SOLUTIONS MANUAL TO ACCOMPANY

ELEMENTS OF PHYSICAL CHEMISTRY

SOLUTIONS MANUAL TO ACCOMPANY

ELEMENTS OF PHYSICAL CHEMISTRY

FOURTH EDITION

Charles A. Trapp
Marshall P. Cady, Jr

OXFORD
UNIVERSITY PRESS

W. H. Freeman and Company
New York

OXFORD
UNIVERSITY PRESS

Great Clarendon Street, Oxford OX2 6DP
Oxford University Press is a department of the University of Oxford.
It furthers the University's objective of excellence in research, scholarship, and education by publishing worldwide in

Oxford New York

Auckland Cape Town Dar es Salaam Hong Kong Karachi
Kuala Lumpur Madrid Melbourne Mexico City Nairobi
New Delhi Shanghai Taipei Toronto

With offices in

Argentina Austria Brazil Chile Czech Republic France Greece
Guatemala Hungary Italy Japan Poland Portugal Singapore
South Korea Switzerland Thailand Turkey Ukraine Vietnam

Oxford is a registered trade mark of Oxford University Press
in the UK and in certain other countries

Published in the United States
by W.H. Freeman and Company, New York

First published 2007

British Library Cataloguing in Publication Data
Data available

Library of Congress Cataloging in Publication Data
Data available

Typeset by Newgen Imaging Systems (P) Ltd., Chennai, India
Printed in Great Britain
on acid-free paper by
Ashford Colour Press Ltd., Gosport, Hampshire

Published, under licence, in the United States and Canada by
W.H. Freeman and Company,
41 Madison Avenue,
New York, NY 10010
www.whfreeman.com

ISBN 978–0–19–928880–9 0–19–928880–1
ISBN (W.H. Freeman) 978–0–7167–3193–1 0–7167–3193–2

10 9 8 7 6 5 4 3 2 1

Preface

This manual provides detailed solutions to all of the discussion questions and exercises in the fourth edition of *Elements of Physical Chemistry* by Peter Atkins and Julio de Paula. We hope that these complete solutions will help in deepening your understanding of physical chemistry. Solutions to exercises carried over from the third edition have been reworked, modified, or corrected when needed.

The solutions to the exercises in this edition rely somewhat more heavily on the mathematical and molecular modelling software that is now generally accessible to physical chemistry students, and this is particularly true for some of the new exercises, which specifically request the use of such software for their solutions. But almost all of the exercises can still be solved with a modern hand-held scientific calculator.

In general, we have adhered rigorously to the rules for significant figures in displaying the final answers. However, when intermediate answers are shown, they are often given with one more figure than would be justified by the data. These excess digits are indicated with an overline.

We have carefully cross-checked the solutions for errors and expect that most have been eliminated. We would be grateful to readers who bring any remaining errors to our attention.

We warmly thank our publishers for their patience in guiding this complex, detailed project to completion.

C. A. T.
M. P. C.

Contents

	Introduction	1
	Answers to discussion questions	1
	Solutions to exercises	2
1	**The properties of gases**	8
	Answers to discussion questions	8
	Solutions to exercises	9
2	**Thermodynamics: the First Law**	22
	Answers to discussion questions	22
	Solutions to exercises	22
3	**Thermochemistry**	35
	Answers to discussion questions	35
	Solutions to exercises	36
4	**Thermodynamics: the Second Law**	50
	Answers to discussion questions	50
	Solutions to exercises	51
5	**Phase equilibria: pure substances**	61
	Answers to discussion questions	61
	Solutions to exercises	62
6	**The properties of mixtures**	72
	Answers to discussion questions	72
	Solutions to exercises	73
7	**Principles of chemical equilibrium**	88
	Answers to discussion questions	88
	Solutions to exercises	89
8	**Consequences of equilibrium**	107
	Answers to discussion questions	107
	Solutions to exercises	108
9	**Electrochemistry**	130
	Answers to discussion questions	130
	Solutions to exercises	131

10 The rates of reactions 153
Answers to discussion questions 153
Solutions to exercises 155

11 Accounting for the rate laws 173
Answers to discussion questions 173
Solutions to exercises 174

12 Quantum theory 189
Answers to discussion questions 189
Solutions to exercises 190

13 Atomic structure 201
Answers to discussion questions 201
Solutions to exercises 202

14 The chemical bond 214
Answers to discussion questions 214
Solutions to exercises 216

15 Metallic, ionic, and covalent solids 231
Answers to discussion questions 231
Solutions to exercises 233

16 Solid surfaces 244
Answers to discussion questions 244
Solutions to exercises 246

17 Molecular interactions 259
Answers to discussion questions 259
Solutions to exercises 260

18 Macromolecules and aggregates 273
Answers to discussion questions 273
Solutions to exercises 276

19 Molecular rotations and vibrations 285
Answers to discussion questions 285
Solutions to exercises 286

20 Electronic transitions and photochemistry 300

Answers to discussion questions 300
Solutions to exercises 303

21 Magnetic resonance 311

Answers to discussion questions 311
Solutions to exercises 313

22 Statistical thermodynamics 322

Answers to discussion questions 322
Solutions to exercises 325

Solutions to box exercises 337

Introduction

Answers to discussion questions

0.1 A gas is a form of matter that fills the container it occupies and is compressible under ordinary atmospheric conditions. It is composed of separated particles in continuous rapid, disordered motion during which particles often travel several diameters before colliding. Consequently, the particles are separated by considerable empty space, a condition that results in compressibility. The interactions between particles are negligibly weak except when they are colliding.

The particles of liquids and solids are in continuous contact with their neighbors, a condition that causes liquids and solids to be incompressible. All particles are in constant motion but they travel only a fraction of a diameter before colliding with a neighbor. The microscopic particles of the liquid phase can slip past each other, a condition that results in a non-rigid fluidity of the macroscopic phase. Microscopic particles of a solid cannot slip past each other. They can only oscillate about an average position within the solid, a condition that results in macroscopic rigidity of the solid. The interactions between particles within the liquid or solid phases are relatively strong.

0.2 The force F acting on any object of mass m equals the mass multiplied by the acceleration a of the object. This is Newton's second law of motion: $F = ma$.

The work done on an object when moving the object against an opposing force equals the opposing force multiplied by the distance over which the object is moved:

Work done on an object = opposing force × distance.

Energy is the capacity to do work. Kinetic energy is the energy that a mass m has because of its speed v: $E_K = \frac{1}{2}mv^2$. The potential energy, E_P, of an object is the energy it possesses due to its position. The potential energy of an object may be due to gravitational, electrical, or magnetic forces. In particular, the 'Coulombic' potential resulting from the presence of electrical charges is especially important in chemistry.

0.3 When the pressures on both sides of an object, for example, a movable piston confined in a cylinder containing a gas, are equal, there is no net force acting upon the object and it will not move in either direction. The object is said to be in mechanical equilibrium. Two objects that exhibit no net flow of energy between them upon contact are in thermal equilibrium. They have the same temperature.

0.4 (a) Extensive; (b) intensive; (c) intensive; (d) intensive; (e) extensive.

Solutions to exercises

0.5 The work done while lifting mass m to a height h near the surface of the Earth is given by the expression: work = force × distance = mass × (acceleration of free fall) × distance = mgh.

$$\text{work} = (65\,\text{kg}) \times (9.81\,\text{m}\,\text{s}^{-2}) \times (3.5\,\text{m}) = 2.2 \times 10^{3}\,\text{kg}\,\text{m}^{2}\,\text{s}^{-2}$$

$$= 2.2 \times 10^{3}\,\text{J} = \boxed{2.2\,\text{kJ}}.$$

0.6 $$E_{\text{K}} = \frac{1}{2}mv^{2} = \frac{1}{2}(58\,\text{g}) \times (30\,\text{m}\,\text{s}^{-1})^{2} \times \left(\frac{1\,\text{kg}}{1000\,\text{g}}\right) = \boxed{26\,\text{J}}.$$

0.7 $$E_{\text{K}} = \frac{1}{2}mv^{2} = \frac{1}{2}(1.5\,\text{t}) \times (50\,\text{km}\,\text{h}^{-1})^{2} \times \left(\frac{1000\,\text{kg}}{1\,\text{t}}\right) \times \left(\frac{1000\,\text{m}}{1\,\text{km}}\right)^{2} \times \left(\frac{1\,\text{h}}{3600\,\text{s}}\right)^{2}$$

$$= 1.4 \times 10^{5}\,\text{J} = \boxed{1.4 \times 10^{2}\,\text{kJ}}.$$

0.8 $$m_{\text{average}} = \frac{M_{\text{average}}}{N_{\text{A}}} = \frac{29\,\text{g}\,\text{mol}^{-1}}{6.022 \times 10^{23}\,\text{mol}^{-1}}\left(\frac{1\,\text{kg}}{1000\,\text{g}}\right) = 4.8 \times 10^{-26}\,\text{kg}.$$

$$E_{\text{K, total}} = N_{\text{A}}\left(\frac{1}{2}mv^{2}\right)_{\text{average}} \approx N_{\text{A}}\left(\frac{1}{2}m_{\text{average}}v_{\text{average}}^{2}\right)$$

$$= \frac{6.022 \times 10^{23}}{2}(4.8 \times 10^{-26}\,\text{kg}) \times (400\,\text{m}\,\text{s}^{-1})^{2}$$

$$\approx 2.3 \times 10^{3}\,\text{J} = \boxed{2.3\,\text{kJ}}.$$

0.9 $$m_{\text{Hg}} = \frac{M_{\text{Hg}}}{N_{\text{A}}} = \frac{200.6\,\text{g}\,\text{mol}^{-1}}{6.0221 \times 10^{23}\,\text{mol}^{-1}}\left(\frac{1\,\text{kg}}{1000\,\text{g}}\right) = 3.331 \times 10^{-25}\,\text{kg}.$$

$$\Delta E_{\text{P}} = m_{\text{Hg}} \times g \times \Delta h = (3.331 \times 10^{-25}\,\text{kg}) \times (9.807\,\text{m}\,\text{s}^{-2}) \times (760.0\,\text{mm}) \times \left(\frac{10^{-3}\,\text{m}}{1\,\text{mm}}\right)$$

$$= \boxed{2.483 \times 10^{-24}\,\text{J}}.$$

0.10 $$E_{\text{min}} = mgh = (25\,\text{g}) \times (9.81\,\text{m}\,\text{s}^{-2}) \times (50\,\text{m}) \times \left(\frac{1\,\text{kg}}{1000\,\text{g}}\right) = \boxed{12\,\text{J}}.$$

0.11 $$E_{\text{gravitational potential}} = -\frac{Gmm_E}{r} = -\frac{Gmm_E}{r_E + h} = -\left(\frac{Gmm_E}{r_E}\right) \times \left(\frac{1}{1+\dfrac{h}{r_E}}\right)$$

$$= -\left(\frac{Gmm_E}{r_E}\right) \times \left(1 + \frac{h}{r_E}\right)^{-1}.$$

Since $h/r_E \ll 1$, the last factor may be expanded in a Taylor expansion series. Second and higher order terms may be discarded because powers of very small fractions produce yet smaller fractions. Using $x = h/r_E$, the Taylor expansion is (Appendix 2)

$$(1+x)^{-1} = 1 - x + x^2 - x^3 + \cdots = 1 - x.$$

Substitution gives

$$E_{\text{gravitational potential}} = -\left(\frac{Gmm_E}{r_E}\right) \times \left(1 - \frac{h}{r_E}\right) = -\frac{Gmm_E}{r_E} + \frac{Gmm_E h}{r_E^2}$$

$$= -\frac{Gmm_E}{r_E} + m \times \left(\frac{Gm_E}{r_E^2}\right) \times h.$$

Inspection of the first term to the right of this expression indicates that it is the gravitational potential of mass m when it is located at position r_E. The second term is the increase in the gravitational potential when the object is lifted from position r_E to height h above r_E. It is the gravitational potential above the surface and it is normally written as [0.2] $E_P = mgh$. Thus,

$$mgh = m \times \left(\frac{Gm_E}{r_E^2}\right) \times h \quad \text{or} \quad \boxed{g = \frac{Gm_E}{r_E^2}}.$$

The values of m_E and r_E (at the equator) are found in the *CRC Handbook of Chemistry and Physics*, so it is possible to calculate the value of g at the equator.

$$g = \frac{(6.67259 \times 10^{-11}\,\text{N m}^2\,\text{kg}^{-2})(5.9763 \times 10^{24}\,\text{kg})}{(6378.077 \times 10^3\,\text{m})^2} \left(\frac{1\,\text{kg m s}^{-2}}{1\,\text{N}}\right)$$

$$= 9.803\,\text{m s}^{-2} \text{ at equator.}$$

0.12 The required fuel energy equals the difference between the gravitational potential at $r = \infty$ and the gravitational potential at $r = r_E$ where r_E is the radius of the Earth.

$$\text{Energy} = \left(-\frac{Gmm_E}{r}\right)_{\text{at } r=\infty} - \left(-\frac{Gmm_E}{r}\right)_{\text{at } r=r_E} = 0 + \frac{Gmm_E}{r_E} = \frac{Gmm_E}{r_E}.$$

$$m = V_E \rho_E = \left(\frac{4\pi(6371 \times 10^3\,\text{m})^3}{3}\right) \times \left(\frac{5.5170\,\text{g}}{\text{cm}^3}\right) \times \left(\frac{1\,\text{cm}}{10^{-2}\,\text{m}}\right)^3 \times \left(\frac{1\,\text{kg}}{10^3\,\text{g}}\right)$$

$$= 5.976 \times 10^{24}\,\text{kg}.$$

$$\text{Energy} = \frac{(6.673\times10^{-11}\,\mathrm{N\,m^2\,kg^{-2}})(185\,\mathrm{kg})(5.976\times10^{24}\,\mathrm{kg})}{6371\times10^{3}\,\mathrm{m}}$$
$$= 1.16\times10^{10}\,\mathrm{J} = \boxed{11.6\,\mathrm{GJ}}.$$

0.13 (a) An infinitesimally small amount of work is done upon an object of mass m when lifting it through the infinitesimal distance $\mathrm{d}r$ against the opposing gravitational attraction F. The work causes an infinitesimal increase $\mathrm{d}E_\mathrm{P}$ in the potential energy of the object.

$$\mathrm{d}E_\mathrm{P} = F\,\mathrm{d}r.$$

$$F = \frac{\mathrm{d}E_\mathrm{P}}{\mathrm{d}r} = \frac{\mathrm{d}}{\mathrm{d}r}\left(-\frac{Gmm_\mathrm{E}}{r}\right) = -Gmm_\mathrm{E}\frac{\mathrm{d}}{\mathrm{d}r}\left(\frac{1}{r}\right) = -Gmm_\mathrm{E}\left(-\frac{1}{r^2}\right).$$

$$F = \frac{Gmm_\mathrm{E}}{r^2}.$$

(b) It is shown in the solution to Exercise 0.11 that $g = \dfrac{Gm_\mathrm{E}}{r_\mathrm{E}^2}$. Consequently, the gravitation attraction at the Earth's surface is given by $\boxed{F = \dfrac{Gmm_\mathrm{E}}{r_\mathrm{E}^2} = mg}$. For an 80 kg person the attraction is

$$F = (80\,\mathrm{kg})\times(9.81\,\mathrm{m\,s^{-2}}) = \boxed{78\,\mathrm{N}}.$$

0.14 Refer to Table 0.1 for pressure conversion factors.

(a) $p = 110\,\mathrm{kPa}\times\dfrac{760\,\mathrm{Torr}}{101.325\,\mathrm{kPa}} = \boxed{824\,\mathrm{Torr}}.$

(b) $p = 0.997\,\mathrm{bar}\times\dfrac{100\,\mathrm{kPa}}{1\,\mathrm{bar}}\times\dfrac{1\,\mathrm{atm}}{101.325\,\mathrm{kPa}} = \boxed{0.984\,\mathrm{atm}}.$

(c) $p = 2.15\times10^4\,\mathrm{Pa}\times\dfrac{1\,\mathrm{kPa}}{10^3\,\mathrm{Pa}}\times\dfrac{1\,\mathrm{atm}}{101.325\,\mathrm{kPa}} = \boxed{0.212\,\mathrm{atm}}.$

(d) $p = 723\,\mathrm{Torr}\times\dfrac{101.325\,\mathrm{kPa}}{760\,\mathrm{Torr}}\times\dfrac{10^3\,\mathrm{Pa}}{1\,\mathrm{kPa}} = \boxed{9.64\times10^4\,\mathrm{Pa}}.$

0.15 $p = g\rho h[0.5] = 9.81\,\mathrm{m\,s^{-2}}\times\dfrac{1.10\,\mathrm{g}}{\mathrm{cm^3}}\times\dfrac{1\,\mathrm{kg}}{10^3\,\mathrm{g}}\times\dfrac{10^6\,\mathrm{cm^3}}{1\,\mathrm{m^3}}\times 11.5\,\mathrm{km}\times\dfrac{10^3\,\mathrm{m}}{1\,\mathrm{km}}$

$= 1.24\times10^8\,\mathrm{kg\,m^{-1}\,s^{-2}} = \boxed{1.24\times10^8\,\mathrm{Pa}}.$

Converting to atmospheres,

$$p = 1.24\times10^8\,\mathrm{Pa}\times\left(\frac{1\,\mathrm{atm}}{1.01325\times10^5\,\mathrm{Pa}}\right) = \boxed{1.22\times10^3\,\mathrm{atm}}.$$

0.16 The thinness of the Martian atmosphere is a result of the low gravitational attraction, which makes it impossible for Mars to maintain an atmosphere. The two factors cannot be separated.

The mass of any given vertical column of gas on Mars would be the same on Earth; therefore, the simple answer to this question would seem to be

$$p\ (\text{Earth}) = 0.0060\,\text{atm} \times \frac{9.81}{3.7} = \boxed{1.6 \times 10^{-2}}\,.$$

The better answer would take into account the ratio of the surface areas of the two planets. The *same* atmosphere would be spread more thinly on Earth.

QUESTION Look up the radii of Mars and Earth and calculate the better answer referred to above.

COMMENT The atmosphere of planets cannot be maintained in an equilibrium state. A complete analysis is quite complicated.

0.17 Identifying p_{ex} in the equation $p = p_{\text{ex}} + \rho g h$ [Derivation 0.1] as the pressure at the top of the straw and p as the atmospheric pressure on the liquid, the pressure difference is $p - p_{\text{ex}} = \rho g h$.

(a) On Earth,

$$p - p_{\text{ex}} = \rho g h = (1.0 \times 10^3\,\text{kg}\,\text{m}^{-3}) \times (9.81\,\text{m}\,\text{s}^{-2}) \times (0.15\,\text{m})$$
$$= \boxed{1.5 \times 10^3\,\text{Pa}} = 0.015\,\text{atm}\,.$$

(b) On Mars,

$$p - p_{\text{ex}} = \rho g h = (1.0 \times 10^3\,\text{kg}\,\text{m}^{-3}) \times (3.7\,\text{m}\,\text{s}^{-2}) \times (0.15\,\text{m})$$
$$= \boxed{5.6 \times 10^2\,\text{Pa}} = 0.0056\,\text{atm}\,.$$

0.18 Using the standard gravitation acceleration of exactly $9.80665\,\text{m}\,\text{s}^{-2}$, the pressure in pascal of 1 mmHg when the mercury density equals $13.5951\,\text{g}\,\text{cm}^{-3}$ (Hg density at 0°C) is [0.5]

$$p = \rho g h = (13.5951\,\text{g}\,\text{cm}^{-3}) \times (9.80665\,\text{m}\,\text{s}^{-2}) \times (0.1\,\text{cm})$$
$$\times (1\,\text{cm}/10^{-2}\,\text{m})^2 \times (1\,\text{kg}/10^3\,\text{g}) = 133.322\,\text{Pa}.$$

This calculated value uses the measured value of the density of mercury and it is not an exact value. It has an uncertainty of ± 0.0001 g cm^{-3} and provides 6 significant figures only. In contrast, 760 Torr is defined to equal 101325 Pa exactly and, consequently,

$$1\,\text{Torr} = 101325\,\text{Pa}/760 = 133.322\,\text{Pa exactly with no uncertainty.}$$

It is apparent that 1 mmHg and 1 Torr are identical under the above conditions of density and standard gravitational acceleration. However, because of some small uncertainty in the density of even the purest sample of mercury, the two are expected to $\boxed{\text{differ by as much as 1 part in } 10^6}$. We conclude that in most practical situations the measurement of pressure in mmHg gives the same result as a measurement in SI units. However, for very precise measurements the unit mmHg, a non-SI unit, should be avoided.

0.19 At $T = 0, \theta/°\text{C} = -273.15$.

We solve the equation for Fahrenheit temperature using $\theta_F/°\text{F} = t_F$

$$-273.15 = \frac{5}{9}(t_F - 32),$$

$$t_F = \left(-273.15 \times \frac{9}{5}\right) + 32 = -459.67.$$

$$\theta_F \text{ (at 0 K)} = \boxed{-459.67°\text{F}}.$$

0.20 The mathematical equation through two points $P_1(x_1, y_1)$ and $P_2(x_2, y_2)$ is for a linear relationship

$$y = \frac{\Delta y}{\Delta x}(x - x_1) + y_1 \quad \text{where} \quad \frac{\Delta y}{\Delta x} = \frac{y_2 - y_1}{x_2 - x_1}.$$

Using $P_1(x_1, y_1) = (-209.9°\text{C}, 0°\text{P}), P_2(x_2, y_2) = (-195.8°\text{C}, 100°\text{P})$ as the two points and θ and θ_P to be temperature in degrees Celsius and degrees Plutonium, respectively, the function $T_P(\theta)$ is

$$\theta_P = \left(\frac{100°\text{P} - 0°\text{P}}{-195.8°\text{C} - (-209.9°\text{C})}\right) \times (\theta + 209.9°\text{C}),$$

$$\boxed{\theta_P = (7.092°\text{P}\,°\text{C}^{-1}) \times (\theta + 209.9°\text{C})}.$$

(a) Substitution of the definition $\theta = (1°\text{C}\,\text{K}^{-1}) \times (T - 273.15\,\text{K})$ where T is Kelvin temperature in the above equation gives

$$\boxed{\theta_P = (7.092°\text{P}\,\text{K}^{-1}) \times (T - 63.25\,\text{K})}.$$

(b) Substitution of the relationship $\theta = (5°\text{C}/9°\text{F}) \times (\theta_F - 32°\text{F})$ where θ_F is Fahrenheit temperature into the top equation gives

$$\boxed{\theta_P = (3.940°\text{P}\,°\text{F}^{-1}) \times (\theta_F + 345.8°\text{F})}.$$

0.21 On the Rankine scale $0°\text{F} = 459.67°\text{R}$ (see the solution to Exercise 0.19) and the degree size is identical to that of the Fahrenheit scale. Hence, the relationship between the Rankine scale and the Fahrenheit scale is

$$T_R = (1\,°\text{R}\,°\text{F}^{-1}) \times \theta_F + 459.67°\text{R}$$

$$T_R(212°\text{F}) = (1\,°\text{R}\,°\text{F}^{-1}) \times (212°\text{F}) + 459.67°\text{R} = \boxed{671.67°\text{R}}.$$

0.22 $$N = 1.0\,\text{g} \times \frac{1\,\text{mol}}{16.1\,\text{kg}} \times \frac{1\,\text{kg}}{10^3\,\text{g}} \times \frac{6.02 \times 10^{23}\,\text{molecules}}{\text{mol}} = \boxed{3.74 \times 10^{19}\,\text{molecules}}.$$

0.23 Let hemoglobin = Hb and myoglobin = Mb.

$$\text{Mass of Hb} = 3 \times 10^8\,\text{molecules} \times \frac{4\,\text{mol Mb}}{1\,\text{mol Hb}} \times \frac{1\,\text{mol Hb}}{6.02 \times 10^{23}\,\text{molecules}} \times \frac{16.1 \times 10^3\,\text{g}}{1\,\text{mol Mb}}$$
$$= 3.\overline{21} \times 10^{-11}\,\text{g}.$$

$$\text{Fraction Hb} = \frac{3.\overline{21} \times 10^{-11}\,\text{g}}{3.33 \times 10^{-11}\,\text{g}} = \boxed{0.9\overline{7}} \text{ or } 97\%.$$

0.24 $m = n \times M$ [0.9] and the molar volume V_m is defined by $V_m = \dfrac{V}{n}$.

$$\text{Therefore, } \rho = \frac{m}{V} = \frac{n \times M}{V} \text{ or } \boxed{\rho = \frac{M}{V_m}}.$$

Chapter 1

The properties of gases

Answers to discussion questions

1.1 An equation of state is an equation that relates the variables that define the state of a system to each other. Boyle, Charles, and Avogadro established these relations for gases at low pressures (perfect gases) by appropriate experiments. Boyle determined how volume varies with pressure ($V \propto 1/p$), Charles how volume varies with temperature ($V \propto T$), and Avogadro how volume varies with amount of gas ($V \propto n$). Combining all of these proportionalities into one, we find

$$V \propto \frac{nT}{p}.$$

Inserting the constant of proportionality, R, yields the perfect gas equation

$$V = \frac{RnT}{P} \quad \text{or} \quad pV = nRT.$$

1.2 The partial pressure of a gas in a mixture of gases is the pressure the gas would exert if it occupied alone the same container as the mixture at the same temperature. It is a limiting law because it holds exactly only under conditions where the gases have no effect upon each other. This can only be true in the limit of zero pressure where the molecules of the gas are very far apart. Hence, Dalton's law holds exactly only for a mixture of perfect gases; for real gases, the law is only an approximation.

1.3 The rms molecular speed c is proportional to $(T/M)^{1/2}$ [1.15]. With the simple molecular kinetic theory proposition that the rates of gaseous diffusion and effusion are proportional to c it is apparent that these rates are proportional to $\sqrt{T}$ and inversely proportional to $\sqrt{M}$, which is Graham's law [1.17].

1.4 The van der Waals equation accounts for repulsive interactions between molecules by supposing that they cause the molecules to behave as small but impenetrable spheres. The nonzero volume of the molecules implies that, instead of moving in a volume V, they are restricted to a smaller volume $V - nb$, where nb is approximately the total volume taken up by the molecules themselves. This argument suggests that the perfect gas law $p = nRT/V$ should be replaced by $p = nRT/(V - nb)$ when repulsions are significant.

The pressure depends on both the frequency of collisions with the walls and the force of each collision. Both the frequency of the collisions and their force are reduced by the attractive forces, which act with a strength proportional to the molar concentration, n/V, of molecules in the sample. Therefore, because the attractive forces reduce both the frequency and the force of the collisions, the pressure is reduced in proportion to the square of this concentration. If the reduction of pressure is written as $-a(n/V)^2$, where a is a positive constant characteristic of each gas, the combined effect of the repulsive and attractive forces is the van der Waals equation of state.

Solutions to exercises

1.5 Solve the perfect gas law $pV = nRT$ [1.2] for pressure.

$$p = \frac{nRT}{V}.$$

$$n = 2.045\,\text{g} \times \frac{1\,\text{mol}\,\text{N}_2}{28.02\,\text{g}} = 0.07298\,\text{mol}, \quad V = 2.00\,\text{dm}^3 = 2.00 \times 10^{-3}\,\text{m}^3.$$

$$p = \frac{nRT}{V} = \frac{(0.07298\,\text{mol})(8.3145\,\text{J}\,\text{K}^{-1}\,\text{mol}^{-1})(294\,\text{K})}{2.00 \times 10^{-3}\,\text{m}^3} = \boxed{89.2\,\text{kPa}}.$$

$(\text{Note:}\ \ 1\,\text{J} = 1\,\text{Pa}\,\text{m}^3)$

1.6 Arranging the perfect gas law [1.2] in a form to solve for pressure,

$$p = \frac{nRT}{V}.$$

$$n = \frac{0.255\,\text{g}}{20.18\,\text{g}\,\text{mol}^{-1}} = 1.26 \times 10^{-2}\,\text{mol}, \quad T = 122\,\text{K}, \quad V = 3.00\,\text{dm}^3 = 3.00 \times 10^{-3}\,\text{m}^3.$$

Therefore,

$$p = \frac{(1.26 \times 10^{-2}\,\text{mol}) \times (8.3145\,\text{J}\,\text{K}^{-1}\,\text{mol}^{-1}) \times 122\,\text{K}}{3.00 \times 10^{-3}\,\text{m}^3} = \boxed{4.26\,\text{kPa}}.$$

1.7 Solve the perfect gas law [1.2] for n.

$$n = \frac{pV}{RT} = \frac{\left(24.5 \times 10^3\,\text{Pa} \times 250.0\,\text{cm}^3 \times \dfrac{10^{-6}\,\text{m}^3}{1\,\text{cm}^3}\right)}{(8.3145\,\text{J}\,\text{K}^{-1}\,\text{mol}^{-1}) \times (292.6\,\text{K})} = 0.00252\,\text{mol}$$

$$= \boxed{2.52 \times 10^{-3}\,\text{mol}}.$$

$(\text{Note:}\ \ 1\,\text{Pa} = 1\,\text{N}\,\text{m}^{-2} = 1\,\text{J}\,\text{m}^{-3})$

1.8 Mass of $CO_2 = 1.04\,\text{kg} - 0.74\,\text{kg} = 0.30\,\text{kg} = 300\,\text{g}$.

$$n\text{ (amount)} = 300\,\text{g} \times \frac{1\,\text{mol}}{44.01\,\text{g}} = 6.82\,\text{mol}, \quad V = 250\,\text{cm}^3 = 2.50 \times 10^{-4}\,\text{m}^3.$$

$$p = \frac{nRT}{V} = \frac{(6.82\,\text{mol}) \times (8.3145\,\text{J}\,\text{K}^{-1}\,\text{mol}^{-1}) \times (293\,\text{K})}{2.50 \times 10^{-4}\,\text{m}^3} = \boxed{6.64 \times 10^4\,\text{kPa}}.$$

1.9 Remember that p is inversely proportional to V at constant temperature. Therefore,

$$p_2 = \frac{V_1}{V_2} \times p_1.$$

$V_1 = 1.00\,\text{dm}^3 = 1.00 \times 10^3\,\text{cm}^3$, $p_1 = 1.00\,\text{atm}$, and $V_2 = 1.00 \times 10^2\,\text{cm}^3$.

$$p_2 = \frac{1.00 \times 10^3\,\text{cm}^3}{100\,\text{cm}^3} \times 1.00\,\text{atm} = \boxed{10.0\,\text{atm}}.$$

1.10 The amount n and V are constant; hence

$$\frac{p_1}{T_1} = \frac{nR}{V} = \frac{p_2}{T_2}. \quad \text{Solving for } p_2 \text{ gives } p_2 = \frac{T_2 p_1}{T_1}.$$

$$p_2 = \frac{973\,\text{K} \times 125\,\text{kPa}}{291\,\text{K}} = \boxed{418\,\text{kPa}}.$$

1.11 At constant temperature eqn. [1.4] becomes $p_1 V_1 = p_2 V_2$.

$$p_2 = \frac{p_1 V_1}{V_2} = \frac{101\,\text{kPa} \times 7.20\,\text{dm}^3}{4.21\,\text{dm}^3} = \boxed{173\,\text{kPa}}.$$

1.12 If we assume constant pressure, then eqn. [1.4] becomes $\dfrac{T_1}{V_1} = \dfrac{T_2}{V_2}$.

$$T_2 = \frac{T_1 V_2}{V_1} = \frac{295.3\,\text{K} \times 100\,\text{cm}^3}{1.00\,\text{dm}^3}\left(\frac{1\,\text{dm}^3}{1 \times 10^3\,\text{cm}^3}\right) = \boxed{29.5\,\text{K}}.$$

1.13 Pressure is constant, so $\dfrac{T_1}{V_1} = \dfrac{T_2}{V_2}$ [1.4].

$T_1 = 340\,\text{K}$ and the volume has increased by 14%, so $V_2 = 1.14\,V_1$.

$$T_2 = \frac{1.14\,V_1}{V_1} \times 340\,\text{K} = 1.14 \times 340\,\text{K} = \boxed{388\,\text{K}}.$$

1.14 (a) $\dfrac{p_1 V_1}{T_1} = \dfrac{p_2 V_2}{T_2}$.

$$V_2 = \frac{(104\,\text{kPa}) \times (2.0\,\text{m}^3) \times (268.2\,\text{K})}{(294.3\,\text{K}) \times (52\,\text{kPa})} = \boxed{3.6\,\text{m}^3}.$$

(b) $$V_2 = \frac{(104\,\text{kPa}) \times (2.0\,\text{m}^3) \times (221.2\,\text{K})}{(294.3\,\text{K}) \times (0.880\,\text{kPa})} = \boxed{178\,\text{m}^3}.$$

1.15 Pressure is inversely proportional to volume. There will be an increase in pressure due to the depth submerged, which can be calculated from $p = \rho g h$ [0.5].

$$V_f = \frac{p_i}{p_f} V_i$$

and total pressure $p_i = 1.0\,\text{atm}$.

$$p_f = 1.0\,\text{atm} + \rho g h.$$

$$\rho g h = (1.025 \times 10^3\,\text{kg m}^{-3}) \times (9.81\,\text{m s}^{-2}) \times 50\,\text{m} = 5.0 \times 10^5\,\text{Pa}.$$

Therefore

$$p_f = (1.01 \times 10^5\,\text{Pa}) + (5.0 \times 10^5\,\text{Pa}) = 6.0 \times 10^5\,\text{Pa},$$

$$V_f = \frac{1.01 \times 10^5\,\text{Pa}}{6.0 \times 10^5\,\text{Pa}} \times 3.0\,\text{m}^3 = \boxed{0.50\,\text{m}^3}.$$

1.16 $pV = nRT$, with n constant, yields

$$\frac{p_f V_f}{T_f} = \frac{p_i V_i}{T_i} \text{ and, solving for } p_f,$$

$$p_f = \frac{p_i V_i T_f}{V_f T_i}.$$

Because $V = \frac{4}{3}\pi r^3$,

$$p_f = \left(\frac{r_i}{r_f}\right)^3 \frac{T_f}{T_i} \times p_i.$$

$$p_f = \left(\frac{1.0\,\text{m}}{3.0\,\text{m}}\right)^3 \times \left(\frac{253\,\text{K}}{293\,\text{K}}\right) \times 1.0\,\text{atm} = \boxed{3.2 \times 10^{-2}\,\text{atm}}.$$

1.17 (a) $V = \dfrac{n_J RT}{p_J}$

There is only one volume, so using the amount of nitrogen and p_J for nitrogen we can calculate the volume.

$$n_{N_2} = \frac{0.225\,\text{g}}{28.02\,\text{g mol}^{-1}} = 8.03 \times 10^{-3}\,\text{mol}, \quad p_{N_2} = 15.2\,\text{kPa}, \quad T = 300\,\text{K}.$$

$$V = \frac{(8.03 \times 10^{-3}\,\text{mol}) \times (8.3145\,\text{J K}^{-1}\,\text{mol}^{-1}) \times (300\,\text{K})}{15.2 \times 10^3\,\text{Pa}} = 1.32 \times 10^{-3}\,\text{m}^3$$

$$= \boxed{1.32\,\text{dm}^3}.$$

(b) $p = \dfrac{nRT}{V}$ where n is the sum of the amounts of each component; therefore

$$n = n_{\mathrm{CH_4}} + n_{\mathrm{Ar}} + n_{\mathrm{N_2}}. \quad n_{\mathrm{CH_4}} = \frac{0.320\,\mathrm{g}}{16.04\,\mathrm{g\,mol^{-1}}} = 2.00 \times 10^{-2}\,\mathrm{mol},$$

$$n_{\mathrm{Ar}} = \frac{0.175\,\mathrm{g}}{39.95\,\mathrm{g\,mol^{-1}}} = 4.38 \times 10^{-3}\,\mathrm{mol},$$

$$n = (2.00 + 0.438 + 0.803) \times 10^{-2}\,\mathrm{mol} = 3.24 \times 10^{-2}\,\mathrm{mol}.$$

Substituting into the equation for p,

$$p = \frac{(3.24 \times 10^{-2}\,\mathrm{mol}) \times (8.3145\,\mathrm{J\,K^{-1}\,mol^{-1}}) \times (300\,\mathrm{K})}{1.32 \times 10^{-3}\,\mathrm{m^3}} = \boxed{61.2\,\mathrm{kPa}}.$$

1.18 $p_{\mathrm{Total}} = p_{\mathrm{A}} + p_{\mathrm{B}} + \cdots \quad [1.6]$

$$760\,\mathrm{Torr} = p_{\mathrm{air}} + p_{\mathrm{water}}$$

$$= p_{\mathrm{air}} + 47\,\mathrm{Torr}.$$

$$p_{\mathrm{air}} = (760 - 47)\,\mathrm{Torr} = \boxed{713\,\mathrm{Torr}}.$$

1.19 To calculate the molar mass of the compound, we need to relate density, temperature, and pressure. Using the perfect gas law,

$$pV = nRT = \left(\frac{m}{M}\right)RT \quad \text{or} \quad M = \left(\frac{m}{V}\right)\frac{RT}{p} = \frac{\rho RT}{p}$$

where ρ is density (i.e. mass/volume).

$$M = \frac{(1.23\,\mathrm{g\,dm^{-3}}) \times (8.3145\,\mathrm{J\,K^{-1}\,mol^{-1}}) \times (330\,\mathrm{K})}{(25.5 \times 10^3\,\mathrm{Pa})}\left(\frac{1\,\mathrm{dm^{-3}}}{1 \times 10^{-3}\,\mathrm{m^3}}\right)$$

$$= \boxed{132\,\mathrm{g\,mol^{-1}}}.$$

1.20 $V_{\mathrm{m}} = \dfrac{V}{n}$ and, using the perfect gas law to calculate n,

$n = \dfrac{pV}{RT}$. Then, convert V units in order to use

$R = 62.364\,\mathrm{dm^3\,Torr\,K^{-1}\,mol^{-1}}$ (Table 1.1).

$V = 250\,\mathrm{cm^3} = 0.250\,\mathrm{dm^3}$.

Substituting into the perfect gas law written above,

$$n = \frac{(152\,\mathrm{Torr}) \times (0.250\,\mathrm{dm^3})}{(62.364\,\mathrm{dm^3\,Torr\,mol^{-1}\,K^{-1}}) \times 298\,\mathrm{K}} = 2.04 \times 10^{-3}\,\mathrm{mol}.$$

The mass of the gas is given, 33.5 mg, so the molar mass,

$$M = \frac{0.0335\,\mathrm{g}}{2.04 \times 10^{-3}\,\mathrm{mol}} = \boxed{16.4\,\mathrm{g\,mol^{-1}}}.$$

1.21 (a) The partial pressures of the gases can be related to their mole fractions. The total amount is found by $n = n_{N_2} + n_{H_2} = (1.0 + 2.0)\ \text{mol} = 3.0\ \text{mol}$.

$$x_{H_2} = \frac{2.0\ \text{mol}}{3.0\ \text{mol}} = 0.67.$$

$$x_{N_2} = \frac{1.0\ \text{mol}}{3.0\ \text{mol}} = 0.33.$$

$$p_J = n_J \frac{RT}{V}.$$

Solving for $\dfrac{RT}{V}$,

$$\frac{RT}{V} = \frac{(8.206 \times 10^{-2}\ \text{dm}^3\ \text{atm}\ \text{K}^{-1}\ \text{mol}^{-1}) \times (273.15\ \text{K})}{22.4\ \text{dm}^3} = 1.00\ \text{atm}\ \text{mol}^{-1}$$

$$p_{H_2} = (2.0\ \text{mol}) \times (1.00\ \text{atm}\ \text{mol}^{-1}) = \boxed{2.0\ \text{atm}}.$$

$$p_{N_2} = (1.0\ \text{mol}) \times (1.00\ \text{atm}\ \text{mol}^{-1}) = \boxed{1.0\ \text{atm}}.$$

(b) The total pressure is the sum of the partial pressures

$$p = p_{H_2} + p_{N_2} = (2.0 + 1.0)\ \text{atm} = \boxed{3.0\ \text{atm}}.$$

1.22 The mean speed $\bar{c}$ equals $(8/3\pi)^{1/2}$ times the root mean square speed c [1.13] and [1.15].

$$\bar{c} = \left(\frac{8}{3\pi}\right)^{1/2} c = \left(\frac{8}{3\pi}\right)^{1/2} \left(\frac{3RT}{M}\right)^{1/2} = \left(\frac{8RT}{\pi M}\right)^{1/2}.$$

(a) $$\bar{c}_{He} = \sqrt{\frac{8 \times (8.3145\ \text{J}\ \text{K}^{-1}\ \text{mol}^{-1}) \times T}{\pi \times (0.00400\ \text{kg}\ \text{mol}^{-1})}} = (72.75\ \text{m}\ \text{s}^{-1}\ \text{K}^{-1/2}) \times T^{1/2}.$$

$$\bar{c}_{He}(77\ \text{K}) = \boxed{638\ \text{m}\ \text{s}^{-1}}.$$

$$\bar{c}_{He}(298\ \text{K}) = \boxed{1.26\ \text{km}\ \text{s}^{-1}}.$$

$$\bar{c}_{He}(1000\ \text{K}) = \boxed{2.30\ \text{km}\ \text{s}^{-1}}.$$

(b) $$\bar{c}_{CH_4} = \sqrt{\frac{8 \times (8.3145\ \text{J}\ \text{K}^{-1}\ \text{mol}^{-1}) \times T}{\pi \times (0.01604\ \text{kg}\ \text{mol}^{-1})}} = (36.33\ \text{m}\ \text{s}^{-1}\ \text{K}^{-1/2}) \times T^{1/2}.$$

$$\bar{c}_{CH_4}(77\ \text{K}) = \boxed{319\ \text{m}\ \text{s}^{-1}}.$$

$$\bar{c}_{CH_4}(298\ \text{K}) = \boxed{627\ \text{m}\ \text{s}^{-1}}.$$

$$\bar{c}_{CH_4}(1000\ \text{K}) = \boxed{1.15\ \text{km}\ \text{s}^{-1}}.$$

1.23 $$\langle v\rangle = \int_0^\infty sF(s)\,\mathrm{d}s = \int_0^\infty s\left\{4\pi\left(\frac{M}{2\pi RT}\right)^{3/2} s^2 \mathrm{e}^{-Ms^2/2RT}\right\}\mathrm{d}s\ [1.16]$$

$$= 4\pi\left(\frac{M}{2\pi RT}\right)^{3/2}\int_0^\infty s^3\mathrm{e}^{-Ms^2/2RT}\,\mathrm{d}s$$

$$= 4\pi\left(\frac{M}{2\pi RT}\right)^{3/2}\left(\frac{1}{2}\right)\left(\frac{2RT}{M}\right)^2 = \left(\frac{8RT}{\pi M}\right)^{1/2}.$$

$$\bar{c} = \langle v\rangle = \left(\frac{8RT}{\pi M}\right)^{1/2}.$$

1.24 $$\langle v^2\rangle = \int_0^\infty s^2F(s)\,\mathrm{d}s = \int_0^\infty s^2\left\{4\pi\left(\frac{M}{2\pi RT}\right)^{3/2} s^2 \mathrm{e}^{-Ms^2/2RT}\right\}\mathrm{d}s\ [1.16]$$

$$= 4\pi\left(\frac{M}{2\pi RT}\right)^{3/2}\int_0^\infty s^4\mathrm{e}^{-Ms^2/2RT}\,\mathrm{d}s$$

$$= 4\pi\left(\frac{M}{2\pi RT}\right)^{3/2}\left(\frac{3}{8}\right)\left(\frac{2RT}{M}\right)^2\pi^{1/2}\left(\frac{2RT}{M}\right)^{1/2} = \frac{3}{2}\left(\frac{2RT}{M}\right) = \frac{3RT}{M}.$$

$$c = \langle v^2\rangle^{1/2} = \left(\frac{3RT}{M}\right)^{1/2}.$$

1.25 $$F(s) = 4\pi\left(\frac{M}{2\pi RT}\right)^{3/2} s^2\mathrm{e}^{-Ms^2/2RT}\quad [1.16].$$

$$\frac{\mathrm{d}F(s)}{\mathrm{d}s} = 4\pi\left(\frac{M}{2\pi RT}\right)^{3/2}\left\{2s\mathrm{e}^{-Ms^2/2RT} + s^2\left(\frac{-2Ms}{2RT}\right)\mathrm{e}^{-Ms^2/2RT}\right\}$$

$$= 4\pi s\left(\frac{M}{2\pi RT}\right)^{3/2}\mathrm{e}^{-Ms^2/2RT}\left(2 - \frac{Ms^2}{RT}\right).$$

The most probable speed c^* is located at the peak of $F(s)$ where $\mathrm{d}F(s)/\mathrm{d}s = 0$. Inspection of the above equation reveals that the last factor must equal zero at the peak.

$$2 - \frac{M(c^*)^2}{RT} = 0 \quad\text{and}\quad \boxed{c^* = \left(\frac{2RT}{M}\right)^{1/2}}.$$

1.26 The Maxwell distribution of speeds is $f = 4\pi\left(\frac{M}{2\pi RT}\right)^{3/2} s^2\mathrm{e}^{-Ms^2/2RT}\Delta s$ [1.16].

At the center of the range, $s = 295\ \mathrm{m\,s^{-1}}$.

$$f = 4\times\pi\times\left(\frac{28.02\times10^{-3}\ \mathrm{kg\,mol^{-1}}}{2\times\pi\times(8.3145\ \mathrm{kg\,m^2\,s^{-2}\,K^{-1}\,mol^{-1}})\times500\ \mathrm{K}}\right)^{3/2}$$

$$\times(295\ \mathrm{m\,s^{-1}})^2\mathrm{e}^{-(28.02\times10^{-3})\,(295)^2/(2\times8.3145\times500)}\times10\ \mathrm{m\,s^{-1}}$$

$$= \boxed{9.06\times10^{-3}}.$$

1.27 The formula for determining the mean free path is $\lambda = \dfrac{RT}{\sqrt{2}N_A\sigma p}$ [1.19]. Solving for p,

$$V = \frac{4}{3}\pi r^3 \text{ Therefore, } r = \left(\frac{3V}{4\pi}\right)^{1/3} = \left(\frac{3(1.0\times 10^{-3}\,\text{m}^3)}{4\pi}\right)^{1/3} = 0.062\,\text{m}.$$

$\lambda = d = 2\text{r} = 0.124\,\text{m}.$

$$p = \frac{RT}{\sqrt{2}N_A\sigma\lambda}.$$

$$p = \frac{(8.3145\,\text{Pa m}^3\,\text{K}^{-1}\,\text{mol}^{-1})\times(298.15\,\text{K})}{\sqrt{2}\times(6.02\times 10^{23}\,\text{mol}^{-1})\times(0.36\times 10^{-18}\,\text{m}^2)\times(0.124\,\text{m})} = \boxed{0.065\,\text{Pa}}.$$

Note: $1\,\text{J} = 1\,\text{Pa m}^3$.

1.28 Solving [1.19] for p with $d = \left(\dfrac{\sigma}{\pi}\right)^{1/2} = \sqrt{\dfrac{0.36\times 10^{-18}\text{m}^2}{\pi}} = 3.39\times 10^{-10}\,\text{m}$,

$\lambda = 10d = 3.39\times 10^{-9}\,\text{m}.$

$$p = \frac{RT}{\sqrt{2}N_A\sigma\lambda}.$$

$$p = \frac{(8.3145\,\text{Pa m}^3\,\text{K}^{-1}\,\text{mol}^{-1})\times(298.15\,\text{K})}{\sqrt{2}\times(6.02\times 10^{23}\,\text{mol}^{-1})\times(0.36\times 10^{-18}\,\text{m}^2)\times(3.39\times 10^{-9}\,\text{m})}$$

$$= \boxed{2.4\times 10^{6}\,\text{Pa}}.$$

1.29 The formula for determining the mean free path is $\lambda = \dfrac{RT}{\sqrt{2}N_A\sigma p}$ [1.19].

$$\lambda = \frac{(8.3145\,\text{J K}^{-1}\,\text{mol}^{-1})\times(217\,\text{K})}{\sqrt{2}\times(6.02\times 10^{23}\,\text{mol}^{-1})\times(0.43\times 10^{-18}\,\text{m}^2)}$$

$$\times \frac{1}{(0.050\,\text{atm})\times(1.013\times 10^{5}\,\text{Pa atm}^{-1})}$$

$$= 973\,\text{nm} = \boxed{0.97\,\mu\text{m}}.$$

1.30 (a) The collision frequency is determined by the relation $z = \dfrac{\sqrt{2}N_A\sigma c p}{RT}$ [1.19] with

$$c = \sqrt{\frac{3RT}{M}}.$$

$$z = 2^{1/2}N_A\sigma\sqrt{\frac{3RT}{M}}\,\frac{p}{RT}.$$

$$z = 2^{1/2} \times (6.02 \times 10^{23}\,\text{mol}^{-1}) \times (0.36 \times 10^{-18}\,\text{m}^2)$$

$$\times \sqrt{\frac{3 \times (8.3145\,\text{J K}^{-1}\,\text{mol}^{-1}) \times 298\,\text{K}}{(39.95\,\text{g mol}^{-1}) \times \left(\frac{1\,\text{kg}}{10^3\,\text{g}}\right)}}$$

$$\times \frac{10 \times 10^5\,\text{Pa}}{(8.3145\,\text{J K}^{-1}\,\text{mol}^{-1}) \times (298\,\text{K})}$$

$$= \boxed{5.3 \times 10^{10}\,\text{s}^{-1}}$$ so, for 1 s, the number of collisions is $\boxed{5.3 \times 10^{10}}$.

(b) Substituting in other pressures gives $\boxed{5.3 \times 10^{9}\text{ collisions}}$.

(c) $\boxed{5.3 \times 10^{4}\text{ collisions}}$.

1.31 Exercise 1.30 shows the calculation of the number of collisions for a single Ar atom. Therefore, we need to calculate the total number of atoms to determine the total number of collisions.

The total number of molecules will be $6.02 \times 10^{23}\,\text{mol}^{-1} \times n$.

Because $n = \dfrac{pV}{RT}$,

$$10\,\text{bar} \times \frac{1\,\text{atm}}{1.013\,\text{bar}} = 9.87\,\text{atm},$$

$$\text{number of molecules} = \frac{6.02 \times 10^{23}\,\text{molecules mol}^{-1} \times pV}{RT}$$

$$= \frac{(6.02 \times 10^{23}\,\text{molecules mol}^{-1}) \times (9.87\,\text{atm}) \times (1.0\,\text{dm}^3)}{(0.08206\,\text{dm}^3\,\text{atm K}^{-1}\,\text{mol}^{-1}) \times (298\,\text{K})}$$

$$= 2.43 \times 10^{23}\text{ molecules.}$$

$$\text{Total number of collisions} = \frac{2.43 \times 10^{23}\text{ molecules} \times (5.3 \times 10^{10}\text{ collisions/molecule})}{2}$$

$$= \boxed{6.4 \times 10^{33}\text{ collisions}}.$$

The answer is divided by two because two molecules collide in one collision event. By substituting in for the other pressures,

(b) $\boxed{6.4 \times 10^{31}\text{ collisions}}$.

(c) $\boxed{6.4 \times 10^{21}\text{ collisions}}$.

1.32 $z = \frac{c}{\lambda}$ [1.18]. We have already calculated λ in problem 1.29. To calculate c use

$$c = \sqrt{\frac{3RT}{M}}\ [1.15]$$

$$= \left(\frac{3 \times (8.3145\,\mathrm{J\,K^{-1}\,mol^{-1}}) \times (217\,\mathrm{K})}{(28.02 \times 10^{-3}\,\mathrm{kg\,mol^{-1}})}\right)^{1/2} = 439\,\mathrm{m\,s^{-1}}.$$

(Note: $1\,\mathrm{J} = 1\,\mathrm{kg\,m^2\,s^{-2}}$)

$$z = \frac{439\,\mathrm{m\,s^{-1}}}{(973\,\mathrm{nm}) \times (10^{-9}\,\mathrm{m/nm})} = \boxed{4.5 \times 10^{8}\,\text{collisions}\,\mathrm{s^{-1}}}.$$

1.33 $\lambda = \frac{RT}{\sqrt{2}N_{\mathrm{A}}\sigma p}$ [1.19]; σ is given as $0.43\,\mathrm{nm^2}$.

Because we want to solve this for several different pressures, calculate first in terms of p so it is easier to substitute into the equation

$$\lambda = \frac{(8.3145\,\mathrm{J\,K^{-1}\,mol^{-1}}) \times (298.15\,\mathrm{K})}{\sqrt{2} \times (6.02 \times 10^{23}\,\mathrm{mol^{-1}}) \times (0.43 \times 10^{-18}\,\mathrm{m^2}) \times p},$$

$$\lambda = \frac{6.8 \times 10^{-3}\,\mathrm{mPa}}{p}.$$

(a) When $p = 10\,\mathrm{bar} = 1.0 \times 10^{6}\,\mathrm{Pa}$,

$$\lambda = \frac{6.8 \times 10^{-3}\,\mathrm{mPa}}{1.0 \times 10^{6}\,\mathrm{Pa}} = 6.8 \times 10^{-9}\,\mathrm{m} = \boxed{6.8\,\mathrm{nm}}.$$

(b) When $p = 10^{3}\,\mathrm{kPa}$,

$$\lambda = \frac{6.8 \times 10^{-3}\mathrm{mPa}}{103 \times 10^{3}\,\mathrm{Pa}} = 0.066 \times 10^{-6}\,\mathrm{m} = 6.6 \times 10^{-8}\,\mathrm{m} = \boxed{68\,\mathrm{nm}}.$$

(c) When $p = 1\,\mathrm{Pa}$, $\lambda = 6.8 \times 10^{-3}\,\mathrm{m} = \boxed{7\,\mathrm{mm}}$.

1.34 $\lambda = \frac{RT}{\sqrt{2}N_{\mathrm{A}}\sigma p}$ [1.19]

At constant V, p varies directly with T:

$$p = \frac{nRT}{V} \text{ and}$$

$$\lambda = \frac{RT}{\sqrt{2}N_{\mathrm{A}}\sigma} \times \frac{V}{nRT} = \frac{V}{\sqrt{2}N_{\mathrm{A}}\sigma n}; \text{ hence } \lambda \text{ is } \boxed{\text{independent of temperature}}.$$

1.35 (a) For a perfect gas $p = \dfrac{nRT}{V}$. $n = 1.0\,\text{mol}$ and $R = 8.2058 \times 10^{-2}\,\text{dm}^3\,\text{atm}\,\text{mol}^{-1}\,\text{K}^{-1}$.

(i) $T = 273.15$ in $22.414\,\text{dm}^3$. Substituting in

$$p = \frac{(1.0\,\text{mol}) \times (8.2058 \times 10^{-2}\,\text{dm}^3\,\text{atm}\,\text{K}^{-1}\,\text{mol}^{-1}) \times (273.15\,\text{K})}{22.414\,\text{dm}^3} = \boxed{1.0\,\text{atm}}.$$

(ii) Using a similar substitution for 1000 K and $100\,\text{cm}^3$

$$p = \frac{(1.0\,\text{mol}) \times (8.2058 \times 10^{-2}\,\text{dm}^3\,\text{atm}\,\text{K}\,\text{mol}^{-1}) \times (1000\,\text{K})}{0.100\,\text{dm}^3} = \boxed{8.3 \times 10^2\,\text{atm}}.$$

(b) For a van der Waals gas, using Table 1.5, $a = 5.507\,\text{dm}^3\,\text{atm}\,\text{mol}^{-2}$, $b = 0.0651\,\text{dm}^3\,\text{mol}^{-1}$, the van der Waals equation is

$$p = \frac{nRT}{V - nb} - a\left(\frac{n}{V}\right)^2$$ [1.23 a]. For conditions (i) substitution gives

$$p = \frac{(1.0\,\text{mol}) \times (8.2058 \times 10^{-2}\,\text{dm}^3\,\text{atm}\,\text{K}^{-1}\,\text{mol}^{-1}) \times (273.15\,\text{K})}{22.414\,\text{dm}^3 - (1.0\,\text{mol}) \times (0.0651\,\text{dm}^3\,\text{mol}^{-1})}$$

$$- 5.507\,\text{dm}^6\,\text{atm}\,\text{mol}^{-2}\left(\frac{1.0\,\text{mol}}{22.414\,\text{dm}^3}\right)^2,$$

$$p = \boxed{0.99\,\text{atm}}.$$

At conditions (ii) $p = \boxed{1.8 \times 10^3\,\text{atm}}$.

1.36 For a perfect gas $p = \dfrac{nRT}{V}$.

$$n = 10.00\,\text{g}\,CO_2 \times \frac{1\,\text{mol}\,CO_2}{44.01\,\text{g}} = 0.2272\,\text{mol}.$$

$$p = \frac{(0.2272\,\text{mol}) \times (0.08206\,\text{dm}^3\,\text{atm}\,\text{mol}^{-1}\,\text{K}^{-1}) \times (298.1\,\text{K})}{0.100\,\text{dm}^3} = \boxed{55.6\,\text{atm}}.$$

For a van der Waals gas:

$$p = \frac{nRT}{V - nb} - a\left(\frac{n}{V}\right)^2 \quad [1.23\text{a}].$$

$$a = 3.610\,\text{dm}^6\,\text{atm}\,\text{mol}^{-2}.$$

$b = 0.0429\,\mathrm{dm^3\,mol^{-1}}$.

$$p = \frac{(0.2272\,\mathrm{mol}) \times (0.08206\,\mathrm{dm^3\,atm\,mol^{-1}\,K^{-1}})(298.1\,\mathrm{K})}{(0.100\,\mathrm{dm^3}) - (0.2272\,\mathrm{mol} \times 0.0429\,\mathrm{dm^3\,mol^{-1}})}$$

$$- 3.610\,\mathrm{dm^6\,atm\,mol^{-2}} \times \left(\frac{0.2272\,\mathrm{mol}}{0.100\,\mathrm{dm^3}}\right)^2$$

$$= 43.10\,\mathrm{atm} \times 101.325\,\mathrm{kPa/atm} = \boxed{4.36\,\mathrm{MPa}}.$$

1.37 Solve the van der Waals equation for p, giving $p = \dfrac{RT}{V_m - b} - \dfrac{a}{V_m^2}$ [1.23a] with $V_m = \dfrac{V}{n}$.

$$p = \frac{RT}{V_m\left(1 - \dfrac{b}{V_m}\right)} - \frac{a}{V_m^2}.$$

$$\frac{1}{1 - \dfrac{b}{V_m}} = 1 + \frac{b}{V_m} + \left(\frac{b}{V_m}\right)^2 + \cdots$$

because $\dfrac{1}{1-x} = 1 + x - x^2 + \cdots$.

$$p = \frac{RT}{V_m}\left(1 + \frac{b}{V_m} + \frac{b^2}{V_m^2} + \cdots\right) - \frac{a}{V_m^2}.$$

$$p = \frac{RT}{V_m}\left[1 + \left(b - \frac{a}{RT}\right)\frac{1}{V_m} + \frac{b}{V_m^2} + \cdots\right].$$

Comparing this expression to equation 1.22 with $V_m = \dfrac{V}{n}$,

$$p = \frac{RT}{V_m}\left(1 + \frac{B}{V_m} + \frac{C}{V_m^2} + \cdots\right),$$

we see $\boxed{B = b - \dfrac{a}{RT} \text{ and } C = b^2}$.

1.38 Because $C = 1200\,\mathrm{cm^6\,mol^{-2}}$, $b = C^{1/2} = \boxed{34.6\,\mathrm{cm^3\,mol^{-1}}}$.

$$a = RT(b - B)$$

$$= (8.206 \times 10^{-2}\,\mathrm{dm^3\,atm\,mol^{-1}\,K^{-1}}) \times (273\,\mathrm{K}) \times (34.6 + 21.7\,\mathrm{cm^3\,mol^{-1}})$$

$$= (22.40\,\mathrm{dm^3\,atm\,mol^{-1}}) \times (56.3 \times 10^{-3}\,\mathrm{dm^3\,mol^{-1}})$$

$$= \boxed{1.26\,\mathrm{dm^6\,atm\,mol^{-2}}}.$$

1.39 From Exercise 1.37,

$$B = b - \frac{a}{RT}.$$

$$B = 0 \text{ when } \frac{a}{RT} = b \text{ or } T = \frac{a}{bR}.$$

For CO_2

$$T = \frac{3.610\ \mathrm{dm^6\ atm\ mol^{-2}}}{0.0429\ \mathrm{dm^3\ mol^{-1}} \times 0.08206\ \mathrm{dm^3\ atm\ K^{-1}\ mol^{-1}}} = \boxed{1.03 \times 10^3\ \mathrm{K}}.$$

1.40 (a) Using the substitution $V_m = V/n$, equation [1.23a] becomes

$$p = \frac{RT}{V_m - b} - \frac{a}{V_m^2}.$$

$$\left(\frac{\partial p}{\partial V_m}\right)_T = -\frac{RT}{(V_m - b)^2} + \frac{2a}{V_m^3}.$$

$$\left(\frac{\partial^2 p}{\partial V_m^2}\right)_T = \frac{2RT}{(V_m - b)^3} - \frac{6a}{V_m^4}.$$

Evaluating these three equations at the critical point (p_c, V_c, T_c), where the first and second derivatives equal zero on the isotherm, yields three independent equations.

(1) $p_c = \dfrac{RT_c}{V_c - b} - \dfrac{a}{V_c^2}.$

(2) $-\dfrac{RT_c}{(V_c - b)^2} + \dfrac{2a}{V_c^3} = 0 \quad \text{or} \quad V_c^3 = \dfrac{2a(V_c - b)^2}{RT_c}.$

(3) $\dfrac{RT_c}{(V_c - b)^3} - \dfrac{3a}{V_c^4} = 0 \quad \text{or} \quad V_c^4 = \dfrac{3a(V_c - b)^3}{RT_c}.$

Division of (3) by (2) and solving for V_c yields

(4) $\boxed{V_c = 3b}.$

Substitution of (4) into (2) and solving for T_c yields

(5) $\boxed{T_c = \dfrac{8a}{27bR}}.$

Substitution of (4) and (5) into (1) and simplifying the expression yields

(6) $\boxed{p_c = \dfrac{a}{27b^2}}.$

(b) Evaluation of equation [1.20b] at the critical point and substitution of equations (4)–(6) yields an expression for Z_c.

$$Z_c = \frac{p_c V_c}{RT_c} = \frac{\left(\dfrac{a}{27b^2}\right)(3b)}{R\left(\dfrac{8a}{27bR}\right)} = \frac{3}{8}.$$

$$\boxed{Z_c = \frac{3}{8}}.$$

Chapter 2

Thermodynamics: the First Law

Answers to discussion questions

2.1 At the molecular level, work is a transfer of energy that results in orderly motion of the atoms and molecules in a system; heat is a transfer of energy that results in disorderly motion. See Figures 2.5 and 2.6 of text.

2.2 In an expansion against constant pressure the internal pressure of the system is greater than the constant opposing pressure by a measurable non-infinitesimal amount. In a reversible expansion the internal pressure is only infinitesimally greater than the opposing pressure. Since the opposing pressure is greater in the reversible case and, in fact, is at a maximum, the work generated in the reversible case is the maximum work.

2.3 The difference results from the definition $H = U + PV$; hence $\Delta H = \Delta U + \Delta(PV)$. As $\Delta(PV)$ is not usually zero, except for isothermal processes in a perfect gas, the difference between ΔH and ΔU is a non-zero quantity. As shown in sections 2.6 and 2.7 of the text, ΔH can be interpreted as the heat associated with a process at constant pressure, and ΔU as the heat at constant volume.

2.4 (a) $q = nRT\ln(V_f/V_i)$ limitations: reversible, isothermal expansion of a perfect gas.

(b) $\Delta H = \Delta U + p\Delta V$ limitation: constant pressure process.

(c) $C_{p,m} - C_{V,m} = R$ limitation: perfect gas.

Solutions to exercises

2.5 To calculate the work use the formula $w = mgh$ [2.1].

(a) $w = 1.0\,\text{kg} \times 9.81\,\text{m}\,\text{s}^{-2} \times 10\,\text{m} = \boxed{98\,\text{J}}$.

(b) $w = 1.0\,\text{kg} \times 1.60\,\text{m}\,\text{s}^{-2} \times 10\,\text{m} = \boxed{16\,\text{J}}$.

2.6 Use the formula $w = mgh$ [2.1].

$$w = 0.200\,\text{kg} \times 9.81\,\text{m}\,\text{s}^{-2} \times 20\,\text{m} = \boxed{39\,\text{J}}.$$

2.7
$$w = mgh \text{ [2.1]}$$
$$= 65\,\text{kg} \times 9.81\,\text{m}\,\text{s}^{-2} \times 4.0\,\text{m} = \boxed{2.6\,\text{kJ}}.$$

2.8 We can assume that the entire mass is concentrated at the center of mass, which is at

$$h = \frac{760}{2}\,\text{mm} = 0.380\,\text{m}.$$

Mass = volume × density.

Volume of cylinder = $A \times h$.

$$V = \pi r^2 \times h$$
$$= \pi \times \left(\frac{0.0100\,\text{m}}{2}\right)^2 \times 0.760\,\text{m}$$
$$= 5.97 \times 10^{-5}\text{m}^3.$$

$$\text{mass} = 5.97 \times 10^{-5}\text{m}^3 \times 13.6\,\text{g}\,\text{cm}^{-3}\,\frac{1\,\text{kg}}{10^3\,\text{g}} \times \frac{10^6\,\text{cm}^3}{1\,\text{m}^3}$$
$$= 0.812\,\text{kg}.$$

$$w = mgh = 0.812\,\text{kg} \times 9.81\,\text{m}\,\text{s}^{-2} \times 0.380\,\text{m}$$
$$= 3.03\,\text{kg}\,\text{m}^2\,\text{s}^{-2} = \boxed{3.03\,\text{J}}.$$

2.9 Taking the mean Earth radius $R = 6370.9$ km, the aircraft's distance from the Earth's center is

$$r = R + \text{altitude} = 6370.9\,\text{km} + 11.0\,\text{km} = 6381.9\,\text{km}.$$

The aircraft's mass is

$$m = 100.0\,\not{t}\left(\frac{1000\,\text{kg}}{1\,\not{t}}\right) = 1.000 \times 10^5\,\text{kg}.$$

$$\text{(a) } w = -\int_R^r F_{\text{gravity}}\,\text{d}r = -\int_R^r \left(-\frac{Gmm_{\text{E}}}{r^2}\right)\text{d}r = Gmm_{\text{E}}\int_R^r \left(\frac{1}{r^2}\right)\text{d}r$$

$$= Gmm_{\text{E}} \times \left(-\frac{1}{r}\right)\Bigg]_{R=6370.9\,\text{km}}^{r=6381.9\,\text{km}}$$

$$= Gmm_{\text{E}} \times \left(\frac{1}{r}\right)\Bigg]_{r=6381.9\,\text{km}}^{R=6370.9\,\text{km}}$$

$$= \frac{(6.6726 \times 10^{-11}\,\mathrm{N\,m^2\,kg^{-2}})(1.000 \times 10^5\,\mathrm{kg})(5.9763 \times 10^{24}\,\mathrm{kg})}{10^3}$$

$$\times \left(\frac{1}{6370.9\,\mathrm{m}} - \frac{1}{6381.9\,\mathrm{m}}\right)$$

$$= 1.0789 \times 10^{10}\,\mathrm{N\,m} = 1.0789 \times 10^{10}\,\mathrm{J}$$

$$w = \boxed{10.789\,\mathrm{GJ}}.$$

(b) Equation [2.1] is a very good approximation to the answer of part (a): $w_{\text{approx.}} = mgh$ where h is the altitude.

$w_{\text{approx.}} = mgh = (1.000 \times 10^5\,\mathrm{kg})(9.807\,\mathrm{m\,s^{-2}})(11.0 \times 10^3\,\mathrm{m}) = 10.79\,\mathrm{GJ}.$

(We have extended the altitude to assume 4 significant figures.)

The percentage error is about 1 part per 1000 (i.e. about 0.1%). The percentage error is expressed analytically as follows.

$$\text{Fractional error} = \left|\frac{w - w_{\text{approx.}}}{w}\right| = \left|1 - \frac{w_{\text{approx.}}}{w}\right| = \left|1 - \frac{mgh}{Gmm_{\mathrm{E}}\left(\dfrac{1}{R} - \dfrac{1}{R+h}\right)}\right|$$

$$= \left|1 - \frac{mgh}{Gmm_{\mathrm{E}}\left(\dfrac{h}{R(R+h)}\right)}\right| = \left|1 - \frac{mgR(R+h)}{Gmm_{\mathrm{E}}}\right|$$

$$= \left|1 - \frac{m\left(\dfrac{Gm_{\mathrm{E}}}{R^2}\right)R(R+h)}{Gmm_{\mathrm{E}}}\right| = \left|1 - \left(1 + \frac{h}{R}\right)\right| = \frac{h}{R}.$$

The percentage error is $\left(\dfrac{h}{R}\right) \times 100\% = \left(\dfrac{11.0\,\mathrm{km}}{6381.9\,\mathrm{km}}\right) 100\% = \boxed{0.172\%}$.

2.10 $w = -\displaystyle\int_0^x F(x)\,\mathrm{d}x = -\int_0^x (-kx)\,\mathrm{d}x = k\int_0^x x\,\mathrm{d}x = \left.\frac{kx^2}{2}\right]_{x=0}^{x=x} = \boxed{\frac{kx^2}{2}}$. This function is sketched in Figure 2.1.

2.11 (a) $w_{\text{expansion}} = -p_{\text{ext}}\Delta V = -(1.00 \times 10^5\,\mathrm{Pa}) \times (1.0 \times 10^{-6}\,\mathrm{m^3}) = \boxed{-0.10\,\mathrm{J}}$.

(b) $w_{\text{expansion}} = -p_{\text{ext}}\Delta V = -(1.00 \times 10^5\,\mathrm{Pa}) \times (1.0 \times 10^{-3}\,\mathrm{m^3}) = \boxed{-10.\,\mathrm{J}}$.

In order to calculate the amount of work required to return the system to its original state an exact knowledge of the details of the process is required. Pressures may change; heat may be extracted. Heat is not a function of the state of the system. It depends upon the process. Hence, no numerical value can be calculated without further information.

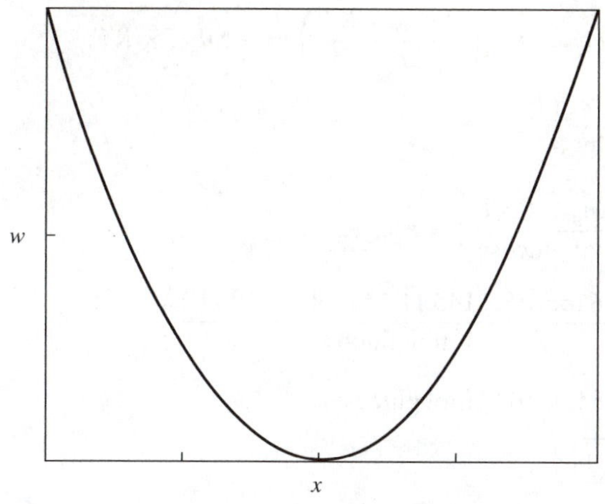

Fig 2.1

2.12 Amount glucose $= 1.0\,\mathrm{g}\,C_6H_{12}O_6 \times \dfrac{1\,\mathrm{mol}}{180.2\,\mathrm{g}} = 5.5\overline{5} \times 10^{-3}$ mol glucose.

The balanced reaction equation for the complete combustion of glucose is written as

$$C_6H_{12}O_6(s) + 6O_2(g) \rightarrow 6\,CO_2(g) + 6\,H_2O$$

In part (a) the net expansion work done on the chemical system, w, is computed for the case in which water is produced as a liquid. In part (b) water is considered to form as a gas. The difference is important because gases have a large molar volume compared to the negligibly small molar volumes of solids and liquids. When a balanced reaction indicates a large net change in the number of moles of gas per reaction, $\Delta n_{\mathrm{gas}} = n_{\mathrm{product\ gases}} - n_{\mathrm{reactant\ gases}}$, the magnitude of w is significantly large. Should there be no change in Δn_{gas} $(= 0)$, the extremely small changes in the volumes of reactant liquids and solids to product liquids and solid can be expressed as $\Delta V \simeq 0$ and, consequently, $w = -p_{\mathrm{ex}}\Delta V\,[2.2] \simeq 0$.

(a) When the reaction water is written as liquid, gaseous carbon dioxide is produced and gaseous oxygen is consumed. The combustion of 1 mol glucose causes the gaseous change

$$\Delta n_{\mathrm{gas}} = n_{\mathrm{product\ gases}} - n_{\mathrm{reactant\ gases}} = 6 - 6 = 0.$$

We conclude that $w = -p_{\mathrm{ex}}\Delta V\,[2.2] \simeq \boxed{0}$.

(b) When the reaction water is written as gas, both gaseous carbon dioxide and gaseous water are produced while gaseous oxygen is consumed. The combustion of 1 mol glucose causes the gaseous change $\Delta n_{\mathrm{gas}} = n_{\mathrm{product\ gases}} - n_{\mathrm{reactant\ gases}} = 6 + 6 - 6 = 6$. Expansion work is significant and it is caused by the appearance of 6 new moles of gas per reaction.

$$\Delta V \text{ per reaction} = \frac{\Delta n_{\mathrm{gas}} \times RT}{p_{\mathrm{ex}}} \quad \text{(perfect gas law [1.2]).}$$

$$w \text{ per reaction} = -p_{ex}\Delta V \ [2.2] = -p_{ex}\left(\frac{\Delta n_{gas} \times RT}{p_{ex}}\right) = -\Delta n_{gas} \times RT.$$

$$w \text{ per mole glucose} = \frac{-\Delta n_{gas} \times RT}{1 \text{ mol glucose}}.$$

$$w \text{ per amount glucose} = \left(\frac{-\Delta n_{gas} \times RT}{1 \text{ mol glucose}}\right) \times n_{glucose}$$

$$= -\left(\frac{(6\,\text{mol})(8.3145\,\text{J}\,\text{K}^{-1}\,\text{mol}^{-1})(293\,\text{K})}{1\,\text{mol glucose}}\right)$$

$$\times\ (5.55 \times 10^{-3}\,\text{mol glucose})$$

$$= \boxed{-81\,\text{J}}.$$

The reaction does 81 J of expansion work as the production of gas pushes on the surrounding atmospheric gases.

2.13 $w = -p_{ex}\Delta V$ [2.2].

To calculate p_{ex} convert to Pa.

$p_{ex} = 1.00 \times 10^5$ Pa.

$\Delta V = 100\,\text{cm}^2 \times 10\,\text{cm} = 1.0 \times 10^3\,\text{cm}^3 = 1.0 \times 10^{-3}\,\text{m}^3$.

$w = -1.00 \times 10^5\,\text{Pa} \times 1.0 \times 10^{-3}\,\text{m}^3 = -1.0 \times 10^2\,\text{Pa}\,\text{m}^3$.

Then, because $1\,\text{Pa}\,\text{m}^3 = 1\,\text{J}$,

$w = \boxed{-1.0 \times 10^2\,\text{J}}$.

The expanding gas does 100 J of work while moving the piston.

2.14 (a) Horizontally (no additional work to raise the piston).

Work done by system = distance × opposing force.

$$w = h \times p_{ex} \times A$$

$$= 155\,\text{cm} \times 105\,\text{kPa} \times 55.0\,\text{cm}^2 \times \frac{1\,\text{m}^3}{10^6\,\text{cm}^3} \times \frac{10^3\,\text{Pa}}{1\,\text{kPa}}$$

$$= 895\,\text{Pa}\,\text{m}^3 = \boxed{895\,\text{J}}.$$

(b) Vertically additional work is required to raise the piston).

Additional work = $h \times mg$ [2.1]

$$w_{additional} = 155\,\text{cm} \times \frac{1\,\text{m}}{100\,\text{cm}} \times 250\,\text{g} \times \frac{1\,\text{kg}}{10^3\,\text{g}} \times 9.81\,\text{m}\,\text{s}^{-2}$$

$$= 3.80\,\text{J}.$$

Total work = $895\,\text{J} + 3.80\,\text{J} = \boxed{899\,\text{J}}$.

2.15 (a) $w = -p_{ex}\Delta V$ [2.2] $= -(30.0 \times 10^3\ \text{Pa}) \times (3.3\ \text{dm}^3) \times \left(\dfrac{1 \times 10^{-3}\ \text{m}^3}{1\ \text{dm}^3}\right) = \boxed{-99\ \text{J}}$.

(b) $w = -nRT \ln \dfrac{V_f}{V_i}$ [2.3].

$$n = \frac{4.50\ \text{g}}{16.04\ \text{g mol}^{-1}} = 0.2805\ \text{mol}.$$

$V_i = 12.7\ \text{dm}^3,\ V_f = (12.7 + 3.3)\ \text{dm}^3 = 16.0\ \text{dm}^3.$

$$w = -0.2805\ \text{mol} \times 8.3145\ \text{J K}^{-1}\ \text{mol}^{-1} \times 310\ \text{K} \times \ln\left(\frac{16.0\ \text{dm}^3}{12.7\ \text{dm}^3}\right) = \boxed{-164\ \text{J}}.$$

2.16 $w = -nRT \ln \dfrac{V_f}{V_i}$ [2.3].

$$nRT = 52.0 \times 10^{-3}\ \text{mol} \times 8.3145\ \text{J K}^{-1}\ \text{mol}^{-1} \times 260\ \text{K}$$

$$= 1.124 \times 10^2\ \text{J}.$$

$$w = -1.124 \times 10^2\ \text{J} \times \ln\left(\frac{100\ \text{cm}^3}{300\ \text{cm}^3}\right) = \boxed{+123\ \text{J}}.$$

2.17 $w = -p_{ex}\Delta V$ [2.2].

$$p_{ex} = 95.2\ \text{bar} \times \frac{10^5\ \text{Pa}}{1\ \text{bar}} = 9.52 \times 10^6\ \text{Pa}.$$

$$\Delta V = -0.550\ \text{dm}^3 \times 0.57 = -0.314\ \text{dm}^3 \times \frac{10^{-3}\ \text{m}^3}{1\ \text{dm}^3} = -3.14 \times 10^{-4}\ \text{m}^3.$$

$$w = (-9.52 \times 10^6\ \text{Pa}) \times (-3.14 \times 10^{-4}\ \text{m}^3) = 2.99 \times 10^3\ \text{Pa m}^3 = \boxed{+2.99\ \text{kJ}}.$$

2.18 $Mg(s) + 2\ HCl(aq) \rightarrow H_2(g) + MgCl_2(aq)$.

$M(Mg) = 24.31\ \text{g mol}^{-1}$.

$$n_{H_2} = n_{Mg} = \frac{12.5\ \text{g}}{24.31\ \text{g mol}^{-1}} = 0.514\ \text{mol}.$$

$w = -p_{ex}\Delta V$ [2.2].

$$V_i = 0, \quad V_f = \frac{nRT}{p_f}, \quad p_f = p_{ex}.$$

$$w = -p_{ex}(V_f - V_i) = -p_{ex} \times \frac{nRT}{p_{ex}} = -nRT.$$

$$w = 0.514\ \text{mol} \times 8.3145\ \text{J K}^{-1}\ \text{mol}^{-1} \times 293.5\ \text{K}$$

$$= -1.25 \times 10^3\ \text{J} = \boxed{-1.25\ \text{kJ}}.$$

2.19 (a) $w = -\int_{V_i}^{V_f} p\,dV = -\int_{V_i}^{V_f} \frac{nRT}{V}\,dV = -\int_{V_i}^{V_f} \frac{nR(T_i - c(V - V_i))}{V}\,dV$ where $c \geq 0$

$$= -nRT_i \int_{V_i}^{V_f} \frac{1}{V}\,dV + nRc\left\{\int_{V_i}^{V_f} dV - V_i \int_{V_i}^{V_f} \frac{1}{V}\,dV\right\}$$

$$= -nR(T_i + cV_i)\int_{V_i}^{V_f} \frac{1}{V}\,dV + nRc\int_{V_i}^{V_f} dV.$$

$$\boxed{w = -nR(T_i + cV_i)\ln\left(\frac{V_f}{V_i}\right) + nRc(V_f - V_i)}.$$

(b) The difference between the work done by the gas during a non-isothermal expansion $W_{\text{non-iso}}$ and the work done by the gas during an isothermal expansion W_{iso} is

$$W_{\text{non-iso}} - W_{\text{iso}} = (-w_{\text{non-iso}}) - (-w_{\text{iso}}) = nR(T_i + cV_i)\ln\left(\frac{V_f}{V_i}\right)$$

$$- nRc(V_f - V_i) - nRT_i \ln\left(\frac{V_f}{V_i}\right)$$

$$= nRcV_i \ln\left(\frac{V_f}{V_i}\right) - nRc(V_f - V_i).$$

To determine whether this difference is positive or negative, consider a slight expansion for which $V_f = V_i + \delta$ where $0 < \delta/V_i \ll 1$. Because of the small value of δ, the Taylor series expansion of the logarithm factor may be truncated after the second-order term.

$$W_{\text{non-iso}} - W_{\text{iso}} = nRcV_i \ln\left(\frac{V_i + \delta}{V_i}\right) - nRc(V_i + \delta_f - V_i)$$

$$= nRcV_i\left\{\ln\left(1 + \frac{\delta}{V_i}\right) - \frac{\delta}{V_i}\right\}$$

$$= nRcV_i\left\{\frac{\delta}{V_i} - \frac{1}{2}\left(\frac{\delta}{V_i}\right)^2 - \frac{\delta}{V_i}\right\} = -\frac{nRc}{2V_i}\delta^2.$$

This expression is always negative. The work done by the gas in a non-isothermal expansion is always less than the work done during an isothermal expansion.

2.20 Exercise 2.19 indicates that the molar difference between the work done by the gas during a non-isothermal expansion, $W_{\text{non-iso}}$ and the work done by the gas during an isothermal expansion, W_{iso}, is

$$\Delta W_m = \frac{W_{\text{non-iso}} - W_{\text{iso}}}{n} = RcV_i\left\{\ln\left(1 + \frac{\delta}{V_i}\right) - \frac{\delta}{V_i}\right\} \quad \text{where} \quad \delta = V_f - V_i.$$

The difference is plotted in Figure 2.2 with $V_i = 25.0\,\text{dm}^3$ and two values of c.

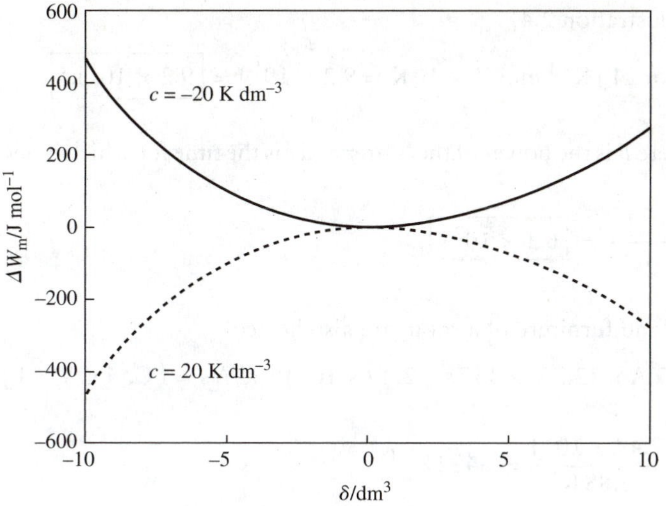

Fig 2.2

2.21 $C = \dfrac{q}{\Delta T}$ [2.4a] $= \dfrac{124\,\text{J}}{5.23\,\text{K}} = \boxed{23.7\,\text{J}\,\text{K}^{-1}}$.

2.22 $q = n\,C_{p,\text{m}}\Delta T$ [see Illustration 2.4].

$n = 250\,\text{g} \times 1\,\text{mol}/18.0\,\text{g} = 13.9\,\text{mol}$.

$C_{p,\text{m}} = 75.3\,\text{J}\,\text{K}^{-1}\,\text{mol}^{-1}$.

$q = 13.9\,\text{mol} \times 75.3\,\text{J}\,\text{mol}^{-1}\,\text{K}^{-1} \times 40\,\text{K} = 4.2 \times 10^4\,\text{J} = \boxed{42\,\text{kJ}}$.

2.23 $q = IVt$ [2.5]

$= 1.34\,\text{A} \times 110\,\text{V} \times 5.0\,\text{min} \times \dfrac{60\,\text{s}}{1\,\text{min}} = \boxed{4.4 \times 10^4\,\text{J}}$.

2.24 $C_V = \dfrac{q_V}{\Delta T}$ [2.11] $= \dfrac{229\,\text{J}}{2.55\,\text{K}} = 89.8\,\text{J}\,\text{K}^{-1}$.

The molar heat capacity at constant pressure is therefore

$$C_{V,\text{m}} = \frac{89.8\,\text{J}\,\text{K}^{-1}}{3.0\,\text{mol}} = \boxed{30\,\text{J}\,\text{K}^{-1}\,\text{mol}^{-1}}.$$

For a perfect gas $C_{p,\text{m}} - C_{V,\text{m}} = R$ [2.17], or

$$C_{p,\text{m}} = C_{V,\text{m}} + R = (30 + 8.3)\,\text{J}\,\text{K}^{-1}\,\text{mol}^{-1} = \boxed{38\,\text{J}\,\text{K}^{-1}\,\text{mol}^{-1}}.$$

2.25 $V = 5.5\,\text{m} \times 6.5\,\text{m} \times 3.0\,\text{m} = 10\overline{7}\,\text{m}^3$.

$$n = \frac{pV}{RT} = \frac{1.0\,\text{atm} \times 10\overline{7}\,\text{m}^3 \times \dfrac{10^3\,\text{dm}^3}{1\,\text{m}^3}}{8.206 \times 10^{-2}\,\text{dm}^3\,\text{atm}\,\text{K}^{-1}\,\text{mol}^{-1} \times 298\,\text{K}} = 4.4 \times 10^3\,\text{mol}.$$

$q = nC_{p,\mathrm{m}}\Delta T$ [Illustration 2.4]

$= 4.4 \times 10^3\,\mathrm{mol} \times 21\,\mathrm{J\,K^{-1}\,mol^{-1}} \times 10\,\mathrm{K} = 9.2 \times 10^5\,\mathrm{J} = \boxed{9.2 \times 10^2\,\mathrm{kJ}}$.

Because $q = P \times t$ where P is the power of the heater and t is the time for which it operates,

$$t = \frac{q}{P} = \frac{9.2 \times 10^5\,\mathrm{J}}{1.5 \times 10^3\,\mathrm{J\,s^{-1}}} = \boxed{6.1 \times 10^2\,\mathrm{s}}.$$

In practice, the walls and furniture of a room are also heated.

2.26 $q = IVt$ [2.5] $= 1.27\,\mathrm{A} \times 12.5\,\mathrm{V} \times 157\,\mathrm{s} = 2.49 \times 10^3\,\mathrm{J}$ $(1\,\mathrm{A\,s} = 1\,\mathrm{C},\ 1\,\mathrm{C\,V} = 1\,\mathrm{J})$

$$\mathrm{C} = \frac{q}{\Delta T}\,[2.4\mathrm{a}] = \frac{2.49 \times 10^3\,\mathrm{J}}{3.88\,\mathrm{K}} = 642\,\mathrm{J\,K^{-1}}.$$

$q = \mathrm{C}\Delta T = 642\,\mathrm{J\,K^{-1}} \times 2.89\,\mathrm{K} = \boxed{1.86 \times 10^3\,\mathrm{J}}$.

2.27 $q = n\,RT\ln\left(\frac{V_\mathrm{f}}{V_\mathrm{i}}\right)$ [2.7] $= 1.00\,\mathrm{mol} \times 8.3145\,\mathrm{J\,K^{-1}\,mol^{-1}} \times 300\,\mathrm{K} \times \ln\left(\frac{30.0}{22.0}\right)$

$= \boxed{773\,\mathrm{J}}$.

2.28 $\Delta U = w + q$ [2.8].

$$w = mgh\,[2.1] = 200\,\mathrm{g} \times \frac{1\,\mathrm{kg}}{10^3\,\mathrm{g}} \times 9.81\,\mathrm{m\,s^{-2}} \times 1.55\,\mathrm{m} = 3.04\,\mathrm{J}.$$

This is the work on the lifted mass; the work done on the animal is $-3.04\,\mathrm{J}$.

$\Delta U_{\mathrm{animal}} = -3.04\,\mathrm{J} + (-5.2\,\mathrm{J}) = \boxed{-8.0\,\mathrm{J}}$.

2.29 $q_{\mathrm{total}} = q_{\mathrm{heater}} + q_{\mathrm{calorimeter}} = q_{\mathrm{heater}}$.

$w = 0$.

$\Delta U = w + q_{\mathrm{total}} = q_{\mathrm{heater}} = IVt$ [2.5] $= 0.01522\,\mathrm{A} \times 12.4\,\mathrm{V} \times 155\,\mathrm{s} = \boxed{29.3\,\mathrm{J}}$.

2.30 For a perfect gas: $\Delta H_\mathrm{m} = \Delta U_\mathrm{m} + R\Delta T$ [2.14b].

Since both ΔU_m and ΔT equal zero for an isothermal change of a perfect gas, $\Delta H_\mathrm{m} = 0$ for an isothermal change of a perfect gas.

2.31 $H = U + pV$ [2.12]; hence the difference between molar enthalpy and molar energy is pV.

For a perfect gas $pV = nRT = RT$ (for $n = 1$).

$RT = 8.3145\,\mathrm{J\,K^{-1}\,mol^{-1}} \times 298.15\,\mathrm{K} = 2.47897 \times 10^3\,\mathrm{J\,mol^{-1}}$

$= 2.479\,\mathrm{kJ\,mol^{-1}}$.

For a real gas, use the virial equation of state [1.22] with B and C written in terms of the van der Waals constants. See the solution to Exercise 1.37 for the required expressions.

$$pV_{\mathrm{m}} = \frac{RT}{V_{\mathrm{m}}}\left(1 + \frac{B}{V_{\mathrm{m}}} + \frac{C}{V_{\mathrm{m}}^2} + \cdots\right).$$

For the purposes of this exercise, we may wish to truncate this power series after the second term and approximate V_{m} as $\dfrac{RT}{p}$; then

$$pV_{\mathrm{m}} = RT\left(1 + \frac{pB}{RT} + \cdots\right), \quad B = \left(b - \frac{a}{RT}\right).$$

$$pV_{\mathrm{m}} = RT + pb - \frac{pa}{RT} + \cdots.$$

Assume $p = 1.000\ \mathrm{atm} = 1.01325 \times 10^5\ \mathrm{Pa}$ in the above expression.

$$pV_{\mathrm{m}} = (2.4790 \times 10^3\ \mathrm{J\,mol^{-1}}) + (1.01325 \times 10^5\ \mathrm{Pa} \times 0.0429\ \mathrm{dm^3\,mol^{-1}} \times \frac{10^{-3}\ \mathrm{m^3}}{1\ \mathrm{dm^3}})$$

$$- \left(\frac{1.01325 \times 10^5\ \mathrm{Pa} \times 3.610\ \mathrm{dm^6\,atm\,mol^{-2}} \times 1.01325 \times 10^5\ \mathrm{Pa/1\,atm} \times 10^{-6}\ \mathrm{m^6/dm^6}}{2.4790 \times 10^3\ \mathrm{J\,mol^{-1}}}\right)$$

$$= [(2.4790 \times 10^3) + (4.37 - 15.0)]\ \mathrm{J\,mol^{-1}}$$

$$= 2.4684 \times 10^3\ \mathrm{J\,mol^{-1}} = \boxed{2.4684\ \mathrm{kJ\,mol^{-1}}}.$$

As this value is less than the perfect gas value, the molar enthalpy is *decreased* by intermolecular forces.

2.32 $q = \boxed{-1.2\ \mathrm{kJ}}$ (heat leaves the sample).

At constant pressure $\Delta H = q$ [2.15b]; hence $\Delta H = \boxed{-1.2\ \mathrm{kJ}}$.

$$C = \frac{q}{\Delta T}\ [2.4a] = \frac{-1.2\ \mathrm{kJ}}{-15\ \mathrm{K}} = \boxed{80\ \mathrm{J\,K^{-1}}}.$$

2.33 $q = C_p\Delta T = nC_{p,\mathrm{m}}\ \ \Delta T = 3.0\ \mathrm{mol} \times (29.4\ \mathrm{J\,K^{-1}\,mol^{-1}}) \times 25\ \mathrm{K} = \boxed{+2.2\ \mathrm{kJ}}$.

$\Delta H = q = \boxed{+2.2\ \mathrm{kJ}}$ (constant pressure).

$\Delta U = \Delta H - \Delta(pV) = \Delta H - \Delta(nRT)$ (perfect gas)

$= \Delta H - nR\Delta T$

$= 2.2\ \mathrm{kJ} - 3.0\ \mathrm{mol} \times 8.3145\ \mathrm{J\,K^{-1}\,mol^{-1}} \times 25\ \mathrm{K}$

$= 2.2\ \mathrm{kJ} - 0.62\ \mathrm{kJ} = \boxed{+1.6\ \mathrm{kJ}}$.

2.34 $C_{V,\mathrm{m}} = C_{p,\mathrm{m}} - R$ [2.17]

$$= 29.14\,\mathrm{J\,K^{-1}\,mol^{-1}} - 8.31\,\mathrm{J\,K^{-1}\,mol^{-1}} = \boxed{20.83\,\mathrm{J\,K^{-1}\,mol^{-1}}}.$$

2.35 (a) $\Delta H_\mathrm{m} = C_{p,\mathrm{m}}\Delta T$ [see Illustration 2.4]

$$= 29.14\,\mathrm{J\,K^{-1}\,mol^{-1}} \times (37°\mathrm{C} - 15°\mathrm{C})[°\mathrm{C} = \mathrm{K}]$$

$$= 29.14\,\mathrm{J\,K^{-1}\,mol^{-1}} \times 22\,\mathrm{K} = \boxed{641\,\mathrm{J\,mol^{-1}}}.$$

(b) $\Delta U_\mathrm{m} = \Delta H_\mathrm{m} - \Delta(pV)$ [2.13a]

$$= \Delta H - \Delta(RT) = \Delta H - R\Delta T.$$

$$\Delta U_m = 641\,\mathrm{J\,mol^{-1}} - (8.31\,\mathrm{J\,K^{-1}\,mol^{-1}} \times 22\,\mathrm{K}) = \boxed{458\,\mathrm{J\,mol^{-1}}}.$$

2.36 $C_{V,\mathrm{m}} = \dfrac{\mathrm{d}U_\mathrm{m}}{\mathrm{d}T}$ at constant volume [[2.11] written in infinitesimal form]

$$= \frac{\mathrm{d}}{\mathrm{d}T}(a + bT + cT^2) = \frac{\mathrm{d}a}{\mathrm{d}T} + \frac{\mathrm{d}(bT)}{\mathrm{d}T} + \frac{\mathrm{d}(cT^2)}{\mathrm{d}T} = 0 + b + 2cT = \boxed{b + 2cT}.$$

2.37 $\Delta H_\mathrm{m} = \displaystyle\int_{T_\mathrm{i}}^{T_\mathrm{f}} C_{p,\mathrm{m}}\mathrm{d}T$ [Box 2.1] $= \displaystyle\int_{T_\mathrm{i}}^{T_\mathrm{f}} \left(a + bT + \frac{c}{T^2}\right)\mathrm{d}T$

$$= \left[aT + \frac{bT^2}{2} - \frac{c}{T}\right]_{T_\mathrm{i}=288.15\,\mathrm{K}}^{T_\mathrm{f}=310.15\,\mathrm{K}}$$

$$= \left\{(44.22\,\mathrm{J\,K^{-1}\,mol^{-1}})(310.15\,\mathrm{K}) + \frac{(8.79\times10^{-3}\,\mathrm{J\,K^{-2}\,mol^{-1}})(310.15\,\mathrm{K})^2}{2} - \frac{(-8.62\times10^{5}\,\mathrm{J\,K\,mol^{-1}})}{310.15\,\mathrm{K}}\right\}$$

$$- \left\{(44.22\,\mathrm{J\,K^{-1}\,mol^{-1}})(288.15\,\mathrm{K}) + \frac{(8.79\times10^{-3}\,\mathrm{J\,K^{-2}\,mol^{-1}})(288.15\,\mathrm{K})^2}{2} - \frac{(-8.62\times10^{5}\,\mathrm{J\,K\,mol^{-1}})}{288.15\,\mathrm{K}}\right\}.$$

$$\boxed{\Delta H_\mathrm{m} = 818\,\mathrm{J\,mol^{-1}}}.$$

2.38 (a) $\Delta H_\mathrm{m}(T) = \displaystyle\int_{T_\mathrm{i}}^{T} C_{p,\mathrm{m}}\,\mathrm{d}T$ [Box 2.1] $= \displaystyle\int_{T_\mathrm{i}}^{T} \left(a + bT + \frac{c}{T^2}\right)\mathrm{d}T$

$$= \left[aT + \frac{bT^2}{2} - \frac{c}{T}\right]_{T_\mathrm{i}=288.15\,\mathrm{K}}^{T}$$

$$= aT + \frac{bT^2}{2} - \frac{c}{T}$$

$$- \left\{ (44.22\,\mathrm{J\,K^{-1}\,mol^{-1}})(288.15\,\mathrm{K}) + \frac{(8.79 \times 10^{-3}\,\mathrm{J\,K^{-2}\,mol^{-1}})(288.15\,\mathrm{K})^2}{2} - \frac{(-8.62 \times 10^5\,\mathrm{J\,K\,mol^{-1}})}{288.15\,\mathrm{K}} \right\}$$

$$\boxed{\Delta H_\mathrm{m}(T) = aT + \frac{bT^2}{2} - \frac{c}{T} - 16.1\,\mathrm{kJ\,mol^{-1}}}.$$

(b) This function is plotted in Figure 2.3.

Note: $\Delta H_\mathrm{m}(T)$ is almost, but not quite, linear over this range.

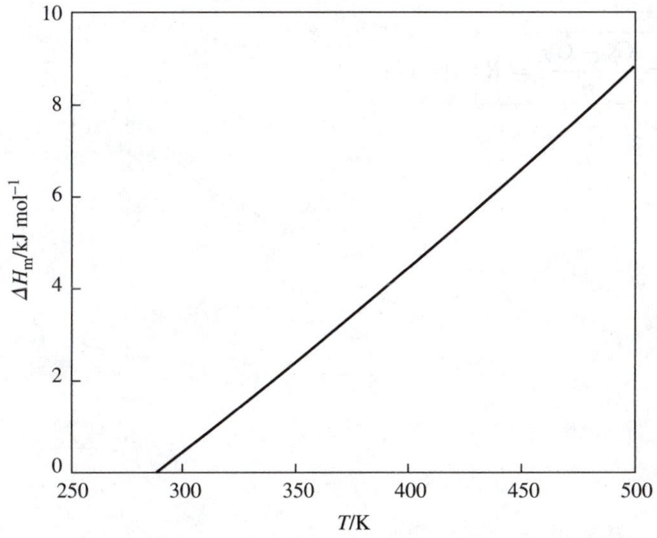

Fig 2.3

2.39 For a perfect gas $V = nRT/p$.

$$\alpha = \frac{1}{V}\left(\frac{\partial V}{\partial T}\right)_p = \frac{p}{nRT}\left(\frac{\partial\left(\frac{nRT}{p}\right)}{\partial T}\right)_p = \left(\frac{p}{nRT}\right)\left(\frac{nR}{p}\right)\left(\frac{\partial T}{\partial T}\right)_p.$$

$$\boxed{\alpha = \frac{1}{T}}.$$

$$\kappa = -\frac{1}{V}\left(\frac{\partial V}{\partial p}\right)_T = -\frac{p}{nRT}\left(\frac{\partial\left(\frac{nRT}{p}\right)}{\partial p}\right)_T$$

$$= -\left(\frac{p}{nRT}\right)(nRT)\left(\frac{\partial\left(\frac{1}{p}\right)}{\partial p}\right)_T = -p\left(-\frac{1}{p^2}\right).$$

$$\boxed{\kappa = \frac{1}{p}}.$$

Thus, $C_p - C_V = \dfrac{\alpha^2 TV}{\kappa} = \dfrac{\left(\frac{1}{T}\right)^2 TV}{\frac{1}{p}} = \dfrac{pV}{T} = nR.$

$$\boxed{C_{p,\mathrm{m}} - C_{V,\mathrm{m}} = \frac{C_p - C_V}{n} = R} \quad [2.17].$$

Chapter 3

Thermochemistry

Answers to discussion questions

3.1 The vaporization of the water is an endothermic transformation that cools the linen and its immediate environment.

3.2 Standard reaction enthalpies can be calculated from a knowledge of the standard enthalpies of formation of all the substances (reactants and products) participating in the reaction. This is an exact method that involves no approximations. The only disadvantage is that standard enthalpies of formation are not known for all substances.

Approximate values can be obtained from mean bond enthalpies. See Example 3.3 for an illustration of the method of calculation. This method is often quite inaccurate, though, because the average values of the bond enthalpies used may not be close to the actual values in the compounds of interest and the method ignores attractive forces between molecules, or is applicable to perfect gases only.

Computer-aided molecular modelling is now the method of choice for estimating standard reaction enthalpies, especially for large molecules with complex three-dimensional structures, but accurate numerical values are still difficult to obtain.

3.3 (a) The standard state of a substance at a specified temperature is its pure form at 1 bar. The reference state of a substance is its most stable state at the specified temperature and 1 bar.

(b) An exothermic compound has a negative standard enthalpy of formation while an endothermic compound has a positive standard enthalpy of formation.

3.4 (a) $\Delta_r H = \Delta_r U + \Delta\nu_{gas} RT$.

Limitations: perfect gas and negligible volume contribution from condensed phases.

(b) $\Delta_r H^\ominus(T') = \Delta_r H^\ominus(T) + \Delta_r C_p^\ominus \times (T' - T)$.

Limitation: negligible dependence of heat capacities upon temperature.

Solutions to exercises

3.5 $H_2(g) + \frac{1}{2}O_2(g) \rightarrow H_2O(l), \quad \Delta_f H(H_2O, l, p).$

$$\Delta_f H(H_2O,l,p) = \Delta_f H^{\ominus}(H_2O,l) + \left(\frac{\partial(\Delta_f H(H_2O,l))}{\partial p}\right)_T \times (p - p^{\ominus})$$

$$= \Delta_f H^{\ominus}(H_2O,l) + \left\{\left(\frac{\partial H_m(H_2O,l)}{\partial p}\right)_T - \left(\frac{\partial H_m(H_2,g)}{\partial p}\right)_T - \frac{1}{2}\left(\frac{\partial H_m(O_2,g)}{\partial p}\right)_T\right\} \times (p - p^{\ominus}).$$

At the low pressures around 1 bar the gases are approximately perfect gases, which means that gas enthalpies do not change with pressure when T is held constant.

$$\Delta_f H(H_2O,l,p) = \Delta_f H^{\ominus}(H_2O,l) + \left(\frac{\partial H_m(H_2O,l)}{\partial p}\right)_T \times (p - p^{\ominus}).$$

The derivative is simplified with the relationship

$$\left(\frac{\partial H_m}{\partial p}\right)_T = V_m - T\left(\frac{\partial V_m}{\partial T}\right)_p \simeq V_m \text{ for a condensed phase.}$$

(See Problem 2.34 in Atkins & de Paula, *Physical Chemistry*, 8th edition, 2006.)

$$\Delta_f H(H_2O,l,p) = \Delta_f H^{\ominus}(H_2O,l) + V_m \times (p - p^{\ominus}).$$

$$\Delta_f H(H_2O,l,p) - \Delta_f H^{\ominus}(H_2O,l) = V_m \times (p - p^{\ominus})$$

where

$$V_m = \left(\frac{1.00\,\text{cm}^3}{1\,\text{g}}\right) \times \left(\frac{18.0\,\text{g}}{1\,\text{mol}}\right) = \frac{18.0 \times 10^{-6}\,\text{m}^3}{1\,\text{mol}}.$$

Evaluating this difference at $p = 1.000\,\text{atm} = 1.013\,\text{bar}$ yields

$$\Delta_f H(H_2O,l,1\,\text{atm}) - \Delta_f H^{\ominus}(H_2O,l) = \frac{18.0 \times 10^{-6}\,\text{m}^3}{1\,\text{mol}} \times (1.013 - 1)\text{bar} \times \left(\frac{10^5\,\text{Pa}}{1\,\text{bar}}\right)$$

$$= \boxed{0.023\,\text{J}\,\text{mol}^{-1}}.$$

3.6 $\Delta_{fus}H^{\ominus} = 2.60\,kJ\,mol^{-1}$ [*Handbook of Chemistry and Physics*].

$$n = \frac{224 \times 10^3\,g}{22.99\,g\,mol^{-1}} = 9.74 \times 10^3\,mol.$$

$$q = n\Delta_{fus}H^{\ominus} = 9.74 \times 10^3\,mol \times 2.60\,kJ\,mol^{-1}$$

$$= \boxed{2.53 \times 10^4\,kJ}.$$

3.7 (a) $q = n\Delta_{vap}H(298.15\,K) = \frac{m}{M}\Delta_{vap}H(298.15\,K)$

$$= \frac{m}{M}\{\Delta_f H(g, 298.15\,K) - \Delta_f H(l, 298.15\,K)\}$$

$$= \left(\frac{1.00 \times 10^3\,g}{18.02\,g\,mol^{-1}}\right) \times (-241.82 - (-285.83))\,kJ\,mol^{-1}$$

$$= \boxed{2.44 \times 10^3\,kJ}\ \text{[Table D1.2]}.$$

(b) $q = n\Delta_{vap}H(373.15\,K) = \frac{m}{M}\Delta_{vap}H(373.15\,K)$

$$= \left(\frac{1.00 \times 10^3 g}{18.02\,g\,mol^{-1}}\right) \times (40.7\,kJ\,mol^{-1})\ \text{[Table 3.1]} = \boxed{2.26 \times 10^3\,kJ}.$$

3.8 The heat supplied to the sample is

$$q = IVt\ [2.5]$$
$$= 0.812\,A \times 11.5\,V \times 303\,s = 2.83 \times 10^3 J.$$

$q = \Delta H = n\Delta_{vap}H^{\ominus}$ (pressure is constant).

$$n = \frac{4.27\,g}{60.04\,g\,mol^{-1}} = 0.0711\,mol.$$

$$\Delta_{vap}H^{\ominus} = \frac{q}{n} = \frac{2.83 \times 10^3\,J}{0.0711\,mol} = 3.98 \times 10^4\,J\,mol^{-1} = \boxed{39.8\,kJ\,mol^{-1}}.$$

3.9 $q = n\Delta_{vap}H^{\ominus} = 1.50\,mol \times 26.0\,kJ\,mol^{-1} = \boxed{+39.0\,kJ}.$

$w = -p_{ex}\Delta V \approx -p_{ex}V(g)$, because $V(g) \gg V(l)$.

$V(g) = \frac{nRT}{p_{ex}}$, because gas is assumed to be perfect.

$$w = -p_{ex}\left(\frac{nRT}{p_{ex}}\right) = -nRT.$$

Substituting into the above equation,

$$w = -1.50\,\text{mol} \times 8.314\,\text{J}\,\text{K}^{-1}\,\text{mol}^{-1} \times 250\,\text{K} = \boxed{-3.12\,\text{kJ}}\,.$$

$$\Delta H = q = \boxed{+39.0\,\text{kJ}}\ \text{(constant pressure)}.$$

$$\Delta U = q + w = +39.0\,\text{kJ} - 3.12\,\text{kJ} = \boxed{35.9\,\text{kJ}}\,.$$

3.10 $n = \dfrac{100\,\text{g ice}}{18.0\,\text{g mol}} = 5.55\,\text{mol}.$

The heat needed to melt 100 g of ice is

$$q_1 = n \times \Delta_{\text{fus}}H^{\ominus}$$
$$= 5.55\,\text{mol} \times 6.01\,\text{kJ}\,\text{mol}^{-1} = 33.4\,\text{kJ}.$$

The heat needed to raise the temperature of the water from 0°C to 100°C is

$$q_2 = 100\,\text{g} \times 4.18\,\text{J}\,\text{K}^{-1}\,\text{g}^{-1} \times 100\,\text{K} = 4.18 \times 10^4\,\text{J} = \boxed{41.8\,\text{kJ}}\,.$$

The heat needed to vaporize the water is

$$q_3 = 5.55\,\text{mol} \times 40.7\,\text{kJ}\,\text{mol}^{-1} = \boxed{226\,\text{kJ}}\,.$$

The total heat is $q = q_1 + q_2 + q_3 = 33.4\,\text{kJ} + 41.8\,\text{kJ} + 226\,\text{kJ} = 301\,\text{kJ}$.

The graph of temperature against time is sketched in Figure 3.1. Note that the length of the liquid + gas, two-phase line is longer than the solid + liquid line in proportion to their $\Delta_{\text{trs}}H$ values.

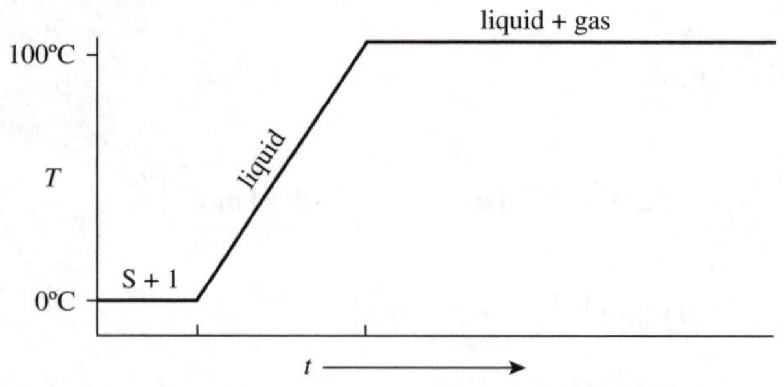

Fig 3.1

3.11 We follow Example 3.2, but with calcium in place of magnesium.

The overall process is

$$\text{Ca(s)} \rightarrow \text{Ca}^{2+}\text{(g)} + 2\text{e}^{-}\text{(g)}.$$

The process is broken down as follows:

				$\Delta H/(\text{kJ/mol})$
Sublimation:	Ca(s)	→	Ca(g)	+178.2
First ionization:	Ca(g)	→	$Ca^{+}(g) + e^{-}(g)$	+590
Second ionization:	$Ca^{+}(g)$	→	$Ca^{2+}(g) + e^{-}(g)$	+1150
Overall:	Ca(s)	→	$Ca^{2+}(g) + 2e^{-}(g)$	+1918

For the 10.0 g sample,

$$q = n\Delta H$$

$$= \left(\frac{10.0\,\text{g}}{40.08\,\text{g mol}^{-1}}\right) \times 1918\,\text{kJ mol}^{-1}$$

$$= \boxed{478\,\text{kJ}}.$$

3.12 $\text{Mg(g)} \rightarrow \text{Mg}^{2+}\text{(g)} + 2\text{e}^{-}\text{(g)}, \quad \Delta\nu_{\text{gas}} = 2.$

$\Delta_{\text{ion}}H = \Delta_{\text{ion}}U + \Delta\nu_{\text{gas}}RT \quad [3.3].$

$$\Delta_{\text{ion}}H - \Delta_{\text{ion}}U = \Delta\nu_{\text{gas}}RT = 2 \times \left(8.3145\,\text{J K}^{-1}\,\text{mol}^{-1}\right) \times (298.15\,\text{K})$$

$$= \boxed{4.96\,\text{kJ mol}^{-1}}.$$

3.13 $\text{Cl(g)} + \text{e}^{-}\text{(g)} \rightarrow \text{Cl}^{-}\text{(g)}, \quad \Delta\nu_{\text{gas}} = -1.$

$\Delta_{\text{eg}}H = \Delta_{\text{eg}}U + \Delta\nu_{\text{gas}}RT \quad [3.3].$

$$\Delta_{\text{eg}}H - \Delta_{\text{eg}}U = \Delta\nu_{\text{gas}}RT = (-1) \times \left(8.3145\,\text{J K}^{-1}\,\text{mol}^{-1}\right) \times (298.15\,\text{K})$$

$$= \boxed{-2.48\,\text{kJ mol}^{-1}}.$$

3.14 $\text{Cl}_2\text{(g)} \rightarrow \text{Cl}^{+}\text{(g)} + \text{Cl}^{-}\text{(g)}.$

$$\Delta_{\text{m}}H = \Delta_{\text{ion}}H(\text{Cl}) + \Delta_{\text{eg}}H(\text{Cl}) + \Delta_{\text{B}}H\left(\text{Cl}_2\right) = \Delta_{\text{ion}}H(\text{Cl}) + \Delta_{\text{eg}}H(\text{Cl}) + 2\Delta_{\text{f}}H(\text{Cl})$$

$$= (1257.5 - 354.8 + 2 \times 121.68)\,\text{kJ mol}^{-1} = \boxed{1146.1\,\text{kJ mol}^{-1}}$$

[Hess's law, [3.4], Table D1.2].

$$\Delta H = n\Delta_{\text{m}}H = 10.0\,\text{g}\left(\frac{1\,\text{mol}}{70.90\,\text{g}}\right)\left(1146.1\,\text{kJ mol}^{-1}\right) = \boxed{162\,\text{kJ}}.$$

3.15 (a) $Cl^-(g) \rightarrow Cl(g) + e^-(g), \quad \Delta_{ion}H(Cl^-) = -\Delta_{eg}H(Cl) = \boxed{354.8\,kJ\,mol^{-1}}$.

(b) $\Delta_{ion}H(Cl^-) = \Delta_{ion}U(Cl^-) + \Delta\nu_{gas}RT \quad [3.3].$

$$\Delta_{ion}U(Cl^-) = \Delta_{ion}H(Cl^-) - \Delta\nu_{gas}RT$$

$$= 354.8\,kJ\,mol^{-1} - 1 \times (8.315\,J\,K^{-1}\,mol^{-1}) \times (298.15\,K)$$

$$= \boxed{352.3\,kJ\,mol^{-1}}.$$

3.16 (a) The mean bond enthalpy is the average of the three enthalpy changes given.

$$\Delta H_B(N{-}H) = \left(\frac{460 + 390 + 314}{3}\right) \times kJ\,mol^{-1}$$

$$= \boxed{388\,kJ\,mol^{-1}}.$$

(b) The bond dissociation energies and enthalpies refer to gas-phase dissociation to atoms. Because of the dissociation, the number of moles of gas-phase particles increases $(\Delta\nu_g > 0)$. Therefore, since

$\Delta H = \Delta U + \Delta\nu_g RT$, ΔU_B (N—H) is expected to be $\boxed{\text{smaller}}$ than ΔH_B (N—H).

3.17 (a) $C_6H_{12}O_6(aq) \rightarrow 2CH_3CH(OH)COOH(aq)$.
The molecular formulas of these compounds are shown.

7 C—H single bonds
5 O—H single bonds
5 C—O single bonds
1 C=O double bond
5 C—C single bonds

```
    HC=O
     |
    HC—OH
     |
  HO—CH
     |
    HC—OH
     |
    HC—OH
     |
    CH2OH
  glucose
```

4 C—H single bonds × 2
2 O—H single bonds × 2
2 C—O single bonds × 2
1 C=O double bond × 2
2 C—C single bonds × 2

```
       O
      //
     C—OH
     |
  HO—C—H
     |
     CH3
 lactic acid
```

$$\Delta_r H = \Sigma\Delta H\ (\text{bonds broken}) + \Sigma\Delta H\ (\text{bonds formed}).$$

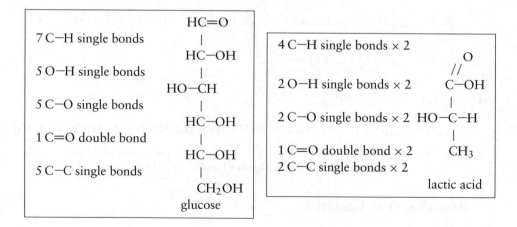

Note that the ΔH for the formation of a bond is the negative of the bond enthalpy.

$$\Sigma\Delta H \text{ (bonds broken)} = 7\Delta H(\text{C—H}) + 5\Delta H(\text{O—H}) + 5\Delta H(\text{C—O})$$

$$+ 1\Delta H(\text{C=O}) + 5\Delta H(\text{C—C})$$

$$= (7 \times 412) + (5 \times 463) + (5 \times 360) + 743 + (5 \times 348)$$

$$= 9482 \text{ kJ mol}^{-1}.$$

$$\Sigma\Delta H\text{(bonds formed)} = -8\Delta H(\text{C—H}) - 4\Delta H(\text{O—H}) - 4\Delta H(\text{C—O})$$

$$- 2\Delta H(\text{C=O}) - 4\Delta H(\text{C—C})$$

$$= -(8\times412)-(4\times463)-(4\times360)-(2\times743)-(4\times348)$$

$$= -9466 \text{ kJ mol}^{-1}.$$

$$\Delta_r H = (9482 - 9466) \text{ kJ mol}^{-1} = \boxed{16 \text{ kJ mol}^{-1}}.$$

The approximate nature of this kind of calculation can be seen by comparing this value of $\Delta_r H$ to the value of $\Delta_r H^\ominus$ calculated from standard enthalpies of formation, which is $\Delta_r H^\ominus = -114 \text{ kJ mol}^{-1}$. Strictly speaking, bond enthalpies apply only to gas-phase processes.

(b) $C_6H_{12}O_6(aq) + 6O_2(g) \rightarrow 6CO_2(g) + 6H_2O(l)$.

$$\Delta_r H = 9482 \text{ kJ mol}^{-1} \text{ [from part (a)]} + 6 \times \Delta H(\text{O=O}) - 12 \times \Delta H(\text{O=CO})$$

$$- 12 \times \Delta H(\text{H—OH})$$

$$= 9482 \text{ kJ mol}^{-1} + (6 \times 497) - (12 \times 799) - (12 \times 492) \text{ kJ mol}^{-1}$$

$$= \boxed{-3028 \text{ kJ mol}^{-1}}.$$

3.18 (a) 1.00 mol N_2 consumed.

$$\Delta H^\ominus = 1.00 \text{ mol} \times (-92.22 \text{ kJ mol}^{-1}) = \boxed{-92.22 \text{ kJ}}.$$

(b) 1.00 mol NH_3 formed.

$$\Delta H^\ominus = \frac{1 \text{ mol N}_2}{2 \text{ mol NH}_3} \times 1 \text{ mol NH}_3 \times (-92.22 \text{ kJ mol}^{-1}) = \boxed{-46.11 \text{ kJ}}.$$

3.19 (a) The standard enthalpy of combustion applies to the combustion of one mole; therefore

$$\Delta_c H^\ominus = \frac{1 \text{ mol}}{2 \text{ mol}} \times (-3120 \text{ kJ mol}^{-1}) = \boxed{-1560 \text{ kJ mol}^{-1}}.$$

(b) $\Delta_r H^\ominus = 3.00\,\text{mol CO}_2 \times \dfrac{2\,\text{mol C}_2\text{H}_6}{4\,\text{mol CO}_2} \times \left(\dfrac{-1560\,\text{kJ}}{\text{mol C}_2\text{H}_6}\right)$

$= \boxed{-2340\,\text{kJ}}$.

3.20 $C_6H_5C_2H_5(l) + \frac{21}{2}O_2(g) \rightarrow 8CO_2(g) + 5\,H_2O(l)$.

$$\Delta_c H^\ominus = 8\Delta_f H^\ominus[CO_2(g)] + 5\,\Delta_f H^\ominus[H_2O(l)] - \Delta_f H^\ominus[C_6H_5C_2H_5(l)]$$

$$= 8 \times (-393.51\,\text{kJ mol}^{-1}) + 5 \times (-285.83\,\text{kJ mol}^{-1}) - (-12.5\,\text{kJ mol}^{-1})$$

$$= \boxed{-4564.7\,\text{kJ mol}^{-1}}.$$

3.21 The reaction wanted is: $C_6H_{10}(l) + H_2(g) \rightarrow C_6H_{12}(l)$.

Use the following reactions:

(1) $C_6H_{10}(l) + 8\frac{1}{2}O_2 \rightarrow 6CO_2(g) + 5\,H_2O(l)$, $\quad\Delta_c H^\ominus = -3752\,\text{kJ mol}^{-1}$,
(2) $C_6H_{12}(l) + 9\,O_2(g) \rightarrow 6CO_2(g) + 6H_2O(l)$, $\quad\Delta_c H^\ominus = -3953\,\text{kJ mol}^{-1}$,
(3) $H_2(g) + \frac{1}{2}O_2(g) \rightarrow H_2O(l)$, $\quad\Delta_f H^\ominus = -286\,\text{kJ mol}^{-1}$.

Now sum eqn (1), eqn (3), and the reverse of eqn (2) to get the desired equation; then

$$\Delta_r H^\ominus = -3752\,\text{kJ mol}^{-1} + 3953\,\text{kJ mol}^{-1} - 286\,\text{kJ mol}^{-1} = \boxed{-85\,\text{kJ mol}^{-1}}.$$

3.22 $3\,C(s) + 3\,H_2(g) + O_2(g) \rightarrow CH_3COOCH_3(l)$, $\quad\Delta_f H^\ominus = -\,442\,\text{kJ mol}^{-1}$.

$\Delta_f U^\ominus = \Delta_f H^\ominus - \Delta(pV)$ [2.13a] $= \Delta_f H^\ominus - \Delta\nu_{gas}RT$, $\quad\Delta\nu_{gas} = -4\,\text{mol}$.

$$\Delta_f U^\ominus = -442\,\text{kJ mol}^{-1} - (-4 \times 8.3145\,\text{J K}^{-1}\,\text{mol}^{-1} \times 298.15\,\text{K} \times 10^{-3}\,\text{kJ J}^{-1})$$

$$= \boxed{-432\,\text{kJ mol}^{-1}}.$$

3.23 $C_{10}H_8(s) + 12\,O_2(g) \rightarrow 10CO_2(g) + 4\,H_2O(l)$, $\quad\Delta_c H^\ominus = -\,5157\,\text{kJ mol}^{-1}$.

The reverse reaction is

$10CO_2(g) + 4\,H_2O(l) \rightarrow C_{10}H_8(s) + 12\,O_2(g)$, $\quad\Delta H^\ominus = 5157\,\text{kJ mol}^{-1}$.

The CO_2 and H_2O can be replaced by adding the following two reactions and using data from Table 3.8 for $\Delta_f H^\ominus(CO_2)$ and $\Delta_f H^\ominus(H_2O)$.

$10\,C(s) + 10\,O_2(g) \rightarrow 10CO_2(g)$, $\quad\Delta H^\ominus = 10 \times (-393.5\,\text{kJ mol}^{-1})$.

$4\,H_2(g) + 2\,O_2(g) \rightarrow 4\,H_2O(l)$, $\quad\Delta H^\ominus = 4 \times (-285.8\,\text{kJ mol}^{-1})$.

Overall,

$$10\,C(s) + 4\,H_2(g) \rightarrow C_{10}H_8(s).$$

$$\Delta_r H^\ominus = (+5157 - 3935 - 1143)\,\text{kJ mol}^{-1} = \boxed{+79\,\text{kJ mol}^{-1}}.$$

3.24 $q = n\Delta_c H^\ominus$ with $\Delta_c H^\ominus = -5157\,\text{kJ mol}^{-1}$ [Table D1.1].

Calculating q,

$$|q| = \frac{320 \times 10^{-3}\,\text{g}}{128.18\,\text{g mol}^{-1}} \times 5157\,\text{J mol}^{-1} = 12.87\,\text{kJ}.$$

$$C = \frac{q}{\Delta T} = \frac{12.87\,\text{kJ}}{3.05\text{K}} = \boxed{4.22\,\text{kJ K}^{-1}}.$$

When phenol is used, $\Delta_c H^\ominus = -3054\,\text{kJ mol}^{-1}$ [Table D1.1].

$$|q| = \frac{100 \times 10^{-3}\,\text{g}}{94.12\,\text{g mol}^{-1}} \times 3054\,\text{J mol}^{-1} = 3.245\,\text{kJ}.$$

$$\Delta T = \frac{q}{C} = \frac{3.245\,\text{kJ}}{4.22\,\text{kJ K}^{-1}} = \boxed{0.769\,\text{K}}.$$

3.25 (a) $q = C\Delta T$.

$$\Delta_c H^\ominus = \frac{q}{n} = \frac{C\Delta T}{n} = \frac{MC\Delta T}{m}$$

where M is the molar mass of glucose and m is the mass of the sample.

$$M = 180.16\,\text{g mol}^{-1}.$$

$$|\Delta_c H^\ominus| = \frac{180.16\,\text{g mol}^{-1} \times 641\,\text{J K}^{-1} \times 7.793\,\text{K}}{0.3212\,\text{g}} = 2802\,\text{kJ mol}^{-1}.$$

Because the combustion is exothermic, $\Delta_c H^\ominus = \boxed{-2.80\,\text{MJ mol}^{-1}}$.

(b) The combustion reaction is

$$C_6H_{12}O_6(s) + 6\,O_2(g) \rightarrow 6CO_2(g) + 6H_2O(l), \quad \Delta n_g = 0.$$

Therefore $\Delta_f U^\ominus = \Delta_c H^\ominus = \boxed{-2.80\,\text{MJ mol}^{-1}}$.

(c) For the enthalpy of formation we combine the following equations:

$$6CO_2(g) + 6\,H_2O(l) \rightarrow C_6H_{12}O_6(s) + 6\,O_2(g), \quad \Delta_r H^\ominus = +2.80\,\text{MJ mol}^{-1},$$

$$6\,C(s) + 6\,O_2(g) \rightarrow 6CO_2(g), \quad \Delta_r H^\ominus = 6 \times \Delta_f H^\ominus[CO_2(g)],$$

$$6\,H_2(g) + 3\,O_2(g) \rightarrow 6\,H_2O(l), \quad \Delta_r H^\ominus = 6 \times \Delta_f H^\ominus[H_2O(l)].$$

The sum of the three equations is

$$6C(s) + 6H_2(g) + 3O_2(g) \rightarrow C_6H_{12}O_6(s).$$

Therefore $\Delta_f H^\ominus$ (glucose) $= (2.80 - 2.36 - 1.72)\,\text{MJ mol}^{-1}$

$$= \boxed{-1.28\,\text{MJ mol}^{-1}}.$$

3.26 (a) On the assumption that the calorimeter is a constant volume bomb calorimeter such as the one described in Figure 2.15 in the main text, the heat released directly equals the internal energy of combustion, $\Delta_c U^\ominus$. Therefore, $\Delta_c U^\ominus = \boxed{-1333\,\text{kJ mol}^{-1}}$.

(b) $HOOCCH{=}CHCOOH(s) + 3O_2(g) \rightarrow 4CO_2(g) + 2H_2O(l)$.

$$\Delta \nu_g = +1.$$

$$\Delta_c H^\ominus = \Delta_c U^\ominus + \Delta \nu_g RT$$

$$= -1333\,\text{kJ mol}^{-1} + (1\,\text{mol} \times 2.5\,\text{kJ mol}^{-1}) = \boxed{-1331\,\text{kJ mol}^{-1}}.$$

(c) $\Delta_c H^\ominus = 4\Delta_f H^\ominus[CO_2(g)] + 2\Delta_f H[H_2O(l)] - \Delta_f H^\ominus$ (fumaric acid)

$$= -1331\,\text{kJ mol}^{-1}.$$

$\Delta_f H^\ominus$ (fumaric acid) $= 4 \times (-393.51\,\text{kJ mol}^{-1})$

$$+ 2 \times (-285.83\,\text{kJ mol}^{-1}) + 1331\,\text{kJ mol}^{-1}$$

$$= \boxed{-815\,\text{kJ mol}^{-1}}.$$

3.27 $AgBr(s) \rightarrow Ag^+(aq) + Br^-(aq)$.

$$\Delta_{soln} H^\ominus = \Delta_f H^\ominus[Ag^+(aq)] + \Delta_f H^\ominus[Br^-(aq)] - \Delta_f H^\ominus[AgBr(s)]$$

$$= [(105.58) + (-121.55) - (-100.37)]\,\text{kJ mol}^{-1} = \boxed{+84.40\,\text{kJ mol}^{-1}}.$$

3.28 Because for $NH_3SO_2(s) \rightarrow NH_3(g) + SO_2(g)$ $\quad \Delta H^\ominus = +40\,\text{kJ mol}^{-1}$,

for $NH_3(g) + SO_2(g) \rightarrow NH_3SO_2(s)$ $\quad \Delta H^\ominus = -40\,\text{kJ mol}^{-1}$.

For the latter reaction

$$\Delta_r H^\ominus = \Delta_f H^\ominus(NH_3SO_2) - \Delta_f H^\ominus(NH_3) - \Delta_f H^\ominus(SO_2) = -40\,\text{kJ mol}^{-1}.$$

Therefore, after solving for $\Delta_f H^\ominus(NH_3SO_2)$,

$$\Delta_f H^\ominus(NH_3SO_2, s) = \Delta_f H^\ominus(NH_3, g) + \Delta_f H^\ominus(SO_2, g) - 40\,\text{kJ mol}^{-1}$$

$$= (-46.11 - 296.83 - 40)\,\text{kJ mol}^{-1} = \boxed{-383\,\text{kJ mol}^{-1}}.$$

3.29 (1) $C(gr) + O_2(g) \rightarrow CO_2(g), \quad \Delta_c H^{\ominus} = -393.5\,kJ\,mol^{-1}.$
(2) $C(diam) + O_2(g) \rightarrow CO_2(g), \quad \Delta_c H^{\ominus} = -395.41\,kJ\,mol^{-1}.$

Subtracting (2) from (1) yields

$$C(gr) \rightarrow C(diam).$$

$$\Delta_{trs}H^{\ominus} = [-393.5 - (-395.41)]\,kJ\,mol^{-1} = \boxed{+1.9\,kJ\,mol^{-1}}.$$

3.30 $\Delta_{trs}U^{\ominus} = w + q = -p_{ex}\Delta V + q.$

$$q = \Delta_{trs}H = +1.9\,kJ\,mol^{-1}.$$

$$p = 150\,kbar = 1.50 \times 10^5\,bar = 1.50 \times 10^{10}\,Pa.$$

For 1 mol graphite,

$$V_{gr} = \frac{12.01\,g/mol}{2.250\,g/cm^3} \times \frac{1\,m^3}{10^6\,cm^3} = 5.338 \times 10^{-6}\,m^3\,mol^{-1}.$$

For 1 mol diamond

$$V_{diam} = \frac{12.01\,g/mol}{3.510\,g/cm^3} \times \frac{1\,m^3}{10^6\,cm^3} = 3.422 \times 10^{-6}\,m^3\,mol^{-1}.$$

$$\Delta V = V_{diam} - V_{gr} = 3.422 \times 10^{-6} m^3 - 5.338 \times 10^{-6} m^3$$

$$= -1.916 \times 10^{-6} m^3\,mol^{-1}.$$

$$-p_{ex}\Delta V = -1.50 \times 10^{10}\,Pa \times (-1.916 \times 10^{-6}\,m^3\,mol^{-1}) = 2.874 \times 10^4\,J\,mol^{-1}$$

$$= 28.74\,kJ\,mol^{-1}.$$

$$\Delta_{trs}U = 28.74\,kJ\,mol^{-1} + 1.9\,kJ\,mol^{-1} = \boxed{+30.6\,kJ\,mol^{-1}}.$$

3.31 $q = n\Delta_c H^{\ominus}$

$$= \frac{1.5\,g}{342.3\,g\,mol} \times (-5645\,kJ\,mol^{-1}) = \boxed{-25\,kJ}.$$

Effective work available is $\approx 25\,kJ \times 0.20 = 5.0\,kJ$

Because $w = mgh$, and $m \approx 68\,kg$,

$$h \approx \frac{5.0 \times 10^3\,J}{68\,kg \times 9.81\,m\,s^{-2}} = \boxed{7.5\,m}.$$

3.32 $C_3H_8(l) \rightarrow C_3H_8(g), \quad \Delta_{vap}H^{\ominus}.$

$C_3H_8(g) + 5\,O_2(g) \rightarrow 3CO_2(g) + 4\,H_2O(l), \quad \Delta_c H^{\ominus}(g).$

(a) $\Delta_c H^{\ominus}(l) = \Delta_{vap}H^{\ominus} + \Delta_c H^{\ominus}(g)$

$$= 15\,\text{kJ mol} - 2220\,\text{kJ mol}^{-1} = \boxed{-2205\,\text{kJ mol}^{-1}}.$$

(b) $\Delta \nu_g = -2[5\ O_2(g)$ replaced with $3\ CO_2(g)]$.

$$\Delta_c U^{\ominus}(l) = \Delta_c H^{\ominus}(l) - (-2)RT$$

$$= -2205\,\text{kJ mol}^{-1} + (2 \times 2.5\,\text{kJ mol}^{-1}) = \boxed{-2200\,\text{kJ mol}^{-1}}.$$

3.33 (a) Exothermic, $\Delta_r H^{\ominus}$ = negative.
(b) Endothermic, $\Delta H^{\ominus}$ = positive.
(c) Endothermic, $\Delta_{vap}H^{\ominus}$ = positive.
(d) Endothermic, $\Delta_{fus}H^{\ominus}$ = positive.
(e) Endothermic, $\Delta_{sub}H^{\ominus}$ = positive.

3.34 (a) $\Delta_r H^{\ominus} = \Delta_f H^{\ominus}(N_2O_4, g) - 2\Delta_f H^{\ominus}(NO_2, g)$

$$= [9.16 - 2 \times 33.18]\,\text{kJ mol}^{-1} = \boxed{-57.20\,\text{kJ mol}^{-1}}.$$

(b) $\Delta_r H^{\ominus} = \frac{1}{2}\Delta_f H^{\ominus}(N_2O_4, g) - \Delta_f H^{\ominus}(NO_2, g)$

$$= \frac{1}{2}(9.16) - 33.18\,\text{kJ mol}^{-1} = \boxed{-28.6\,\text{kJ mol}^{-1}}.$$

(c) $\Delta_r H^{\ominus} = 2 \times \Delta_f H^{\ominus}(HNO_3, aq) + \Delta_f H^{\ominus}(NO, g) - 3 \times \Delta_f H^{\ominus}(NO_2, g) - \Delta_f H^{\ominus}(H_2O, l)$

$$= [2 \times (-207.36) + 90.25 - 3 \times (33.18) - (-285.83)]\,\text{kJ mol}^{-1}$$

$$= \boxed{-138.2\,\text{kJ mol}^{-1}}.$$

(d) $\Delta_r H^{\ominus} = \Delta_f H^{\ominus}(\text{propene, g}) - \Delta_f H^{\ominus}(\text{cyclopropane, g})$

$$= [20.42 - 53.30]\,\text{kJ mol}^{-1} = \boxed{-32.88\,\text{kJ mol}^{-1}}.$$

(e) In order to calculate $\Delta_r H^{\ominus}$ first write the net ionic equation:

$$H^+(aq) + Cl^-(aq) + Na^+(aq) + OH^-(aq) \rightarrow Na^+(aq) + Cl^-(aq) + H_2O(l).$$

Simplifying we obtain

$$H^+(aq) + OH^-(aq) \rightarrow H_2O(l).$$

$$\Delta_r H^{\ominus} = \Delta_f H^{\ominus}(H_2O, l) - \Delta_f H^{\ominus}(H^+, aq) - \Delta_f H^{\ominus}(OH^-, aq)$$

$$= [-285.83 - 0 - (-229.99)]\,\text{kJ mol}^{-1} = \boxed{-55.84\,\text{kJ mol}^{-1}}.$$

3.35 The formation of N_2O_5 is the sum of the three reactions

	$\Delta_r H^\ominus/(\text{kJ mol}^{-1})$
$2NO(g) + O_2(g) \rightarrow 2NO_2(g)$	-114.1
$\frac{1}{2}O_2(g) + 2\,NO_2(g) \rightarrow N_2O_5(g)$	$\frac{1}{2}(-110.2)$
$N_2(g) + O_2(g) \rightarrow 2NO(g)$	180.5
$N_2(g) + \frac{5}{2}O_2(g) \rightarrow N_2O_5(g)$	$+11.3$

Therefore, $\Delta_f H^\ominus[N_2O_5(g)] = \boxed{+11.3\,\text{kJ mol}^{-1}}$.

3.36 We use equations 3.6 and 3.7.

$$\Delta_r H^\ominus(T_2) = \Delta_r H^\ominus[T_1] + \Delta_r C_p \Delta T \quad [3.6].$$

$$\Delta_r C_p = \sum \nu C_{p,\text{m}}\,(\text{products}) - \sum \nu C_{p,\text{m}}\,(\text{reactants}) \quad [3.7]$$

$$= (77.28 - 2 \times 37.20)\,\text{J K}^{-1}\,\text{mol}^{-1} = +2.88\,\text{J K}^{-1}\,\text{mol}^{-1}.$$

$$\Delta_r H^\ominus(373\,\text{K}) = \Delta_r H^\ominus(298\,\text{K}) + \Delta_r C_p \Delta T$$

$$= -57.20\,\text{kJ mol}^{-1} + (2.88\,\text{J K}^{-1} \times 75\,\text{K})$$

$$= (-57.20 + 0.22)\,\text{kJ mol}^{-1} = \boxed{-56.98\,\text{kJ mol}^{-1}}.$$

3.37 $\Delta_{\text{vap}} H^\ominus(T') = \Delta_{\text{vap}} H^\ominus(T) + \Delta_r C_p^\ominus \times (T' - T) \quad [3.6].$

$$\Delta_{\text{vap}} H^\ominus(373\,\text{K}) = 44.01\,\text{kJ mol}^{-1} + (-41.71\,\text{J K}^{-1}\,\text{mol}^{-1}) \times (373\,\text{K} - 298\,\text{K})$$

$$= \boxed{40.88\,\text{kJ mol}^{-1}}.$$

3.38 The sign of $\Delta_r C_p$ in equation 3.6 determines whether or not $\Delta_r H^\ominus$ will increase or decrease with increasing T. A negative value of $\Delta_r C_p$ implies a decrease; a positive value an increase with increasing T.

(a) $\Delta_r C_p = (2 \times 4R) - \left(3 \times \frac{7}{2} R\right) = \frac{-5}{2} R$; therefore $\Delta H^\ominus$ will $\boxed{\text{decrease}}$ with increasing T.

(b) $\Delta_r C_p = 8R - \frac{7}{2} R - \left(3 \times \frac{7}{2} R\right) = -6R$, $\boxed{\text{decrease}}$.

(c) $\Delta_r C_p = 8R + \frac{7}{2} R - 4R - \left(2 \times \frac{7}{2} R\right) = +\frac{1}{2} R$, $\boxed{\text{increase}}$.

3.39 (a) $\Delta_r C_p = 2 \times 9R - 3 \times \frac{7}{2} R = +\frac{15}{2} R$, $\boxed{\text{increase}}$.

(c) $\Delta_r C_p = \frac{7}{2} R + (2 \times 9R) - 4R - \left(2 \times \frac{7}{2} R\right) = +\frac{21}{2} R$, $\boxed{\text{increase}}$.

3.40 $C_6H_{12}O_6(s) + 6O_2(g) \rightarrow 6CO_2(g) + 6H_2O(l)$.

Higher because, as temperature increases, $\Delta_r C_p$ will likely increase because of the large n for products.

3.41 $\nu_A A + \nu_B B + \cdots \rightarrow \nu_P P + \nu_Q Q + \cdots \ \Delta_r U(T)$

Let $\Delta U_m(i)$ be the molar internal energy change of the ith chemical species due to the temperature change $\Delta T = T' - T$. The reaction internal energy change due to the temperature change is $\Delta_r U(T') - \Delta_r U(T)$, which equals the sum of $\nu_{product}\Delta U_m(\text{product})$ minus the sum of $\nu_{reactant}\Delta U_m(\text{reactant})$.

$$\Delta_r U(T') - \Delta_r U(T) = \nu_P \Delta U_m(P) + \nu_Q \Delta U_m(Q) + \cdots$$

$$- \{\nu_A \Delta U_m(A) + \nu_B \Delta U_m(B) + \cdots\}$$

$$= \Sigma \nu \Delta U_m(\text{products}) - \Sigma \nu \Delta U_m(\text{reactants})$$

$$\Delta_r U(T') = \Delta_r U(T) + \Sigma \nu \Delta U_m(\text{products}) - \Sigma \nu \Delta U_m(\text{reactants})$$

Substitution of $\Delta U_m(i) = C_{V,m}(i)\ \Delta T$ [2.11] gives

$$\Delta_r U(T') = \Delta_r U(T) + \Sigma \nu C_{V,m}(\text{products})\ \Delta T - \Sigma \nu C_{V,m}(\text{reactants})\ \Delta T$$

$$= \Delta_r U(T) + \{\Sigma \nu C_{V,m}(\text{products}) - \Sigma \nu C_{V,m}(\text{reactants})\}\Delta T$$

or

$$\boxed{\Delta_r U(T') = \Delta_r U(T) + \Delta_r C_V \times (T' - T)} \quad \text{where}$$

$$\Delta_r C_V = \Sigma \nu C_{V,m}(\text{products}) - \Sigma \nu C_{V,m}(\text{reactants})$$

3.42 $\nu_A A + \nu_B B + \cdots \rightarrow \nu_P P + \nu_Q Q + \cdots \ \Delta_r H(T)$.

Let $dH_m(i)$ be the infinitesimal molar internal energy change of the ith chemical species due to the infinitesimal temperature change dT. The infinitesimal reaction internal energy change due to the temperature change is $d\Delta_r H(T)$, which equals the sum of $\nu_{product} dH_m(\text{product})$ minus the sum of $\nu_{reactant} dH_m(\text{reactant})$.

$$d\Delta_r H(T) = \nu_P dH_m(P) + \nu_Q dH_m(Q) + \cdots - \{\nu_A dH_m(A) + \nu_B dH_m(B) + \cdots\}$$

$$= \Sigma \nu dH_m(\text{products}) - \Sigma \nu dH_m(\text{reactants})$$

$$= \Sigma \nu dH_m(\text{products}) - \Sigma \nu dH_m(\text{reactants}).$$

Substitution of $dH_m(i) = C_{p,m}(i)\ dT$ [2.16 as infinitesimal expression] gives

$$d\Delta_r H(T) = \Sigma \nu C_{p,m}(\text{products})\ dT - \Sigma \nu C_{p,m}(\text{reactants})\ dT$$

$$= \{\Sigma \nu C_{p,m}(\text{products}) - \Sigma \nu C_{p,m}(\text{reactants})\}\ dT$$

or

$$\mathrm{d}\Delta_r H(T) = \Delta_r C_p \,\mathrm{d}T \quad \text{where } \Delta_r C_p = \Sigma \nu C_{p,\mathrm{m}}(\text{products}) - \Sigma C_{p,\mathrm{m}}(\text{reactants}).$$

Integration between T and T' gives

$$\int_T^{T'} \mathrm{d}\Delta_r H(T) = \int_T^{T'} \Delta_r C_p \,\mathrm{d}T.$$

$$\Delta_r H(T') - \Delta_r H(T) = \int_T^{T'} \Delta_r C_p \,\mathrm{d}T \text{ or } \Delta_r H(T') = \Delta_r H(T) + \int_T^{T'} \Delta_r C_p \,\mathrm{d}T.$$

Case 1. If $\Delta_r C_p$ is either temperature independent or negligibly dependent upon temperature over the temperature range, the integral on the right simplifies to Kirchhoff's law.

$$\Delta_r H(T') = \Delta_r H(T) + \int_T^{T'} \Delta_r C_p \,\mathrm{d}T = \Delta_r H(T) + \Delta_r C_p \int_T^{T'} \mathrm{d}T$$

$$= \Delta_r H(T) + \Delta_r C_p \times (T' - T).$$

Case 2. If $\Delta_r C_p$ is temperature dependent, care must be taken with the integral on the right. For example, if $\Delta_r C_p = a + bT + c/T^2$ the integral is

$$\int_T^{T'} \Delta_r C_p \,\mathrm{d}T = \int_T^{T'} \left(a + bT + \frac{c}{T^2}\right) \mathrm{d}T$$

$$= a\int_T^{T'} \mathrm{d}T + b\int_T^{T'} T\,\mathrm{d}T + c\int_T^{T'} \frac{1}{T^2}\,\mathrm{d}T$$

$$= aT\Big|_T^{T'} + \frac{bT^2}{2}\Big|_T^{T'} - \frac{c}{T}\Big|_T^{T'}$$

$$= a(T' - T) + \frac{b}{2}(T'^2 - T^2) - c\left(\frac{1}{T'} - \frac{1}{T}\right)$$

and the reaction enthalpy is

$$\Delta_r H(T') = \Delta_r H(T) + \int_T^{T'} \Delta_r C_p \,\mathrm{d}T$$

$$= \boxed{\Delta_r H(T) + a(T' - T) + \frac{b}{2}(T'^2 - T^2) - c\left(\frac{1}{T'} - \frac{1}{T}\right)}.$$

Chapter 4

Thermodynamics: the Second Law

Answers to discussion questions

4.1 $\Delta S_{\text{total}} = \Delta S_{\text{isolated system}} > 0$ (the universe is an isolated system) is a statement of the Second Law of thermodynamics. It is applicable to all macroscopic changes such as heat process, chemical reactions, and phase transitions. It establishes criteria for spontaneity of changes and processes within an isolated system.

$\mathrm{d}G \leq 0$ is valid under conditions of constant temperature and pressure, with no additional work other than pressure–volume work. The statement originates with the Gibbs energy definition, $G = H - TS$, and application of the Second Law. The inequality $\mathrm{d}G < 0$ is the criterion for spontaneous change of processes at constant T and constant p processes within non-isolated systems, including chemical reactions. The equality $\mathrm{d}G = 0$ is the criterion for equilibrium with respect to constant T and constant p changes and processes within non-isolated systems.

4.2 A liquid and its vapor are in equilibrium at the boiling point. Since the process is reversible under this condition, $\Delta S = q_{\text{rev}}/T$ [4.1] $= \Delta H/T$ [2.15b, constant p] and the entropy of vaporization is given by $\Delta_{\text{vap}}S_{\text{m}} = \Delta_{\text{vap}}H_{\text{m}}/T_{\text{b}}$.

Trouton's rule states that the ratio $\Delta_{\text{vap}}H/T_{\text{b}}$, and thus the vaporization entropy, is a constant. Explore the origin of the constancy by considering that the vaporization entropy has two components, only one of which depends upon liquid phase properties. They are the molar entropy of the liquid and the molar entropy of the gas: $\Delta_{\text{vap}}S_{\text{m}} = S_{\text{m}}(\text{g}) - S_{\text{m}}(\text{l})$. Under ordinary conditions, the value of $S_{\text{m}}(\text{g})$ is expected to be identical for all gases and relatively large with respect to $S_{\text{m}}(\text{l})$ because gases behave as perfect gases for which molecular volume and intermolecular forces are negligibly small. This allows completely random, very high entropy molecular motion, which is independent of molecular properties. In addition to the large $S_{\text{m}}(\text{g})$ value, should the value of $S_{\text{m}}(\text{l})$ be either negligibly small or a constant value for a series of compounds, $\Delta_{\text{vap}}S_{\text{m}}$, and the ratio $\Delta_{\text{vap}}H/T_{\text{b}}$, will be a constant. Trouton's rule is followed. Small, non-polar molecules provide examples that meet these conditions. The relative absence of molecular order in their liquid states gives relatively small and constant $S_{\text{m}}(\text{l})$ values. Thus, bromine, carbon tetrachloride, and cyclohexane have approximately identical vaporization entropies ($\sim 85\,\mathrm{J\,K^{-1}\,mol^{-1}}$, Table 4.1).

Exceptions to Trouton's rule include liquids in which the interactions between molecules result in the liquid being less disordered than the random jumble of molecules in something like carbon tetrachloride. This includes liquids in which hydrogen bonding creates local order, as in water and small alcohols. It also includes liquid metals in which the metallic bond creates atomic organization, as in mercury.

It is also interesting to explore the origin of Trouton's rule with a careful analysis of the vaporization enthalpy (heat). Energy in the form of heat supplied to a liquid manifests itself as an increase in thermal motion. This is an increase in the kinetic energy of molecules. When the kinetic energy of the molecules is sufficient to overcome the attractive energy that hold them together the liquid vaporizes. The enthalpy of vaporization is the heat required to accomplish this at constant pressure. It seems reasonable that, the greater the enthalpy of vaporization, the greater the kinetic energy required, and the greater the temperature needed to achieve this kinetic energy. Hence, we expect that $\Delta_{\mathrm{vap}}H$ is proportional to T_{b}, which implies that their ratio is a constant.

4.3 We must remember that the Second Law of thermodynamics states only that the total entropy of both the system (here, the molecules organizing themselves into cells) and the surroundings (here, the medium) must increase in a naturally occurring process. It does not state that entropy must increase in a portion of the universe that interacts with its surroundings. In this case, the cells grow by using chemical energy from their surroundings (the medium) and in the process the increase in the entropy of the medium outweighs the decrease in entropy of the system. Hence, the Second Law is not violated.

4.4 (a) $\boxed{\text{Positive}}$, due to greater disorder in the product, though the difference may not be large.

(b) $\boxed{\text{Negative}}$, less disorder (smaller number of moles of gas) in the product.

(c) $\boxed{\text{Positive}}$, two new substances are formed, resulting in greater disorder on the product side.

Solutions to exercises

4.5 $$\Delta S_{\mathrm{sur}} = \frac{q_{\mathrm{sur}}}{T}\,[4.7] = \frac{120\,\mathrm{J}}{293\,\mathrm{K}} = \boxed{0.410\,\mathrm{J\,K^{-1}}}.$$

4.6 (a) We assume that the ice melts reversibly under the conditions described; therefore

$$\Delta S_{\mathrm{ice}} = \frac{q_{\mathrm{rev}}}{T}\,[4.1] = \frac{33\,\mathrm{kJ}}{273\,\mathrm{K}} = +0.12\,\mathrm{kJ\,K^{-1}}.$$

(b) $$\Delta S_{\mathrm{sur}} = \frac{q_{\mathrm{sur}}}{T}\,[4.7] = \frac{-33\,\mathrm{kJ}}{273\,\mathrm{K}} = \boxed{-0.12\,\mathrm{kJ\,K^{-1}}}.$$

Note. Because this process is reversible, the total entropy change is zero.

4.7 $q = nC_{p,\mathrm{m}}\Delta T.$

$$q = \frac{1.25\times10^3\,\mathrm{g}}{26.98\,\mathrm{g\,mol^{-1}}}\times 24.35\,\mathrm{J\,K^{-1}\,mol^{-1}}\times(-40\,\mathrm{K})$$

$$= \boxed{-45.1\,\mathrm{kJ}}\,.$$

$$\Delta S = C_p \ln\frac{T_2}{T_1} = nC_{p,\mathrm{m}}\ln\frac{T_2}{T_1}$$

$$= \frac{1.25\times10^3\,\mathrm{g}}{26.98\,\mathrm{g\,mol^{-1}}}\times 24.35\,\mathrm{J\,K^{-1}\,mol^{-1}}\times\ln\frac{260\,\mathrm{K}}{300\,\mathrm{K}} = \boxed{-161\,\mathrm{J\,K^{-1}}}\,.$$

4.8 For the first step, melting 100 g ice:

$$\Delta_{\mathrm{fus}}S = \frac{\Delta_{\mathrm{fus}}H}{T_{\mathrm{fus}}} = \frac{6.01\,\mathrm{kJ\,mol^{-1}}}{273\,\mathrm{K}}\times 100\,\mathrm{g}\times\frac{1\,\mathrm{mol}}{18.0\,\mathrm{g}}$$

$$= \boxed{122\,\mathrm{J\,K^{-1}}}\,.$$

For the second step, heating the water:

$$\Delta S = C_V \ln\frac{T_{\mathrm{f}}}{T_{\mathrm{i}}} = 4.18\,\mathrm{J\,K^{-1}\,g^{-1}}\times 100\,\mathrm{g}\times\ln\frac{373}{273} = \boxed{130\,\mathrm{J\,K^{-1}}}\,.$$

For the third step, vaporization:

$$\Delta_{\mathrm{vap}}S = \frac{\Delta_{\mathrm{vap}}H}{T_{\mathrm{b}}} = \frac{40.7\,\mathrm{kJ\,mol^{-1}}}{373\,\mathrm{K}}\times 100\,\mathrm{g}\times\frac{1\,\mathrm{mol}}{18.0\,\mathrm{g}} = \boxed{606\,\mathrm{J\,K^{-1}}}\,.$$

$$\Delta S_{\mathrm{total}} = (122+130+606)\,\mathrm{J\,K^{-1}} = \boxed{858\,\mathrm{J\,K^{-1}}}\,.$$

(a) A graph of temperature vs. time (Figure 4.1) would show a constant 273 K temperature until all the ice had melted. Temperature would increase until the boiling point, 373 K, was reached. Temperature would again remain constant until all the liquid was vaporized.

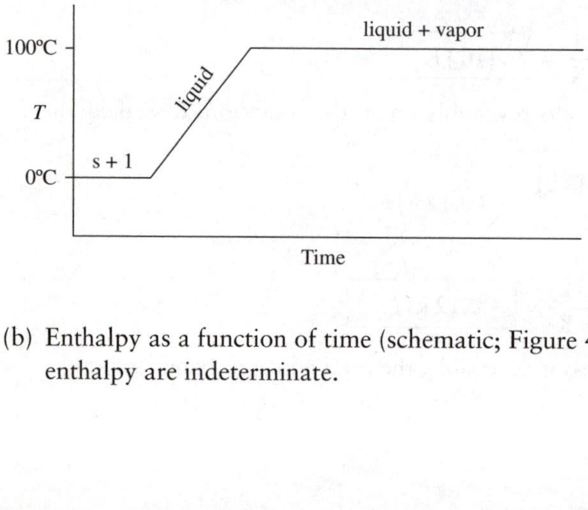

Fig 4.1

(b) Enthalpy as a function of time (schematic; Figure 4.2). Note that absolute values of enthalpy are indeterminate.

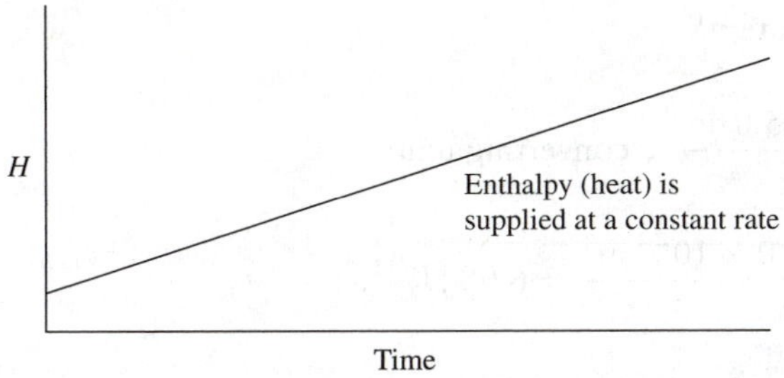

Fig 4.2

(c) Entropy as a function of time (schematic; Figure 4.3).

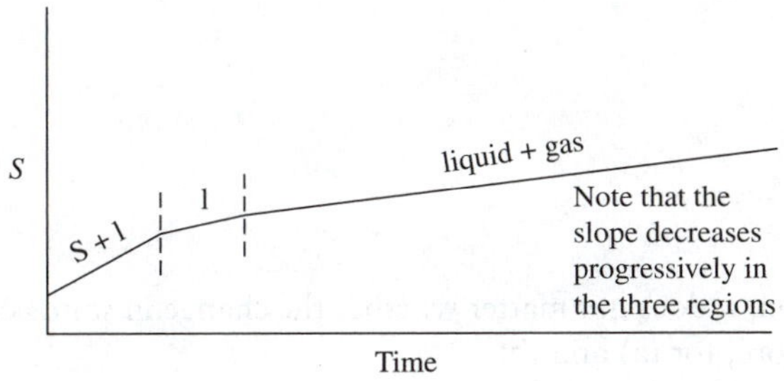

Fig 4.3

The graph of entropy against time does not look much different from that of enthalpy against time. The reason is that $\Delta t \propto \Delta H$. Therefore

$$\frac{\Delta S}{\Delta t} \propto \frac{\Delta S}{\Delta H}.$$

In the three regions, we see that this ratio progressively decreases

$$\frac{\Delta S}{\Delta H} = \frac{122\,\mathrm{J\,K^{-1}}}{33.3\,\mathrm{kJ}} = 3.66\,\mathrm{J\,K^{-1}\,kJ^{-1}}\ (\mathrm{s+l\ region}).$$

$$\frac{\Delta S}{\Delta H} = \frac{130\,\mathrm{J\,K^{-1}}}{41.8\,\mathrm{kJ}} = 3.17\,\mathrm{J\,K^{-1}\,kJ^{-1}}\ (\text{liquid region}).$$

$$\frac{\Delta S}{\Delta H} = \frac{606\,\mathrm{J\,K^{-1}}}{226\,\mathrm{kJ}} = 2.68\,\mathrm{J\,K^{-1}\,kJ^{-1}}\ (\text{liquid + gas region}).$$

4.9 $\Delta S = nR \ln \frac{V_f}{V_i}$ [4.2] (assume gas is perfect)

$$= 8.31\,\mathrm{J\,K^{-1}\,mol^{-1}} \times \ln\left(\frac{4.5\,\mathrm{dm^3}}{1.5\,\mathrm{dm^3}}\right) = \boxed{+9.1\,\mathrm{J\,K^{-1}\,mol^{-1}}}.$$

4.10 $\Delta S = nR \ln \frac{V_f}{V_i}$ [4.2] $= -\,10.0\,\mathrm{J\,K^{-1}}$.

Use the perfect gas law to calculate nR.

$$nR = \frac{p_i V_i}{T_i} = \frac{1.00\,\text{atm} \times 15.0\,\text{dm}^3}{250\,\text{K}}, \text{ converting units}$$

$$= \frac{1.013 \times 10^5\,\text{Pa} \times 15.0 \times 10^{-3}\,\text{m}^3}{250\,\text{K}} = 6.08\,\text{J}\,\text{K}^{-1}.$$

$$\ln \frac{V_f}{V_i} = \frac{\Delta S}{nR} = \frac{-10.0\,\text{J}\,\text{K}^{-1}}{6.8\,\text{J}\,\text{K}^{-1}} = -1.64.$$

Therefore, $V_i = 15.0\,\text{dm}^3$ and

$$V_f = \boxed{2.91\,\text{dm}^3}.$$

4.11 $\Delta S = nR \ln \frac{V_f}{V_i}$ [4.2].

Because entropy is a state function, it does not matter whether the change in state occurs reversibly or irreversibly. Therefore, for (a) and (b),

substitute $V = \frac{nRT}{p}$ into the expression for ΔS.

$$\Delta S = nR \ln \frac{p_i}{p_f}$$

$$= \frac{25\,\text{g}}{16.04\,\text{g}\,\text{mol}^{-1}} \times 8.3145\,\text{J}\,\text{K}^{-1}\,\text{mol}^{-1} \times \ln \frac{185}{2.5} = \boxed{56\,\text{J}\,\text{K}^{-1}}.$$

4.12 $\Delta S = C_p \ln \frac{T_f}{T_i} = nC_{p,m} \ln \frac{T_f}{T_i}$ [4.3]

$$= \frac{100\,\text{g}}{18.02\,\text{g}\,\text{mol}^{-1}} \times 75.5\,\text{J}\,\text{K}^{-1}\,\text{mol}^{-1} \times \ln\left(\frac{310\,\text{K}}{293\,\text{K}}\right) = \boxed{23.6\,\text{J}\,\text{K}^{-1}}.$$

4.13 Because entropy changes depend only on the initial and final states, it does not matter if the change is accomplished in one or more than one step. Therefore, calculate the change in two steps.

ΔS for compression:

$$\Delta S = nR \ln \frac{V_f}{V_i} \quad [4.2]$$

$$= 1\,\text{mol} \times 8.3145\,\text{J}\,\text{K}^{-1}\,\text{mol}^{-1} \times \ln\left(\frac{0.500\,\text{dm}^3}{2.0\,\text{dm}^3}\right) = -11.5\,\text{J}\,\text{K}^{-1}.$$

ΔS for heating:

$$\Delta S = nC_{V,\mathrm{m}} \ln \frac{T_\mathrm{f}}{T_\mathrm{i}} \quad [4.3], \; C_{V,\mathrm{m}} = \frac{3}{2}R$$

$$= 1\,\mathrm{mol} \times \frac{3}{2} \times 8.3145\,\mathrm{J\,K^{-1}\,mol^{-1}} \times \ln\left(\frac{400\,\mathrm{K}}{300\,\mathrm{K}}\right) = +3.59\,\mathrm{J\,K^{-1}}.$$

$$\Delta S_\mathrm{total} = (-11.5 + 3.59)\,\mathrm{J\,K^{-1}} = \boxed{-7.9\,\mathrm{J\,K^{-1}}}.$$

4.14 $V_\mathrm{f} = 2V_\mathrm{i}$.

$$\Delta S = nR \ln \frac{V_\mathrm{f}}{V_\mathrm{i}} = nR \ln 2, \text{ for the first step (expansion).}$$

$$\Delta S = nC_{V,\mathrm{m}} \ln \frac{T_\mathrm{f}}{T_\mathrm{i}} \text{ for the second step (cooling).}$$

$\Delta S =$ for the second step is the negative of ΔS for the first step; therefore

$$nR \ln 2 = -n \times R \ln\left(\frac{T_\mathrm{f}}{T_\mathrm{i}}\right).$$

$$\ln\left(\frac{T_\mathrm{f}}{T_\mathrm{i}}\right) = -\frac{2}{3} \ln 2 = -0.4621.$$

$$T_\mathrm{f} = \boxed{0.630\,T_\mathrm{i}}.$$

4.15 Entropy changes occur in steps 1 and 3 and are the negatives of each other. Temperature changes in steps 2 and 4 are the negatives of each other. See Figure 4.4.

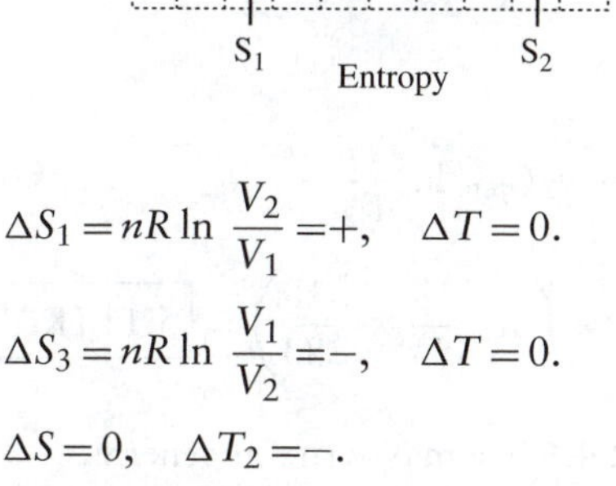

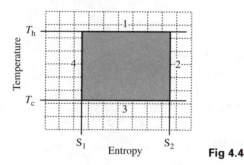

Fig 4.4

$$\text{Step 1: } \Delta S_1 = nR \ln \frac{V_2}{V_1} = +, \quad \Delta T = 0.$$

$$\text{Step 3: } \Delta S_3 = nR \ln \frac{V_1}{V_2} = -, \quad \Delta T = 0.$$

Step 2: $\Delta S = 0, \quad \Delta T_2 = -.$

Step 4: $\Delta S = 0, \quad \Delta T_4 = +.$

$$\Delta S_1 = -\Delta S_3, \quad \Delta T_2 = -\Delta T_4.$$

4.16 We use the formula of Derivation 4.4.

$$S_m(T)-S_m(0)=\frac{1}{3}C_{V,m}(T)=\frac{1}{3}\times 1.2\times 10^{-3}\,\mathrm{J\,K^{-1}\,mol^{-1}}=4.0\times 10^{-4}\,\mathrm{J\,K^{-1}\,mol^{-1}}.$$

As $S_m(0)$ for the pure crystalline substance KCl is expected to be zero,

$$S_m(T)=\boxed{4.0\times 10^{-4}\,\mathrm{J\,K^{-1}\,mol^{-1}}}.$$

4.17 $$\Delta S=\int_{T_i}^{T_f}\frac{C}{T}\,dT\ [\text{Derivation 4.2}]=\int_{T_i}^{T_f}\left(\frac{a+bT+\dfrac{c}{T^2}}{T}\right)dT$$

$$=a\int_{T_i}^{T_f}\frac{1}{T}\,dT+b\int_{T_i}^{T_f}dT+c\int_{T_i}^{T_f}\frac{1}{T^3}\,dT=a\ln(T)\Big|_{T_i}^{T_f}+bT\Big|_{T_i}^{T_f}-\frac{c}{2T^2}\Big|_{T_i}^{T_f}.$$

$$\Delta S=a\ln\left(\frac{T_f}{T_i}\right)+b(T_f-T_i)-\frac{c}{2}\left(\frac{1}{T_f^2}-\frac{1}{T_i^2}\right).$$

4.18 First find the common final temperature, T_f, by noting that the heat lost by the hot sample is gained by the cold sample.

$$n_1C_{p,m}(T_f-T_{i1})=n_2C_{p,m}(T_f-T_{i2}).$$

Solving for T_f,

$$T_f=\frac{n_1T_{i1}\times n_2T_{i2}}{n_1\times n_2}.$$

$$\text{Because } n_1=n_2=\frac{100\,\mathrm{g}}{18.02\,\mathrm{g\,mol^{-1}}}=5.55\,\mathrm{mol},$$

$$T_f=\frac{1}{2}(353\,\mathrm{K}+283\,\mathrm{K})=318\,\mathrm{K}.$$

The total entropy change is therefore

$$\Delta S_{total}=\Delta S_1+\Delta S_2=n_1C_{p,m}\ln\frac{T_f}{T_{i1}}+n_2C_{p,m}\ln\frac{T_f}{T_{i2}}$$

$$=5.55\,\mathrm{mol}\times 75.5\,\mathrm{J\,K^{-1}\,mol^{-1}}\times\left(\ln\frac{318}{353}\times\frac{318}{283}\right)=\boxed{5.11\,\mathrm{J\,K^{-1}}}.$$

4.19 In a manner similar to equations 4.5 and 4.6, we may write in general,

$$\Delta_{trs}S=\frac{\Delta_{trs}H}{T_{trs}};\ \text{therefore } \Delta_{trs}S=\frac{+1.9\,\mathrm{kJ\,mol^{-1}}}{2000\,\mathrm{K}}=\boxed{0.95\,\mathrm{J\,K^{-1}\,mol^{-1}}}.$$

4.20 (a) $\Delta_{vap}S = \dfrac{\Delta_{vap}H}{T_b}$ [4.6]

$$= \frac{29.4 \times 10^3\,\mathrm{J\,mol^{-1}}}{334.88\,\mathrm{K}} = \boxed{+87.8\,\mathrm{J\,K^{-1}\,mol^{-1}}}.$$

(b) Because the vaporization process can be accomplished reversibly,

$\Delta S_{total} = 0$; hence

$$\Delta S_{sur} = \boxed{-87.8\,\mathrm{J\,K^{-1}\,mol^{-1}}}.$$

4.21 $\Delta_{fus}C_p^{\ominus} = C_p^{\ominus}(\mathrm{l}) - C_p^{\ominus}(\mathrm{s})$ [3.7]

$$= (28 - 19)\,\mathrm{J\,K^{-1}\,mol^{-1}} = 9\,\mathrm{J\,K^{-1}\,mol^{-1}}.$$

$$\Delta_{fus}S^{\ominus}(T') = \Delta_{fus}S^{\ominus}(T) + \int_T^{T'} \frac{\Delta_{fus}C_p^{\ominus}}{T}\,\mathrm{d}T$$

[Derivation 4.2 modified for a constant pressure transition]

$$= \Delta_{fus}S^{\ominus}(T) + \Delta_{fus}C_p^{\ominus} \int_T^{T'} \frac{1}{T}\,\mathrm{d}T = \Delta_{fus}S^{\ominus}(T) + \Delta_{fus}C_p^{\ominus} \ln(T)\big|_T^{T'}$$

$$= \Delta_{fus}S^{\ominus}(T) + \Delta_{fus}C_p^{\ominus} \ln\left(\frac{T'}{T}\right).$$

At $T = T_f$ the transition is reversible and $\Delta_{fus}S^{\ominus}(T_f) = \dfrac{\Delta_{fus}H^{\ominus}(T_f)}{T_f}$ [4.5].

$$\Delta_{fus}S^{\ominus}(T') = \frac{\Delta_{fus}H^{\ominus}(T_f)}{T_f} + \Delta_{fus}C_p^{\ominus} \ln\left(\frac{T'}{T_f}\right).$$

$$\Delta_{fus}S^{\ominus}(298\,\mathrm{K}) = \frac{32\,\mathrm{kJ\,mol^{-1}}}{419\,\mathrm{K}} + (9\,\mathrm{J\,K^{-1}\,mol^{-1}}) \times \ln\left(\frac{298\,\mathrm{K}}{419\,\mathrm{K}}\right)$$

$$= \boxed{73\,\mathrm{J\,K^{-1}\,mol^{-1}}}.$$

4.22 (a) According to Trouton's rule, which applies well to hydrocarbons such as octane,

$$\Delta_{vap}S = \boxed{+85\,\mathrm{J\,K^{-1}\,mol^{-1}}}.$$

(b) $\Delta_{vap}S = \dfrac{\Delta_{vap}H}{T_b} = +85\,\mathrm{J\,K^{-1}\,mol^{-1}}$.

$$\Delta_{vap}H = \Delta_{vap}S \times T_b = +85\,\mathrm{J\,K^{-1}\,mol^{-1}} \times 399\,\mathrm{K} = \boxed{+34\,\mathrm{kJ\,K^{-1}\,mol^{-1}}}.$$

4.23 (a) $\Delta_r S^\ominus = 2S_m^\ominus(CH_3COOH, l) - 2S_m^\ominus(CH_3CHO, g) - S_m^\ominus(O_2, g)$

$$= [(2 \times 159.8) - (2 \times 250.3) - 205.14]\,J\,K^{-1}\,mol^{-1}$$

$$= \boxed{-386.1\,J\,K^{-1}\,mol^{-1}}.$$

(b) $\Delta_r S^\ominus = 2S_m^\ominus(AgBr, s) + S_m^\ominus(Cl_2, g) - 2S_m^\ominus(AgCl, s) - S_m^\ominus(Br_2, l)$

$$= [(2 \times 107.1) + 223.07 - (2 \times 96.2) - 152.23]\,J\,K^{-1}\,mol^{-1}$$

$$= \boxed{+92.6\,J\,K^{-1}\,mol^{-1}}.$$

(c) $\Delta_r S^\ominus = S_m^\ominus(HgCl_2, s) - S_m^\ominus(Hg, l) - S_m^\ominus(Cl_2, g)$

$$= (146.0 - 76.02 - 223.07)\,J\,K^{-1}\,mol^{-1}$$

$$= \boxed{-153.1\,J\,K^{-1}\,mol^{-1}}.$$

(d) $\Delta_r S^\ominus = S_m^\ominus(Zn^{2+}, aq) + S_m^\ominus(Cu, s) - S_m^\ominus(Zn, s) - S_m^\ominus(Cu^{2+}, aq)$

$$= (-112.1 + 33.15 - 41.63 + 99.6)\,J\,K^{-1}\,mol^{-1}$$

$$= \boxed{-21.0\,J\,K^{-1}\,mol^{-1}}.$$

(e) $\Delta_r S^\ominus = 12S_m^\ominus(CO_2, g) + 11S_m^\ominus(H_2O, l) - S_m^\ominus(C_{12}H_{22}O_{11}, s) - 12S_m^\ominus(O_2, g)$

$$= [(12 \times 213.74) + (11 \times 69.91) - 360.2 - (12 \times 205.14)]\,J\,K^{-1}\,mol^{-1}$$

$$= \boxed{+512.0\,J\,K^{-1}\,mol^{-1}}.$$

4.24 $\Delta S = nC_{p,m} \ln \dfrac{T_f}{T_i}$ for each substance in each reaction; therefore

$\Delta_r S = \Delta_r C_p \ln \dfrac{T_f}{T_i}$ for each reaction.

(a) $\Delta_r C_p = (2\,mol \times 4R) - \left(3\,mol \times \dfrac{7}{2}R\right) = -\dfrac{5}{2}R\,mol.$

$$\Delta_r S = -\frac{5}{2}R\,mol \times \ln\left(\frac{283\,K}{273\,K}\right) = \boxed{-0.75\,J\,K^{-1}}.$$

(b) $\Delta_r C_p = (2\,mol \times 4R) + \left(1\,mol \times \dfrac{7}{2}R\right) - (1\,mol \times 4R) - \left(2\,mol \times \dfrac{7}{2}R\right)$

$$= +\frac{1}{2}R\,mol.$$

$$\Delta_r S = +\frac{1}{2}R\,mol \times \ln\left(\frac{283\,K}{273\,K}\right) = \boxed{+0.15\,J\,K^{-1}}.$$

4.25 $\Delta_c H^{\ominus} = -2808\,\mathrm{kJ\,mol^{-1}} = q_p = q_{rev} = q_{body}$.

$$\Delta S^{\ominus} = \frac{q_{rev}}{T} = \frac{-2808\,\mathrm{kJ\,mol^{-1}} \times 100\,\mathrm{g} \times \dfrac{1\,\mathrm{mol}}{180\,\mathrm{g}}}{273\,\mathrm{K} + 37\,\mathrm{K}} = \boxed{-5.03\,\mathrm{kJ\,K^{-1}}}.$$

NOTE The above calculation uses the value of $\Delta_c H^{\ominus}$ at 25°C. The value at 37°C should not be much different and can be calculated from knowledge of the heat capacities of all of the substances involved in the reaction.

4.26 (a) $\Delta G = \Delta H - T\Delta S$ [4.14]

$$= -125\,\mathrm{kJ\,mol^{-1}} - 310\,\mathrm{K} \times (-126\,\mathrm{J\,K^{-1}\,mol^{-1}}) = \boxed{-86\,\mathrm{kJ\,mol^{-1}}}.$$

(b) Yes, ΔG is negative.

(c) $\Delta G = -T\Delta S_{total}$ [4.15].

$$\Delta S_{total} = -\frac{\Delta G}{T} = -\left(\frac{-86\,\mathrm{kJ\,mol^{-1}}}{310\,\mathrm{K}}\right) = \boxed{+0.28\,\mathrm{kJ\,K^{-1}\,mol^{-1}}}.$$

4.27 $\Delta G = w'_{max} = -2828\,\mathrm{kJ\,mol^{-1}}$, so the maximum work that can be done is $2828\,\mathrm{kJ\,mol^{-1}}$. We will assume that we will be able to extract the maximum work from the reaction.

$$w = mgh = 65\,\mathrm{kg} \times 9.81\,\mathrm{m\,s^{-2}} \times 10\,\mathrm{m}$$

$$= 6.4 \times 10^3\,\mathrm{J} = 6.4\,\mathrm{kJ}.$$

$$\text{Amount}(n) = \frac{6.4\,\mathrm{kJ}}{2828\,\mathrm{kJ\,mol^{-1}}} = 2.3 \times 10^{-3}\,\mathrm{mol}.$$

$$\text{Mass of glucose} = 2.3 \times 10^{-3}\,\mathrm{mol} \times 180\,\mathrm{g\,mol^{-1}} = \boxed{0.41\,\mathrm{g}}.$$

4.28 (a) $\boxed{\text{Yes}}$, coupling the two reactions can give a net ΔG that is negative; hence the overall process is spontaneous. For example, for one mole of glutamate and one mole of ATP,

$$\Delta G = (14.2 - 31)\,\mathrm{kJ\,mol^{-1}} = -17\,\mathrm{kJ\,mol^{-1}}.$$

(b) The minimum amount of ATP required is $\dfrac{1\,\mathrm{mol} \times (-14.2\,\mathrm{kJ\,mol^{-1}})}{-31\,\mathrm{kJ\,mol^{-1}}}$

$$= \boxed{0.46\,\mathrm{mol\,ATP}}.$$

4.29 For the synthesis, $\Delta G = +42\,\mathrm{kJ\,mol^{-1}}$; hence at least $-42\,\mathrm{kJ}$ would need to be provided by the ATP in order to make ΔG overall negative.

$$\text{Amount}(n) \text{ of ATP} = \frac{-42\,\mathrm{kJ}}{-31\,\mathrm{kJ\,mol^{-1}}} = 1.35\,\mathrm{mol\,ATP}.$$

$$1.35\,\mathrm{mol} \times 6.02 \times 10^{23}\,\mathrm{mol^{-1}} = \boxed{8.1 \times 10^{23}\text{ molecules of ATP}}.$$

4.30 $n(\mathrm{ATP}) = \dfrac{10^6}{6.02 \times 10^{23}\ \mathrm{mol}^{-1}} = 1.7 \times 10^{-18}\ \mathrm{mol}.$

$$\Delta G = 1.7 \times 10^{-18}\ \mathrm{mol\,s}^{-1} \times (-31\ \mathrm{kJ\,mol}^{-1}) = -5.3 \times 10^{-17}\ \mathrm{kJ\,s}^{-1}$$

$$= -5.3 \times 10^{-14}\ \mathrm{J\,s}^{-1}.$$

$$\text{Power density of cell} = \frac{\Delta G \text{ of cell per second}}{\text{volume of cell}}.$$

$$V_{\mathrm{cell}} = \frac{4}{3}\pi r^3 = \frac{4}{3}\pi(10 \times 10^{-6}\ \mathrm{m})^3 = 4.2 \times 10^{-15}\ \mathrm{m}^3.$$

$$\text{Power density of cell} = \frac{5.3 \times 10^{-14}\ \mathrm{J\,s}^{-1}}{4.2 \times 10^{-15}\ \mathrm{m}^3} = \boxed{13\ \mathrm{W\,m}^{-3}}.$$

$$\text{Power density of battery} = \frac{15\ \mathrm{W}}{100\ \mathrm{cm}^3 \times 10^{-6}\ \mathrm{m}^3/\mathrm{cm}^3} = \boxed{150\ \mathrm{kW\,m}^{-3}}.$$

The $\boxed{\text{battery}}$ has the greater power density.

Chapter 5

Phase equilibria: pure substances

Answers to discussion questions

5.1 Consider two phases of a system, labelled α and β. The phase with the lower molar Gibbs energy under the given set of conditions is the more stable phase. First, consider the variation of the molar Gibbs energy of each phase with temperature at a fixed pressure by comparing the equation 5.3 expressions

$$\frac{\Delta G_\alpha}{\Delta T} = -S_\alpha \quad \text{and} \quad \frac{\Delta G_\beta}{\Delta T} = -S_\beta, \text{ at constant } p.$$

They clearly show that, if S_β is larger in magnitude than S_α, then ΔG_β decreases to a greater extent than ΔG_α as temperature increases. β phase becomes the more stable phase at higher temperature.

Second, consider the variation of the molar Gibbs energy of each phase with pressure at a fixed temperature by comparing the equation [5.1] expressions

$$\frac{\Delta G_\alpha}{\Delta p} = V_\alpha \quad \text{and} \quad \frac{\Delta G_\beta}{\Delta p} = V_\beta, \text{ at constant } T.$$

These equations clearly show that, if V_β is larger in magnitude than V_α, then ΔG_β increases to a greater extent than ΔG_α as pressure increases. The β phase becomes the unstable phase at higher pressure; α phase becomes the stable phase.

5.2 (a) Attractive interactions tend to decrease the pressure of a gas relative to its perfect value for the same volume. We may qualitatively use equation 5.1 to decide that the molar Gibbs energy will be $\boxed{\text{lowered}}$ relative to its 'perfect' value.

(b) Repulsive interactions have the opposite effect on the pressure of a gas, so we may qualitatively decide that they will $\boxed{\text{raise}}$ the molar Gibbs energy relative to its 'perfect' value.

5.3 The Clapeyron equation is exact and applies rigorously to all first-order phase transitions. It shows how pressure and temperature vary with respect to each other (temperature or pressure) along the phase boundary line, and in that sense, it defines the phase boundary line.

The Clausius–Clapeyron equation serves the same purpose, but it is not exact; its derivation involves approximations, in particular the assumptions that the perfect gas law holds and that the volume of condensed phases can be neglected in comparison to the volume of the gaseous phase. It applies only to phase transitions between the gaseous state and condensed phases.

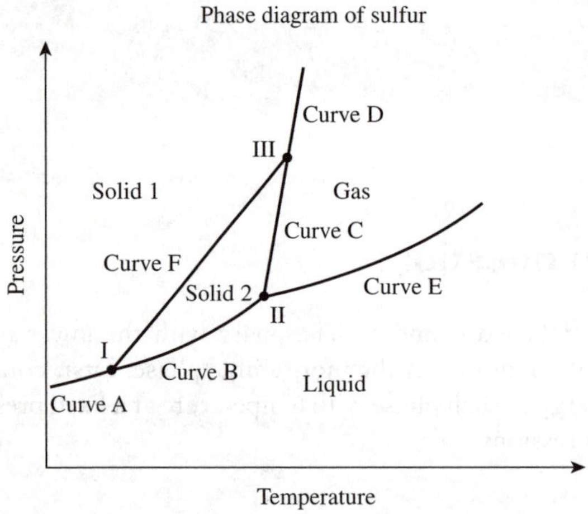

Fig 5.1

5.4 $C = 1$ for the sulfur phase diagram and the phase rule is $F = C - P + 2 = 3 - P$.

See Figure 5.1. In the areas, off any line segment, labelled Solid 1, Solid 2, Liquid, and Gas there is one phase and $F = 2$. Both T and p may be independently varied within these regions.

Curve segments A and B are sublimation curves between crystal form 1 of sulfur and the gas phase, and crystal form 2 of sulfur and the gas phase, respectively. On these equilibrium curves $P = 2$ and $F = 1$. There is only one independent variable, which means that once either T or P is set the other variable has a unique value determined by the equilibrium criteria. For identical reasons curve segments C, D, E, and F also have only one independent variable. Curves C and D are fusion curves in which the liquid is in equilibrium with either crystal form 2 or crystal form 1, respectively. Curve E is the vapor pressure curve. Curve F represents the points at which the two crystal forms are in equilibrium.

Points I, II, and III are triple points at which $P = 3$ and $F = 0$. There is no independent variable at these points because triple points are fixed by equilibrium criteria.

Solutions to exercises

5.5 The substance with the lower molar Gibbs energy is the more stable; therefore, rhombic sulfur is the more stable.

5.6 $\boxed{\text{No}}$, the application of pressure tends to favor the substance with the smaller molar volume (higher density; see Exercise 5.1). Therefore, rhombic sulfur becomes even more stable relative to monoclinic sulfur as the pressure increases. This is easily seen in equation 5.1, $\Delta G_m = V_m \Delta p$.

ΔG is less positive for rhombic sulfur than for monoclinic sulfur so, relative to monoclinic sulfur, the Gibbs energy of rhombic sulfur becomes more negative,

5.7 (a) We assume that the molar volume of water is approximately constant with respect to variation in pressure. Then

$$V_m = \frac{18.02\,\text{g mol}^{-1}}{1.03\,\text{g cm}^{-3}} = 17.5\,\text{cm}^3\,\text{mol}^{-1}$$

$$= 1.75 \times 10^{-5}\,\text{m}^3\,\text{mol}^{-1}.$$

$$\Delta p = g\rho h[0.5] \quad [\Delta p = p_{\text{trench}} - p_{\text{surface}}]$$

$$= 9.81\,\text{m s}^{-2} \times 1.03\,\text{g cm}^{-3} \times \frac{1\,\text{kg}}{10^3\,\text{g}} \times \frac{10^6\,\text{cm}^3}{\text{m}^3} \times 11.5 \times 10^3\,\text{m}$$

$$= 1.16 \times 10^8\,\text{Pa} = 116\,\text{MPa}.$$

$$\Delta G_m = V_m \Delta p = 1.75 \times 10^{-5}\,\text{m}^3\,\text{mol}^{-1} \times 1.16 \times 10^8\,\text{Pa}$$

$$= 2.03 \times 10^3\,\text{J mol}^{-1} = \boxed{+2.03\,\text{kJ mol}^{-1}}.$$

(b) The pressure at the bottom of the mercury column is 1.000 atm = 1.013×10^5 Pa.

$$\Delta p = 1.013 \times 10^5\,\text{Pa} - 0.160\,\text{Pa} \approx 1.013 \times 10^5\,\text{Pa}.$$

$$V_m = \frac{200.6\,\text{g mol}^{-1}}{13.6\,\text{g cm}^3} = 14.8\,\text{cm}^3\,\text{mol} = 1.48 \times 10^{-5}\,\text{m}^3\,\text{mol}^{-1}.$$

$$\Delta G_m = V_m \Delta p = 1.48 \times 10^{-5}\,\text{m}^3\,\text{mol}^{-1} \times 1.013 \times 10^5\,\text{Pa}$$

$$= \boxed{+1.50\,\text{J mol}^{-1}}.$$

5.8 $\Delta G_m = V_m \Delta p.$

$$V_m = \frac{891.51\,\text{g mol}^{-1}}{0.95\,\text{g cm}^{-3}} = 938\,\text{cm}^3\,\text{mol}^{-1} = 9.4 \times 10^{-4}\,\text{m}^3\,\text{mol}^{-1}.$$

$$\Delta p = g\rho h = 9.81\,\text{m s}^{-2} \times 1.03 \times 10^3\,\text{kg m}^{-3} \times 2.0 \times 10^3\,\text{m} = 2.0 \times 10^7\,\text{Pa}.$$

$$\Delta G_m = 9.4 \times 10^{-4}\,\text{m}^3\,\text{mol}^{-1} \times 2.0 \times 10^7\,\text{Pa}$$

$$= +1.9 \times 10^4\,\text{J mol}^{-1} = \boxed{+19\,\text{kJ mol}^{-1}}.$$

5.9 $\Delta G_m = RT \ln \frac{p_f}{p_i}$ [5.2b].

(a) $\Delta G_m = 8.3145\,\mathrm{J\,K^{-1}\,mol^{-1}} \times 293\,\mathrm{K} \times \ln\left(\frac{2.0\,\mathrm{bar}}{1.0\,\mathrm{bar}}\right)$

$= 1.7 \times 10^3\,\mathrm{J\,mol^{-1}} = \boxed{+1.7\,\mathrm{kJ\,mol^{-1}}}$.

(b) $\Delta G_m = 8.3145\,\mathrm{J\,K^{-1}\,mol^{-1}} \times 293\,\mathrm{K} \times \ln\left(\frac{0.00027\,\mathrm{atm}}{1.0\,\mathrm{atm}}\right)$

$= -2.0 \times 10^4\,\mathrm{J\,mol^{-1}} = \boxed{-2.0\,\mathrm{kJ\,mol^{-1}}}$.

5.10 At these pressures, water vapor may be considered a perfect gas; therefore $p_i V_i = p_f V_f$ and

$$\frac{p_f}{p_i} = \frac{V_i}{V_f}.$$

$$\Delta G_m = RT \ln \frac{p_f}{p_i} = RT \ln \frac{V_i}{V_f}$$

$$= 8.3145\,\mathrm{J\,K^{-1}\,mol^{-1}} \times 473\,\mathrm{K} \times \ln\left(\frac{300\,\mathrm{cm^3}}{100\,\mathrm{cm^3}}\right)$$

$$= +4.3 \times 10^3\,\mathrm{J\,mol^{-1}} = \boxed{+4.3\,\mathrm{kJ\,mol^{-1}}}.$$

5.11 The van der Waals equation of state [1.23b] for one mole of gas is

$$\left(p + \frac{a}{V_m^2}\right)(V_m - b) = RT[V_m = \text{molar volume}].$$

After neglecting attractive effects, it becomes

$$p(V_m - b) = RT.$$

(a) Solving for V_m we get

$$V_m = \frac{RT + bp}{p} = +\frac{RT}{p} + b.$$

Note that the first term is the molar volume of a perfect gas. Following Derivation 5.2 gives

$$\Delta G_m = \int_{p_i}^{p_f} V_m\,dp = \int_{p_i}^{p_f} \frac{RT}{p}\,db + \int_{p_i}^{p_f} b\,dp$$

$$= \boxed{RT \ln \frac{p_f}{p_i} + b(p_f - p_i)} = \text{perfect value } + b(p_f - p_i).$$

(b) For $p_f > p_i$, the change is $\boxed{\text{greater}}$ for this gas by the amount $b(p_f - p_i)$.

(c) Perfect gas value $= RT \ln \frac{p_f}{p_i}$; assume $T = 298.15\,\text{K}$.

$$RT \ln \frac{p_f}{p_i} = 8.3145\,\text{J K}^{-1}\,\text{mol}^{-1} \times 298.15\,\text{K} \times \ln\left(\frac{10.0\,\text{atm}}{1.0\,\text{atm}}\right)$$

$$= 5.71 \times 10^3\,\text{J} = 5.71\,\text{kJ}.$$

$$b(p_f - p_i) = 0.0429\,\text{dm}^3\,\text{mol}^{-1} \times \frac{1\,\text{m}^3}{10^3\,\text{dm}^3} \times (10 - 1)\,\text{atm} \times \frac{1.013 \times 10^5\,\text{Pa}}{1.0\,\text{atm}}$$

$$= 39\,\text{J}.$$

The imperfect gas value is then $5.71 \times 10^3\,\text{J} + 39\,\text{J} = 5.75 \times 10^3\,\text{J} = 5.75\,\text{kJ}$
The percentage difference is approximately

$$\frac{39\,\text{J}}{5.7 \times 10^3\,\text{J}} \times 100\% = \boxed{0.68\%}.$$

5.12 For the transition S (rhombic) $\rightarrow$ S (monoclinic),

$$\Delta G_m(298\,\text{K}) = +0.33\,\text{kJ mol}^{-1} \quad \text{and} \quad \Delta S_m(298\,\text{K}) = (32.6 - 31.8)\,\text{J K}^{-1}\,\text{mol}^{-1}$$

$$= 0.8\,\text{J K}^{-1}\,\text{mol}^{-1}.$$

(a) We expect that an increase in temperature can result in monoclinic sulfur being more stable than rhombic sulfur provided that at temperature T the value of $\Delta G_m(T)$ for the reaction is equal to or less than zero because the transition becomes spontaneous to the right at that point.

(b) $\Delta G_m = \Delta H_m - T\Delta S_m$ and

$$\Delta(\Delta G_m) = \Delta G_m(T) - \Delta G_m(298\,\text{K}) = -\Delta G_m(298\,\text{K}) = -0.33\,\text{kJ mol}^{-1}$$

when the transition becomes spontaneous to the right.

We will assume that ΔH_m, and ΔS_m are roughly independent of temperature. We need $\Delta(\Delta G_m)$ to be $-0.33\,\text{kJ mol}^{-1}$ as a result of the change in temperature.

That is,

$$-(T_f - T_i)\Delta S_m = -0.33\,\text{kJ mol}^{-1}.$$

Solve for T_f, $T_f = \dfrac{0.33 \times 10^3\,\text{J mol}^{-1}}{0.8\,\text{J K}^{-1}\,\text{mol}^{-1}} + 298.15\,\text{K}.$

$$T_f = \boxed{7 \times 10^2\,\text{K}}.$$

5.13 $\Delta G_m = -S_m \Delta T$ at constant p and small temperature changes [5.3].

$$\Delta G_m = -S_m(T_f - T_i) = -173.3\,\mathrm{J\,K^{-1}\,mol^{-1}} \times 30\,\mathrm{K}$$

$$= -5.2 \times 10^3\,\mathrm{J\,mol^{-1}} = \boxed{-5.2\,\mathrm{kJ\,mol^{-1}}}.$$

5.14 The slope of a graph of G_m against T is $-S_m$, that is, $\dfrac{\Delta G_m}{\Delta T} = -S_m$ [5.3].

The slopes in all phases are negative, because S_m is always positive, but

$$\left|\frac{\Delta G_m}{\Delta T}(\mathrm{g})\right| > \left|\frac{\Delta G_m}{\Delta T}(\mathrm{l})\right| > \left|\frac{\Delta G_m}{\Delta T}(\mathrm{s})\right|$$

because $S_m(\mathrm{g}) > S_m(\mathrm{l}) > S_m(\mathrm{s})$.

Therefore, a graph of G_m against T appears as in Figure 5.2 below. Absolute values of G_m are not known, but ΔG_m in each phase could be calculated as illustrated in Exercise 5.13.

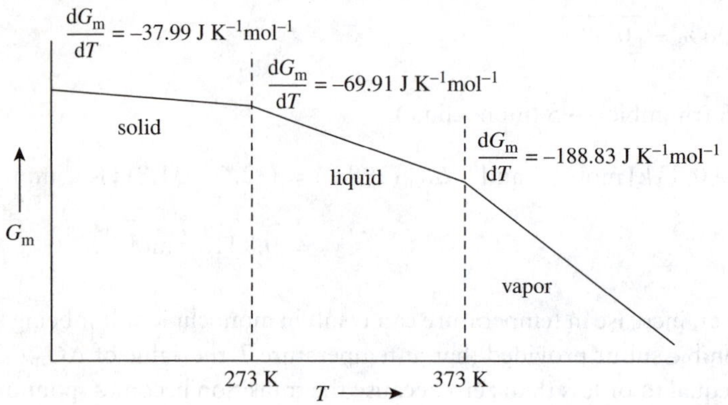

Fig 5.2

Note the discontinuous change in slopes at the transition temperatures.

5.15 Use the perfect gas law to calculate the amount and the mass.

$$n = \frac{pV}{RT},\quad n = \frac{m}{M},\quad m = \frac{pVM}{RT},\quad V = 6.0\,\mathrm{m} \times 5.3\,\mathrm{m} \times 3.2\,\mathrm{m} = 101.76\,\mathrm{m^3}$$

$$= 10\bar{2} \times 10^3\,\mathrm{dm^3}.$$

(a) $$m = \frac{3.2\,\mathrm{kPa} \times (10\bar{2} \times 10^3\,\mathrm{dm^3}) \times (18.02\,\mathrm{g\,mol^{-1}})}{(8.3145\,\mathrm{dm^3\,kPa\,K^{-1}\,mol^{-1}}) \times (298.15\,\mathrm{K})} = \boxed{2.37\,\mathrm{kg}}.$$

(b) $$m = \frac{14\,\mathrm{kPa} \times (10\bar{2} \times 10^3\,\mathrm{dm^3}) \times (78.11\,\mathrm{g\,mol^{-1}})}{(8.3145\,\mathrm{dm^3\,kPa\,K^{-1}\,mol^{-1}}) \times (298.15\,\mathrm{K})} = \boxed{45\,\mathrm{kg}}.$$

(c) $$m = \frac{(0.23 \times 10^{-3}\,\mathrm{kPa}) \times (10\bar{2}\,\mathrm{dm^3}) \times (200.59\,\mathrm{g\,mol^{-1}})}{(8.3145\,\mathrm{dm^3\,kPa\,K^{-1}\,mol^{-1}}) \times (298.15\,\mathrm{K})} = \boxed{1.9 \times 10^{-3}\,\mathrm{g}}.$$

5.16 (a) The Clapeyron equation for the solid–liquid phase boundary is

$$\frac{\mathrm{d}p}{\mathrm{d}T} = \frac{\Delta_{\mathrm{fus}}H}{T_{\mathrm{fus}}\Delta_{\mathrm{fus}}V} \text{ [5.4b].}$$

$$\Delta_{\mathrm{fus}}V = V_{\mathrm{m}}(\mathrm{l}) - V_{\mathrm{m}}(\mathrm{s}) = M\left(\frac{1}{\rho_{\mathrm{l}}} - \frac{1}{\rho_{\mathrm{s}}}\right)$$

$$= 18.02\,\mathrm{g\,mol^{-1}}\left(\frac{1}{0.99984\,\mathrm{g\,cm^{-3}}} - \frac{1}{0.91671\,\mathrm{g\,cm^{-3}}}\right)$$

$$= -1.634\,\mathrm{cm^3\,mol^{-1}} = -1.634\times10^{-6}\,\mathrm{m^3\,mol^{-1}}.$$

$$\frac{\mathrm{d}p}{\mathrm{d}T} = \frac{6.008\times10^{3}\,\mathrm{J\,mol^{-1}}}{273.15\,\mathrm{K}\times(-1.634\times10^{-6}\,\mathrm{m^3\,mol^{-1}})}$$

$$= -1.346\times10^{7}\,\mathrm{Pa\,K^{-1}} = \boxed{-134.6\,\mathrm{bar\,K^{-1}}}.$$

The slope is very steep!

(b) $\dfrac{\Delta p}{\Delta T} = -134.6\,\mathrm{bar\,K^{-1}}$.

For $\Delta T = -1\,\mathrm{K}$, $\Delta p = 134.6\,\mathrm{bar}$. Consequently, $p = p_{\mathrm{i}} + \Delta p = 1.0\,\mathrm{bar} + 134.6\,\mathrm{bar}$

$$= \boxed{135.6\,\mathrm{bar}}.$$

5.17 Starting with the differential form of the Clausius–Clapeyron equation [Derivation 5.5, perfect gas],

$$\frac{\mathrm{d}\ln p}{\mathrm{d}T} = \frac{\Delta_{\mathrm{vap}}H}{RT^2} \quad \text{or} \quad \mathrm{d}\ln p = \frac{\Delta_{\mathrm{vap}}H}{RT^2}\,\mathrm{d}T,$$

integrate between the points (p, T) and (p', T').

$$\int_{\ln p}^{\ln p'} \mathrm{d}\ln p = \int_T^{T'} \frac{\Delta_{\mathrm{vap}}H}{RT^2}\,\mathrm{d}T = \int_T^{T'} \frac{a+bT}{RT^2}\,\mathrm{d}T = \frac{1}{R}\left\{a\int_T^{T'}\frac{1}{T^2}\,\mathrm{d}T + b\int_T^{T'}\frac{1}{T}\,\mathrm{d}T\right\}.$$

$$\ln\left(\frac{p'}{p}\right) = \frac{1}{R}\left\{a\left(\frac{-1}{T}\right)\Bigg|_T^{T'} + b\ln\left(\frac{T'}{T}\right)\right\},$$

$$\ln\left(\frac{p'}{p}\right) = \frac{1}{R}\left\{a\left(\frac{1}{T} - \frac{1}{T'}\right) + b\ln\left(\frac{T'}{T}\right)\right\},$$

$$\boxed{\ln p' = \ln p + \frac{1}{R}\left\{a\left(\frac{1}{T} - \frac{1}{T'}\right) + b\ln\left(\frac{T'}{T}\right)\right\}}.$$

5.18 (a) $\log(p/\text{kPa}) = A - B/T$ [5.7]

For hexane $A = 6.849$, $B = 1655\,\text{K}$ from Table 5.1.

$\Delta_{\text{vap}}H = BR\ln(10)$ [see Example 5.1].

$$\Delta_{\text{vap}}H = 2.303 \times 8.3145\,\text{J K}^{-1}\,\text{mol}^{-1} \times 1655\,\text{K}$$

$$= \boxed{31.69\,\text{kJ mol}^{-1}}.$$

This value is about 10% different from the value for hexane at its boiling point, which is $28.85\,\text{kJ mol}^{-1}$.

(b) $$T_\text{b} = \frac{\Delta_{\text{vap}}H(T_\text{b})}{\Delta_{\text{vap}}S}\ [4.6] \simeq \frac{31.69\,\text{kJ mol}^{-1}}{85\,\text{J K}^{-1}\,\text{mol}^{-1}}\ \text{(Trouton's rule)} = \boxed{37\bar{3}\,\text{K}}.$$

Once again, this estimate is about 10% high. The normal boiling point of hexane is 342 K.

5.19 $$\log\left(\frac{p}{\text{kPa}}\right) = A - \frac{B}{T} \quad \text{where} \quad A = 7.455,\ B = 2047\,\text{K for methylbenzene [Table 5.1]}.$$

Since $760\,\text{Torr} = 101.325\,\text{kPa}$,

$$\log\left(\left\{\frac{p}{\text{kPa}}\right\}\left\{\frac{101.325\,\text{kPa}}{760\,\text{Torr}}\right\}\right) = A - \frac{B}{T},$$

$$\log\left(\left\{\frac{p}{\text{Torr}}\right\}\left\{\frac{101.325}{760}\right\}\right) = A - \frac{B}{T},$$

$$\log\left(\frac{p}{\text{Torr}}\right) + \log\left(\frac{101.325}{760}\right) = A - \frac{B}{T},$$

$$\log\left(\frac{p}{\text{Torr}}\right) = A - \log\left(\frac{101.325}{760}\right) - \frac{B}{T},$$

$$\log\left(\frac{p}{\text{Torr}}\right) = A' - \frac{B}{T} \quad \text{where} \quad A' = A - \log\left(\frac{101.325}{760}\right)$$

$$= 7.455 - \log\left(\frac{101.325}{760}\right) = \boxed{8.330}.$$

B is unchanged.

5.20 We use

$$\ln\frac{p'}{p} = \frac{\Delta_{\text{vap}}H}{R}\left(\frac{1}{T} - \frac{1}{T'}\right)\ [5.6]$$

with $T = 293\,\text{K}$, $p = 160\,\text{mPa}$, and $T' = 323\,\text{K}$; then solve for p'.

$$\ln\frac{p'}{p} = \frac{59.30 \times 10^3\,\text{J mol}^{-1}}{8.3145\,\text{J K}^{-1}\,\text{mol}^{-1}}\left(\frac{1}{293\,\text{K}} - \frac{1}{323\,\text{K}}\right) = 2.261.$$

$$\frac{p'}{p} = 9.59.$$

$$p' = 9.59 \times 160\,\text{mPa} = 1.53 \times 10^3\,\text{mPa} = \boxed{1.53\,\text{Pa}}.$$

5.21 We use

$$\ln\frac{p'}{p} = \frac{\Delta_{\text{vap}}H}{R}\left(\frac{1}{T} - \frac{1}{T'}\right) \quad [5.6]$$

with $p' = 101.3\,\text{kPa}$, $p = 50.0\,\text{kPa}$, $= 365.7\,\text{K}$, and $T' = 388.4\,\text{K}$; then solve for $\Delta_{\text{vap}}H$.

$$\ln\frac{101.3}{50.0} = \frac{\Delta_{\text{vap}}H}{8.315\,\text{J K}^{-1}\,\text{mol}^{-1}}\left(\frac{1}{365.7\,\text{K}} - \frac{1}{388.4\,\text{K}}\right).$$

$$0.706 = 1.922 \times 10^{-5}\,\text{J}^{-1}\,\text{mol} \times \Delta_{\text{vap}}H.$$

$$\Delta_{\text{vap}}H = \boxed{36.7\,\text{kJ mol}^{-1}}.$$

5.22 We use

$$\ln\frac{p'}{p} = \frac{\Delta_{\text{vap}}H}{R}\left(\frac{1}{T} - \frac{1}{T'}\right) \quad [5.6].$$

There are two ways to proceed with the estimate of T_b. In method 1, we (i) look up the experimental value of $\Delta_{\text{vap}}H$, (ii) use one measurement of p at T, and (iii) solve the above equation for the normal boiling point at $p' = 1\,\text{atm}$. In method 2, we (i) substitute $\Delta_{\text{vap}}H = T_b\Delta_{\text{vap}}S \approx (85\,\text{J K}^{-1}\,\text{mol}^{-1}) \times T_b$, which combines equation 4.6 and Trouton's rule, (ii) use a pair of p–T measurements, and (iii) solve the above equation for T_b.

Method 1. Substitute $p' = 1\,\text{atm} = 1.013 \times 10^5\,\text{Pa}$, $p = 2.0 \times 10^4\,\text{Pa}$, $T = 308\,\text{K}$, $\Delta_{\text{vap}}H = 30.8\,\text{kJ mol}^{-1}$ [Table 3.1], and then solve for $T_b = T'$.

$$\ln\left(\frac{1.013 \times 10^5\,\text{Pa}}{2.0 \times 10^4\,\text{Pa}}\right) = \frac{30.8 \times 10^3\,\text{J mol}^{-1}}{8.3145\,\text{J K}^{-1}\,\text{mol}^{-1}}\left(\frac{1}{308\,\text{K}} - \frac{1}{T_b}\right).$$

$$1.622 = 3704\,\text{K}\left(\frac{1}{308\,\text{K}} - \frac{1}{T_b}\right).$$

$$T_b = \boxed{356\,\text{K}}.$$

Method 2. Substitute $p' = 50.0 \times 10^3$ Pa, $T' = 332$ K, $p = 2.0 \times 10^4$ Pa, $T = 308$ K, $\Delta_{vap}H = (85\,\mathrm{J\,K^{-1}\,mol^{-1}}) \times T_b$, and then solve for T_b.

$$\ln\left(\frac{50.0 \times 10^3\,\mathrm{Pa}}{2.0 \times 10^4\,\mathrm{Pa}}\right) = \frac{(85\,\mathrm{J\,K^{-1}\,mol^{-1}})T_b}{8.3145\,\mathrm{J\,K^{-1}\,mol^{-1}}}\left(\frac{1}{308\,\mathrm{K}} - \frac{1}{332\,\mathrm{K}}\right).$$

$$T_b = \frac{0.916}{2.40 \times 10^{-3}\,\mathrm{K^{-1}}} = \boxed{382\,\mathrm{K}}.$$

The normal boiling point of benzene (353 K) gives evidence that Method 1 is more accurate.

5.23 The vapor pressure of ice at – 5°C is 3.9×10^{-3} atm, or 3.0 Torr (*Handbook of Chemistry and Physics*). Since the partial pressure of water is lower (2 Torr), the frost will sublime. A partial pressure of 3.0 Torr or more will ensure that the frost remains.

5.24 (a) The volume decreases as the vapor is cooled from 400 K, at constant pressure, in a manner described by the perfect gas equation

$$V = \frac{nRT}{p},$$

that is, V is a linear function of T. This continues until 373 K is reached where the vapor condenses to a liquid and there is a large decrease in volume. As the temperature is lowered further to 273 K, liquid water freezes to ice. Only a small decrease in volume occurs in the liquid as temperature is decreased, and a small (~9%) increase in volume occurs when the liquid freezes. Water remains as a solid at 260 K.

(b) The cooling curve appears roughly as sketched in Figure 5.3 below. The vapor and solid phases show a steeper rate of decline than for the liquid phase due to their smaller heat capacities. The temperature halt in the liquid plus vapor region is longer than for the liquid plus solid region due to its larger heat of transition.

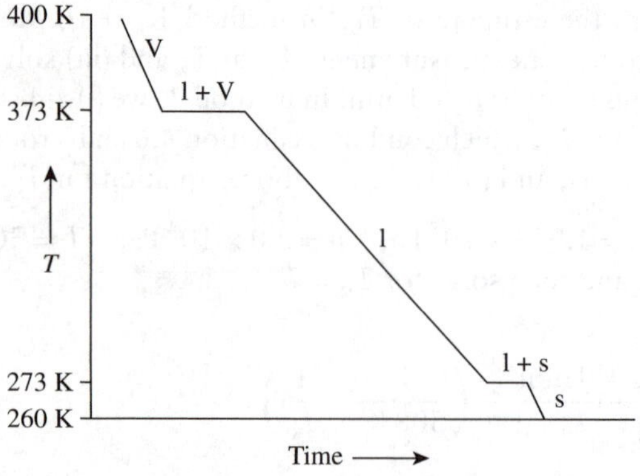

Fig 5.3

5.25 Cooling from, say, 400 K will cause a decrease in volume of gaseous water until 273.16 K is reached, at which temperature liquid and solid water will appear. All three phases will remain in equilibrium until the temperature falls below 273.16 K.

5.26 (a) The gaseous sample expands. (b) The sample contracts but remains gaseous because 320 K is greater than the critical temperature. (c) The gas contracts and forms a liquid-like substance without the appearance of a discernible surface. As the temperature lowers further to the solid phase boundary line, solid carbon dioxide forms in equilibrium with the liquid. At 210 K the sample has become all solid. (d) The solid expands slightly as the pressure is reduced and sublimes when the pressure reaches about 5 atm. (e) The gas expands as it is heated at constant pressure.

5.27 The slope of the He-II/He-I phase boundary line appears to be negative everywhere. That is,

$$\frac{\mathrm{d}p}{\mathrm{d}T} = \frac{\Delta_{\mathrm{trs}}H_{\mathrm{m}}}{T_{\mathrm{trs}}\Delta_{\mathrm{trs}}V_{\mathrm{m}}} = -.$$

If we assume that $\Delta_{\mathrm{trs}}H_{\mathrm{m}}$ for He-II $\rightarrow$ He-I is positive, it implies that $\Delta_{\mathrm{trs}}V_{\mathrm{m}}$ is negative, so He-I is $\boxed{\text{expected to be more dense}}$ than He-II. This argument, which would work well for normal fluids, fails in the case of He. The transition to the superfluid shows no measurable $\Delta_{\mathrm{trs}}H_{\mathrm{m}}$ or $\Delta_{\mathrm{trs}}V_{\mathrm{m}}$. So at the transition, there is $\boxed{\text{no difference in density}}$ between the two forms of He.

Chapter 6

The properties of mixtures

Answers to discussion questions

6.1 At equilibrium, the chemical potentials of any component in both the liquid and vapor phases must be equal. This is justified with the equilibrium criterion that under constant temperature and pressure conditions, with no additional work, $\Delta G = 0$. Consider the relationships $\mathrm{d}G_{\mathrm{m,i}(\alpha)} = V_{\mathrm{m,i}(\alpha)}\,\mathrm{d}p - S_{\mathrm{m,i}(\alpha)}\mathrm{d}T + \mu_{\mathrm{i}}(\alpha)\mathrm{d}n_{\mathrm{i}}$ and $\mathrm{d}G_{\mathrm{m,i}(\beta)} = V_{\mathrm{m,i}(\beta)}\mathrm{d}p - S_{\mathrm{m,i}(\beta)}\mathrm{d}T + \mu_{\mathrm{i}}(\beta)\,\mathrm{d}n_{\mathrm{i}}$ for a chemical species 'i' that is present in the two phases α and β (see Derivations 5.3 and 6.2). The terms with $\mathrm{d}p$ and $\mathrm{d}T$ on the right side of these expressions equal zero for constant p and constant T processes, and the near equilibrium transformation $\mathrm{i}(\beta) \rightleftharpoons \mathrm{i}(\alpha)$ is such a process. Subtraction and application of the equilibrium criteria gives $\mathrm{d}G_{\mathrm{i}} = \mathrm{d}G_{\mathrm{m,i}(\alpha)} - \mathrm{d}G_{\mathrm{m,i}(\beta)} = \{\mu_{\mathrm{i}}(\alpha) - \mu_{\mathrm{i}}(\beta)\}\mathrm{d}n_{\mathrm{i}} = 0$ or $\mu_{\mathrm{i}}(\alpha) - \mu_{\mathrm{i}}(\beta) = 0$ at equilibrium. The chemical potential of each chemical species must be equal in the liquid (α) and vapor (β) phases or in any two phases that are at equilibrium: $\boxed{\mu_{\mathrm{i}}(\alpha) = \mu_{\mathrm{i}}(\beta)}$.

6.2 All the colligative properties (properties that depend only on the number of solute particles present, not their chemical identity) are a result of the lowering of the chemical potential of the solvent due to the presence of the solute. This reduction takes the form $\mu_{\mathrm{A}} = \mu_{\mathrm{A}}^* + RT\ln x_{\mathrm{A}}$ or $\mu_{\mathrm{A}} = \mu_{\mathrm{A}}^* + RT\ln a_{\mathrm{A}}$, depending on whether or not the solution can be considered ideal. The lowering of the chemical potential results in a freezing point depression and a boiling point elevation as illustrated in Figures 6.16 and 6.17 of the text. Both of these effects can be explained by the lowering of the vapor pressure of the solvent in solution due to the presence of the solute. The solute molecules get in the way of the solvent molecules, reducing their escaping tendency.

6.3 The activity of a solute is that property that determines how the chemical potential of the solute varies from its value in a specified reference state. This is seen from the relationship $\mu = \mu^{\ominus} + RT\ln a$, where $\mu^{\ominus}$ is the value of the chemical potential in the reference state. The reference state is either the hypothetical state where the pure solute obeys Henry's law (if the solute is volatile) or the hypothetical state where the solute at unit molality obeys Henry's law (if the solute is involatile). The activity of the solute is defined as that physical property which makes the above relationship true. It can be interpreted as an effective concentration.

6.4 The osmotic pressure, Π, method (see Example 6.4) for determination of polymer molar mass involves measurement of Π for a series of successively more dilute mass concentrations c_{polymer}. The extrapolated intercept at $c_{\text{polymer}} = 0$ of a Π/c_{polymer} against c_{polymer} plot equals RT/M_{polymer}. Consequently, $M_{\text{polymer}} = RT/\text{intercept}$.

Solutions to exercises

6.5 $$\text{Mass} = 250.0\,\text{cm}^3 \times \frac{1\,\text{dm}^3}{10^3\,\text{cm}^3} \times \frac{0.112\,\text{mol}}{1\,\text{dm}^3} \times \frac{180.16\,\text{g}}{\text{mol}} = \boxed{5.04\,\text{g}}.$$

6.6 $$\text{Mass} = 250.0\,\text{g} \times \frac{1\,\text{kg}}{10^3\,\text{g}} \times \frac{0.112\,\text{mol}}{1\,\text{kg}} \times \frac{180.16\,\text{g}}{\text{mol}} = \boxed{5.04\,\text{g}}.$$

6.7 $$\text{Mass} = 25.00\,\text{cm}^3 \frac{1\,\text{dm}^3}{10^3\,\text{cm}^3} \times \frac{0.245\,\text{mol}}{\text{dm}^3} \times \frac{75.07\,\text{g}}{\text{mol}} = \boxed{0.460\,\text{g}}.$$

6.8 $$0.134\,m = \frac{0.134\,\text{mol}}{1\,\text{kg water}}.$$

$$n_{H_2O} = \frac{1000\,\text{g}}{18.02\,\text{g mol}^{-1}} = 55.5\,\text{mol}.$$

$$x_{\text{alanine}} = \frac{0.134\,\text{mol}}{0.134\,\text{mol} + 55.5\,\text{mol}} = \boxed{2.41 \times 10^{-3}}.$$

6.9 $$x_{\text{sucrose}} = 0.124 = \frac{n_{\text{sucrose}}}{n_{\text{sucrose}} + n_{H_2O}} = \frac{m/M_{\text{sucrose}}}{m/M_{\text{sucrose}} + \dfrac{100\,\text{g}}{18.02\,\text{g mol}^{-1}}}.$$

Solve the above equation for mass (m).

$$0.124\left(\frac{m}{M_{\text{sucrose}}} + 5.55\,\text{mol}\right) = \frac{m}{M_{\text{sucrose}}}.$$

$$0.124 \times 5.55\,\text{mol} = \frac{m}{M_{\text{sucrose}}}(1 - 0.124).$$

$$m = \frac{0.124 \times 5.55\,\text{mol} \times M_{\text{sucrose}}}{0.876} \left[M_{\text{sucrose}} = \frac{342.30\,\text{g}}{\text{mol}}\right].$$

$$\text{Mass} = \boxed{269\,\text{g sucrose}}.$$

6.10 Let 1 = 1-propanol and 2 = 2-propanol.

$$x_1 = \frac{n_1}{n_1 + n_2}; \quad x_2 = \frac{n_2}{n_1 + n_2}.$$

Because both alcohols have the same molar mass, $n_1 = n_2$ and $x_1 = x_2$.

Because $x_1 + x_2 = 1$, $x_1 = x_2 = \boxed{0.500}$.

6.11 Let p = 1-propanol and b = 1-butanol.

$$x_p = \frac{n_p}{n_p + n_b}; \quad x_b = \frac{n_b}{n_p + n_b}.$$

$$n_p = \frac{40.0\,\text{g}}{60.10\,\text{g mol}^{-1}} = 0.665\,\text{mol}.$$

$$n_b = \frac{60.0\,\text{g}}{74.13\,\text{g mol}^{-1}} = 0.809\,\text{mol}.$$

$$x_p = \frac{0.665\,\text{mol}}{0.665\,\text{mol} \times 0.809\,\text{mol}} = \boxed{0.451}.$$

$$x_p + x_b = 1; \text{ therefore } x_b = 1 - 0.451 = \boxed{0.549}.$$

6.12 Let A denote propanone and C denote trichloromethane. The mass of the sample is then

$$n_A M_A + n_C M_C = m. \qquad (1)$$

We also know that

$$x_A = \frac{n_A}{n_A + n_C} \text{ and rearranging we get } (x_A - 1)n_A + x_A n_C = 0.$$

Because $x_C = 1 - x_A$, we can substitute to obtain

$$-x_C n_A + x_A n_C = 0. \qquad (2)$$

Upon solving equations (1) and (2) simultaneously we obtain

$$n_A = \frac{x_A}{x_C} \times n_C, \quad n_C = \frac{m x_C}{x_A M_A + x_C M_C}.$$

Because $x_C = 0.4693$, $x_A = 1 - x_C = 0.5307$.

$$n_C = \frac{0.4693 \times 1000\,\text{g}}{(0.5307 \times 58.08) + (0.4693 \times 119.37)\,\text{g mol}^{-1}} = 5.505\,\text{mol}.$$

$$n_A = \frac{0.5307}{0.4693} \times 5.404 = 6.111\,\text{mol}.$$

The total volume, $V = n_A V_A + n_C V_C$ and

$$V = (6.111\,\text{mol} \times 74.166\,\text{cm}^3\,\text{mol}^{-1}) \times (5.404 \times 80.235\,\text{cm}^3\,\text{mol}^{-1})$$

$$= \boxed{886.8\,\text{cm}^3}.$$

6.13 Let E = ethanol, W = water

$$n_E = 50.0\,\text{cm}^3 \times 0.789\,\text{g}\,\text{cm}^{-3} \times \frac{1\,\text{mol}}{46.07\,\text{g}} = 0.856\,\text{mol}.$$

$$n_W = 50.0\,\text{cm}^3 \times 1.000\,\text{g}\,\text{cm}^{-3} \times \frac{1\,\text{mol}}{18.02\,\text{g}} = 2.775\,\text{mol}.$$

$$x_E = \frac{n_E}{n_E + n_W} = \frac{0.856}{0.856 + 2.775} = 0.236.$$

$$x_W = 1 - 0.236 = 0.764.$$

From Figure 6.1 in the main text we may roughly estimate the partial molar volumes as $V_E = 54.5\,\text{cm}^3/\text{mol}$, $V_W = 17.9\,\text{cm}^3/\text{mol}$. Then

$$V = n_E V_E + n_W V_W$$

$$= 0.856\,\text{mol} \times 54.5\,\text{cm}^3\,\text{mol}^{-1} + 2.775\,\text{mol} \times 17.9\,\text{cm}^3\,\text{mol}^{-1}$$

$$= \boxed{96\,\text{cm}^3}.$$

6.14 Figure 6.1 is a plot of the partial molar volume of ethanol solutions at 25°C.

$$V_{\text{ethanol}}/(\text{cm}^3\,\text{mol}^{-1}) = 54.6664 - 0.72788\,b + 0.084768\,b^2 \quad \text{where}$$

b is the magnitude of molality.

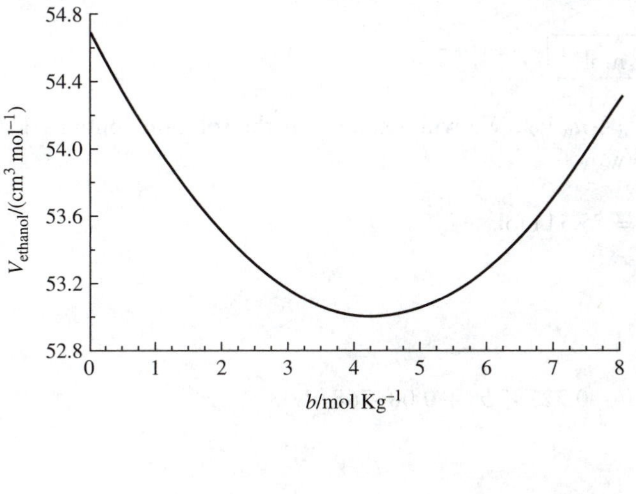

Fig 6.1

Examination of the plot shows that the minimum occurs at

$b \cong \boxed{4.3\ \text{mol kg}^{-1}}$ with $V_{\text{ethanol}} = \boxed{53.1\ \text{cm}^3\ \text{mol}^{-1}}$. This minimum may also be obtained by taking the derivative of V_{ethanol} with respect to b, setting this equal to zero, and solving for b.

$$\frac{dV_{\text{ethanol}}}{db} = -0.72788 + (2 \times 0.084768\ b) = 0.$$

$$b = \boxed{4.29\ \text{mol kg}^{-1}}, \quad V_{\text{ethanol}} = \boxed{53.1\ \text{cm}^3\ \text{mol}^{-1}}.$$

Not surprisingly, the agreement is almost exact. The plotting software used (PSI plot) is very good. To convert from molality to mole fraction we proceed as in Example 6.1 and consider the mass of solvent to be 1 kg. If the solvent is water

$$n_{\text{water}} = \frac{1000\ \text{g/kg}}{18.02\ \text{g mol}^{-1}} = 55.5\ \text{mol kg}^{-1}.$$

$$x_{\text{ethanol}} = \frac{b_{\text{ethanol}}}{b_{\text{ethanol}} + n_{\text{water}}} = \frac{4.29\ \text{mol kg}^{-1}}{(4.29 + 55.5)\ \text{mol kg}^{-1}} = \boxed{0.072}.$$

6.15 $V_{\text{ethanol}} = c_0 + c_1 b + c_2 b^2.$

$$\frac{dV_{\text{ethanol}}}{db} = c_1 + 2c_2 b.$$

The minimum occurs at $b = b_{\text{min}}$ where $\dfrac{dV_{\text{ethanol}}(b_{\text{min}})}{db} = c_1 + 2c_2 b_{\text{min}} = 0.$

$$b_{\text{min}} = \frac{-c_1}{2c_2} = \frac{-(-0.72788\ \text{cm}^3\ \text{kg mol}^{-2})}{2(0.084768\ \text{cm}^3\ \text{kg}^2\ \text{mol}^{-3})} = \boxed{4.2934\ \text{mol kg}^{-1}}.$$

$$V_{\text{min}} = 54.6664\ \text{cm}^3\ \text{mol}^{-1} - 0.72788\ \text{cm}^3\ \text{kg mol}^{-2}(4.2934\ \text{mol kg}^{-1})$$

$$+ 0.084768\ \text{cm}^3\ \text{kg}^2\ \text{mol}^{-3}(4.2934\ \text{mol kg}^{-1})^2$$

$$= \boxed{51.5570\ \text{cm}^3\ \text{mol}^{-1}}.$$

6.16 $V = n_E V_E + n_w V_w = b V_E + n_w V_w$. We will assume that the solution contains 1.000 kg of water and solve for V_W.

$$n_w = \frac{1000\ \text{g}}{18.02\ \text{g mol}^{-1}} = 55.51\ \text{mol}.$$

$$V_w = \frac{V - bV_E}{n_w}.$$

$$bV_E/\text{cm}^3 = 54.6664\, b - 0.72788\, b^2 + 0.084768\, b^3.$$

$$(V - bV_E)/\text{cm}^3 = 1002.93 - 0.36394\,b^2 + 0.72788\,b^2 + 0.028256\,b^3 - 0.084768\,b^3$$
$$= 1002.93 + 0.36394\,b^2 - 0.056512\,b^3.$$

$$V_w/\text{cm}^3\,\text{mol}^{-1} = \boxed{18.067 + 6.556 \times 10^{-3}b^2 - 1.018 \times 10^{-3}b^3}.$$

From an examination of Figure 6.2, we see that the $\boxed{\text{maximum occurs at } b \approx 4.3\,\text{mol kg}^{-1}}$ in agreement with the minimum in V_{ethanol}.

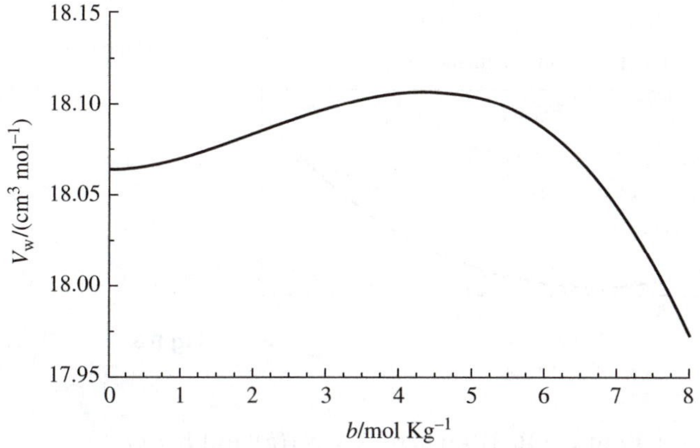

Fig 6.2

6.17 The moles of ethanol (EtOH) in 1 kg of water, n_{EtOH}, is related to the molality b by the expression $b = \dfrac{n_{\text{EtOH}}}{1\,\text{kg}}$. Consequently, $\dfrac{db}{dn_{\text{EtOH}}} = \dfrac{1}{1\,\text{kg}} = 1\,\text{kg}^{-1}$.

$$V = c_0 + c_1 b + c_2 b^2 + c_3 b^3 \quad \text{where}$$

$$c_0 = 1002.93\,\text{cm}^3, \quad c_1 = 54.6664\,\text{cm}^3\,\text{mol}^{-1}\,\text{kg},$$

$$c_2 = -0.36394\,\text{cm}^3\,\text{mol}^{-2}\,\text{kg}^2, \quad c_3 = 0.028256\,\text{cm}^3\,\text{mol}^{-3}\,\text{kg}^3.$$

$$V_{\text{EtOH}} = \frac{dV}{dn_{\text{EtOH}}} = \frac{dV}{db} \times \frac{db}{dn_{\text{EtOH}}} = \frac{dV}{db} \times (1\,\text{kg}^{-1})$$
$$= \frac{d(c_0 + c_1 b + c_2 b^2 + c_3 b^3)}{db} \times (1\,\text{kg}^{-1}) = (c_1 + 2c_2 b + 3c_3 b^2) \times (1\,\text{kg}^{-1}).$$

This equation is used to calculate the points for a plot of V_{EtOH} as a function of b. Alternatively, a plot of V_{EtOH} as a function of x_{EtOH} may be prepared by calculating V_{EtOH} over a range of b values with the above equation and calculating corresponding value of x_{EtOH} at each b value with

$$x_{\text{EtOH}} = \frac{b \times (1\,\text{kg})}{b \times (1\,\text{kg}) + (10^3/18.02)\text{mol}} \quad \text{[Example 6.1].}$$

Sample points for a $V_{\text{EtOH}}(x_{\text{EtOH}})$ plot are in the table and a complete plot appears in Figure 6.3.

b/m	x_{EtOH}	$V_{\text{EtOH}}/\text{cm}^3\,\text{mol}^{-1}$
0	0	54.6664
1	0.017701	54.02312
2	0.034786	53.54904

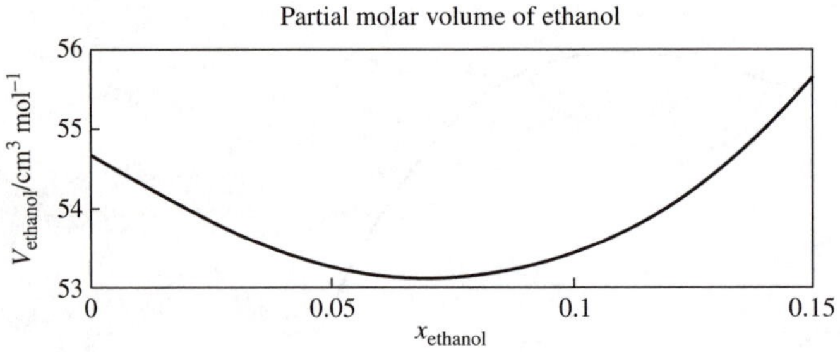

Fig 6.3

6.18 (a) $\Delta G_{\text{m}} = RT\,(x_{\text{A}} \ln x_{\text{A}} + x_{\text{B}} \ln x_{\text{B}})$ [6.10] where $\text{A} = \text{N}_2(\text{g})$ and $\text{B} = \text{O}_2(\text{g})$.

$$\Delta G_{\text{m}} = 2.4790\,\text{kJ}\,\text{mol}^{-1}\{0.78\ln(0.78) + 0.22\ln(0.22)\}$$

$$= \boxed{-1.31\,\text{kJ}\,\text{mol}^{-1}}.$$

Because ΔG_{m} is negative, the mixing is $\boxed{\text{spontaneous}}$.

(b) $\Delta S_{\text{m}} = -R(x_{\text{A}} \ln x_{\text{A}} + x_{\text{B}} \ln x_{\text{B}})$ [6.11b]

$$= \boxed{+4.38\,\text{J}\,\text{K}^{-1}\,\text{mol}^{-1}}.$$

6.19 $\Delta G_{\text{m}} = RT(x_{\text{A}} \ln x_{\text{A}} + x_{\text{B}} \ln x_{\text{B}} + x_{\text{C}} \ln x_{\text{C}})$ where $\text{A} = \text{N}_2(\text{g})$, $\text{B} = \text{O}_2(\text{g})$, and $\text{C} = \text{Ar}(\text{g})$.

$$\Delta G_{\text{m}} = 2.4790\,\text{kJ}\,\text{mol}^{-1}\{0.780\ln(0.780) + 0.210\ln(0.210) + 0.0096\ln(0.0096)\}$$

$$= \boxed{-1.40\,\text{kJ}\,\text{mol}^{-1}}.$$

Because the change in ΔG_{m} is negative upon the addition of argon, the mixing is $\boxed{\text{spontaneous}}$.

$$\delta S_{\text{m}} = -R(x_{\text{A}} \ln x_{\text{A}} + x_{\text{B}} \ln x_{\text{B}} + x_{\text{C}} \ln x_{\text{C}})$$

$$= \boxed{+4.71\,\text{J}\,\text{K}^{-1}\,\text{mol}^{-1}}.$$

6.20 $p_J = x_J p_J^*$ [6.12].

$$x_{\text{toluene}} = \frac{n_{\text{toluene}}}{n_{C_{60}} + n_{\text{toluene}}}.$$

$$n_{C_{60}} = \frac{1.23\,\text{g}}{720.6\,\text{g mol}^{-1}} = 1.71 \times 10^{-3}\,\text{mol}.$$

$$n_{\text{toluene}} = \frac{100\,\text{g}}{92.14\,\text{g mol}^{-1}} = 1.085\,\text{mol}.$$

$$x_{C_{60}} = \frac{1.085\,\text{g}}{(1.71 \times 10^{-3}\,\text{mol}) + 1.085\,\text{mol}} = 0.9984.$$

$$p_{\text{toluene}} = 0.9984 \times 5.00\,\text{kPa} = \boxed{4.99\,\text{kPa}}.$$

6.21 Let us assume that $1.000\,\text{dm}^3$ of seawater contains roughly 1000 g of water. Then

$$n_{\text{water}} = \frac{1000\,\text{g}}{18.02\,\text{g mol}^{-1}} = 55.5\,\text{mol}.$$

$$n_{\text{solutes}} = 2 \times 0.05\,\text{mol} = 1.00\,\text{mol}.$$

$$x_{\text{water}} = \frac{n_{\text{water}}}{n_{\text{water}} + n_{\text{solutes}}} = \frac{55.5}{56.5} = 0.982.$$

$$p_{\text{water}} = x_{\text{water}}\, p^*_{\text{water}} \quad [6.12] = 0.982 \times 2.338\,\text{kPa} = \boxed{2.30\,\text{kPa}}.$$

6.22 Check whether p_B/x_B [6.16] is equal to a constant (K_B).

x	0.005	0.012	0.019
p/x	6.4×10^3	6.4×10^3	6.4×10^3 kPa

Hence, $\boxed{K_B \sim 6.4 \times 10^3\,\text{kPa}}$.

6.23 $x_B = \dfrac{p_B}{K_B}$ [6.16].

Convert Torr to kPa. 760 Torr = 101.3 kPa.

1 Torr = 0.1333 kPa.

$$x_B = \frac{55\,\text{kPa}}{(8.6 \times 10^4\,\text{Torr}) \times 0.1333\,\text{kPa Torr}^{-1}} = \boxed{4.8 \times 10^{-3}}.$$

6.24 $p_{H_2} = [H_2]K_{H_2}$ [6.17]

$$= \left(\frac{1.0 \times 10^{-3}\,\text{mol}}{1 \times 10^{-3}\,\text{m}^3}\right) \times (121.2\,\text{kPa m}^3\,\text{mol}^{-1}) \text{ [Table 6.1]} = \boxed{121.2\,\text{kPa}}.$$

6.25 $[\mathrm{CO_2}] = \dfrac{p_{\mathrm{CO_2}}}{K_{\mathrm{CO_2}}}$ [6.17], $K_{\mathrm{CO_2}} = 2.937\,\mathrm{kPa\,m^3\,mol^{-1}}$.

(a) $[\mathrm{CO_2}] = \dfrac{4.0\,\mathrm{kPa}}{2.937\,\mathrm{kPa\,m^3\,mol^{-1}}} = 1.3\bar{6}\,\mathrm{mol\,m^{-3}} = \boxed{1.3\bar{6}\,\mathrm{mmol\,dm^{-3}}}$.

(b) $[\mathrm{CO_2}] = \dfrac{100.\,\mathrm{kPa}}{2.937\,\mathrm{kPa\,m^3\,mol^{-1}}} = 34.0\,\mathrm{mol\,m^{-3}} = \boxed{34.0\,\mathrm{mmol\,dm^{-3}}}$.

6.26 $[\mathrm{J}] = \dfrac{p_{\mathrm{J}}}{K_{\mathrm{J}}}$ Henry's law [6.17]

$= \dfrac{x_{\mathrm{J}}(\mathrm{gas}) \times p}{K_{\mathrm{J}}}$ [1.7] and we assume that $p = p^{\ominus} = 1.00\,\mathrm{bar} = 100\,\mathrm{kPa}$.

$[\mathrm{N_2}] = \dfrac{0.78 \times (100\,\mathrm{kPa})}{155\,\mathrm{kPa\,m^3\,mol^{-1}}}$ [Table 6.1] $= 0.50\,\mathrm{mol\,m^{-3}} = 0.50\,\mathrm{mmol\,dm^{-3}}$.

$[\mathrm{O_2}] = \dfrac{0.21 \times (100\,\mathrm{kPa})}{74.68\,\mathrm{kPa\,m^3\,mol^{-1}}}$ [Table 6.1] $= 0.28\,\mathrm{mol\,m^{-3}} = 0.28\,\mathrm{mmol\,dm^{-3}}$.

The magnitudes of molarity and molality concentrations are equal in very dilute solutions such as these. Consequently, $\boxed{b_{\mathrm{N_2}} = 0.50\,\mathrm{mmol\,kg^{-1}}}$ and $b_{\mathrm{O_2}} = \boxed{0.28\,\mathrm{mmol\,kg^{-1}}}$.

6.27 Assume $1\,\mathrm{dm^3}$ of water has a mass of roughly 1 kg.

$$x_{\mathrm{CO_2}} = \frac{p_{\mathrm{CO_2}}}{K_{\mathrm{CO_2}}} = \frac{3.0\,\mathrm{atm} \times 760\,\mathrm{Torr\,atm^{-1}}}{1.25 \times 10^6\,\mathrm{Torr}} = 1.82 \times 10^{-3}.$$

$$x_{\mathrm{CO_2}} = \frac{n_{\mathrm{CO_2}}}{n_{\mathrm{CO_2}} + n_{\mathrm{H_2O}}}.$$

Solving for

$$n_{\mathrm{CO_2}} = \frac{x_{\mathrm{CO_2}} n_{\mathrm{H_2O}}}{1 - x_{\mathrm{CO_2}}}$$

$$= \frac{(1.82 \times 10^{-3}) \times 1000\,\mathrm{g}/18.02\,\mathrm{g\,mol^{-1}}}{1 - 1.82 \times 10^{-3}} = 0.101\,\mathrm{mol}.$$

$$\text{Molarity} = \frac{0.101\,\mathrm{mol}}{1\,\mathrm{dm^3}} = \boxed{0.101\,\mathrm{mol\,dm^{-3}}}.$$

6.28 $p = p_{\mathrm{A}} + p_{\mathrm{B}} = x p_{\mathrm{A}}^* + x p_{\mathrm{B}}^* = x p_{\mathrm{A}}^* + (1 - x_{\mathrm{A}}) p_{\mathrm{B}}^*$.

Solving for x_{A},

$$x_{\mathrm{A}} = \frac{p - p_{\mathrm{B}}^*}{p_{\mathrm{A}}^* - p_{\mathrm{B}}^*}.$$

For boiling under 0.50 atm (380 Torr) pressure, the combined vapor pressure must be 380 Torr; hence if A = toluene and B = *o*-xylene

$$x_A = \frac{380-150}{400-150} = \boxed{0.920}, \qquad x_B = \boxed{0.080}.$$

The composition of the vapor is given by

$$y_A = \frac{p_A}{p} = \frac{x_A - p_A^*}{p_B^* + (p_A^* - p_B^*)x_A} = \frac{0.920 \times 400}{150 + [(400-150) \times 0.920]} = \boxed{0.968}$$

and $y_B = 1 - 0.968 = \boxed{0.032}$.

6.29 Let B denote benzene and A the solute; then

$$p_B = x_B p_B^* \quad \text{and} \quad x_B = \frac{n_B}{n_A + n_B}.$$

Hence

$$p_B = \frac{n_B p_B^*}{n_A + n_B}$$

which solves to

$$n_A = \frac{n_B(p_B^* - p_B)}{p_B}.$$

Then, because $n_A = m_A/M_A$, where m_A is the mass of A present,

$$M_A = \frac{m_A p_B}{n_B(p_B^* - p_B)} = \frac{m_A M_B p_B}{m_B(p_B^* - p_B)}.$$

$$M_A = \frac{(0.125\,\text{g}) \times (78.11\,\text{g mol}^{-1}) \times (51.2\,\text{kPa})}{5.00\,\text{g} \times (53.0 - 51.2)\,\text{kPa}} = \boxed{55.5\,\text{g mol}^{-1}}.$$

6.30 Assume 150 cm^3 of water has a mass of 0.150 kg.

$$\Delta T = K_f b_B \quad [6.22] = 1.86\,\text{K kg mol}^{-1} \times \frac{7.5\,\text{g}}{342.3\,\text{g mol}^{-1} \times 0.150\,\text{kg}} = 0.27\,\text{K}.$$

The freezing point will be approximately $\boxed{-0.27°\text{C}}$.

6.31 $\Delta T = K_f b_B$ [6.22], where b_B is the molality of B and is given by

$$b_B = \frac{n_B}{\text{mass of CC1}_4\text{ in kg}} = \frac{28.0\,\text{g}}{M \times 0.750\,\text{kg}}$$

$$= \frac{37.3\,\text{g/kg}}{M} = \frac{\Delta T}{K_f}.$$

Solve for M,

$$M = \frac{37.3\,\text{g/kg} \times 30\,\text{K kg mol}^{-1}}{5.40\,\text{K}} = \boxed{207\,\text{g mol}^{-1}}.$$

6.32 $K = \frac{[A_2]}{[A]^2}$ and let the initial number of moles $A = n$.

At equilibrium $n_{A2} = fn$, $n_A = (1 - 2f)\,n$, and the total amount of solute is $(1 - f)\,n$.

Therefore, if the volume is V,

$$K = \frac{fnV}{(1-2f)^2n^2} = \frac{f}{(1-2f)^2c} \quad \text{where} \quad c = n/V.$$

Vapor pressure, p is $p = x_{\text{solvent}}p^*$.

$$p = x_s p^* = \frac{n_s p^*}{n_A + n_{A_2} + n_s} = \frac{n_s p^*}{(1-f)n + n_s}.$$

$n_s = Vr$ with $r = \rho/M$, ρ = density of solvent.

$$p = \frac{rp^*}{(1-f)c + r}; \text{ rearranging } f = 1 - \frac{r(p^* - p)}{cP} \text{ and, finally.}$$

$$\boxed{K = \frac{1 - \dfrac{r(p^*-p)}{cp}}{c\left(1 - \dfrac{2r(p^*-p)}{cp}\right)^2}}.$$

6.33 $\Pi V = n_B RT$ [6.23a].

$$\frac{n_B}{V} = M_B \text{ (molarity)} \approx b\rho[\rho = \text{density}]$$

with $\rho = 10^3\,\text{kg m}^{-3}$ for dilute aqueous solutions.

Then

$$b \approx \frac{n_B}{V\rho} = \frac{\Pi}{RT\rho}.$$

$$\Delta T = K_f b_B \approx K_f \times \frac{\Pi}{RT\rho}.$$

Therefore, with $K_f = 1.86\,\text{K kg mol}^{-1}$ (Table 6.3),

$$\Delta T = \frac{(1.86\,\text{K kg mol}^{-1}) \times (120 \times 10^3\,\text{Pa})}{(8.3145\,\text{J K}^{-1}\,\text{mol}^{-1}) \times (300\,\text{K}) \times (1.00 \times 10^3\,\text{kg m}^{-3})} = 0.089\,\text{K}.$$

Therefore, the solution will freeze at about $\boxed{-0.09°\text{C}}$.

6.34 Our strategy is to avoid assuming that these solutions behave as ideal-dilute solutions and to analyze the data as illustrated in Example 6.4. The method of analysis is suggested by the equation

$$\frac{\Pi}{c} = \frac{RT}{M} + \left(\frac{RTB}{M^2}\right)c \quad \text{where} \quad c = m/V.$$

This says that a plot of Π/c against c has an intercept equal to RT/M and a slope equal to RTB/M^2 where B is the osmotic virial coefficient. We draw up a table to calculate Π/c values, prepare a plot to check linearity, and perform a linear regression analysis of the plot with a scientific calculator. See Figure 6.4. The molar mass is given by $M = RT/\text{intercept}$.

The virial coefficient is given by $B = \text{slope} \times M^2/RT$.

$c/(\text{g dm}^{-3})$	2.042	6.613	9.521	12.602
Π/Pa	58.3	188.2	270.8	354.6
$\Pi/c/(\text{Pa g}^{-1}\text{ dm}^3)$	28.55	28.46	28.44	28.14

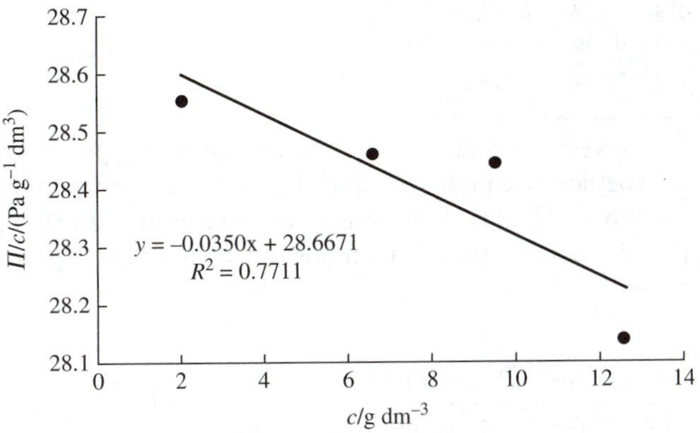

Fig 6.4

$$M = \frac{(8.3145\,\text{J mol}^{-1}\,\text{K}^{-1})(298\,\text{K})}{28.67\,\text{Pa g}^{-1}\,\text{dm}^3}\left(\frac{\text{dm}^3}{10^{-3}\,\text{m}^3}\right) = \boxed{86.4\,\text{kg mol}^{-1}}.$$

$$B = \frac{(-0.0350\,\text{Pa g}^{-2}\,\text{dm}^6)(8.64\times10^4\,\text{g mol}^{-1})^2}{(8.3145\,\text{J mol}^{-1}\,\text{K}^{-1})(298\,\text{K})}\left(\frac{10^{-3}\,\text{m}^3}{\text{dm}^3}\right) = -105\,\text{mol}^{-1}\,\text{dm}^3.$$

Note. The low value of the linear regression correlation coefficient (R^2) indicates that the apparent linearity is an artifact of experimental uncertainty of the individual data points. If this is confirmed, the data would be interpreted as indicating that B is too small to be confidently resolved with the above analysis and that the polymer solutions are behaving as ideal-dilute solutions within the limits of experimental data. In this case, the data should be analyzed with the van't Hoff equation [6.23b] in the form: $M = RTc/\Pi$. That is, the molar mass equals the mean of RTc/Π for the data set. This analysis gives $(c/\Pi)_{\text{mean}} = 3.522\times10^{-2}\,\text{g dm}^{-3}\,\text{Pa}^{-1}$ and, therefore,

$$M = (8.3145\,\text{J K}^{-1}\,\text{mol}^{-1})\times(298.15\,\text{K})\times(3.522\times10^{-2}\,\text{g dm}^{-3}\,\text{Pa}^{-1})\times\left(\frac{1\,\text{dm}^3}{10^{-3}\,\text{m}^3}\right)$$

$$= 87.\overline{31}\,\text{kg mol}^{-1}.$$

6.35 Our strategy is to avoid assuming that these solutions behave as ideal-dilute solutions and to analyze the data as illustrated in Example 6.4. The method of analysis is suggested by the equation

$$\frac{\Pi}{c} = \frac{RT}{M} + \left(\frac{RTB}{M^2}\right)c \quad \text{where} \quad c = m/V.$$

$\Pi = \rho g h$ [hydrostatic pressure] so

$$\frac{h}{c} = \frac{\Pi}{\rho g c} = \left(\frac{RT}{\rho g M}\right) + \left(\frac{RTB}{\rho g M^2}\right)c.$$

This says that a plot of h/c against c has an intercept equal to $RT/\rho g M$ and a slope equal to $RTB/\rho g M^2$ where B is the osmotic virial coefficient. We draw up a table to calculate h/c values

$c/(\text{mg cm}^{-3})$	3.221	4.618	5.112	6.722
h/cm	5.746	8.238	9.119	11.990
$h/c/(\text{mg}^{-1}\,\text{cm}^4)$	1.784	1.783	1.784	1.784

Inspection of the h/c values reveals that they are a constant for this experimental set. This implies that the virial coefficient equals zero and that these enzyme solutions are behaving as ideal-dilute solutions. The last term in the above equation vanishes giving the van't Hoff equation [6.23b]. Solving for M (assuming a density of $1.000\,\text{g cm}^{-3}$),

$$M = \frac{RT}{\rho g \times (h/c)}$$

$$= \frac{(8.3145\,\text{J K}^{-1}\,\text{mol}^{-1}) \times (293.15\,\text{K})}{(1.000\,\text{g cm}^{-3}) \times (9.807\,\text{m s}^{-2}) \times (1.784 \times 10^3\,\text{g}^{-1}\,\text{cm}^4)} \times \left(\frac{1\,\text{cm}}{10^{-2}\,\text{m}}\right)$$

$$= \boxed{13.9\bar{3}\,\text{kg mol}^{-1}}.$$

6.36 The data are plotted in Figure 6.5. From tie line (a) on the graph, the vapor in equilibrium with liquid of composition $x_T = 0.250$ has $y_T = \boxed{0.36}$. From tie line (b), for $x_O = 0.250$, $x_T = 0.750$, $y_T = \boxed{0.81}$.

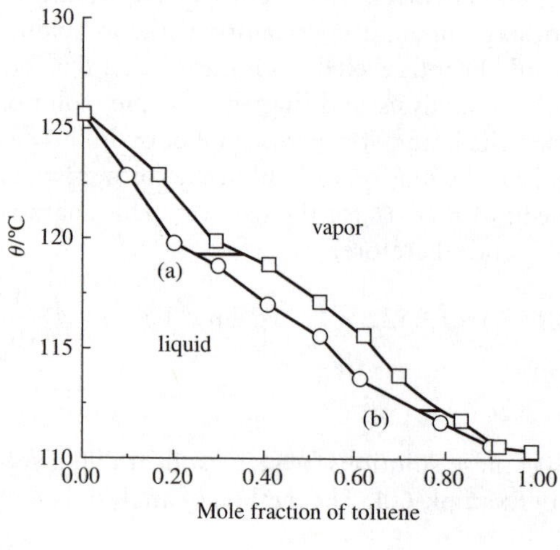

Fig 6.5

6.37 The phase diagram of the NH_3/N_2H_4 system is sketched in Figure 6.6.

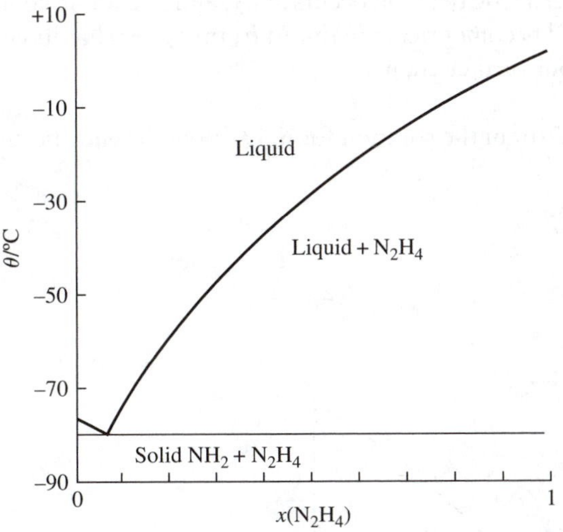

Fig 6.6

6.38 Refer to Figure 6.35 of the text. At b_3 there are two phases with compositions $x_A = 0.18$ and $x_A = 0.70$; their abundances are in the ratio 0.13 [lever rule]. Because $C = 2$ and $P = 2$, we have $F = 2$ (such as p and x). On heating, the phases merge, and the single-phase region is encountered. Then $F = 3$ (such as p, T, and x). The liquid comes into equilibrium with its vapor when the isopleth cuts the phase line. At this temperature, and for all points up to b_1, $C = 2$ and $P = 2$, implying that $F = 2$. The whole sample is a vapor above b_1.

6.39 The phase diagrams and cooling curves are shown in Figure 6.7.

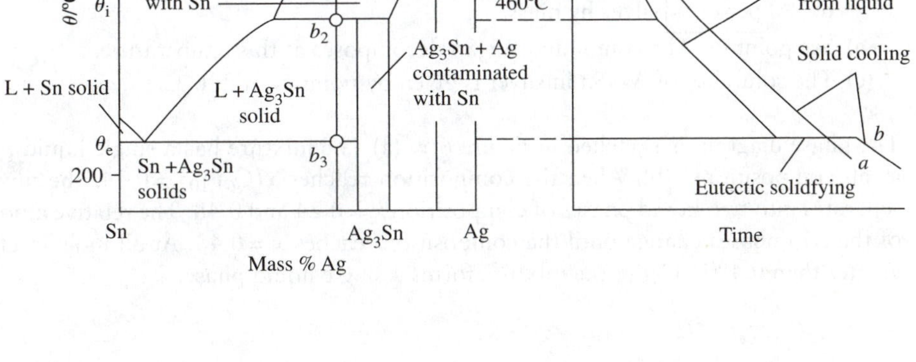

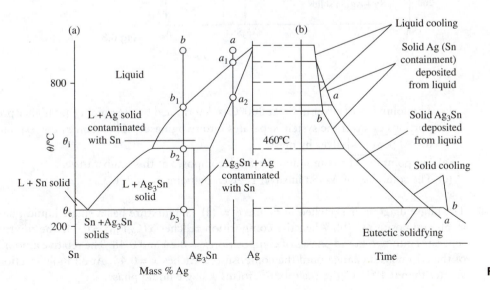

Fig 6.7

(a) Solid silver with dissolved tin begins to precipitate at a_1, and the sample solidifies completely at a_2. (b) Solid silver with dissolved tin begins to precipitate at b_1, and the liquid becomes richer in Sn. The peritectic reaction occurs at b_2, and as cooling continues Ag_3Sn is precipitated and the liquid becomes richer in tin. At b_3 the system has its eutectic composition (e) and freezes without further change.

6.40 The curves are shown in Figure 6.7(b) in the solution for 6.39. Note the eutectic halt for the isopleth b.

6.41 Refer to Figure 6.8.

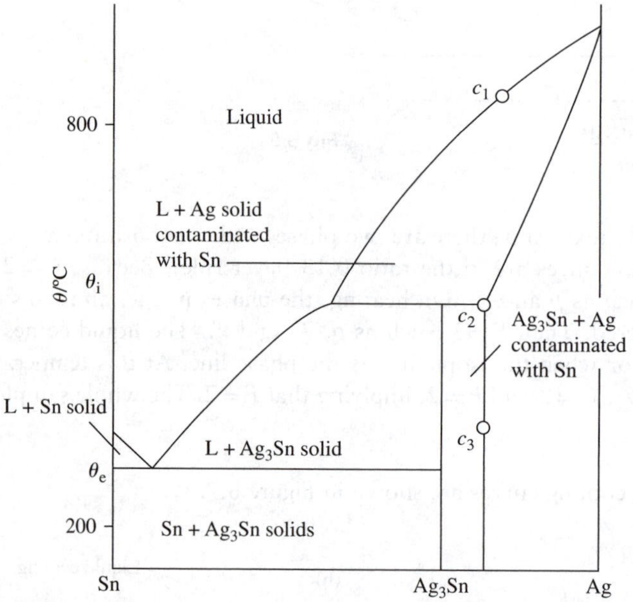

Fig 6.8

(a) The solubility of silver in tin at 800°C is determined by the point c_1 [at higher proportions of silver the system separates into two phases]. The point c_1 corresponds to 80 per cent silver by mass.
(b) See point c_2. The compound Ag_3Sn decomposes at this temperature.
(c) The solubility of Ag_3Sn in silver is given by point c_3 at 300°C.

6.42 The phase diagram is sketched in Figure 6.9. (a) The mixture has a single liquid phase at all compositions. (b) When the composition reaches $x(C_6F_{10}) = 0.24$, the mixture separates into two liquid phases of composition $x = 0.24$ and 0.48. The relative amounts of the two phases change until the composition reaches $x = 0.48$. At all mole fractions greater than 0.48 in C_6F_{14} the mixture forms a single liquid phase.

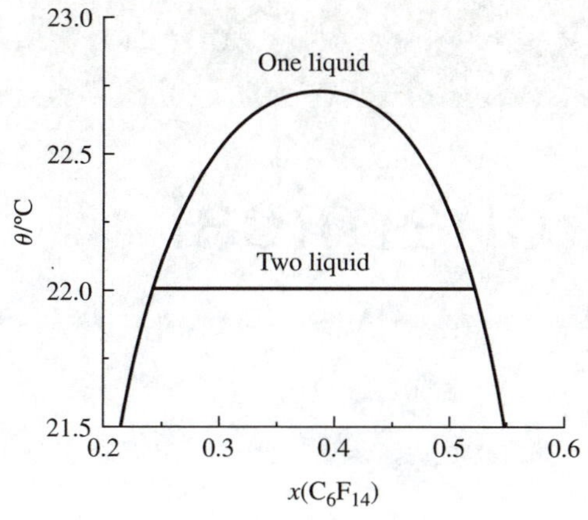

Fig 6.9

6.43 (a) No, the region of stability of the molten-globule form does not extend below 0.1 concentration of denaturant.

(b) The native form converts to the molten-globule form at $T \approx 0.65$ and finally to the unfolded form at $T \approx 0.85$.

6.44 At roughly 34°C, solid begins to form as the state point enters the two-phase region. Within the two-phase region, the proportion of liquid and solid can be determined by the lever rule. As the temperature is lowered through the two-phase region, the proportion of solid increases until, at roughly 20°C, the system becomes totally solid.

Chapter 7

Principles of chemical equilibrium

Answers to discussion questions

7.1 The position of equilibrium is always determined by the condition that the reaction quotient, Q, must equal the equilibrium constant, K. If the mixing in of an additional amount of reactant or product destroys that equality, then the reacting system will shift in such a way as to restore the equality. That implies that some of the added reactant or product must be removed by the reacting system and the amounts of other components will also be affected. These adjustments restore the concentrations to their (new) equilibrium values.

7.2 A non-spontaneous, endergonic reaction may be driven forward by a spontaneous, exergonic reaction that can supply the requisite reaction Gibbs energy. The total Gibbs energy change must be exergonic. This is accomplished in many coupled biochemical reactions. The exergonic reaction gives up a portion of its Gibbs energy not as heat but to the conversion of a low potential biochemical intermediate to a high potential one. The high potential species carries the energy to the endergonic reaction and in the process of releasing its chemical energy to the endergonic reaction it returns to its low potential form. The energy-carrying intermediate effectively 'couples' the two reactions and the energy transfers often occur on the surfaces of enzymes. Adenosine diphosphate (ADP) and adenosine triphosphate (ADP) is an example set of coupling intermediates:

$$\mathrm{ATP(aq)} + \mathrm{H_2O(l)} \rightarrow \mathrm{ADP(aq)} + \mathrm{P_i^-(aq)} + \mathrm{H^+(aq)}, \quad \Delta_r G^\ominus = -31\,\mathrm{kJ\,mol^{-1}}.$$

7.3 (1) Response to change in pressure. The equilibrium constant is independent of pressure, but the individual partial pressures can change as the total pressure changes. This will happen when there is a difference, Δn_g, between the sums of the number of moles of gases on the product and reactant sides of the chemical equation. The requirement of an unchanged equilibrium constant implies that the side with the smaller number of moles of gas will be favored as pressure increases.

(2) Response to change in temperature. Equation 7.15 shows that K decreases with increasing temperature when the reaction is exothermic; thus the reaction shifts to the

left, the opposite occurs in endothermic reactions. See Section 7.9 for a more detailed discussion.

7.4 The van't Hoff equation, written as $\ln K' - \ln K = \dfrac{\Delta_r H^\ominus}{R}\left(\dfrac{1}{T} - \dfrac{1}{T'}\right)$, is valid over small temperature ranges in which neither $\Delta_r H^\ominus$ nor $\Delta_r S^\ominus$ vary much with temperature.

Solutions to exercises

7.5 (a) $Q = \dfrac{[\text{G}][\text{P}_\text{i}]}{[\text{G6P}]}$. H_2O doesn't appear in these expressions because it is the solvent.

(b) $Q = \dfrac{[\text{Gly—Ala}]}{[\text{Gly}][\text{Ala}]}$.

(c) $Q = \dfrac{[\text{MgATP}^{2-}]}{[\text{Mg}^{2+}][\text{ATP}^{4-}]}$.

(d) $Q = \dfrac{p_{\text{CO}_2}^6}{p_{\text{O}_2}^5[\text{CH}_3\text{COCOOH}]^2}$.

7.6 $\frac{1}{2}\text{N}_2(\text{g}) + \frac{3}{2}\text{H}_2(\text{g}) \rightarrow \text{NH}_3(\text{g}), \quad \Delta_r G = \Delta_f G^\ominus + RT\ln Q.$

$$Q = \frac{p_{\text{NH}_3}}{p_{\text{N}_2}^{1/2}p_{\text{H}_2}^{3/2}} = \frac{4.0}{(3.0)^{1/2}(1.0)^{3/2}} = \frac{4.0}{\sqrt{3.0}}.$$

Therefore,

$$\Delta_r G = -16.45\,\text{kJ mol}^{-1} + RT\ln\left(\frac{4.0}{\sqrt{3.0}}\right) = -16.45\,\text{kJ mol}^{-1} + 2.07\,\text{kJ mol}^{-1}$$

$$= \boxed{-14.38\,\text{kJ mol}^{-1}}.$$

Because $\Delta_r G < 0$, the spontaneous direction of the reaction is toward the products.

7.7 (a) $K = \dfrac{a_{\text{COCl}}a_{\text{Cl}}}{a_{\text{CO}}a_{\text{Cl}_2}} = \dfrac{p_{\text{COCl}}p_{\text{Cl}}}{p_{\text{CO}}p_{\text{Cl}_2}}$.

(b) $K = \dfrac{a_{\text{SO}_3}^2}{a_{\text{O}_2}a_{\text{SO}_2}^2} = \dfrac{p_{\text{SO}_3}^2}{p_{\text{O}_2}p_{\text{SO}_3}^2}$.

(c) $K = \dfrac{a_{\text{HBr}}^2}{a_{\text{H}_2}a_{\text{Br}_2}^2} = \dfrac{p_{\text{HBr}}^2}{p_{\text{H}_2}p_{\text{Br}_2}}$.

(d) $K = \dfrac{a_{\text{O}_2}^3}{a_{\text{O}_3}^2} = \dfrac{p_{\text{O}_2}^3}{p_{\text{O}_3}^2}$.

7.8 $K_f = \dfrac{[C]}{[A][B]} = 0.224.$

$$K_r = \frac{[A][B]}{[C]} = \frac{1}{K_f} = \frac{1}{0.224} = \boxed{4.46}.$$

7.9 $K = \dfrac{[C]^2}{[A][B]} = 3.4 \times 10^4.$

(a) $K' = \dfrac{[C]^4}{[A]^2[B]^2} = K^2 = (3.4 \times 10^4)^2 = \boxed{1.2 \times 10^9}.$

(b) $K'' = \dfrac{[C]}{[A]^{1/2}[B]^{1/2}} = K^{1/2} = (3.4 \times 10^4)^{1/2} = \boxed{1.8 \times 10^2}.$

7.10 $\Delta_r G^\ominus = -RT \ln K$ [7.8]

$$= -8.315\,\mathrm{J\,K^{-1}\,mol^{-1}} \times 400\,\mathrm{K} \times \ln 2.07$$

$$= -2.42 \times 103\,\mathrm{J\,mol^{-1}} = \boxed{-2.42\,\mathrm{kJ\,mol^{-1}}}.$$

7.11 $\Delta_r G^\ominus = -RT \ln K.$

Therefore, $K = e^{-\Delta G^\ominus/RT} = e^{+3.67\times10^3\,\mathrm{J\,mol^{-1}}/8.3145\,\mathrm{J\,K^{-1}\,mol^{-1}}\times 400\,\mathrm{K}} = \boxed{3.01}.$

7.12 $\Delta_r G^\ominus = -RT \ln K$ [7.8] or $K = e^{-\Delta_r G^\ominus/RT}.$

$$\frac{K_{r1}}{K_{r2}} = \frac{e^{-\Delta_{r1} G^\ominus/RT}}{e^{-\Delta_{r2} G^\ominus/RT}} = e^{-(\Delta_{r1} G^\ominus - \Delta_{r2} G^\ominus)/RT}.$$

Let $\Delta_{r1} G^\ominus = -200\,\mathrm{kJ\,mol^{-1}}$ and $\Delta_{r2} G^\ominus = -100\,\mathrm{kJ\,mol^{-1}}$; then

$$\frac{K_{r1}}{K_{r2}} = e^{-(-200\times10^3\,\mathrm{J\,mol^{-1}} - (-100\times10^3\,\mathrm{J\,mol^{-1}}))/(8.3145\,\mathrm{J\,K^{-1}\,mol^{-1}})(300\,\mathrm{K})}$$

$$= e^{(100\times10^3\,\mathrm{J\,mol^{-1}})/(8.3145\,\mathrm{J\,K^{-1}\,mol^{-1}})(300\,\mathrm{K})} = \boxed{2.58 \times 10^{17}}.$$

COMMENT This is an enormous difference. Because of the exponential relation between K and $\Delta_r G^\ominus$ even small differences in $\Delta_r G^\ominus$ can make large differences in K.

7.13 $K_1 = 10 \times K_2.$

$$\Delta_r G_2^\ominus = -RT \ln(0.10 \times K_1) = -RT \ln K_1 - RT \ln(0.10)$$

$$= -300\,\mathrm{kJ\,mol^{-1}} - (2.478\,\mathrm{kJ\,mol^{-1}})(-2.303)$$

$$= \boxed{-294\,\mathrm{kJ\,mol^{-1}}}.$$

7.14 $\Delta_r G_2^\ominus = -RT \ln K = 0.$

$\ln K = 0.$

$K = \boxed{1}.$

7.15 Let glucose-1-phosphate = G1P, glucose-6-phosphate = G6P, and glucose-3- phosphate = G3P.

$$\ln K = \frac{-\Delta_r G^{\oplus}}{RT} \quad \text{or} \quad K = e^{-\Delta_r G^{\oplus}/RT} \text{ (The biological standard state, } \oplus\text{, has pH} = 7.)$$

$$RT(T = 310\,\text{K}) = 8.3145\,\text{J}\,\text{K}^{-1}\text{mol}^{-1} \times 310\,\text{K} = 2.577\,\text{kJ}\,\text{mol}^{-1}.$$

$$K(\text{G1P}) = \exp\left(\frac{21\,\text{kJ}\,\text{mol}^{-1}}{2.577\,\text{kJ}\,\text{mol}^{-1}}\right) = \boxed{3.5 \times 10^3}.$$

$$K(\text{G6P}) = \exp\left(\frac{14\,\text{kJ}\,\text{mol}^{-1}}{2.577\,\text{kJ}\,\text{mol}^{-1}}\right) = \boxed{2.3 \times 10^2}.$$

$$K(\text{G3P}) = \exp\left(\frac{9.2\,\text{kJ}\,\text{mol}^{-1}}{2.577\,\text{kJ}\,\text{mol}^{-1}}\right) = \boxed{36}.$$

7.16 $\text{ATP(aq)} + H_2O\text{(l)} \rightarrow \text{ADP(aq)} + P_i^-\text{(aq)} + H^+\text{(aq)}.$

The standard value of the Gibbs energy quoted applies to the state where $a_{H^+} = 10^{-7}$ and all other activities are 1 (i.e. the biological standard state, $\oplus$, has pH = 7).

Hence $Q^{\oplus} = 1 \times 10^{-7}$.

$$\Delta_r G^{\oplus} = -30.5\,\text{kJ}\,\text{mol}^{-1} = \Delta_r G^{\ominus} + RT \ln Q^{\oplus}.$$

$$\Delta_r G = \Delta_r G^{\ominus} + RT \ln Q = \Delta_r G^{\oplus} + RT \ln\left(\frac{a_{\text{ADP}} a_{P_i^-}}{a_{\text{ATP}}}\right)$$

where we have used

$$Q = Q^{\oplus}\left(\frac{a_{\text{ADP}} a_{P_i^-}}{a_{\text{ATP}}}\right).$$

We assume that all activities equal concentrations (all γ's = 1).

(a) $\Delta_r G = -30.5\,\text{kJ}\,\text{mol}^{-1} + 2.577\,\text{kJ}\,\text{mol}^{-1} \times \ln(1.0 \times 10^{-3})$

$$= \boxed{-48.3\,\text{kJ}\,\text{mol}^{-1}}.$$

(b) $\Delta_r G = -30.5\,\text{kJ}\,\text{mol}^{-1} + 2.577\,\text{kJ}\,\text{mol}^{-1} \times \ln(1.0 \times 10^{-6})$

$$= \boxed{-66.1\,\text{kJ}\,\text{mol}^{-1}}.$$

7.17 $\mu_J = \mu_J^{\ominus} + RT \ln a_J \quad [7.3].$

Assume $a_{Na^+} = [Na^+]$.

$$\Delta G = \mu_{Na^+}\text{ (outside)} - \mu_{Na^+}\text{ (inside)}$$

$$= RT\ln(140) - RT\ln(10) = RT\ln\left(\frac{140}{10}\right)$$

$$= 8.3145\,\text{J}\,\text{K}^{-1}\,\text{mol}^{-1} \times 310\,\text{K} \times \ln 14$$

$$= 6.8 \times 10^3\,\text{J}\,\text{mol}^{-1} = \boxed{6.8\,\text{kJ}\,\text{mol}^{-1}}.$$

7.18 $\Delta_r G^\ominus = \Delta_r H^\ominus - T\Delta_r S^\ominus = 0$ when $\Delta_r H^\ominus = T\Delta_r S^\ominus$.

Therefore the decomposition temperature (when $K = 1$) is

$$T = \frac{\Delta_r H^\ominus}{\Delta_r S^\ominus}.$$

(a) $CaCO_3(s) \rightarrow CaO(s) + CO_2(g)$.

$$\Delta_r H^\ominus = [-635.09 - 393.51 - (-1206.9)]\,\text{kJ}\,\text{mol}^{-1} = \boxed{+178.3\,\text{kJ}\,\text{mol}^{-1}}.$$

$$\Delta_r S^\ominus = [39.75 + 213.74 - 92.9]\,\text{J}\,\text{K}^{-1}\,\text{mol}^{-1} = \boxed{+160.6\,\text{J}\,\text{K}^{-1}\,\text{mol}^{-1}}.$$

$$T = \frac{178.3 \times 10^3\,\text{J}\,\text{mol}^{-1}}{160.6\,\text{J}\,\text{K}^{-1}\,\text{mol}^{-1}} = \boxed{1110\,\text{K}\,(837°\text{C})}.$$

(b) $CuSO_4 \cdot 5H_2O(s) \rightleftharpoons CuSO_4(s) + 5H_2O(g)$.

$$\Delta_r H^\ominus = [-771.36 + 5 \times (-241.82) - (-2279.7)]\,\text{kJ}\,\text{mol}^{-1}$$

$$= +229.2\,\text{kJ}\,\text{mol}^{-1}.$$

$$\Delta_r S^\ominus = [109 + (5 \times 188.83) - 300.4]\,\text{J}\,\text{K}^{-1}\,\text{mol}^{-1}$$

$$= 753\,\text{J}\,\text{K}^{-1}\,\text{mol}^{-1}.$$

Therefore,

$$T = \frac{299.2 \times 10^3\,\text{J}\,\text{mol}^{-1}}{753\,\text{J}\,\text{K}^{-1}\,\text{mol}^{-1}} = \boxed{397\,\text{K}\,(124°\text{C})}.$$

7.19 We use the van't Hoff equation [7.15]

$$\ln K' - \ln K = \frac{\Delta_r H^\ominus}{R}\left(\frac{1}{T} - \frac{1}{T'}\right).$$

Substituting $\ln K = -\dfrac{\Delta_r G^\ominus}{RT}$ we obtain

$$\frac{\Delta_r G^\ominus}{T'} - \frac{\Delta_r G^\ominus}{T} = \Delta_r H^\ominus\left(\frac{1}{T'} - \frac{1}{T}\right).$$

$T = 1280\,\text{K}$, T' = temperature at which $K' = 1$. When $K' = 1$, $\ln K' = 0$, and $\Delta_r G^{\ominus\prime} = 0$. This occurs when

$$-\frac{\Delta_r G^{\ominus}}{T} = \Delta_r H^{\ominus}\left(\frac{1}{T'} - \frac{1}{T}\right).$$

We solve for T'.

$$\frac{1}{T'} = \frac{1}{T} - \frac{\Delta_r G^{\ominus}}{T\Delta_r H^{\ominus}} = \frac{1}{T}\left(1 - \frac{\Delta_r G^{\ominus}}{\Delta_r H^{\ominus}}\right)$$

$$= \frac{1}{1280\,\text{K}}\left(1 - \frac{33\,\text{kJ mol}^{-1}}{224\,\text{kJ mol}^{-1}}\right) = 6.66 \times 10^{-4}\,\text{K}^{-1}.$$

$$T' = \boxed{1.5 \times 10^3\,\text{K}}.$$

7.20 $I_2(g) \rightarrow 2I(g)$, $\Delta\nu_{gas} = 1$, $K = 0.26$ at $1000\,\text{K}$.

$$K_c = K \times \left(\frac{c^{\ominus}RT}{p^{\ominus}}\right)^{-\Delta\nu_{gas}} \quad [7.13a]$$

$$= 0.26 \times \left(\frac{(1\,\text{mol dm}^{-3})(0.08314\,\text{dm}^3\,\text{bar K}^{-1}\,\text{mol}^{-1})(1000\,\text{K})}{1\,\text{bar}}\right)^{-1}$$

$$= \boxed{0.0031}.$$

7.21 G6P $\rightarrow$ F6P, $Q = [\text{F6P}]/[\text{G6P}]$ [Illustration 7.1], $\Delta_r G^{\ominus} = +1.7\,\text{kJ mol}^{-1}$ [Example 7.1].

$$f = \frac{[\text{F6P}]}{[\text{F6P}] + [\text{G6P}]}\ [\text{Example 7.1}] = \frac{1}{1 + \dfrac{[\text{G6P}]}{[\text{F6P}]}} = \frac{1}{1 + \dfrac{1}{Q}}.$$

Solving for Q gives

$$Q = \frac{f}{1 - f}$$

$$\Delta_r G = \Delta_r G^{\ominus} + RT \ln Q = \Delta_r G + RT \ln\left(\frac{f}{1-f}\right) \quad [\text{where}\quad T = 298\,\text{K}]$$

$$= 1.7\,\text{kJ mol}^{-1} + (2.479\,\text{kJ mol}^{-1}) \times \ln\left(\frac{f}{1-f}\right).$$

Figure 7.1 gives a plot of $\Delta_r G$ against f. When $\Delta_r G < 0$, the reaction proceeds spontaneously to the right until $\Delta_r G = 0$ at the equilibrium value of f, i.e. $f_{eq} = 0.33$ (Example 7.1). When $\Delta_r G > 0$, the reaction proceeds spontaneously to the left until $\Delta_r G = 0$ at the equilibrium value of f.

7.22 We look up $\Delta_f G^{\ominus}$ for each compound and note the sign. (a) −, exergonic; (b) +, endergonic; (c) +, endergonic; (d) −, exergonic.

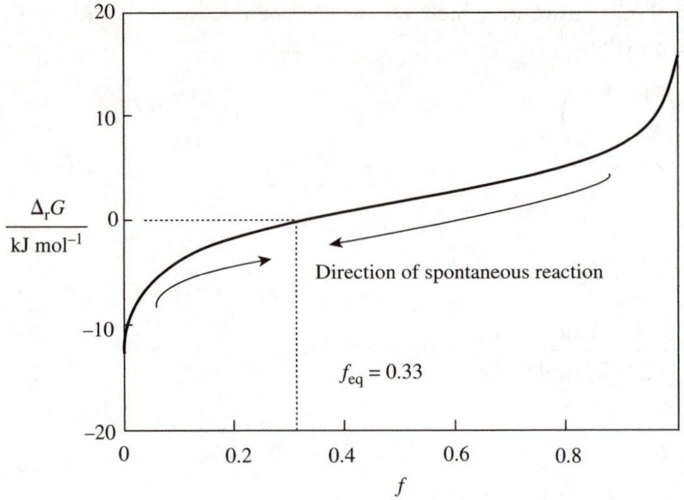

Fig 7.1

7.23 In each case, calculate $\Delta_r S^\ominus$ and $\Delta_r H^\ominus$ from data in Table D1.2 and then calculate $\Delta_r G^\ominus$ from $\Delta_r G^\ominus = \Delta_r H^\ominus - T\Delta_r S^\ominus$.

(a) $\Delta_r S^\ominus = (94.6 - 186.91 - 192.45)\,\mathrm{J\,K^{-1}\,mol^{-1}} = -284.8\,\mathrm{J\,K^{-1}\,mol^{-1}}$.

$\Delta_r H^\ominus = (-314.43 + 92.31 + 46.11)\,\mathrm{kJ\,mol^{-1}} = -176.01\,\mathrm{kJ\,mol^{-1}}$.

$\Delta_r G^\ominus = [-176.01 - 298 \times (-0.2848)]\,\mathrm{kJ\,mol^{-1}} = \boxed{-91.14\,\mathrm{kJ\,mol^{-1}}}$.

(b) $\Delta_r S^\ominus = (4 \times 28.33 + 3 \times 41.84 - 2 \times 50.92 - 3 \times 18.83)\,\mathrm{J\,K^{-1}\,mol^{-1}}$

$= +80.51\,\mathrm{J\,K^{-1}\,mol^{-1}}$.

$\Delta_r H^\ominus = [3 \times (-910.93) - 2 \times (-1675.7)]\,\mathrm{kJ\,mol^{-1}} = +618.6\,\mathrm{kJ\,mol^{-1}}$.

$\Delta_r G^\ominus = [+618.6 - 298 \times (0.08051)]\,\mathrm{kJ\,mol^{-1}} = \boxed{+594.6\,\mathrm{kJ\,mol^{-1}}}$.

(c) $\Delta_r S^\ominus = (60.29 + 130.684 - 27.28 - 205.79)\,\mathrm{J\,K^{-1}\,mol^{-1}} = -42.10\,\mathrm{J\,K^{-1}\,mol^{-1}}$.

$\Delta_r H^\ominus = [-100.0 - (-20.63)]\,\mathrm{kJ\,mol^{-1}} = -79.4\,\mathrm{kJ\,mol^{-1}}$.

$\Delta_r G^\ominus = [-79.47 - 298 \times (-0.04210)]\,\mathrm{kJ\,mol^{-1}} = \boxed{-66.8\,\mathrm{kJ\,mol^{-1}}}$.

(d) $\Delta_r S^\ominus = (27.28 + 2 \times 205.79 - 52.93 - 2 \times 130.684)\,\mathrm{J\,K^{-1}\,mol^{-1}}$.

$= +124.56\,\mathrm{J\,K^{-1}\,mol^{-1}}$.

$\Delta_r H^\ominus = [2 \times (-20.63) - (-178.2)]\,\mathrm{kJ\,mol^{-1}} = +136.9\,\mathrm{kJ\,mol^{-1}}$.

$\Delta_r G^\ominus = (+136.9 - 298 \times 0.12456)\,\mathrm{kJ\,mol^{-1}} = \boxed{+99.8\,\mathrm{kJ\,mol^{-1}}}$.

(e) $\Delta_r S^\ominus = (156.9 + 2 \times 130.684 - 2 \times 109.6 - 205.79)\,\mathrm{J\,K^{-1}\,mol^{-1}}$

$= -6.7\,\mathrm{J\,K^{-1}\,mol^{-1}}.$

$\Delta_r H^\ominus = [-813.99 - 2 \times (-187.78) - (-20.63)]\,\mathrm{kJ\,mol^{-1}}$

$= -417.80\,\mathrm{kJ\,mol^{-1}}.$

$\Delta_r G^\ominus = [-417.80 - 298 \times (-6.7 \times 10^{-3})]\,\mathrm{kJ\,mol^{-1}}$

$= \boxed{-415.80\,\mathrm{kJ\,mol^{-1}}}.$

7.24 In each case, calculate $\Delta_r G^\ominus$ from the values of $\Delta_f G^\ominus$ found in Tables D1.1 and D1.2. Then, from $\Delta_r G^\ominus = -RT \ln K$ decide which reactions have $K > 1$. If $\ln K > 0$, then $K > 1$.

$\ln K > 0$, if $\Delta_r G^\ominus < 0$. So if $\Delta_r G^\ominus < 0$, $K > 1$.

(a) $\Delta_r G^\ominus = 2\Delta_f G^\ominus(\mathrm{CH_3COOH, l}) - 2\Delta_f G^\ominus(\mathrm{CH_3CHO, g})$

$= [2 \times (-389.9) - 2 \times (-128.86)]\,\mathrm{kJ\,mol^{-1}}$

$= \boxed{-522.1\,\mathrm{kJ\,mol^{-1}},\ K > 1}.$

(b) $\Delta_r G^\ominus = 2\Delta_f G^\ominus(\mathrm{AgBr, s}) - 2\Delta_f G^\ominus(\mathrm{AgCl, s})$

$= [2 \times (-96.90) - 2 \times (-109.79)]\,\mathrm{kJ\,mol^{-1}}$

$= \boxed{+25.78\,\mathrm{kJ\,mol^{-1}},\ K < 1}.$

(c) $\Delta_r G^\ominus = \Delta_f G^\ominus(\mathrm{HgCl_2, s}) = \boxed{-178.6\,\mathrm{kJ\,mol^{-1}},\ K > 1}.$

(d) $\Delta_r G^\ominus = \Delta_f G^\ominus(\mathrm{Zn^{2+}, aq}) - \Delta_f G^\ominus(\mathrm{Cu^{2+}, aq})$

$= [-147.06 - 65.49]\,\mathrm{kJ\,mol^{-1}} = \boxed{-212.55\,\mathrm{kJ\,mol^{-1}},\ K > 1}.$

(e) $\Delta_r G^\ominus = 12\Delta_f G^\ominus(\mathrm{CO_2, g}) + 11\Delta_f G^\ominus(\mathrm{H_2O, l}) - \Delta_f G^\ominus(\mathrm{C_{12}H_{22}O_{11}, s})$

$= [12 \times (-394.36) + 11 \times (-237.13) - (-1543)]\,\mathrm{kJ\,mol^{-1}}$

$= \boxed{-5798\,\mathrm{kJ\,mol^{-1}},\ K > 1}.$

7.25 (a) $\Delta_c H^\ominus = -890\,\mathrm{kJ\,mol^{-1}}.$

$$n(\mathrm{CH_4}) = \frac{1.0 \times 10^3\,\mathrm{g}}{16.04\,\mathrm{g\,mol^{-1}}} = 62\,\mathrm{mol}.$$

$q(\text{heat}) = -890\,\mathrm{kJ\,mol^{-1}} \times 62\,\mathrm{mol} = \boxed{5.5 \times 10^4\,\mathrm{kJ}}.$

(b) $\Delta_r G^\ominus$ = maximum non-expansion work.

$$\Delta_r G^\ominus = 2 \times \Delta_f G^\ominus(H_2O, l) + \Delta_f G^\ominus(CO_2, g) - \Delta_f G^\ominus(CH_4, g)$$
$$= [2 \times (-237.13) - 394.36 + 50.72]\,\text{kJ mol}^{-1}$$
$$= -817.90\,\text{kJ mol}^{-1}.$$

For $n(CH_4) = 62$ mol,

$$\Delta_r G^\ominus = -817.90\,\text{kJ mol}^{-1} \times 62\,\text{mol} = \boxed{-5.1 \times 10^4\,\text{kJ}}.$$

7.26 (a) The value for $\Delta_c H^\ominus$ given in Table D1.1 for glucose applies to the formation of 6 $H_2O(l)$. The value for the formation of 6 $H_2O(g)$ is

$$\Delta_c H^\ominus(H_2O, g) = \Delta_c H^\ominus(H_2O, l) + 6 \times \Delta_{vap} H^\ominus(H_2O, l)$$
$$= -2808\,\text{kJ mol}^{-1} + 6 \times 44\,\text{kJ mol}^{-1}$$
$$= -2544\,\text{kJ mol}^{-1}.$$

$$n = 1.0 \times 10^3\,\text{g} \times \frac{1\,\text{mol}}{180.16\,\text{g}} = 5.55\,\text{mol}.$$

$$\text{Heat} = -2544\,\text{kJ mol}^{-1} \times 5.55\,\text{mol} = \boxed{-1.4 \times 10^4\,\text{kJ}}.$$

(b) Non-expansion work.

$$\Delta G = \Delta H - T\Delta S.$$

$$\Delta S = 6S_m^\ominus(H_2O, g) + 6S_m^\ominus(CO_2, g) - 6S_m^\ominus(O_2, g) - S_m^\ominus(C_6H_{12}O_6, s)$$
$$= (6 \times 188.83\,\text{J K}^{-1}\,\text{mol}^{-1}) + (6 \times 213.74\,\text{J K}^{-1}\,\text{mol}^{-1})$$
$$- (6 \times 205.14\,\text{J K}^{-1}\,\text{mol}^{-1}) - (212\,\text{J K}^{-1}\,\text{mol}^{-1})$$
$$= 972\,\text{J K}^{-1}\,\text{mol}^{-1}.$$

$$\Delta G = -2544\,\text{kJ mol}^{-1} - (298\,\text{K}) \times (0.972\,\text{kJ K}^{-1}\,\text{mol}^{-1})$$
$$= -2.83 \times 10^3\,\text{kJ mol}^{-1}.$$

$$\Delta G = -2.83 \times 10^3\,\text{kJ mol}^{-1} \times 5.55\,\text{mol}$$
$$= \boxed{-1.57 \times 10^4\,\text{kJ or } 1.57 \times 10^4\,\text{kJ of non-expansion work}}.$$

7.27 In order to answer the question about energy effectiveness we need to compare $\Delta_r G^\ominus$ for the combustion of sucrose and glucose. The value for 1.0 kg of glucose was determined to be -1.57×10^4 kJ in Exercise 7.26.

Here we calculate $\Delta_r G^\ominus$ for the combustion of sucrose.

(a) Water vapor.

$$C_{12}H_{22}O_{11}(s) + 12O_2(g) \rightarrow 12CO_2(g) + 11H_2O(g).$$

$$\Delta_r G^\ominus = 12 \times \Delta_f G^\ominus(CO_2, g) + 11 \times \Delta_f G^\ominus(H_2O, g) - \Delta_f G^\ominus(\text{sucrose})$$

$$= [12 \times (-394.36) + 11 \times (-228.57) - (-1543)]\,\text{kJ mol}^{-1}$$

$$= -5704\,\text{kJ mol}^{-1}.$$

$$n(\text{sucrose}) = \frac{1000\,\text{g}}{342.30\,\text{g mol}^{-1}} = 2.92\,\text{mol}.$$

$$n \times \Delta_r G^\ominus = 2.92\,\text{mol} \times (-5704\,\text{kJ mol}^{-1}) = -1.67 \times 10^4\,\text{kJ}.$$

Non-expansion work = $\boxed{1.67 \times 10^4\,\text{kJ}}$.

Expansion work: $w = -p_{\text{ex}}\Delta V, \quad p = p_{\text{ex}} = 1.00\,\text{atm}.$

$$\Delta V = \frac{RT}{p}\Delta\nu_{\text{gas}}, \qquad \Delta\nu_{\text{gas}} = 2.29\,\text{mol} \times 11 = 31.9\,\text{mol}.$$

$$w = -RT\Delta\nu_{\text{gas}} = -2.478\,\text{kJ mol}^{-1} \times 31.9\,\text{mol}$$

$$= -79.6\,\text{kJ}$$

$$= 79.6\,\text{kJ expansion work done.}$$

Total work done $= 1.67 \times 10^4\,\text{kJ} + 79.0\,\text{kJ} = \boxed{1.68 \times 10^4\,\text{kJ}}$.

(b) Liquid water.

From Exercise 7.24(e), $\Delta_r G^\ominus = -5798\,\text{kJ mol}^{-1}$.

Non-expansion work = $\boxed{1.69 \times 10^4\,\text{kJ}}$.

Expansion work = $\boxed{0}$ $(\Delta\nu_{\text{gas}} = 0)$

Total work = $\boxed{1.69 \times 10^4\,\text{kJ}}$.

7.28 To calculate the standard free energy of formation, we need to calculate the entropy of formation and the enthalpy of formation.

$$6C(s) + 3H_2(g) + \frac{1}{2}O_2(g) \rightarrow C_6H_5OH(s).$$

$$\Delta_f G^\ominus = \Delta_f H^\ominus - T\Delta_f S^\ominus.$$

$$\Delta_f S^\ominus = S_m^\ominus(C_6H_5OH, s) - 6S_m^\ominus(C, s) - 3S_m^\ominus(H_2, g) - \frac{1}{2}S_m^\ominus(O_2, g)$$

$$= \left(144.0 - 6 \times 5.740 - 3 \times 130.68 - \frac{1}{2} \times 205.14\right) \mathrm{J\,K^{-1}\,mol^{-1}}$$

$$= -385.05\,\mathrm{J\,K^{-1}\,mol^{-1}}.$$

$$C_6H_5OH(s) + 7O_2(g) \rightarrow 6CO_2(g) + 3H_2O(1).$$

$$\Delta_r H^\ominus = 6\Delta_f H^\ominus(CO_2, g) + 3\Delta_f H^\ominus(H_2O, l) - \Delta_f H^\ominus(C_6H_5OH, s).$$

Rearranging,

$$\Delta_f H^\ominus(C_6H_5OH, s) = 6\Delta_f H^\ominus(CO_2, g) + 3\Delta_f H^\ominus(H_2O, l) - \Delta_c H^\ominus$$

$$= [6 \times (-393.51) + 3(-285.83) - (-3054)]\,\mathrm{kJ\,mol^{-1}}$$

$$= -164.55\,\mathrm{kJ\,mol^{-1}}.$$

Therefore,

$$\Delta_f G^\ominus = -164.55\,\mathrm{kJ\,mol^{-1}} - 298.15\,\mathrm{K} \times (-0.38505\,\mathrm{kJ\,K^{-1}\,mol^{-1}})$$

$$= \boxed{-49.8\,\mathrm{kJ\,mol^{-1}}}.$$

7.29 $CH_4(g) + 2O_2(g) \rightarrow CO_2(g) + 2H_2O(l).$

$w'_{max} = \Delta_r G$ [4.16].

Assume that $\Delta_r G = \Delta_r G^\ominus$ and calculate $\Delta_r G$ in the usual manner from $\Delta_r G^\ominus$ data given in Table D1.2. We obtain

$$\Delta_r G^\ominus = -817.90\,\mathrm{kJ\,mol^{-1}}.$$

Therefore, the maximum non-expansion work that can be obtained is $\boxed{817.90\,\mathrm{kJ\,mol^{-1}}}$.

7.30 We follow the procedure illustrated in Example 7.3. Here, however, protons occur as reactants.

$$Q = \frac{a_{\text{lactate}}a_{\text{NAD}^+}}{a_{\text{pyruvate}}a_{\text{NADH}}a_{\text{H}^+}} = \frac{1 \times 1}{1 \times 1 \times 10^{-7}} = 1 \times 10^7$$

Therefore,

$$\Delta_r G^\oplus = \Delta_r G^\ominus + RT\ln(1 \times 10^7)$$

$$= -66.6\,\mathrm{kJ\,mol^{-1}} + (8.3145 \times 10^{-3}\,\mathrm{kJ\,K^{-1}\,mol^{-1}} \times 310\,\mathrm{K} \times 16.12)$$

$$= \boxed{-25.0\,\mathrm{kJ\,mol^{-1}}}.$$

7.31 The reaction is $AMP(aq) + H_2O(l) \rightarrow A(aq) + P_i^-(aq) + H^+(aq)$.

Therefore, as in Example 7.3, the relationship between the standard state values of $\Delta_r G$ is

$$\Delta_r G^{\ominus} = \Delta_r G^{\oplus} - RT\ln(1 \times 10^{-7})$$
$$= -14\,\text{kJ mol}^{-1} - [8.31 \times 10^{-3}\,\text{kJ K}^{-1}\,\text{mol}^{-1} \times 298\,\text{K} \times (-16.1 \times 2)]$$
$$= \boxed{+26\,\text{kJ mol}^{-1}}.$$

7.32 $$\ln K = -\frac{\Delta_r G^{\ominus}}{RT} = -\frac{\Delta_r H^{\ominus}}{RT} + \frac{\Delta_r S^{\ominus}}{R} \text{ [Derivation 7.2].}$$

For many reactions both $\Delta_r H^{\ominus}$ and $\Delta_r S^{\ominus}$ are weakly dependent upon temperature and a data plot of $\ln K$ against $1/T$ is approximately linear. In this case, the equation indicates that a linear regression fit of the plot will yield a slope equal to $-\Delta_r H^{\ominus}/R$ and an intercept equal to $\Delta_r S^{\ominus}/R$. Consequently, $\Delta_r H^{\ominus}$ and $\Delta_r S^{\ominus}$ may often be determined by (i) measuring K over a range of temperatures, (ii) checking linearity of the $\ln K$ against $1/T$ plot, and (iii) performing a linear regression fit of the plot. The thermodynamic properties are determined with the expressions $\Delta_r H^{\ominus} = -R \times \text{slope}$ and $\Delta_r S^{\ominus} = R \times \text{intercept}$.

7.33 See Problem 7.32 for the methodology for finding both $\Delta_r H^{\ominus}$ and $\Delta_r S^{\ominus}$ using the available data. Then, fill out a table that includes T and K data columns as well as $1/T$ and $\ln K$ transformations. Prepare a $\ln K$ against $1/T$ plot and visually check its linearity.

T/K	K	$1000\,\text{K}/T$	$\ln K$
300	4.00×10^{31}	3.33	72.77
500	4.00×10^{18}	2.00	42.83
1000	5.1×10^{8}	1.00	20.05

The plot (Figure 7.2) is linear with a slope equal to 22.6×10^3 K and an intercept equal to -2.46.

$$\Delta_r H^{\ominus} = -R \times \text{slope} = -(8.3145\,\text{J K}^{-1}\,\text{mol}^{-1}) \times (22.6 \times 10^3\,\text{K}) = \boxed{-188\,\text{kJ mol}^{-1}}.$$

$$\Delta_r S^{\ominus} = R \times \text{intercept} = (8.3145\,\text{J K}^{-1}\,\text{mol}^{-1}) \times (-2.46) = \boxed{-20.5\,\text{J K}^{-1}\,\text{mol}^{-1}}.$$

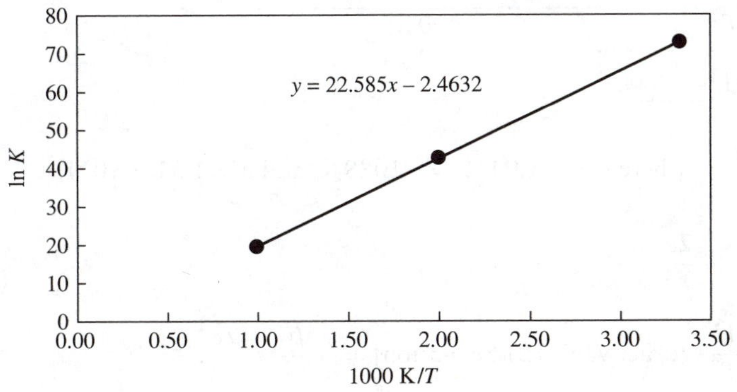

Fig 7.2

7.34 First substitute in and solve for K at 390 K and 410 K from the equation given; thus

$$\ln K = -1.04 - \frac{1088\,\mathrm{K}}{390} + \frac{1.51\times10^5\,\mathrm{K}^2}{(390)^2},$$

$$\ln K = -2.84.$$

Solving for K' where $T = 410\,\mathrm{K}$,

$$\ln K' = -2.80.$$

Use the equation

$$\ln K' - \ln K = -\frac{\Delta_r H}{R}\left(\frac{1}{T'} - \frac{1}{T}\right).$$

Substitution gives

$$0.04 = -\frac{\Delta_r H^\ominus}{8.3145\,\mathrm{J\,K^{-1}\,mol^{-1}}}\left(\frac{1}{410} - \frac{1}{390}\right).$$

Solving for $\Delta_r H^\ominus$ gives

$$\Delta_r H^\ominus = \boxed{2.66\,\mathrm{kJ\,mol^{-1}}}.$$

$$\Delta_r G^\ominus = -RT\ln K$$

$$= RT \times \left(1.04 + \frac{1088}{T} - \frac{1.51\times10^5}{T^2}\right)$$

$$= RT \times \left(1.04 + \frac{1088}{400} - \frac{1.51\times10^5}{(400)^2}\right) = +9.36\,\mathrm{kJ\,mol^{-1}}.$$

$$\Delta_r G^\ominus = \Delta_r H^\ominus - T\Delta_r S^\ominus.$$

Therefore,

$$\Delta_r S^\ominus = \frac{\Delta_r H^\ominus - \Delta_r G^\ominus}{T} = \frac{2.66\,\mathrm{kJ\,mol^{-1}} - 9.36\,\mathrm{kJ\,mol^{-1}}}{400\,\mathrm{K}}$$

$$= \boxed{-16.8\,\mathrm{J\,K^{-1}\,mol^{-1}}}.$$

7.35 $\ln K = a + \frac{b}{T} + \frac{c}{T^2}$ where $a = -1.04$, $b = -1088\,\mathrm{K}$, and $c = 1.51\times10^5\,\mathrm{K}^2$.

$$\frac{\mathrm{d}(\ln K)}{\mathrm{d}T} = -\frac{b}{T^2} - \frac{2c}{T^3}.$$

$$\frac{\mathrm{d}(\ln K)}{\mathrm{d}T} = -\frac{\Delta_r H^\ominus}{RT^2}\ \text{[exact van't Hoff equation]} = -\frac{b}{T^2} - \frac{2c}{T^3}.$$

$$\Delta_r H^{\ominus} = -RT^2 \times \left(-\frac{b}{T^2} - \frac{2c}{T^3}\right) = R\left(b + \frac{2c}{T}\right)$$

$$= -9.046\,\text{kJ mol}^{-1} + \frac{2.51 \times 10^3\,\text{kJ K mol}^{-1}}{T}.$$

The reaction enthalpy is plotted in Figure 7.3 in the temperature range 300 K to 700 K.

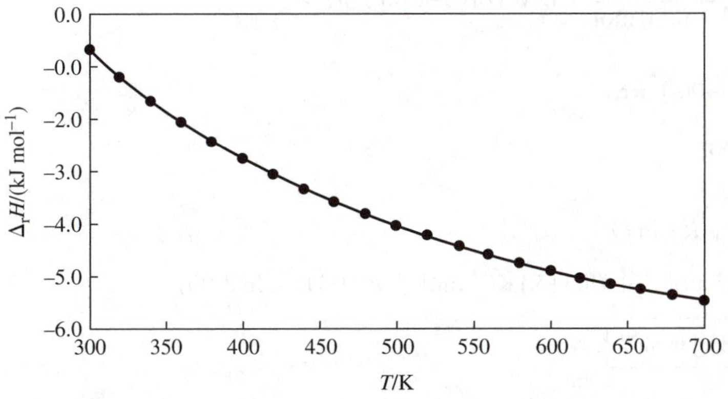

Fig 7.3

7.36 $K' = \left(\frac{c^{\ominus}RT'}{p^{\ominus}}\right)^{\Delta\nu_{\text{gas}}} K'_c$ [7.13 at temperature T'].

$$\ln K' = \ln\left\{\left(\frac{c^{\ominus}RT'}{p}\right)^{\Delta\nu_{\text{gas}}} K'_c\right\} = \ln\left(\frac{c^{\ominus}RT'}{p^{\ominus}}\right)^{\Delta\nu_{\text{gas}}} + \ln K'_c.$$

Likewise, at temperature T, $\ln K = \ln\left\{\left(\frac{c^{\ominus}RT}{p^{\ominus}}\right)^{\Delta\nu_{\text{gas}}} K_c\right\} = \ln\left(\frac{c^{\ominus}RT}{p^{\ominus}}\right)^{\Delta\nu_{\text{gas}}} + \ln K_c.$

Subtracting the expression at T from the expression at T' and setting the result equal to [7.15],

$$\ln K' - \ln K = \ln\left(\frac{c^{\ominus}RT'}{p^{\ominus}}\right)^{\Delta\nu_{\text{gas}}} - \ln\left(\frac{c^{\ominus}RT}{p^{\ominus}}\right)^{\Delta\nu_{\text{gas}}} + \ln K'_c - \ln K_c$$

$$= \frac{\Delta_r H^{\ominus}}{R}\left(\frac{1}{T} - \frac{1}{T'}\right).$$

$$\ln K'_c - \ln K_c - \ln\left(\frac{T}{T'}\right)^{\Delta\nu_{\text{gas}}} = \frac{\Delta_r H^{\ominus}}{R}\left(\frac{1}{T} - \frac{1}{T'}\right).$$

$$\boxed{\ln K'_c - \ln K_c = \ln\left(\frac{T}{T'}\right)^{\Delta\nu_{\text{gas}}} + \frac{\Delta_r H^{\ominus}}{R}\left(\frac{1}{T} - \frac{1}{T'}\right)}.$$

7.37 Let B = borneol, I = isoborneol.

For B → I,

$$Q = \frac{p_I}{p_B},$$

$$p_B = x_B p$$

$$= \frac{0.15\,\text{mol}}{0.15\,\text{mol} + 0.30\,\text{mol}} \times 600\,\text{Torr} = 200\,\text{Torr},$$

$$p_I = p - p_B = 400\,\text{Torr},$$

$$Q = \frac{p_I}{p_B} = 2.00,$$

$$\Delta_r G = \Delta_r G^\ominus + RT \ln Q$$

$$= +9.4\,\text{kJ mol}^{-1} + (8.315\,\text{J K}^{-1}\,\text{mol}^{-1} \times 503\,\text{K} \times \ln 2.00)$$

$$= \boxed{+12.3\,\text{kJ mol}^{-1}}.$$

7.38 $$K = \frac{p_I}{p_B} = \frac{x_I}{x_B}\;[p_I = x_I p, p_B = x_B p] = \frac{1 - x_I}{x_B}.$$

Rearranging, $x_B = \dfrac{1}{1+K} = \dfrac{1}{1+0.106} = \boxed{0.904}$.

$$x_I = \boxed{0.096}.$$

Note that the other information given in the exercise is superfluous.

The initial amounts of the isomers are

$$n_B = \frac{7.50\,\text{g}}{M}, \quad n_I = \frac{140\,\text{g}}{M}, \quad n = \frac{21.50\,\text{g}}{M}.$$

The total amount remains the same, but at equilibrium

$$\frac{n_B}{n} = x_B = 0.904 \quad \text{and} \quad x_I = 0.096.$$

7.39

$$\begin{array}{llll} N_2(g) \;+ & 3H_2(g) & \rightleftharpoons 2NH_3(g) & \\ (1.00 - x) & (4.00 - 3x) & \quad 2x & \text{partial pressure at equilibrium} \end{array}$$

$$K = \frac{p^2_{NH_3}}{p_{N_2}p^3_{NH_2}} = \frac{(2x)^2}{(1.00 - x)(4.00 - 3x)^3} = 89.8.$$

The expression above needs to be expanded and the resulting equation solved for x.

$$2424.6x^4 - 12123x^3 + 22625.6x^2 - 18678.4x + 5747.2 = 0.$$

This equation is most readily solved with mathematical software on a computer. With the use of Mathcad, the root found is 0.968.

$$p_{NH_3} = 2 \times 0.968 = 1.936\,\text{bar}.$$

$$p_{N_2} = 1.00 - 0.968 = 0.032\,\text{bar}.$$

$$p_{H_2} = 4.00 - (3 \times 0.968) = 1.096\,\text{bar}.$$

Total $p = 3.064\,\text{bar}$.

$$x_{NH_3} = \frac{1.936}{3.064} = \boxed{0.632}.$$

$$x_{N_2} = \frac{0.032}{3.064} = \boxed{0.010}.$$

$$x_{H_2} = \frac{1.096}{3.064} = \boxed{0.358}.$$

7.40 $I_2(g) \rightleftharpoons 2I(g), \quad K = \frac{[I]^2}{[I_2]}.$

Initially, $[I_2] = \left(\frac{1.00\,g\,I_2}{1.0\,dm^3}\right) \times \left(\frac{1\,mol}{254\,g}\right) = 3.94 \times 10^{-3}\,mol\,dm^{-3}$ and, at equilibrium,

$$[I_2] = \left(\frac{0.83\,g\,I_2}{1.0\,dm^3}\right) \times \left(\frac{1\,mol}{254\,g}\right) = 3.27 \times 10^{-3}\,mol\,dm^{-3}$$

$$\text{and } [I] = 2 \times (\text{mol } I_2 \text{ reacted}) = 2 \times (3.94 \times 10^{-3} - 3.27 \times 10^{-3})\,mol\,dm^{-3}$$

$$= 1.34 \times 10^{-3}\,mol\,dm^{-3}.$$

Therefore,

$$K = \frac{(1.34 \times 10^{-3})^2}{(3.3 \times 10^{-3})} = \boxed{5.4 \times 10^{-4}}.$$

7.41 $SbCl_5(g) \rightleftharpoons SbCl_3(g) + Cl_2(g).$

$$K = \frac{p_{SbCl_3}p_{Cl_2}}{p_{SbCl_5}} = 3.5 \times 10^{-4}$$

$$= \frac{(0.20) \times p_{Cl_2}}{(0.15)} = 3.5 \times 10^{-4}.$$

$$p_{Cl_2} = \boxed{2.6 \times 10^{-4}\,\text{bar}}.$$

7.42 (a) $K = 0.36$ is assumed to be the equilibrium constant in terms of pressures. Below we refer to this as K_p and we need to convert from K_p to K in terms of concentrations. The conversion is

$$K = \frac{K_p}{(RT)^{\Delta\nu_{gas}}}$$

where $\Delta\nu_{gas}$ is the change in the number of moles of gas between reactants and products. In this case $\Delta\nu_{gas} = +1$.

$$[PCl_5] = \frac{2.0\,g}{0.250\,dm^3} \times \frac{1\,mol}{208.22\,g} = 3.8 \times 10^{-2}\,mol\,dm^{-3}.$$

$$PC1_5(g) \rightleftharpoons Cl_3(g) + Cl_2(g)$$

Initial conc./(mol dm^{-3})	3.8×10^{-2}	0	0
Change	$-x$	$+x$	$+x$
Equil.	$(3.8 \times 10^{-2} - x)$	x	x

$$K = \frac{K_P}{(RT)^{\Delta\nu_{gas}}} = \frac{0.36}{(0.0821) \times (400)^1} = 1.1 \times 10^{-2}$$

$$= \frac{x^2}{3.8 \times 10^{-2} - x} = 1.1 \times 10^{-2}.$$

Solving the quadratic,

$$x^2 + (1.1 \times 10^{-2})x - 4.2 \times 10^{-4} = 0,$$

$$x = [PCl_3] = [Cl_2] = \boxed{1.6 \times 10^{-2}\,mol\,dm^{-3}}.$$

$$[PCl_5] = 3.9 \times 10^{-2} - x = \boxed{2.3 \times 10^{-2}\,mol\,dm^{-3}}.$$

(b) $$\%\text{ decomposition} = \frac{\text{amount decomposed}}{\text{initial amount}} \times 100\%$$

$$= \frac{1.6 \times 10^{-2}\,mol\,dm^{-3}}{3.8 \times 10^{-2}\,mol\,dm^{-3}} \times 100\% = \boxed{42\%}.$$

7.43 $$N_2(g) + 3H_2(g) \rightleftharpoons 2NH_3(g)$$

Initial pressure / atm	0.020	0.020	0
Change	$-x$	$-3x$	$+2x$
Equil.	$0.020 - x$	$0.020 - 3x$	$2x$

$$K = \frac{p^2_{NH_3}}{p_{N_2}p^2_{NH_3}} = 0.036.$$

Because at this temperature $K \ll 1$, we may, at least initially, make the approximation that x and $3x$ are small compared to 0.020. Then, in a second approximation we may get a better value.

$$0.036 = \frac{4x^2}{(0.02)(0.02)^3}.$$

Solving for x yields

$$x = 3.8 \times 10^{-5}.$$

Then $0.036 = \dfrac{4x^2}{(0.019962) \times (0.019886)^3}$.

Solving again for x yields

$x = 3.8 \times 10^{-5}$.

$2x = p_{NH_3} = \boxed{7.6 \times 10^{-5}\ \text{bar}}$.

$p_{N_2} = 0.020\ \text{bar} - 3.8 \times 10^{-5}\ \text{bar} \cong \boxed{0.020\ \text{bar}}$.

$p_{N_2} = 0.020\ \text{bar} - 3 \times 3.8 \times 10^{-5}\ \text{bar} \cong \boxed{0.020\ \text{bar}}$.

7.44 $K = \dfrac{p_{NO_2}^2}{p_{N_2O_4}p^\ominus}\ [p^\ominus = 1\ \text{bar}]$.

	$N_2O_4(g)$ $\rightleftharpoons$	$2NO_2(g)$
At equilibrium	$1 - \alpha$	2α

$p_{NO_2} = x_{NO_2}\, p = 2\alpha p$ and, similarly, $p_{N_2O_2} = (1 - \alpha)p$.

$K_p = \dfrac{(2\alpha p)^2}{(1 - \alpha)p} = \dfrac{4\alpha^2 p}{(1 - \alpha)}$

When $\alpha \ll 1$ then $K \approx 4\alpha^2 p/p^\ominus$ and $\alpha \propto \dfrac{1}{\sqrt{(p/p^\ominus)}}$.

7.45 $U(s) + \dfrac{3}{2} H_2(g) \rightleftharpoons UH_3(s)$.

$K = \dfrac{1}{p_{H_2}^{3/2}}, \quad p_{H_2} = p.$

Pressures here are unitless pressures relative to the standard pressure or 1 bar.

$\ln K = \ln p^{-3/2} = -\dfrac{3}{2} \ln p.$

$\Delta_r G^\ominus = -RT \ln K = \dfrac{3}{2} \ln p.$

$= \dfrac{3}{2} \times 8.3145\ \text{J K}^{-1}\ \text{mol}^{-1} \times 500\ \text{K} \times \ln \dfrac{1.04\ \text{Torr}}{750\ \text{Torr bar}^{-1}}$

$= \boxed{-41.0\ \text{kJ mol}^{-1}}$.

7.46 Determine whether $\Delta_r H^\ominus > 0$ at 298 K using $\Delta_f H^\ominus$ values.

(a) Two high-energy radicals form in this reaction. It is very likely to be endothermic.

(b) $\Delta_r H^\ominus = 2 \times (-395.72) - 2 \times (-296383) = \boxed{-197.78\ \text{kJ mol}^{-1}}$

(c) $\Delta_r H^\ominus = 2 \times (-36.4) - (30.91) = \boxed{-103.71\ \text{kJ mol}^{-1}}$

(d) $\Delta_r H^\ominus = -2 \times (142.7) = \boxed{+285.4\ \text{kJ mol}^{-1}}$

Because (b) and (c) are exothermic, an increase in temperature favors the reactants; $\boxed{\text{(a) and (d)}}$ are endothermic, and an increase in temperature favors the products.

7.47 $\ln \dfrac{K'}{K} = \dfrac{\Delta_r H^\ominus}{R}\left(\dfrac{1}{T} - \dfrac{1}{T'}\right).$

Solving for $\Delta_r H^\ominus$,

$$\Delta_r H^\ominus = \frac{R\ln(K'/K)}{\left(\dfrac{1}{T} - \dfrac{1}{T'}\right)}.$$

$T' = 308\,\text{K}$; hence with $K'/K = k$,

$$\Delta_r H^\ominus = \frac{8.3145\,\text{J}\,\text{K}^{-1}\,\text{mol}^{-1} \times \ln k}{\left(\dfrac{1}{298\,\text{K}} - \dfrac{1}{308\,\text{K}}\right)} = 76\,\text{kJ}\,\text{mol}^{-1} \times \ln k.$$

Therefore,

(a) $k = 2$, $\Delta_r H^\ominus = 76\,\text{kJ}\,\text{mol}^{-1} \times \ln 2 = \boxed{+53\,\text{kJ}\,\text{mol}^{-1}}.$

(b) $k = 1/2$, $\Delta_r H^\ominus = 76\,\text{kJ}\,\text{mol}^{-1} \times \ln(1/2) = \boxed{-53\,\text{kJ}\,\text{mol}^{-1}}.$

7.48 $NH_4Cl(s) \rightleftharpoons NH_3(g) + HCl(g).$

$p = p_{NH_3} + p_{HCl} = 2p_{NH_3} p_{NH_3} = p_{HCl}$ [all ps are ps relative to $p^\ominus = 1\,\text{bar}$].

(a) $K = p_{NH_3} p_{HCl} = p_{NH_3}^2 = \dfrac{1}{4}p^2$

At 427°C(700 K), $K = \dfrac{1}{4}\left(\dfrac{608\,\text{kPa}}{100\,\text{kPa}\,\text{bar}^{-1}}\right)^2 = \boxed{9.24}.$

At 459°C(732 K), $K = \dfrac{1}{4}\left(\dfrac{1115\,\text{kPa}}{100\,\text{kPa}\,\text{bar}^{-1}}\right)^2 = \boxed{31.08}.$

(b) $\Delta_r G^\ominus = -RT\ln K$

$= -8.3145\,\text{J}\,\text{K}^{-1}\,\text{mol}^{-1} \times 700\,\text{K} \times \ln 9.24$

$= -12.9\,\text{kJ}\,\text{mol}^{-1}$ (at 427°C).

(c) $\Delta_r H^\ominus \approx \dfrac{R\ln(K'/K)}{\left(\dfrac{1}{T} - \dfrac{1}{T'}\right)}$

$$\approx \frac{8.3145\,\text{J}\,\text{K}^{-1}\,\text{mol}^{-1} \times \ln \dfrac{31.08}{9.24}}{\left(\dfrac{1}{700\,\text{K}} - \dfrac{1}{732\,\text{K}}\right)} = \boxed{+161\,\text{kJ}\,\text{mol}^{-1}}.$$

(d) $\Delta_r S^\ominus = \dfrac{\Delta_r H^\ominus - \Delta_r G^\ominus}{T} = \dfrac{161\,\text{kJ}\,\text{mol}^{-1} - (-12.9\,\text{kJ}\,\text{mol}^{-1})}{700\,\text{K}}$

$= \boxed{+248\,\text{J}\,\text{K}^{-1}\,\text{mol}^{-1}}.$

Chapter 8

Consequences of equilibrium

Answers to discussion questions

8.1 (a) Figure 8.4 of the text illustrates the typical pH curve for the titrations of a weak acid with a strong base. Prior to reaching the stoichiometric point, when titrant volumes are small, the slope of the curve is positive but of small magnitude and an important inflection point is observed, which characterizes the pK_a of the weak acid. pH changes are small in this region because HA can both neutralize titrant and dissociate to maintain pH. Near the stoichiometric point the curve slope is very large because each addition of even a small drop of base causes a large pH change. The remaining amount of weak acid is not sufficient to maintain pH constancy through dissociation. At the stoichiometric point, $[A^-] = [HA]_{initial} V_{initial}/V$. After the stoichiometric point the curve levels off as the mixture approaches the pH of the titrant. If the acid is extremely weak, there is no large increase in pH near the stoichiometric point.

(b) Figure 8.5 of the text illustrates the typical pH curve for the titrations of a weak base with a strong acid. Prior to reaching the stoichiometric point, when titrant volumes are small, the slope of the curve is negative but of small magnitude and an important inflection point is observed, which characterizes the pK_b of the weak base. pH changes are small in this region because the base can both neutralize titrant and dissociate to maintain pH. Near the stoichiometric point the curve slope is very steep because each addition of even a small drop of acid causes a large pH change. The remaining amount of weak base is not sufficient to maintain pH constancy through dissociation. After the stoichiometric point the curve levels off as the mixture approaches the pH of the titrant. If the base is extremely weak, there is no large drop in pH near the stoichiometric point.

8.2 An acid buffer is a solution of approximately equal concentrations of a weak acid and its salt. This solution maintains some constancy in the acidic pH range through its ability to neutralize both a small amount of strong base and a small amount of strong acid. A base buffer, a solution of equal concentrations of a weak base and its salt, does much the same thing but in the base pH range.

Indicators are weak acids that in their undissociated acid form have one color, and in their dissociated anion form another. In acidic solution, the indicator exists in the predominantly acid form (one color), in basic solution in the predominantly anion form (the other color). The ratio of the two forms is very pH sensitive because of the small value of pK_a of the indicator, so the color change can occur very rapidly with change in pH. The indicator dye for an acid/base titration is chosen with care to match the pH at which color change occurs with the stoichiometric point.

8.3 (a) The relationship $pH = \frac{1}{2}(pK_{a1} + pK_{a2})$ [8.10] is limited to the condition that the solution of salt MHA, where HA^- is an amphiprotic anion, has an initial ('preparation') concentration that is dilute to the extent that it is valid to replace activities with concentrations in equilibrium expressions. Additionally (see Derivation 8.1), the preparation concentration of HA^- must be (i) much larger than either the equilibrium concentration $[H_2A]$ or the equilibrium concentration $[A^-]$, and (ii) much larger than the product $[H_2A]K_{a1}$. These conditions cause $[H_2A] \approx [A^-]$.

(b) The Henderson–Hasselbalch equation, $pH = pK_a - \log\left(\frac{[\text{acid}]}{[\text{base}]}\right)$ [8.11], is limited to the condition that the weak acid solution has an initial ('preparation') concentration that is dilute to the extent that it is valid to replace activities with concentrations in equilibrium expressions. When preparing a solution by mixing a preparative concentration of HA, c_{HA}, with a preparative concentration of conjugate base salt MA, c_{MA}, the approximations $[\text{acid}] \approx c_{HA}$ and $[\text{base}] \approx c_{MA}$ are limited to the conditions that both c_{HA} and c_{MA} are much larger than $[H_3O^+]$. To see this, draw up a concentration table and use it to closely examine the equilibrium constant K_a.

	HA	A^-	H_3O^+
Initial molar concentration/(mol dm^{-3})	c_{HA}	c_{MA}	0
Change to reach equilibrium/(mol dm^{-3})	$-x$	$+x$	$+x$
Equilibrium concentration/(mol dm^{-3})	$c_{HA} - x$	$c_{MA} + x$	$+x$

$$K_a = \frac{a_{H_3O^+}a_{A^-}}{a_{HA}} \simeq \frac{[H_3O^+][A^-]}{[HA]} \simeq \frac{[H_3O^+](c_{MA} + x)}{(c_{HA} - x)}.$$

$$K_a \approx [H_3O^+]\left(\frac{c_{MA}}{c_{HA}}\right) \quad \text{and} \quad pH \approx pK_a - \log\left(\frac{c_{HA}}{c_{MA}}\right) \quad \text{provided that } x \ll c_{HA}, c_{MA}.$$

8.4 The common-ion effect is the phenomenon in which the solubility of a sparingly soluble salt is reduced by a second salt when the two share a common ion. For example, the solubility of lead(II) sulfate is reduced by sodium sulfate.

Solutions to exercises

8.5 (a)

conjugate (acid$_1$ ↔ base$_1$)

$$H_2SO_4 + H_2O \rightleftharpoons H_3O^+ + HSO_4^-.$$

acid$_1$ base$_2$ acid$_2$ base$_1$

conjugate (base$_2$ ↔ acid$_2$)

(b)

conjugate (acid$_1$ ↔ base$_1$)

$$HF + H_2O \rightleftharpoons H_3O^+ + F^-.$$

acid$_1$ base$_2$ acid$_2$ base$_1$

conjugate (base$_2$ ↔ acid$_2$)

(c) conjugate

$C_6H_5NH_3^+ + H_2O \rightleftharpoons H_3O^+ + C_6H_5NH_2$.
acid$_1$ base$_2$ acid$_2$ base$_1$

conjugate

(d) conjugate

$H_2PO_4^- + H_2O \rightleftharpoons H_3O^+ + HPO_4^{2-}$.
acid$_1$ base$_2$ acid$_2$ base$_1$

conjugate

(e) conjugate

$HCOOH + H_2O \rightleftharpoons H_3O^+ + HCO_2^-$.
acid$_1$ base$_2$ acid$_2$ base$_1$

conjugate

(f) conjugate

$NH_2NH_3^+ + H_2O \rightleftharpoons H_3O^+ + NH_2NH_2$.
acid$_1$ base$_2$ acid$_2$ base$_1$

conjugate

8.6 (a) $CH_3CH(OH)COOH + H_2O \rightleftharpoons CH_3CH(OH)COO^- + H_3O^+$.

(b) $HOOC(CH_2)_2CH(NH_2)COOH + H_2O \rightleftharpoons HOOC(CH_2)_2CH(NH_2)COO^- + H_3O^+$,

$HOOC(CH_2)_2CH(NH_2)COO^- + H_2O \rightleftharpoons {}^-OOC(CH_2)_2CH(NH_2)COO^- + H_3O^+$.

(c) $NH_2CH_2COOH + H_2O \rightleftharpoons NH_2CH_2COO^- + H_3O^+$.

(d) $HOOCCOOH + H_2O \rightleftharpoons HOOCCOO^- + H_3O^+$.

$HOOCCOO^- + H_2O \rightleftharpoons {}^-OOCCOO^- + H_3O^+$.

8.7 (a) $K_w = 2.5 \times 10^{-14} = [H_3O^+][OH^-] = x^2$.

$$[H_3O^+] = \sqrt{2.5 \times 10^{-14}} = 1.6 \times 10^{-7}\ \text{mol dm}^{-3}.$$

$$pH = -\log[H_3O^+] = \boxed{6.80}.$$

(b) $[OH^-] = [H_3O^+] = 1.6 \times 10^{-7}\ \text{mol dm}^{-3}$.

$$pOH = -\log[OH^-] = \boxed{6.80}.$$

8.8 $K_{w,D} = [D_3O^+][OD^-] = 1.35 \times 10^{-15}$.

(a) $D_2O + D_2O \rightleftharpoons D_3O^+ + OD^-$.

(b) $K_{w,D} = [D_3O^+][OD^-] = 1.35 \times 10^{-15}$ $\quad pK_{w,D} = -\log K_{w,D} = \boxed{14.870}$.

(c) $[D_3O^+] = [OD^-] = \sqrt{1.35 \times 10^{-15}} = \boxed{3.67 \times 10^{-8}\ \text{mol dm}^{-3}}$.

(d) $pD = -\log(3.67 \times 10^{-8}) = \boxed{7.45} = pOD$.

(e) $pD + pOD = pK_w(D_2O) = \boxed{14.870}$.

8.9 $\ln K' - \ln K = \dfrac{\Delta_r H^\ominus}{R}\left(\dfrac{1}{T} - \dfrac{1}{T'}\right)$ van't Hoff equation [7.15].

Use the relationship $\ln(x) = \ln(10) \times \log(x) = 2.303 \log(x)$ in order to introduce log terms and subsequently pK terms into the van't Hoff equation.

$$\log K' - \log K = \frac{\Delta_r H^\ominus}{\ln 10 \times R}\left(\frac{1}{T} - \frac{1}{T'}\right),$$

$$-\log K - (-\log K') = \frac{\Delta_r H^\ominus}{\ln 10 \times R}\left(\frac{1}{T} - \frac{1}{T'}\right),$$

$$pK - pK' = \frac{\Delta_r H^\ominus}{\ln 10 \times R}\left(\frac{1}{T} - \frac{1}{T'}\right).$$

K' can be considered as a reference equilibrium constant at the specific temperature T' while K is the equilibrium constant at the variable temperature T. This places the working equation in the linear form

$$pK = \text{slope} \times \frac{1}{T} + \text{intercept}$$

where the slope equals $\boxed{\dfrac{\Delta_r H^\ominus}{\ln 10 \times R}}$ and the intercept equals $pK' - \dfrac{\Delta_r H^\ominus}{\ln 10 \times RT'}$.

8.10 $\text{pH} = -\log a_{H_3O^+}[8.1] \simeq -\log([H_3O^+])$ and $\text{pOH} = pK_w - \text{pH}[8.7] = 14 - \text{pH}$.

	$[H_3O^+]/(\text{mol dm}^{-3})$	pH	pOH
(a)	1.50×10^{-5}	4.82	9.18
(b)	1.50×10^{-3}	2.82	11.18
(c)	5.10×10^{-14}	13.29	0.71
(d)	5.01×10^{-5}	4.30	9.70

8.11 (a) Amount (moles) $H_3O^+ = (0.0250\ \text{dm}^3) \times (0.144\ \text{mol dm}^{-3}) = 3.60 \times 10^{-3}$ mol.
Amount (moles) $OH^- = (0.0250\ \text{dm}^3) \times (0.125\ \text{mol dm}^{-3}) = 3.12 \times 10^{-3}$ mol.
Excess $H_3O^+ = (3.60 \times 10^{-3} - 3.12 \times 10^{-3}) = 0.48 \times 10^{-3}\ \text{mol}\ H_3O^+$.

$$[H_3O^+] = \frac{4.8 \times 10^{-4}\ \text{mol}}{0.0500\ \text{dm}^3} = \boxed{9.60 \times 10^{-3}\ \text{mol dm}^{-3}}.$$

$$\text{pH} = -\log(9.60 \times 10^{-3}) = \boxed{2.02}.$$

(b) Amount of $H_3O^+ = (0.0250\ \text{dm}^3) \times (0.15\ \text{mol dm}^{-3}) = 3.75 \times 10^{-3}\ \text{mol}\ H_3O^+$.
Amount of $OH^- = (0.0350\ \text{dm}^3) \times (0.15\ \text{mol dm}^{-3}) = 5.25 \times 10^{-3}\ \text{mol}\ OH^-$.
Excess $OH^- = (5.25 \times 10^{-3} - 3.75 \times 10^{-3}) = 1.50 \times 10^{-3}\ \text{mol}\ OH^-$.

$$[OH^-] = \frac{1.50 \times 10^{-3}\ \text{mol}}{0.060\ \text{dm}^3} = \boxed{0.025\ \text{mol dm}^{-3}}.$$

$$\text{pOH} = -\log(0.025) = 1.60.$$

$$\text{pH} = 14.00 - 1.30 = \boxed{12.40}.$$

(c) Amount of $H_3O^+ = (0.0212\,dm^3) \times (0.22\,mol\,dm^{-3}) = 4.7 \times 10^{-3}\,mol\,H_3O^+$.
Amount of $OH^- = (0.0100\,dm^3) \times (0.30\,mol\,dm^{-3}) = 3.0 \times 10^{-3}\,mol\,OH^-$.

$$\text{Concentration of excess } H_3O^+ = \frac{1.7 \times 10^{-3}\,mol}{0.031\,dm^3} = 5.5 \times 10^{-2}\,M.$$

$$pH = \boxed{1.26}.$$

8.12 The general rule is:

(1) salt of strong acid and strong base is neutral;
(2) salt of strong acid and weak base is acidic;
(3) salt of weak acid and strong base is basic;
(4) salt of weak acid and weak base is often close to neutral.

(a) Acidic; $NH_4^+(aq) + H_2O(l) \rightleftharpoons H_3O^+(aq) + NH_3(aq)$.
(b) Basic; $H_2O(l) + CO_3^{2-}(aq) \rightleftharpoons HCO_3^-(aq) + OH^-(aq)$.
(c) Basic; $H_2O(l) + F^-(aq) \rightleftharpoons HF(aq) + OH^-(aq)$.
(d) Neutral.
(e) Acidic; $[Al(H_2O)_6]^{3+}(aq) + H_2O(l) \rightleftharpoons [Al(H_2O)_5OH]^{2+}(aq) + H_3O^+(aq)$.
(f) Acidic; $[Co(H_2O)_6]^{2+}(aq) + H_2O(l) \rightleftharpoons [Co(H_2O)_5OH]^+(aq) + H_3O^+(aq)$.

8.13 (a) Calculate the molar concentration of $KC_2H_3O_2$.

$$(8.4\,g) \times \left(\frac{1\,mol}{98.15\,g}\right) \times \left(\frac{1}{0.250\,dm^3}\right) = 0.342\,mol\,dm^{-3}.$$

Conc./(mol dm^{-3})	$H_2O(l)$	+	$C_2H_3O_2^-(aq)$	$\rightleftharpoons$	$HC_2H_3O_2(aq)$	+	$OH^-(aq)$
Initial	—		0.342		0		0
Change	—		$-x$		$+x$		$+x$
Equilibrium	—		$0.342 - x \cong 0.342$		x		x

$$K_b = \frac{K_w}{K_a} = \frac{1.0 \times 10^{-14}}{1.8 \times 10^{-5}} \approx \frac{x^2}{0.342}.$$

$$[OH^-] = 1.4 \times 10^{-5}\,mol\,dm^{-3}.$$

$$pOH = 4.85 \quad pH = 14.00 - 4.85 = \boxed{9.15}.$$

(b) $$(3.75\,g) \times \left(\frac{1\,mol}{97.9\,g}\right) \times \left(\frac{1}{0.100\,dm^3}\right) = 0.383\,mol\,dm^{-3}\,NH_4Br.$$

Conc./(mol dm^{-3})	$NH_4^+(aq)$	+	$H_2O(l)$	$\rightleftharpoons$	$NH_3(aq)$	+	$H_3O^+(aq)$
Initial	0.383		—		0		0
Change	$-x$		—		$+x$		$+x$
Equilibrium	$0.383 - x \cong 0.383$		—		x		x

$$K_a = \frac{1.0 \times 10^{-14}}{1.8 \times 10^{-5}} \approx \frac{x^2}{0.383}.$$

$[H_3O^+] = 1.46 \times 10^{-5}\ \text{mol dm}^{-3}$.

$\text{pH} = -\log(1.46 \times 10^{-5}) = \boxed{4.84}$.

(c) HBr is a strong acid and is therefore essentially completely ionized in aqueous solution. Therefore none of the Br^- is protonated.

8.14 (a) $K_a = \dfrac{[H_3O^+][L^-]}{[HL]}$, HL = lactic acid.

If $[L^-] = [HL]$, then $K_a = [H_3O^+]$.

$-\log K_a = -\log[H_3O^+]$.

$pK_a = \text{pH} = 3.08$.

$K_a = \boxed{8.3 \times 10^{-4}}$.

(b)

Conc./(mol dm^{-3})	HL(aq)	+	H_2O(l)	⇌	H_3O^+(aq)	+	L^{-1}(aq)
Initial	x		—		—		$2x$
Change	$-y$		—		$+y$		$+y$

$$K_a = \frac{[H_3O^+][L^-]}{[HL]} = \frac{[y][y+x]}{[2x-y]} \approx \frac{[y][x]}{[2x]} = 8.3 \times 10^{-4}.$$

$y = 2(8.3 \times 10^{-4}) = 1.66 \times 10^{-3}\ \text{mol dm}^{-3} = [H_3O^+]$.

$\text{pH} = \boxed{2.78}$.

8.15 Figure 8.1 is the titration curve of a strong base ($Ba(OH)_2$) with a strong acid (HCl). This curve looks roughly like Figure 8.3 of the text turned upside down.

Initial $\text{pH} = 14.00 - (-\log 0.30) = \boxed{13.48}$.

At the stoichiometric point for a strong acid and strong base the pH = 7.0.

Volume of HCl at the stoichiometric point:

$$\text{Amount (moles) } OH^- = 0.025\ \text{dm}^3 \times 2 \times 0.15\ \text{mol dm}^{-3} = 7.5 \times 10^{-3}\ \text{mol}$$

$$= \text{amount (moles) } H^+ = \text{amount (moles) HCl}.$$

$$\text{Volume (HCl)} = \frac{7.5 \times 10^{-3}\ \text{mol}}{0.22\ \text{mol dm}^{-3}} = 0.034\ \text{dm}^3 = \boxed{34\ \text{cm}^3}.$$

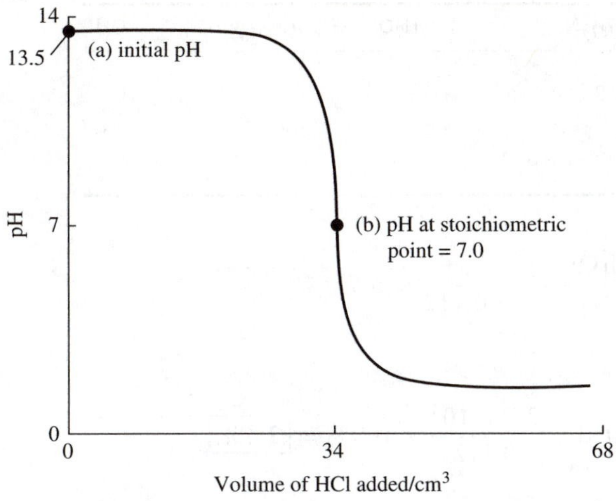

Fig 8.1

8.16 (a)

Conc./(mol dm^{-3})	C_6H_5COOH	+ H_2	⇌ H_3O^+	+ $C_6H_5COO^-$
Initial	0.250	—	0	0
Change	$-x$	—	$+x$	$+x$
Equilibrium	$0.250 - x \cong 0.250$	—	x	x

$$K_a \approx \frac{x^2}{0.250} = 6.5 \times 10^{-5}.$$

$$x = 4.0 \times 10^{-3}.$$

$$\text{Percentage deprotonated} = \frac{4.0 \times 10^{-3}}{0.250} \times 100\% = \boxed{1.6\%}.$$

(b)

Conc./(mol dm^{-3})	H_2O	+ NH_2NH_2	⇌ $NH_2NH_3^+$	+ OH^-
Initial	—	0.150	0	0
Change	—	$-x$	$+x$	$+x$
Equilibrium	—	$0.150 - x \cong 0.150$	x	x

$$K_b = \frac{[NH_2NH_3^+][OH^-]}{[NH_2NH_2]} \approx \frac{x^2}{0.150} = 1.7 \times 10^{-6}.$$

$$x = 5.0 \times 10^{-4}.$$

$$\text{Percentage protonated} = \frac{5.0 \times 10^{-4}}{0.150} \times 100\% = \boxed{0.33\%}.$$

(c)

Conc./(mol dm^{-3})	$(CH_3)_3N$	+ H_2O	$\rightleftharpoons$ $(CH_3)_3NH^+$	+ OH^-
Initial/(mol dm^{-3})	0.112	—	0	0
Change	$-x$	—	$+x$	$+x$
Equilibrium	$0.112 - x \cong 0.112$	—	x	x

$$K_b = \frac{[(CH_3)_3NH][OH^-]}{[(CH_3)_3N]} \approx \frac{x^2}{0.112} = 6.5 \times 10^{-5}.$$

$x = 2.7 \times 10^{-3}$.

$$\text{Percentage protonated} = \frac{2.7 \times 10^3}{0.112} \times 100\% = \boxed{2.4\%}.$$

8.17 See the solutions to previous exercises for detailed information on setting up the equilibrium calculations for weak acids.

(a) $$K_a = 8.4 \times 10^{-4} = \frac{[H_3O^+][CH_3CH(OH)CO_2^-]}{[CH_3CH(OH)COOH]} = \frac{x^2}{0.120 - x} \approx \frac{x^2}{0.120}.$$

Assuming x is small relative to 0.120,

$x = [H_3O^+] = 0.010\,M$.

$\text{pH} = -\log(0.010) = \boxed{2.00}$.

$\text{pOH} = 14.00 - 2.00 = \boxed{12.00}$.

$$\text{Fraction deprotonated} = \frac{0.010}{0.120} = \boxed{0.083}.$$

Without the approximation, the quadratic equation must be solved to give $[H_3O^+] = 9.6 \times 10^{-3}M$, pH = 2.02, and pOH = 11.98. The fraction deprotonated is 0.080 which is not much different from the approximate result.

(b) $$8.4 \times 10^{-4} = \frac{x^2}{1.4 \times 10^{-4} - x}.$$

$x^2 + (8.4 \times 10^{-4})x - 1.18 \times 10^{-7} = 0$

$x = 1.22 \times 10^{-4}M$ [negative root is not possible].

$\text{pH} = -\log(1.22 \times 10^{-4}) = \boxed{3.91}$.

$\text{pOH} = 14.00 - 3.91 = \boxed{10.09}$.

$$\text{Fraction deprotonated} = \frac{1.22 \times 10^{-4}}{1.4 \times 10^{-4}} = \boxed{0.87}.$$

(c) $K_a = \dfrac{[H_3O^+][C_6H_5SO_3^-]}{[C_6H_5SO_3H]} = \dfrac{x^2}{(0.10) - x} = 0.20.$

$$x^2 + 0.20x - 0.02 = 0.$$

$$x = \frac{-(0.2) \pm \sqrt{(0.2)^2 - 4 \times (-0.02)}}{2} = 0.073.$$

$$\text{pH} = -\log(0.073) = \boxed{1.14}.$$

$$\text{pOH} = 14.0 - 1.1 = \boxed{12.86}.$$

$$\text{Fraction deprotonated} = \frac{0.073}{0.10} = \boxed{0.73}.$$

8.18 Glycine can accept one proton on its nitrogen atom and donate one from its carboxyl group.

$$pK_{a1} = 2.35, \quad pK_{a2} = 9.60.$$

We follow the procedure of Example 8.4. The three species present are H_2Gly^+, HGly, and Gly^-. The equilibria are:

$$H_2Gly^+(aq) + H_2O(l) \rightleftharpoons H_3O^+(aq) + HGly(aq).$$

$$K_{a1} = \frac{[H_3O^+][HGly]}{[H_2Gly^+]} = \frac{H[HGly]}{[H_2Gly^+]} \quad H = [H_3O^+]$$

$$HGly + H_2O(l) \rightleftharpoons H_3O^+(aq) + Gly^-.$$

$$K_{a2} = \frac{H[Gly^-]}{[HGly]}.$$

Total concentration $= G = [H_2Gly^+] + [HGly] + [Gly^-]$.

$[Gly^-] = K_{a2}[HGly]/H = K_{a2}K_{a1}[H_2Gly^+]/H^2$ because

$[HGly] = K_{a1}[H_2Gly^+]/H$.

Then $G = [H_2Gly^+] + K_{a1}[H_2Gly^+]/H + K_{a2}K_{a1}[H_2Gly^+]/H^2$.

The fractions are

$$f_1 = f(H_2Gly^+) = \frac{[H_2Gly^+]}{G} = \frac{1}{1 + K_{a1}/H + K_{a2}K_{a1}/H^2}$$

$$= \frac{H^2}{H^2 + K_{a1}H + K_{a2}K_{a1}} = \frac{H^2}{K}$$

where $K = H^2 + K_{a1}H + K_{a2}K_{a1}$.

Similarly, we find

$$f_2 = \mathrm{f(HGly)} = \frac{HK_{a1}}{K},$$

$$f_3 = \mathrm{f(Gly^-)} = \frac{K_{a1}K_{a2}}{K}.$$

These fractions are plotted in Figure 8.2 against $\mathrm{pH} = -\log H$. These plots were produced with Mathcad.

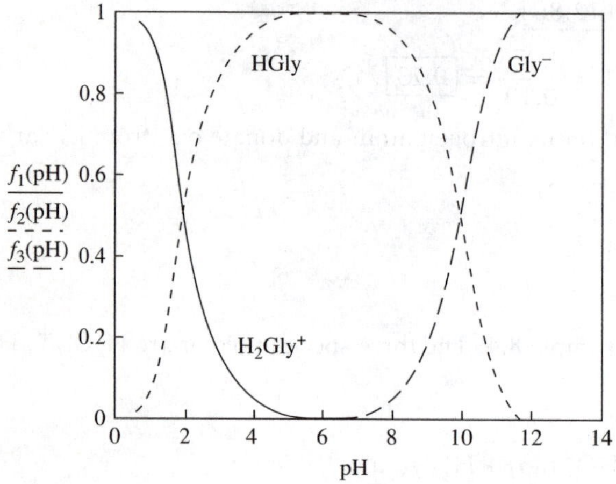

Fig 8.2

8.19 Tyrosine can accept one proton on its nitrogen atom and donate two protons, one from the carboxyl group, and another from the hydroxyl group. The formal analysis is very similar to that of Example 8.4, but with the following values for the equilibrium constants:

$$\mathrm{p}K_{a1} = 2.20 \quad \mathrm{p}K_{a2} = 9.11 \quad \mathrm{p}K_{a3} = 10.07.$$

The equations of Example 8.4 were placed in a Mathcad worksheet with the above values for the equilibrium constants. The resulting graph is shown in Figure 8.3. The legend is as follows:

f_1(pH) = fractional composition of H_3Tyr^+.

f_2(pH) = fractional composition of H_2Tyr.

f_3(pH) = fractional composition of $HTyr^-$.

f_4(pH) = fractional composition of Tyr^{2-}.

Figure 8.3 should be compared to Figure 8.1 in the text.

8.20 Derive the fractional composition of each protonated species by analogy to those of oxalic acid in Example 8.4. Since oxalic acid is a diprotic acid and lysine may be triprotonated in very acid solutions, draw up a comparative table of the fractional composition

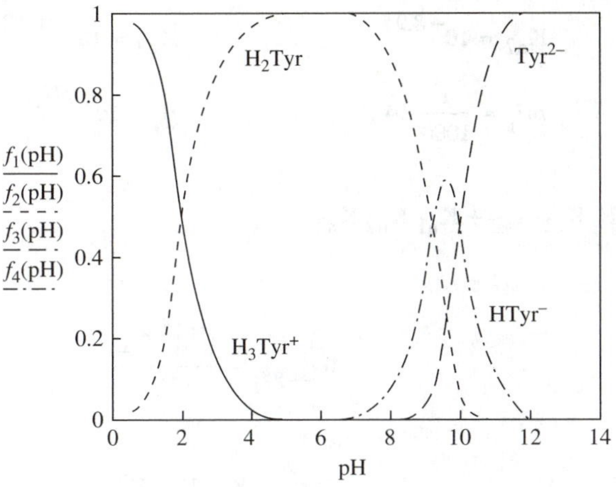

Fig 8.3

numerators for each species. Begin with a deduction of the numerator of the fractional composition for the totally deprotonated species and proceed to deduction of the numerator of the fractional composition for the maximally protonated species. The deduction is an analogy to the equations that appear in Example 8.4. The denominator of the fractional compositions equals the sum of these numerators.

Species	Fractional composition numerator	Species	Fractional composition numerator
$C_2O_4^{2-}$	$K_{a1}K_{a2}$	Lys^-	$K_{a1}K_{a2}K_{a3}$
$HC_2O_4^-$	$[H_3O^+]\,K_{a1}$	HLys	$[H_3O^+]\,K_{a1}K_{a2}$
$H_2C_2O_4$	$[H_3O^+]^2$	H_2Lys^+	$[H_3O^+]^2K_{a1}$
		H_3Lys^{2+}	$[H_3O^+]^3$
Fractional composition denominator	$[H_3O^+]^2+[H_3O^+]\,K_{a1}$ $+K_{a1}K_{a2}$		$[H_3O^+]^3+[H_3O^+]^2K_{a1}$ $+[H_3O^+]K_{a1}K_{a2}$ $+K_{a1}K_{a2}K_{a3}$

The table gives formulas for the fractional composition for every species of a lysine solution. For example,

$$f_{\mathrm{HLys}}=\frac{[H_3O^+]K_{a1}K_{a2}}{[H_3O^+]^3+[H_3O^+]^2K_{a1}+[H_3O^+]K_{a1}K_{a2}+K_{a1}K_{a2}K_{a3}}.$$

A speciation diagram (Figure 8.4) is prepared over the pH range of 0 to 14 by calculating the hydronium ion concentration at each point in the range with the expression $[H_3O^+]=10^{-pH}$. The fractional composition of each species is calculated at each $[H_3O^+]$ with the equations given in the table. The data editor of a scientific graphing calculator serves as an efficient calculation environment. Alternatively, software such as Microsoft Excel or Mathcad are especially effective for preparing the calculations and graph. A Mathcad worksheet is shown below.

$K_{a1} := 10^{-2.18}$ $\quad K_{a2} := 10^{-8.95}$ $\quad K_{a3} := 10^{-10.53}$

$i := 0..1000$ $\quad pH_i := \frac{i}{1000} \cdot 14$ $\quad H_i := 10^{-pH_i}$

$$D_i := (H_i)^3 + (H_i)^2 \cdot K_{a1} + H_i \cdot K_{a1} \cdot K_{a2} + K_{a1} \cdot K_{a2} \cdot K_{a3}$$

$$f_{H3Lys_i} := \frac{(H_i)^3}{D_i} \qquad f_{H2Lys_i} := \frac{(H_i)^2 \cdot K_{a1}}{D_i}$$

$$f_{HLys_i} := \frac{H_i \cdot K_{a1} \cdot K_{a2}}{D_i} \qquad f_{Lys_i} := \frac{K_{a1} \cdot K_{a2} \cdot K_{a3}}{D_i}$$

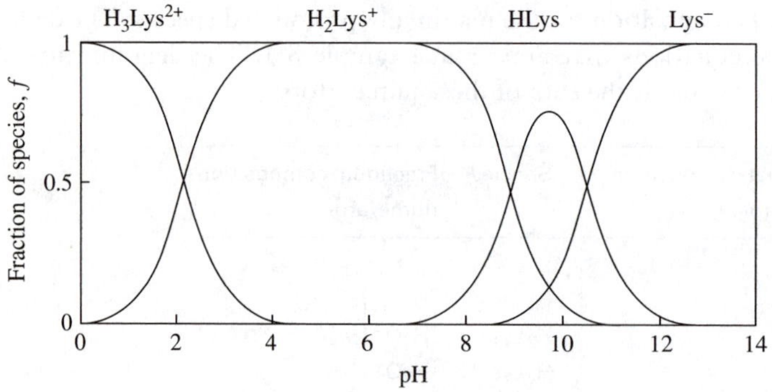

Fig 8.4

8.21

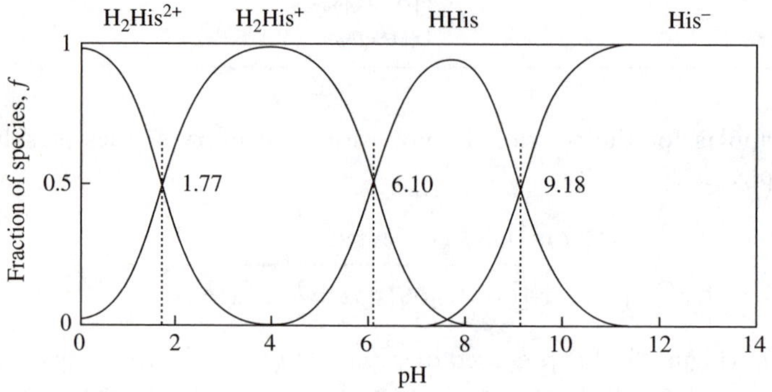

Fig 8.5

8.22 For the case of an amphiprotic salt we use the relation

$$\mathrm{pH} = \frac{1}{2}(\mathrm{p}K_{a1} + \mathrm{p}K_{a2}) \quad [8.10].$$

Note that in this case it is not necessary to specify the concentration of the salt.

For oxalic acid [Table 8.2], $pK_{a1} = 1.23$ and $pK_{a2} = 4.19$. Therefore,

$$pH = \frac{1}{2}(1.23 + 4.19) = \boxed{2.71}.$$

8.23 (a)

$Conc./(mol\,dm^{-3})$	$B(OH)_3$	+	$2H_2O$	$\rightleftharpoons$	H_3O^-	+	$B(OH)_4^-$
Initial	1.0×10^{-4}		—		0		0
Change	$-x$		—		$+x$		$+x$
Equilibrium	$1.0 \times 10^{-4} - x$		—		x		x

$$K_a = 7.2 \times 10^{-10} = \frac{[H_3O^+][B(OH)_4^-]}{[B(OH)_3]} = \frac{x^2}{1.0 \times 10^{-4} - x} \approx \frac{x^2}{1.0 \times 10^{-4}}.$$

$$x = [H_3O^+] = 2.7 \times 10^{-7}\ mol\,dm^{-3}.$$

$$pH = -\log(2.7 \times 10^{-7}) = 6.57.$$

NOTE This value of $[H_3O^+]$ is not much different from the value for pure water, $1.0 \times 10^{-7}\ mol\,dm^{-3}$; hence, it is at the lower limit of safely ignoring the contribution to $[H_3O^+]$ from the autoprotolysis of water. The exercise should be solved by simultaneously considering both equilibria.

$Conc./(mol\,dm^{-3})$	$B(OH)_3$	+	$2H_2O$	$\rightleftharpoons$	H_3O^+	+	$B(OH)_4^-$
Equilibrium	$1.0 \times 10^{-4} - x$		—		x		y
			$2H_2O$	$\rightleftharpoons$	H_3O^+	+	$(OH)^-$
Equilibrium			—		x		z

Because there are now two contributions to $[H_3O^+]$, $[H_3O^+]$ is no longer equal to $[B(OH)_4]$, nor is it equal to $[OH^-]$, as in pure water. To avoid a cubic equation, x will again be ignored relative to $1.0 \times 10^{-4}\ mol\,dm^{-3}$. This approximation is justified by the approximate calculation above and because K_a is very small relative to 1.0×10^{-4}. Let a = initial concentration of $B(OH)_3$, then

$$K_a = 7.2 \times 10^{-10} = \frac{xy}{a - x} \approx \frac{xy}{a} \quad \text{or} \quad y = \frac{aK_a}{x}.$$

$$K_w = 1.0 \times 10^{-14} = xz.$$

Electroneutrality requires $x = y + z$ or $z = x - y$; hence $K_w = xz = x(x - y)$.

Substituting for y from above,

$$x \times \left(x - \frac{aK_a}{x}\right) = K_w,$$

$$x^2 - aK_a = K_w,$$

$$x^2 = K_w + aK_a,$$

$$x = \sqrt{K_w + aK_a} = \sqrt{1.0 \times 10^{-14} + (1.0 \times 10^{-4} \times 7.2 \times 10^{-10})},$$

$$x = 2.9 \times 10^{-7}\ \text{mol dm}^{-3} = [H_3O^+].$$

$$\text{pH} = -\log(2.9 \times 10^{-7}) = \boxed{6.54}.$$

This value is slightly, but measurably, different from the value 6.57 obtained by ignoring the contribution to $[H_3O^+]$ from water.

(b) In this case, the second ionization can safely be ignored; $K_{a2} \ll K_{a1}$.

Conc./(mol dm^{-3})	H_3PO_4	+	H_2O	⇌	H_3O^+	+	$H_2PO_4^-$
Initial	0.015		—		0		0
Change	$-x$		—		$+x$		$+x$
Equilibrium	$0.015 - x$		—		x		x

$$K_{a1} = 7.6 \times 10^{-3} = \frac{x^2}{0.015 - x}.$$

$$x^2 + 7.6 \times 10^{-3}x - 1.14 \times 10^{-4} = 0.$$

$$x = [H_3O^+] = \frac{-7.6 \times 10^{-3} + \sqrt{(7.6 \times 10^{-3} + 4.56 \times 10^{-4}}}{2}$$

$$= 7.6 \times 10^{-3}\ \text{mol dm}^{-3}.$$

$$\text{pH} = -\log(7.6 \times 10^{-3}) = \boxed{2.12}.$$

(c) In this case, second ionization can safely be ignored; $K_{a2} \ll K_{a1}$.

Conc./(mol dm^{-3})	H_2SO_3	+	H_2O	⇌	H_3O^+	+	HSO_3^-
Initial	0.1		—		0		0
Change	$-x$		—		$+x$		$+x$
Equilibrium	$0.1 - x$		—		x		x

$$K_{a1} = 1.5 \times 10^{-2} = \frac{x^2}{0.10 - x}.$$

$$x^2 + 1.5 \times 10^{-2}x - 1.5 \times 10^{-3} = 0.$$

$$x = [H_3O^+] = 0.032\,\text{mol dm}^{-3}.$$

$$\text{pH} = -\log(0.032) = \boxed{1.49}.$$

8.24 At half neutralization, pH = pK_a. This is the pH at which the buffering action is best $\boxed{(\text{pH} = 8.3)}$.

8.25 (a) $\text{pH} = \text{p}K_a - \log \frac{[\text{acid}]}{[\text{base}]}$ [8.11].

$$7.00 = 2.20 - \log \frac{[\text{acid}]}{[\text{base}]}; \quad \text{therefore } \frac{[\text{acid}]}{[\text{base}]} = \boxed{1.6 \times 10^{-5}}.$$

(b) $\log \frac{[\text{acid}]}{[\text{base}]} = 0; \quad \text{therefore } \frac{[\text{acid}]}{[\text{base}]} = \boxed{1}.$

(c) $\log \frac{[\text{acid}]}{[\text{base}]} = 0.7; \quad \text{therefore } \frac{[\text{acid}]}{[\text{base}]} = \boxed{5}.$

8.26 (a)

Conc./(mol dm^{-3})	$H_2C_2O_4$	+	H_2O	$\rightleftharpoons$	H_3O^+	+	$HC_2O_4^-$
Initial	0.15		—		0		0
Change	$-x$		—		$+x$		$+x$
Equilibrium	$0.15 - x$		—		x		x

$$K_{a1} = 5.9 \times 10^{-2}, \quad K_{a2} = 6.5 \times 10^{-5}.$$

The second ionization can be ignored in the calculation of $[H_3O^+]$, but not in the calculation of $[C_2O_4^-]$.

$$K_{a1} = 5.9 \times 10^{-2} = \frac{x^2}{0.15 - x}.$$

$$x^2 + (5.9 \times 10^{-2})x - (8.85 \times 10^{-3}) = 0.$$

$$x = \boxed{0.069\,\text{mol dm}^{-3}} = [H_3O^+].$$

$$[OH^-] = \boxed{1.4 \times 10^{-13}\,\text{mol dm}^{-3}}.$$

$$[H_2C_2O_4] = 0.15 - 0.069 = \boxed{0.08\,\text{mol dm}^{-3}}.$$

Conc./(mol dm^{-3})	$HC_2O_4^-$	+ H_2O	⇌ H_3O^+	+ $C_2O_4^{2-}$
Initial	0.069	—	0.069	0
Change	$-x$	—	$+x$	$+x$
Equilibrium	$0.069-x$	—	$0.069+x$	x

$$K_{a2} = 6.5 \times 10^{-5} = \frac{(0.069+x)x}{(0.069-x)} \approx x \quad \text{[because } x \text{ is small]}.$$

$$x = [C_2O_4^{2-}] = 6.5 \times 10^{-5}\ \text{mol dm}^{-3} \quad \text{and}$$

$$[HC_2O_4^-] = 0.069 - x = 0.069 - 0.000065 = \boxed{0.069\ \text{mol dm}^{-3}}.$$

(b) Similarly to part (a), K_{a2} can be ignored in the first part of the calculation and

$$K_{a1} = 1.3 \times 10^{-7} = \frac{x^2}{0.065-x} \approx \frac{x^2}{0.065}.$$

$$[H_2S] = \boxed{0.065\ \text{mol dm}^{-3}}.$$

$$x = [H_3O^+] = [HS^-] = \boxed{9.2 \times 10^{-5}\ \text{mol dm}^{-3}}.$$

$$[OH^-] = \frac{1.0 \times 10^{-14}}{9.2 \times 10^{-5}} = \boxed{1.1 \times 10^{-10}\ \text{mol dm}^{-3}}.$$

For the second ionization, $K_{a2} = 7.1 \times 10^{-15} = x = [S^{2-}]$.

$$[S^{2-}] = \boxed{7.1 \times 10^{-15}\ \text{mol dm}^{-3}}.$$

8.27 (a)

Conc./(mol dm^{-3})	$CH_3COOH(aq)$	+ $H_2O(l)$	⇌ $H_3O^+(aq)$	+ $CH_3CO_2^-(aq)$
Initial	0.10	—	0	0
Change	$-x$	—	$+x$	$+x$
Equilibrium	$0.10-x$	—	x	x

$$K_a = \frac{[H_3O^+][CH_3CO_2^-]}{[CH_3COOH]} = 1.8 \times 10^{-5} = \frac{x^2}{0.010-x} \approx \frac{x^2}{0.010}.$$

$$x^2 = 1.8 \times 10^{-6}.$$

$$x = 1.3 \times 10^{-3}\ \text{mol dm}^{-3} = [H_3O^+].$$

$$\text{Initial pH} = -\log(1.3 \times 10^{-3}) = \boxed{2.89}.$$

(b) Moles of $CH_3COOH = (0.0250\ \text{dm}^3) \times (0.10\ \text{M})$

$$= 2.5 \times 10^{-3}\ \text{mol } CH_3COOH.$$

Moles of NaOH $= (0.0100\ \mathrm{dm}^3) \times (0.10\ \mathrm{M}) = 1.0 \times 10^{-3}\ \mathrm{mol\ OH^-}$.

So, $\dfrac{1.5 \times 10^{-3}\ \mathrm{mol\ CH_3COOH}}{0.0350\ \mathrm{dm}^3} = 4.29 \times 10^{-2}\ \mathrm{mol\ dm^{-3}}$

and $\dfrac{1.0 \times 10^{-3}\ \mathrm{mol\ CH_3CO_2^-}}{0.0350\ \mathrm{dm}^3} = 2.86 \times 10^{-2}\ \mathrm{mol\ dm^{-3}}$.

Then, consider equilibrium, $K_a = \dfrac{[\mathrm{H_3O^+}][\mathrm{CH_3CO_2^-}]}{[\mathrm{CH_3COOH}]}$.

Con./(mol dm^{-3})	$CH_3COOH(aq)$	+	$H_2O(l)$	$\rightleftharpoons$	$H_3O^+(aq)$	+	$CH_3CO_2^-(aq)$
Initial conc.	4.29×10^{-2}		—		0		2.86×10^{-2}
Change	$-x$		—		$+x$		$+x$
Equilibrium	$4.29 \times 10^{-2} - x$		—		x		$2.86 \times 10^{-2} + x$

$$1.8 \times 10^{-5} = \frac{(x)(x + 2.86 \times 10^{-2})}{(4.29 \times 10^{-2} - x)}; \quad \text{assume} + x \text{ and} - x \text{ negligible.}$$

$$[\mathrm{H_3O^+}] = x = 2.7 \times 10^{-5}\ \mathrm{mol\ dm^{-3}} \quad \text{and} \quad \mathrm{pH} = -\log(2.7 \times 10^{-5}) = \boxed{4.57}.$$

(c) Because acid and base concentrations are equal, their volumes are equal at the stoichiometric point. Therefore, 25.0 cm^3 NaOH are required to reach the stoichiometric point and 12.5 cm^3 NaOH are required to reach halfway to the stoichiometric point.

(d) At the half stoichiometric point, pH $= \mathrm{p}K_a$ and pH $= \boxed{4.74}$.

(e) $\boxed{25.0\ \mathrm{cm}^3}$; see part (c).

(f) The final pH is that of 0.050 M $NaCH_3CO_2$.

Conc./(mol dm^{-3})	$H_2O(l)$	+	$CH_3CO_2^-(aq)$	$\rightleftharpoons$	$CH_3COOH(aq)$	+	$OH^-(aq)$
Initial	—		0.050		0		0
Change	—		$-x$		$+x$		$+x$
Equilibrium	—		$0.050 - x$		x		x

$$K_b = \frac{K_w}{K_a} = \frac{1.00 \times 10^{-14}}{1.8 \times 10^{-5}} = 5.6 \times 10^{-10} = \frac{x^2}{0.050 - x} \approx \frac{x^2}{0.050}.$$

$$x^2 = 2.8 \times 10^{-11}.$$

$$x = 5.3 \times 10^{-6}\ \mathrm{mol\ dm^{-3}} = [\mathrm{OH^-}].$$

$$\mathrm{pOH} = 5.28, \quad \mathrm{pH} = 14.00 - 5.28 = \boxed{8.72}.$$

8.28 $K_a = \dfrac{[H_3O^+][CH_3CO_2^-]}{[CH_3COOH]}$.

$$pH = pK_a + \log \frac{[CH_3CO_2^-]}{[CH_3COOH]}.$$

(a) $pH = pK_a + \log \dfrac{[0.10]}{[0.10]} = pK_a = \boxed{4.74}$.

(b) 3.3 mmol NaOH = 3.3×10^{-3} mol OH^- [strong base] produces 3.3×10^{-3} mol CH_3COO^- from CH_3COOH.

Initially n (CH_3COOH) = $n(CH_3COO^-)$ = 0.10 mol dm^{-3} × 0.100 dm^3 = 1.0×10^{-2} mol.

After adding NaOH,

$$[CH_3COOH] = \frac{1.0 \times 10^{-2} - 3.3 \times 10^{-3}}{0.10\ dm^3} = 6.7 \times 10^{-2}\ mol\ dm^{-3},$$

$$[CH_3COO^-] = \frac{1.0 \times 10^{-2} + 3.3 \times 10^{-3}}{0.10\ dm^3} = 0.13\ mol\ dm^{-3},$$

$$pH = 4.74 + \log \frac{0.13}{0.067} = \boxed{5.03}.$$

Change in pH = $\boxed{0.29}$.

(c) 6.0 mmol HNO_3 = 6.0×10^{-3} mmol H_3O^+ [strong acid] produces 6.0×10^{-3} mol CH_3COOH from CH_3COO^-.
After adding HNO_3 [see (b) above],

$$[CH_3COOH] = \frac{(1.0 \times 10^{-2}) + (6.0 \times 10^{-3})\ mol}{0.100\ dm^3} = 0.16\ mol\ dm^{-3},$$

$$[CH_3COO^-] = \frac{(1.0 \times 10^{-2}) - (6.0 \times 10^{-3})\ mol}{0.100\ dm^3} = 4.0 \times 10^{-2}\ mol\ dm^{-3},$$

$$pH = 4.74 + \log \frac{4.0 \times 10^{-2}}{0.16} = 4.74 - 0.60 = \boxed{4.14}.$$

Change in $\boxed{pH = -0.60}$.

8.29 The rule of thumb we use is that the effective range of a buffer is roughly within plus or minus one pH unit of the pK_a of the acid. Therefore,

(a) $pK_a = 3.08$; pH range, $\boxed{2\text{–}4}$.

(b) $pK_a = 4.19$; pH range, $\boxed{3\text{–}5}$.

(c) $pK_{a3} = 12.68$; pH range, $\boxed{11.5\text{–}13.5}$.

(d) $pK_{a2} = 7.21$; pH range, $\boxed{6\text{–}8}$.

(e) $pK_b = 7.97$; $pK_a = 6.03$; pH range, $\boxed{5\text{–}7}$.

8.30 At the halfway point, pH = pK_a = $\boxed{4.66}$ and $K_a = \boxed{2.19 \times 10^{-5}}$.

$$K_a = 2.19 \times 10^{-5} = \frac{x^2}{(0.015 - x)} \approx \frac{x^2}{(0.015)}.$$

$$x = 5.7 \times 10^{-4}\,\text{M} = [H_3O^+].$$

$$\text{pH} = \boxed{3.24}.$$

8.31 (a) $NH_4^+ + H_2O \rightleftharpoons NH_3 + H_3O^+$.

$$K_a = \frac{[NH_3][H_3O^+]}{[NH_4^+]} = 5.6 \times 10^{-10}$$

$$= \frac{x^2}{0.15 - x} \approx \frac{x^2}{0.15}.$$

$$x = 9.2 \times 10^{-6}\,\text{M} = [H_3O^+].$$

$$\text{pH} = -\log(9.2 \times 10^{-6}) = \boxed{5.04}.$$

(b) $CH_3COO^- + H_2O \rightleftharpoons CH_3COOH + OH^-$.

$$K_b = 5.6 \times 10^{-10} = \frac{x^2}{0.15 - x} \approx \frac{x^2}{0.15}.$$

$$x = 9.2 \times 10^{-5}\,\text{M} = [OH^-].$$

$$\text{pOH} = -\log(9.2 \times 10^{-5}) = 5.04.$$

$$\text{pH} = 14 - 5.04 = \boxed{8.96}.$$

(c) Use the reverse of the above reaction.

$$K_a = 1.8 \times 10^{-5} = \frac{x^2}{0.15 - x} \approx \frac{x^2}{0.15}.$$

$$x = 1.6 \times 10^{-3}\,\text{M} = [H_3O^+].$$

$$\text{pH} = -\log(1.6 \times 10^{-3}) = \boxed{2.80}.$$

8.32 At the stoichiometric point the solution will consist of the lactate ion, which is a weak base, and Na^+ ions. To calculate the pH we first need to calculate the total volume of solution at the stoichiometric point.

Amount (moles) lactate ion = $0.02500\,\text{dm}^3 \times 0.100\,\text{mol dm}^{-3} = 2.50 \times 10^{-3}\,\text{mol}$.

Volume of base added = $(2.500 \times 10^{-3}\,\text{mol})/(0.175\,\text{mol dm}^{-3}) = 0.0143\,\text{dm}^3$.

Total volume = $0.02500\,\text{dm}^3 + 0.0143\,\text{dm}^3 = 0.0393\,\text{dm}^3$.

[Lactate ion] = $(2.50 \times 10^{-3}\,\text{mol})/(0.0393\,\text{dm}^3) = 0.0636\,\text{mol dm}^{-3}$.

$$K_b = 1.2 \times 10^{-11} = \frac{[\text{lactic acid}][\text{OH}^-]}{[\text{lactate ion}]} = \frac{x^2}{0.0636 - x} \approx \frac{x^2}{0.0636}$$

$$x = [\text{OH}^-] = 8.7 \times 10^{-7}\ \text{M}.$$

$$\text{pOH} = -\log(8.7 \times 10^{-7}) = 6.06.$$

$$\text{pH} = 14.00 - 6.06 = \boxed{7.94}.$$

8.33 The initial pH is calculated from

$$K_b = \frac{[\text{CH}_3\text{COOH}][\text{OH}^-]}{[\text{CH}_3\text{CO}_2^-]} \approx \frac{x^2}{0.10} = 5.6 \times 10^{-10}.$$

$$x = [\text{OH}^-] = 7.5 \times 10^{-6}, \quad \text{pOH} = 5.12, \quad \text{pH} = 8.88.$$

Use the Henderson–Hasselbalch equation to calculate the other pHs for specific concentrations of CH_3COOH.

When $[\text{CH}_3\text{COOH}] = [\text{NaCH}_3\text{CO}_2] = 0.10\ \text{M}$, $\text{pH} = \text{p}K_a = 4.74$.

When $[\text{CH}_3\text{COOH}] = \frac{1}{2}[\text{NaCH}_3\text{CO}_2] = 0.05\ \text{M}$, $\text{pH} = 5.04$.

When $[\text{CH}_3\text{COOH}] = \frac{1}{4}[\text{NaCH}_3\text{CO}_2] = 0.025\ \text{M}$, $\text{pH} = 5.34$.

When $[\text{CH}_3\text{COOH}] = \frac{1}{10}[\text{NaCH}_3\text{CO}_2] = 0.010\ \text{M}$, $\text{pH} = 5.74$.

When $[\text{CH}_3\text{COOH}] = 2[\text{NaCH}_3\text{CO}_2] = 0.20\ \text{M}$, $\text{pH} = 4.44$.

A rough sketch is shown in Figure 8.6.

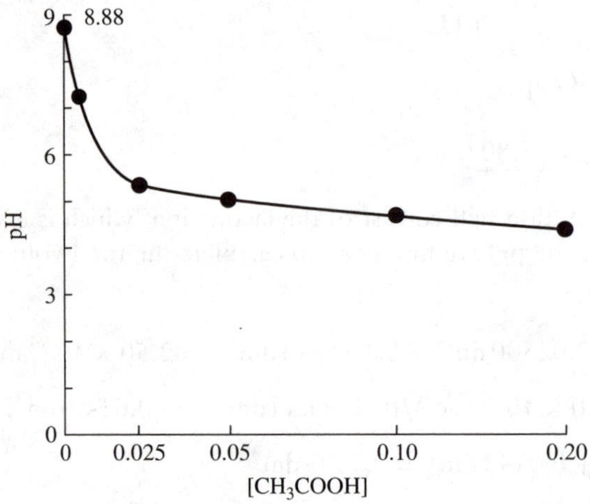

Fig 8.6

8.34 Choose a buffer system in which the conjugate acid has a pK_a close to the desired pH. Therefore,

(a) H_3PO_4 and NaH_2PO_4;
(b) NaH_2PO_4 and Na_2HPO_4 or $NaHSO_3$ and Na_2SO_3.

8.35 (a) $K_s = [Ag^+][I^-]$.
(b) $K_s = [Hg_2^{+2}][S^{2-}]$.
(c) $K_s = [Fe^{3+}][OH^-]^3$.
(d) $K_s = [Ag^+]^2[CrO_4^{2-}]$.

8.36 (a) $K_s = [Ba^{2+}][SO_4^{2-}] = S^2 = 1.1 \times 10^{-10}$.

$$S = \boxed{1.0 \times 10^{-5}\ \text{mol dm}^{-3}}.$$

(b) $K_s = [Ag^+]^2[CO_3^{2-}] = (2S)^2(S) = 4S^3 = 6.2 \times 10^{-12}$.

$$S = \boxed{1.2 \times 10^{-4}\ \text{mol dm}^{-3}}.$$

(c) $K_s = [Fe^{3+}][OH^-]^3 = (S)(3S)^3 = 27S^4 = 2.0 \times 10^{-39}$.

$$S = \boxed{9.3 \times 10^{-11}\ \text{mol dm}^{-3}}.$$

(d) $K_s = [Hg_2^{2+}][Cl^-]^2 = S(2S)^2 = 4S^3 = 1.3 \times 10^{-18}$.

$$S = \boxed{16.9 \times 10^{-7}\ \text{mol dm}^{-3}}.$$

8.37 (a)

Conc./(mol dm^{-3})	$AgBr(s)$	$\rightleftharpoons$	$Ag^+(aq)$	+	$Br^-(aq)$
Initial	0		0		0.0014
Change	—		$+S$		$+S$
Equilibrium	—		S		$0.0014 + S$

$$K_s = [Ag^+][Br^-] = 7.7 \times 10^{-13} = S \times (0.0014 + S) \approx S \times (.0014).$$

$$S = \boxed{5.5 \times 10^{-10}\ \text{mol dm}^{-3}} = [Ag^+]$$

= molar solubility of AgBr in 0.0014 M NaBr.

(b)

Conc./(mol dm^{-3})	$MgCO_3(s)$	$\rightleftharpoons$	$Mg^{2+}(aq)$	+	$CO_3^{2-}(aq)$
Initial	0		0		1.1×10^{-5}
Change	—		$+S$		$+S$
Equilibrium	—		$+S$		$1.1 \times 10^{-5} + S$

$$K_s = [Mg^{2+}][CO_3^{2-}] = S \times (1.1 \times 10^{-5} + S) = 1.0 \times 10^{-5}.$$

Assume 1.1×10^{-5} negligible compared to S.

$$S^2 = 1.0 \times 10^{-5}.$$

$$S = \boxed{3.2 \times 10^{-3}\ \text{mol dm}^{-3}}.$$

(c)

Conc./(mol dm^{-3})	$PbSO_4$(s)	⇌	Pb^{2+}(aq)	+	SO_4^{2-}(aq)
Initial	0		0		0.10
Change	—		$+S$		$+S$
Equilibrium	—		S		$0.10 + S$

$$K_s = [Pb^{2+}][SO_4^{2-}] = 1.6 \times 10^{-8} = (S) \times (0.10 + S) = 0.10(S)$$

$$S = \boxed{1.6 \times 10^{-7}\ \text{mol dm}^{-3}}.$$

(d)

Conc./(mol dm^{-3})	$Ni(OH)_2$(s)	⇌	Ni^{2+}(aq)	+	$2OH^-$(aq)
Initial	0		2.7×10^{-5}		0
Change	—		$+S$		$+2S$
Equilibrium	—		$2.7 \times 10^{-5} + S$		$2S$

$$K_s = [Ni^{2+}][OH^-]^2 = 6.5 \times 10^{-18} = (S + 2.7 \times 10^{-5}) \times (2S)^2.$$

Assume S in $(S + 2.7 \times 10^{-5})$ is negligible.

$$6.5 \times 10^{-18} = (2.7 \times 10^{-5})4S^2$$

$$S = \boxed{12.5 \times 10^{-7}\ \text{mol dm}^{-3}}.$$

8.38 $HgCl_2(s) \rightleftharpoons Hg^{2+}(aq) + 2Cl^-(aq), \quad K = [Hg^{2+}][Cl^-]^2.$

$[Cl^-] = 2 \times [Hg^{2+}]$; therefore $K = 4[Hg^{2+}]^3$

and the solubility of the salt is

$$S = [Hg^{2+}] = \left(\frac{1}{4}K\right)^{1/3}\ \text{mol dm}^{-3}.$$

From $\Delta_r G^\ominus = \Delta_f G^\ominus(Hg^{2+}) + 2\Delta_f G^\ominus(Cl^-) - \Delta_f G^\ominus(HgCl_2)$

$$= +164.40 + 2 \times (-131.23) - (-178.6)\ \text{kJ mol}^{-1}$$

$$= 80.54\ \text{kJ mol}^{-1},$$

$$\ln K = \frac{-\Delta_r G^{\ominus}}{RT} = \frac{-80.54 \times 10^3\,\mathrm{J\,mol^{-1}}}{8.3145\,\mathrm{J\,K^{-1}\,mol^{-1}} \times 298.15\,\mathrm{K}} = -32.49.$$

Therefore $K = 7.758 \times 10^{-15}$ and $S = \boxed{1.25 \times 10^{-5}\,\mathrm{mol\,dm^{-3}}}$.

8.39 (a) $AgCl(s) \rightleftharpoons Ag^+(aq) + Cl^-(aq)$.

$$K_S = a_{Ag^+}a_{Cl^-} = S^2 \quad \text{where } S = [Ag^+] = [Cl^-].$$

$$\ln K_S = \ln S^2 = 2\ln S \text{ at temperature } T.$$

$$\ln K'_S = \ln (S')^2 = 2\ln S' \text{ at temperature } T'.$$

Subtracting the expression at temperature T from the expression at temperature T' gives:

$$\ln K'_S - \ln K_S = 2\ln S' - 2\ln S = 2\ln\left(\frac{S'}{S}\right)$$

$$= \frac{\Delta_r H^{\ominus}}{R}\left(\frac{1}{T} - \frac{1}{T'}\right) \quad \text{van't Hoff equation [7.15].}$$

$$\ln\left(\frac{S'}{S}\right) = \frac{\Delta_r H^{\ominus}}{2R}\left(\frac{1}{T} - \frac{1}{T'}\right)$$

$$\boxed{\frac{S'}{S} = \mathrm{e}^{\frac{\Delta_r H^{\ominus}}{2R}\left(\frac{1}{T} - \frac{1}{T'}\right)}}.$$

If $T' > T$ and $\Delta_S H^{\ominus} > 0$, then $S' > S$ because $\frac{1}{T} - \frac{1}{T'} > 0$

and e to a positive exponent is greater than 1. Solubility increases.

If $T' > T$ and $\Delta_S H^{\ominus} < 0$, then $S' < S$ because $\frac{1}{T} - \frac{1}{T'} > 0$

and e to a negative exponent is less than 1. Solubility decreases.

(b) $\Delta_S H^{\ominus}$ is evaluated for AgCl(s) using the information of Table D1.2.

$$\Delta_S H^{\ominus}(\mathrm{AgCl}) = \Delta_f H^{\ominus}(\mathrm{Ag^+(aq)}) + \Delta_f H^{\ominus}(\mathrm{Cl^-(aq)}) - \Delta_f H^{\ominus}(\mathrm{AgCl(s)})$$
$$= \{105.58 + (-167.16) - (-127.07)\}\,\mathrm{kJ\,mol^{-1}} = 65.49\,\mathrm{kJ\,mol^{-1}}.$$

The solubility of silver chloride $\boxed{\text{increases}}$ as the temperature increases.

Chapter 9

Electrochemistry

Answers to discussion questions

9.1 The Debye–Hückel theory is a theory of the activity coefficients of ions in solution. It is the Coulombic (electrostatic) interaction of the ions in solution with each other and also the interaction of the ions with the solvent that is responsible for the deviation of their activity coefficients from the ideal value of 1. The electrostatic ion–ion interaction is the stronger of the two and is fundamentally responsible for the deviation. Because of this interaction, there is a build-up of charge of opposite sign around any given ion in the overall electrically neutral solution. The energy, and hence the chemical potential of any given ion, is lowered because of the existence of this ionic atmosphere. The lowering of the chemical potential below its ideal value is identified with a non-zero value of $RT \ln \gamma_{\pm}$. This non-zero value implies that $\gamma_{\pm}$ will have a value different from unity, which is its ideal value.

9.2 According to the Grotthus mechanism, there is an effective motion of a proton that involves the rearrangement of bonds in a group of water molecules. However, the actual mechanism is still highly contentious. Attention now focuses on the $H_9O_4^+$ unit in which the nearly trigonal planar H_3O^+ ion is linked to three strongly solvating H_2O molecules. This cluster of atoms is itself hydrated, but the hydrogen bonds in the secondary sphere are weaker than in the primary sphere. It is envisaged that the rate-determining step is the cleavage of one of the weaker hydrogen bonds of this secondary sphere (Figure 9.4 of the text). After this bond cleavage has taken place, and the released molecule has rotated through a few degrees (a process that takes about 1 ps), there is a rapid adjustment of bond lengths and angles in the remaining cluster, to form an $H_5O_2^+$ cation of structure $H_2O \cdots H^+ \cdots OH_2$. Shortly after this reorganization has occurred, a new $H_9O_4^+$ cluster forms as other molecules rotate into a position where they can become members of a secondary hydration sphere, but now the positive charge is located one molecule to the right of its initial location. According to this model, there is no coordinated motion of a proton along a chain of molecules, simply a very rapid hopping between neighboring sites, with low activation energy. The model is consistent with the observation that the molar conductivity of protons increases as the pressure is raised, for increasing pressure ruptures the hydrogen bonds in water.

9.3 A galvanic cell uses a spontaneous chemical reaction to generate a potential difference and deliver an electric current to an external device. An electrolytic cell uses an external potential difference to drive a chemical reaction in the cell that is by itself non-spontaneous. In their essential features, these two kinds of cells can be considered opposites of each other, in the sense that an electrolytic cell can be thought of as a galvanic cell operating in the reverse direction. For some electrochemical cells, this is easy to accomplish. We say they are rechargeable. The most common example is the lead-acid battery used in automobiles. For many other cells, however, this kind of reversibility cannot be achieved. A fuel cell, like the galvanic cell, uses a spontaneous chemical reaction to generate a potential difference and deliver an electric current to an external device. Unlike the galvanic cell, the fuel cell must receive reactants from an external storage tank. See Box 9.2 for greater detail.

9.4 The starting point is the Nernst equation for the cell [9.14], $E = E^{\ominus} - \dfrac{RT}{\nu F}\ln Q$.

Measurement of E, with knowledge of the reaction quotient, Q, would seem to provide a straightforward determination of $E^{\ominus}$. But a problem arises because the calculation of Q requires not only knowledge of the concentrations of the species involved in the cell reaction but also of their activity coefficients. These coefficients are not usually available, so the calculation cannot be directly completed. However, at very low concentrations, the Debye–Hückel limiting law for the coefficients holds. The procedure then is to substitute the Debye–Hückel law for the activity coefficients into the specific form of the Nernst equation for the cell under investigation and then to take measurements of E as a function of concentration. From an extrapolation of the data to zero concentration where the law holds exactly, the standard potential of the cell can be obtained.

Solutions to exercises

For notational simplicity, we have used both the molality concentration expression $a_J = \gamma_J b_J / b^{\ominus}$ where $b^{\ominus} = 1\,\text{mol kg}^{-1}$ [9.1a] and $a_J = \gamma_J b_J$ [9.1b] where b_J in the latter expression is the unitless magnitude of molality. The convention of eqn 9.1b is most often used in calculations of ionic strength while the convention of eqn 9.1a appears in Nernst equation computations.

9.5 $$I = \frac{1}{2}\left(z_+^2 b_+ + z_-^2 b_-\right) \quad [9.5].$$

Let the preparation molality be b. Determination of solution ionic strength requires the deduction of z_+^2, b_+, z_-^2, and b_-. These values are substituted into the above equation.

(a) $\text{KCl(s)} \xrightarrow{\text{water}} \text{K}^+\text{(aq)} + \text{Cl}^-\text{(aq)}$.

$$z_+^2 = (+1)^2 = 1, \quad b_+ = b, \quad z_+^2 b_+ = b.$$

$$z_-^2 = (-1)^2 = 1, \quad b_- = b, \quad z_-^2 b_- = b.$$

$$I = \frac{1}{2}(b + b) = \boxed{b}.$$

(b) $FeCl_3(s) \xrightarrow{\text{water}} Fe^{3+}(aq) + 3Cl^-(aq).$

$$z_+^2 = (+3)^2 = 9, \quad b_+ = b, \quad z_+^2 b_+ = 9b.$$

$$z_-^2 = (-1)^2 = 1, \quad b_- = 3b, \quad z_-^2 b_- = 3b.$$

$$I = \frac{1}{2}(9b + 3b) = \boxed{6b}.$$

(c) $CuSO_4(s) \xrightarrow{\text{water}} Cu^{2+}(aq) + SO_4^{2-}(aq).$

$$z_+^2 = (+2)^2 = 4, \quad b_+ = b, \quad z_+^2 b_+ = 4b.$$

$$z_-^2 = (-2)^2 = 4, \quad b_- = b, \quad z_-^2 b_- = 4b.$$

$$I = \frac{1}{2}(4b + 4b) = \boxed{4b}.$$

9.6 $I = I_{KCl} + I_{CuSO_4} = b_{KCl} + 4b_{CuSO_4}$ [see Exercise 9.5].

$$= (0.10) + 4 \times (0.20) = \boxed{0.90}.$$

COMMENT Note that the ionic strength of a solution of more than one electrolyte may be calculated by summing the ionic strengths of each electrolyte considered as a separate solution, as in the solution to this exercise, or by summing the product $\frac{1}{2} b_J z_J^2$ for each individual ion, as in the definition of [9.5].

9.7 $I_{KNO_3} = b_{KNO_3} = 0.150.$

Therefore, the ionic strengths of the added salts must be 0.100 to result in a total of 0.250.

(a) $I_{Ca(NO_3)_2} = \frac{1}{2}(2^2 + 2) \times b = 3b.$

Therefore, the solution should be made

$$\frac{1}{3}\left(0.100\,\text{mol kg}^{-1}\right) = 0.0333\,\text{mol kg}^{-1} \text{ in } Ca(NO_3)_2.$$

The mass that should be added to 500 g of the KNO_3 solution is therefore

$$(0.500\,\text{kg}) \times \left(0.0333\,\text{mol kg}^{-1}\right) \times \left(164\,\text{g mol}^{-1}\right) = \boxed{2.73\,\text{g}}.$$

(b) $I_{NaCl} = b$; therefore, with $b = 0.100\,\text{mol kg}^{-1}$,

$$(0.500\,\text{kg}) \times \left(0.100\,\text{mol kg}^{-1}\right) \times \left(58.4\,\text{g mol}^{-1}\right) = \boxed{2.92\,\text{g}}.$$

9.8 $\gamma_\pm = (\gamma_+^p \gamma_-^q)^{1/s} \quad s = p + q$ [9.3b].

For $CaCl_2$, $p = 1, q = 2, s = 3,$ $\boxed{\gamma_\pm = (\gamma_+ \gamma_-^2)^{1/3}}$.

9.9 These concentrations are sufficiently dilute for the Debye–Hückel limiting law to give a good approximate value for the mean ionic activity coefficient. Hence,

$$\log \gamma_\pm = -\,|z_+z_-|\,AI^{1/2} \quad [9.4].$$

$$I = \frac{1}{2}\sum_i z_i^2 b_i \quad \text{(see Illustration 9.2)}$$

$$= \frac{1}{2}\{(4\times 0.010) + (1\times 0.020) + (1\times 0.030) + (1\times 0.030)\} = \boxed{0.060}.$$

$$\log \gamma_\pm = -2\times 1\times 0.509\times (0.060)^2 = -0.249\overline{4}.$$

$$\gamma_\pm = 10^{-0.249\overline{4}} = 0.56\overline{3} = \boxed{0.56}.$$

9.10 $$\log \gamma_\pm = -\frac{A\,|z_+z_-|\,I^{1/2}}{1+BI^{1/2}} \quad [9.6 \text{ with } C=0].$$

Solving for B,

$$B = -\left(\frac{1}{I^{1/2}} + \frac{A\,|z_+z_-|}{\log \gamma_\pm}\right).$$

For HBr, $I = b$ and $|z_+z_-| = 1$, so

$$B = -\left(\frac{1}{b^{1/2}} + \frac{0.509}{\log \gamma_\pm}\right).$$

Hence, draw up the accompanying table.

b	0.0050	0.0100	0.0200
$\gamma_\pm$	0.930	0.907	0.879
B	2.01	2.01	2.02

The constancy of B indicates that the mean activity coefficient of HBr obeys the extended Debye–Hückel law very well.

9.11 The basis for the solution is Kohlrausch's law of independent migration of ions [9.8]. Switching counterions does not affect the mobility of the remaining other ion at infinite dilution.

$$\Lambda_m^o(\text{KCl}) = \lambda_+(\text{K}^+) + \lambda_-(\text{Cl}^-)\ [9.9] = 14.99\,\text{mS m}^2\,\text{mol}^{-1}.$$

$$\Lambda_m^o(\text{KNO}_3) = \lambda(\text{K}^+) + \lambda(\text{NO}_3^-) = 14.50\,\text{mS m}^2\,\text{mol}^{-1}.$$

$$\Lambda_m^o(\text{AgNO}_3) = \lambda(\text{Ag}^+) + \lambda(\text{NO}_3^-) = 13.34\,\text{mS m}^2\,\text{mol}^{-1}.$$

Hence, $\Lambda_m^o(\text{AgCl}) = \Lambda_m^o(\text{AgNO}_3) + \Lambda_m^o(\text{KCl}) - \Lambda_m^o(\text{KNO}_3)$

$$= (13.34 + 14.99 - 14.50)\,\text{mS m}^2\,\text{mol}^{-1} = \boxed{13.83\,\text{mS m}^2\,\text{mol}^{-1}}.$$

9.12 Molar ionic conductivity is related to mobility u by

$$\lambda = zuF$$
$$= 1 \times 7.91 \times 10^{-8}\,\mathrm{m^2\,s^{-1}\,V^{-1}} \times 96485\,\mathrm{C\,mol^{-1}}$$
$$= \boxed{7.63 \times 10^{-3}\,\mathrm{S\,m^2\,mol^{-1}}}.$$

9.13 $s = u\mathcal{E}$ [9.10], $\mathcal{E} = \dfrac{\Delta\phi}{l}$.

Therefore,

$$s = u\left(\frac{\Delta\phi}{l}\right) = (7.92 \times 10^{-8}\,\mathrm{m^2\,s^{-1}\,V^{-1}}) \times \left(\frac{35.0\,\mathrm{V}}{8.00 \times 10^{-3}\,\mathrm{m}}\right)$$
$$= 3.47 \times 10^{-4}\,\mathrm{m\,s^{-1}}, \quad \text{or} \quad \boxed{347\,\mathrm{mm\,s^{-1}}}.$$

9.14 $\Lambda_\mathrm{m} = \Lambda_\mathrm{m}^\circ - Kc^{1/2}$ [9.8], $\Lambda_\mathrm{m} = \dfrac{\kappa}{c}$ [9.7] $= \dfrac{C}{cR} = \dfrac{0.2063\,\mathrm{cm^{-1}}}{cR}$.

$C = \kappa^* R^*$, where κ^* and R^* are the conductivity and resistance, respectively, of a standard solution.

We draw up the following table using $1\,\mathrm{M} = 1\,\mathrm{mol\,dm^{-3}}$.

c/M	0.0005	0.001	0.005	0.010	0.020	0.050
$(c/\mathrm{M})^{1/2}$	0.224	0.032	0.071	0.100	0.141	0.224
R/Ω	3314	1669	342.1	174.1	89.08	37.14
$\Lambda_\mathrm{m}/(\mathrm{mS\,m^2\,mol^{-1}})$	12.45	12.36	12.06	11.85	11.58	11.11

The values of Λ_m are plotted against $c^{1/2}$ in Figure 9.1.

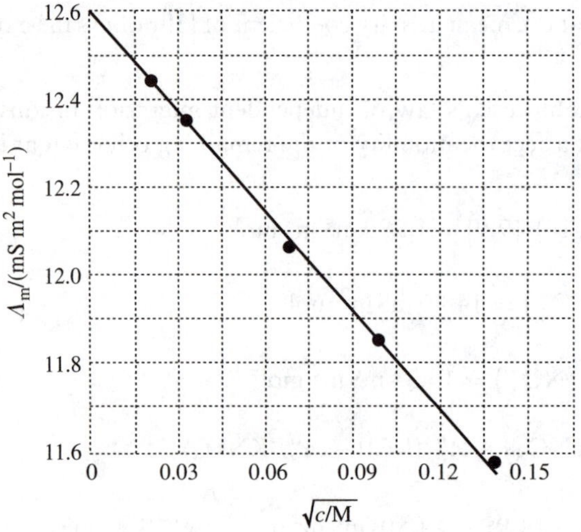

Fig 9.1

(a) The plot of Λ_m against $c^{1/2}$ is linear. Hence, Kohlrausch's law is obeyed.

(b) The limiting value is $\Lambda_m^o = \boxed{12.6\,\text{mS}\,\text{m}^2\,\text{mol}^{-1}}$. The slope is -7.30; hence $K = \boxed{7.30\,\text{mS}\,\text{m}^2\,\text{mol}^{-1}\,\text{M}^{-1/2}}$.

$$\Lambda_m = (5.01 + 7.68)\,\text{mS}\,\text{m}^2\,\text{mol}^{-1} - (+7.30\,\text{mS}\,\text{m}^2\,\text{mol}^{-1}) \times (0.010)^{1/2}$$

$$= \boxed{11.96\,\text{mS}\,\text{m}^2\,\text{mol}^{-1}}.$$

(c) For $c = 10\,\text{mmol}\,\text{dm}^{-3}$ solution, the table above gives

(i) $\Lambda_m = \boxed{11.96\,\text{mS}\,\text{m}^2\,\text{mol}^{-1}}$.

(ii) $\kappa = c\Lambda_m = (10\,\text{mol}\,\text{m}^{-3}) \times (11.96\,\text{mS}\,\text{m}^2\,\text{mol}^{-1}) = 119.6\,\text{mS}\,\text{m}^2\,\text{m}^{-3}$ $= \boxed{119.6\,\text{mS}\,\text{m}^{-1}}$.

(iii) $R = \dfrac{C}{\kappa} = \dfrac{20.63\,\text{m}^{-1}}{119.6\,\text{mS}\,\text{m}^{-1}} = \boxed{172.5\,\Omega}$.

9.15 $c = \dfrac{\kappa}{\Lambda_m}$ [9.7] $\approx \dfrac{\kappa}{\Lambda_m^o}$ [c small, conductivity of water allowed for in the data]

$$c \approx \frac{1.887 \times 10^{-6}\,\text{S}\,\text{cm}^{-1}}{138.3\,\text{S}\,\text{cm}^2\,\text{mol}^{-1}} \text{ [Exercise 9.11]}$$

$$\approx 1.36 \times 10^{-8}\,\text{mol}\,\text{cm}^{-3} = \text{solubility} = \boxed{1.36 \times 10^{-5}\,\text{M}}.$$

9.16 $$\alpha = \frac{\Lambda_m}{\Lambda_m^o} = \frac{\Lambda_m}{\lambda_+ + \lambda_-} \text{ [9.9]} = \frac{3.83\,\text{mS}\,\text{m}^2\,\text{mol}^{-1}}{(34.96 + 5.46)\,\text{mS}\,\text{m}^2\,\text{mol}^{-1}\text{ [Table 9.1]}} = 0.0948.$$

$$K_a = \frac{[\text{H}_3\text{O}^+][\text{HCOO}^-]}{[\text{HCOO}^-]} = \frac{(\alpha c_{\text{HCOOH}})^2}{c_{\text{HCOOH}} - \alpha c_{\text{HCOOH}}} \text{ [See Example 9.1]} = \frac{\alpha^2 c_{\text{HCOOH}}}{1-\alpha}$$

$$= \left\{\frac{(0.0948)^2}{1 - 0.0948}\right\}(0.020) = 1.99 \times 10^{-4}.$$

$$\text{p}K_a = -\log K_a = -\log(1.99 \times 10^{-4}) = \boxed{3.70}.$$

9.17 $\boxed{\text{Yes}}$, NADH is oxidized (it loses electrons); pyruvate, $CH_3COCO_2^-$, is reduced (it gains electrons).

9.18 (1) $NAD^+ + 2e^- + H^+ \rightarrow NADH$,

(2) $CH_3COCO_2^- + 2e^- + 2H^+ \rightarrow CH_3CH(OH)CO_2^-$.

The reaction in Exercise 9.17 is then obtained as (2) – (1);

$CH_3COCO_2^- + NADH + H^+ \rightarrow CH_3CH(OH)CO_2^- + NAD^+$.

9.19 (3) $CH_3CH_2OH + NAD^+ \rightarrow CH_3CHO + NADH + H^+$,

(1) $NAD^+ + 2e^- + H^+ \rightarrow NADH$,

(2) $CH_3CHO + 2e^- + 2H^+ \rightarrow CH_3CH_2OH$.

Reaction (3) is obtained as (1) – (2). The reaction quotients for each of these reactions are:

$$Q_{(1)} = \frac{[NADH]}{[NAD^+][H^+]},$$

$$Q_{(2)} = \frac{[CH_3CH_2OH]}{[CH_3CHO][H^+]^2},$$

$$Q_{(3)} = \frac{[CH_3CHO][NADH][H^+]}{[CH_3CH_2OH][NAD^+]}.$$

It is seen that $Q_{(3)} = \dfrac{Q_{(1)}}{Q_{(2)}}$.

9.20 (1) cystine(aq) + $2H^+$(aq) + $2e^-$ $\rightarrow$ 2 cysteine(aq),

(2) $O_2(g) + 4H^+(aq) + 4e^- \rightarrow 2H_2O$.

The overall reaction is obtained as (2) – 2 × (1) and is

4 cysteine(aq) +O_2 (g) $\rightarrow$ 2cystine(aq) + $2H_2O$(l).

9.21 (1) $NADP^+(aq) + H^+(aq) + 2e^- \rightarrow NADPH(aq)$.

Subtraction of this half-reaction from the overall reaction yields

$2fd_{red}(aq) + H^+(aq) - 2e^- \rightarrow 2fd_{ox}(aq)$,

which may be rearranged to

(2) $2fd_{ox}(aq) + 2e^- \rightarrow 2fd_{red}(aq) + H^+(aq)$.

Then the overall reaction is obtained as (1) – (2), with 2 electrons transferred.

9.22 The half-reactions are

(1) $NAD^+(aq) + H^+(aq) + 2e^- \rightarrow NADH(aq)$,

(2) $O_2(g) + 4H^+(aq) + 4e^- \rightarrow 2H_2O(l)$.

In order to cancel electrons we rewrite (1) as (1) $2NAD^+(aq) + 2H^+(aq) + 4e^- \rightarrow 2NADH(aq)$.

The overall reaction is then (2 – 1):

$$2NADH(aq) + O_2(g) + 2H^+(aq) \rightarrow 2NAD^+(aq) + 2H_2O(l).$$

The standard potential for the overall reaction is then

$$E^{\oplus} = E_2^{\oplus} - E_1^{\oplus} \; [9.17]$$
$$= +0.82\,V - (-0.32\,V) = +1.14\,V.$$

$$\Delta_r G^{\ominus} = -\nu F E^{\ominus} \quad [9.13]$$
$$= -4 \times 96\,485\,\mathrm{C\,mol^{-1}} \times (+1.14\,\mathrm{V})$$
$$= \boxed{-440\,\mathrm{kJ\,mol^{-1}}}.$$

9.23 The Nernst equation applies to half-cells as well as cells.

$$E_i = E^{\ominus} - \frac{RT}{\nu F} \ln Q_i \quad [9.14]$$
$$= E^{\ominus} - \frac{25.7\,\mathrm{mV}}{\nu} \ln Q_i.$$
$$E_f = E^{\ominus} - \frac{25.7\,\mathrm{mV}}{\nu} \ln Q_f.$$

The half-reaction for the electrode is

$$2H^+(aq) + 2e^- \rightarrow H_2(g) \qquad E^{\ominus} = 0.$$
$$Q = \frac{p_{H_2}}{[H^+]^2} = \frac{1.45}{[H^+]^2}.$$
$$\Delta E = E_f - E_i = -\frac{25.7\,\mathrm{mV}}{2} \ln\left(\frac{Q_f}{Q_i}\right)$$
$$= -25.7\,\mathrm{mV} \ln \frac{[H^+]_i}{[H^+]_f}$$
$$= 25.7\,\mathrm{mV} \ln \frac{[H^+]_f}{[H^+]_i}$$
$$= 25.7\,\mathrm{mV} \ln \left(\frac{25.0}{5.0}\right)$$
$$= \boxed{41\,\mathrm{mV}}.$$

9.24 The potential of the hydrogen electrode as a function of pH is obtained from the Nernst equation for the electrode (see the solution to Exercise 9.23 and Section 9.10) and is

$$E = -59.2\,\mathrm{mV} \times \mathrm{pH} \quad [E^{\ominus} = 0 \text{ for this electrode}],$$
$$\Delta E = -59.2\,\mathrm{mV} \times (\mathrm{pH}_2 - \mathrm{pH}_1).$$

For lactic acid $K_a = 8.4 \times 10^{-4} = \dfrac{[H^+][L^-]}{[HL]} \cong \dfrac{x^2}{[HL]}$.

For $[HL] = 5.0 \times 10^{-3}\,\mathrm{mol\,dm^{-3}}$, $x = 2.0 \times 10^{-3}\,\mathrm{mol\,dm^{-3}} = [H^+]$, $\mathrm{pH}_1 = 2.70$.

For $[HL] = 2.5 \times 10^{-2}\,\mathrm{mol\,dm^{-3}}$, $x = 4.6 \times 10^{-3}\,\mathrm{mol\,dm^{-3}} = [H^+]$, $\mathrm{pH}_2 = 2.34$.

$$\Delta E = -59.2\,\mathrm{mV} \times (2.34 - 2.70) = \boxed{21\,\mathrm{mV}}.$$

9.25 R: $Cl_2\,(g) + 2e^- \rightarrow (aq)$, $\quad E^{\ominus} = +1.36\,\mathrm{V}$,

L: $Mn^{2+}(aq) + 2e^- \rightarrow Mn\,(s)$, $\quad E^{\ominus} = ?$.

The cell corresponding to these half-reactions is

$$\mathrm{Mn|MnCl_2\,(aq)\,|Cl_2\,(g)\,|Pt}, \qquad E^\ominus = 1.36\,\mathrm{V} - E^\ominus\left(\mathrm{Mn^{2+}, Mn}\right).$$

Hence, $E^\ominus(\mathrm{Mn^{2+}, Mn}) = 1.36\,\mathrm{V} - 2.54\,\mathrm{V} = \boxed{-1.18\,\mathrm{V}}$.

9.26 (a) R: $\mathrm{Ag^+(aq}, b_\mathrm{R}) + \mathrm{e^-} \rightarrow \mathrm{Ag(s)}$,

L: $\mathrm{Ag^+(aq}, b_\mathrm{L}) + \mathrm{e^-} \rightarrow \mathrm{Ag(s)}$,

R – L: $\mathrm{Ag^+(aq}, b_\mathrm{R}) \rightarrow \mathrm{Ag^+(aq}, b_\mathrm{L})$.

(b) R: $\mathrm{2H^+(aq) + 2e^-} \rightarrow \mathrm{H_2(g}, p_\mathrm{R})$,

L: $\mathrm{2H^+(aq) + 2e^-} \rightarrow \mathrm{H_2(g}, p_\mathrm{L})$,

R – L: $\mathrm{H_2(g, p_L)} \rightarrow \mathrm{H_2(g}, p_\mathrm{R})$.

(c) R: $\mathrm{MnO_2(s) + 4H^+(aq) + 2e^-} \rightarrow \mathrm{Mn^{2+}(aq) + 2H_2O(l)}$,

L: $\mathrm{[Fe(CN)_6]^{3-} + e^-} \rightarrow \mathrm{[Fe(CN)_6]^{4-}}$,

R – L: $\mathrm{MnO_2(s) + 4H^+(aq) + 2[Fe(CN)_6]^{4-}(aq)} \rightarrow \mathrm{Mn^{2+}(aq)}$
$\mathrm{+\,2[Fe(CN)_6]^{3-}(aq) + 2H_2O(l)}$.

(d) R: $\mathrm{Br_2(l) + 2e^-} \rightarrow \mathrm{2Br^-(aq)}$,

L: $\mathrm{Cl_2(g) + 2e^-} \rightarrow \mathrm{2Cl^-(aq)}$,

R – L: $\mathrm{Br_2(l) + 2Cl^-(aq)} \rightarrow \mathrm{Cl_2(g) + 2Br^-(aq)}$.

(e) R: $\mathrm{Sn^{4+}(aq) + 2e^-} \rightarrow \mathrm{Sn^{2+}(aq)}$,

L: $\mathrm{2Fe^{3+}(aq) + 2e^-} \rightarrow \mathrm{2Fe^{2+}(aq)}$,

R – L: $\mathrm{Sn^{4+}(aq) + 2Fe^{2+}(aq)} \rightarrow \mathrm{Sn^{2+}(aq) + 2Fe^{3+}(aq)}$.

(f) R: $\mathrm{MnO_2(s) + 4H^+(aq) + 2e^-} \rightarrow \mathrm{Mn^{2+}(aq) + 2H_2O(l)}$,

L: $\mathrm{Fe^{2+}(aq) + 2e^-} \rightarrow \mathrm{Fe(s)}$,

R – L: $\mathrm{Fe(s) + MnO_2(s) + 4H^+(aq)} \rightarrow \mathrm{Fe^{2+}(aq) + Mn^{2+}(aq) + 2H_2O(l)}$.

9.27 (a) $E = E^\ominus - \dfrac{RT}{F} \ln \dfrac{b_\mathrm{L}}{b_\mathrm{R}}$.

(b) $E = E^\ominus - \dfrac{RT}{2F} \ln \dfrac{p_\mathrm{R}}{p_\mathrm{L}}$.

In the following Nernst equations involving ions in aqueous solution we have replaced activities with molar concentrations.

(c) $E = E^\ominus - \dfrac{RT}{2F} \ln \dfrac{[\mathrm{Mn^{2+}}][\mathrm{Fe(CN)_6^{3-}}]^2}{[\mathrm{H^+}]^4[\mathrm{Fe(CN)_6^{4-}}]^2}$.

(d) $E = E^\ominus - \dfrac{RT}{2F} \ln \dfrac{p_{\mathrm{Cl_2}}[\mathrm{Br^-}]^2}{[\mathrm{Cl^-}]^2}$.

(e) $E = E^{\ominus} - \dfrac{RT}{2F} \ln \dfrac{[Sn^{2+}][Fe^{3+}]^2}{[Sn^{4+}][Fe^{2+}]^2}$.

(f) $E = E^{\ominus} - \dfrac{RT}{2F} \ln \dfrac{[Fe^{2+}][Mn^{2+}]}{[H^+]^4}$.

9.28 (a) R: $PbSO_4(s)+2e^- \rightarrow Pb(s) + SO_4^{2-}(aq)$ −0.36 V.

L: $Fe^{2+}(aq) + 2e^- \rightarrow Fe(s)$ −0.44 V,

The cell is:

$Fe(s)|FeSO_4(aq)|PbSO_4(s)|Pb(s)$ $\boxed{\nu = 2}$.

(b) R: $Hg_2Cl_2(s)+2e^- \rightarrow 2Hg(l)+2Cl^-(aq)$ +0.27 V,

L: $2H^+(aq) + 2e^- \rightarrow H_2(g)$ 0.

The cell is:

$Pt|H_2(g)|HC1(aq)|Hg_2Cl_2(s)|Hg(l)$ $\boxed{\nu = 2}$.

(c) R: $O_2(g)+4H^+(aq)+4e^- \rightarrow 2H_2O(l)$ +1.23 V,

L: $4H^+(aq)+4e^- \rightarrow 2H_2(g)$ 0.

The cell is:

$Pt|H_2(g)|H^+(aq), H_2O|O_2(g)\ Pt|$ $\boxed{\nu = 4}$.

(d) R: $O_2(g)+2H^+(aq)+2e^- \rightarrow H_2O_2(aq)$ +0.695 V,

L: $2H^+(aq) + 2e^- \rightarrow H_2(g)$ 0.

The cell is

$Pt|H_2(g)|H^+(aq),H_2O_2(aq)|O_2(g)|Pt$ $\boxed{\nu = 2}$.

(e) R: $I_2(s) + 2e^- \rightarrow 2I^-(aq)$ +0.54 V,

L: $2H^+(aq)+2e^- \rightarrow H_2(g)$ 0.

The cell is

$Pt|H_2(g)|H^+(aq), I^-(aq)|I_2(s)|Pt$

or more simply

$Pt|H_2(g)|HI(aq)|I_2(s)|Pt$ $\boxed{\nu = 2}$.

(f) R: $Cu^+(aq) + e^- \rightarrow Cu(s)$ +0.52 V,

L: $Cu^{2+}(aq) + e^- \rightarrow Cu^+(aq)$ +0.15 V.

$Pt|CuCl_2(aq)||CuCl(aq)|Cu(s)$ $\boxed{\nu = 1}$.

9.29 (a) This reaction is analyzed in Exercise 9.19. $\boxed{\nu = 2}$.

A possible cell arrangement is

$Pt|CH_3CH_2OH(aq), CH_3CHO(aq), H^+(aq)||H^+(aq), NAD^+(aq), NADH(aq)|Pt$.

(b) R: $Mg^{2+}(aq) + 2e^- \rightarrow Mg(s)$,

L: $MgATP^{2-} + 2e^- \rightarrow Mg(s) + ATP^{4-}$,

$Mg(s)|ATP^{4-}(aq), MgATP^{2-}(aq)||Mg^{2+}(aq)|Mg(s)$ $\boxed{\nu = 2}$.

(c) R: $CH_3COCO_2^-(aq) + 2e^- + 2H^+(aq) \rightarrow CH_3CH(OH)CO_2^-(aq)$,

L: Cyt-c(ox, aq) + $e^- \rightarrow$ Cyt-(red, aq).

The overall reaction is obtained from R − 2 × L $\boxed{\nu = 2}$.

Pt|Cyt-c(red, aq), Cty-c (ox, aq)||$H^+(aq)$, $CH_3CH(OH)CO_2^-(aq)$, $CH_3COCO_2^-(aq)$|Pt.

9.30 See the solution to Exercise 9.26.

(a) $E^\ominus_{cell} = \boxed{0}$ (same electrode on right and left).

(b) $E^\ominus_{cell} = \boxed{0}$ (same electrode on right and left).

(c) $E^\ominus_{cell} = E^\ominus_R - E^\ominus_L = 1.23\,V - 0.36\,V = \boxed{+0.87\,V}$.

(d) $E^\ominus_{cell} = E^\ominus_R - E^\ominus_L = 1.09\,V - 1.36\,V = \boxed{-0.27\,V}$.

(e) $E^\ominus_{cell} = E^\ominus_R - E^\ominus_L = 0.15\,V - 0.77\,V = \boxed{-0.62\,V}$.

(f) $E^\ominus_{cell} = E^\ominus_R - E^\ominus_L = 1.23\,V - (-0.44\,V) = \boxed{+1.67\,V}$.

9.31 See the solution to Exercise 9.28.

(a) $E^\ominus_{cell} = -0.36\,V - (-0.44\,V) = \boxed{+0.08\,V}$.

(b) $E^\ominus_{cell} = +0.27\,V - 0\,V = \boxed{+0.27\,V}$.

(c) $E^\ominus_{cell} = 1.23\,V - 0\,V = \boxed{+1.23\,V}$.

(d) $E^\ominus_{cell} = +0.695\,V - 0\,V = \boxed{+0.695\,V}$.

(e) $E^\ominus_{cell} = +0.54\,V - 0\,V = \boxed{+0.54\,V}$.

(f) $E^\ominus_{cell} = +0.52\,V - 0.15\,V = \boxed{+0.37\,V}$.

9.32 See the solution to Exercise 9.29.

(a) $E^\ominus_{cell} = E^\ominus_{NAD^+, NADH} - E^\ominus_{CH_3CHO, CH_3CH_2OH}$.

(b) $E^\ominus_{cell} = E^\ominus_{Mg^{2+}, Mg} - E^\ominus_{Mg, MgATP^{4-}}$.

(c) $E^\ominus_{cell} = E^\ominus_{CH_3COCO_2^-, CH_3CH(OH)CO_2} - E^\ominus_{Cyt-c(ox), Cyt-c(red)}$.

9.33 $\frac{1}{2}N_2(g) + \frac{3}{2}H_2(g) \rightarrow NH_3(g)$.

$\Delta_f G^\ominus = -16.45\,kJ\,mol^{-1}$ [See Table D1.2].

The reaction is exergonic and spontaneous. It can be the basis of a fuel cell that yields a maximum non-expansion work equal to $\boxed{16.45\,\text{kJ}\,\text{mol}^{-1}}$ [9.12].

$$(100.\,\text{g}) \times \left(\frac{1\,\text{mol}}{17.03\,\text{g}}\right) \times \left(16.45\,\text{kJ}\,\text{mol}^{-1}\right) = \boxed{96.6\,\text{kJ}}.$$

9.34 $MnO_4^- + 8H^+ + 5e^- \rightarrow Mn^{2+} + 4H_2O,$ $\qquad E^\ominus = 1.51\,\text{V}.$

(a) Use the Nernst equation to determine the reduction potential for pH 7.00, keeping $a_{MnO_4^-}$ and $a_{Mn^{2+}} = 1.00$,

$$E = E^\ominus - \frac{25.7\,\text{mV}}{\nu}\ln Q.$$

$$\ln Q = \ln\frac{1}{a_{H^+}^8} = 8 \times 2.303 \times \text{pH}.$$

$$E = 1.51\,\text{V} - \frac{25.7\,\text{mV} \times 8 \times 2.303 \times 6.00}{5} = \boxed{+0.94\,\text{V}}.$$

(b) In general, $\boxed{E = 1.51 - 0.0947\,\text{pH}}$.

9.35 $Pt(s) \mid H_2(g) \mid HCl(aq) \mid AgCl(s) \mid Ag(s).$

The cell reaction is $\frac{1}{2}H_2(g) + AgCl(s) \rightarrow HCl(aq) + Ag(s).$

$$E = E^\ominus(\text{AgCl/Ag,Cl}^-) - \frac{RT}{\nu F}\ln Q \qquad [\text{Nernst equation [9.14]},\ \nu = 1]$$

$$= E^\ominus - \frac{RT}{F}\ln\left(\frac{a_{H^+}a_{Cl^-}}{a_{H_2}}\right) = E^\ominus - \frac{RT}{F}\ln\left(a_{H^+}a_{Cl^-}\right)$$

$$(\text{i.e., } a_{H_2} = p_{H_2}/p^\ominus = 1\,\text{bar}/p^\ominus = 1)$$

$$= E^\ominus - \frac{RT}{F}\ln\left\{(\gamma_\pm b/b^\ominus) \times (\gamma_\pm b/b^\ominus)\right\} \qquad [9.1a] \text{ and } [9.3a]; \quad b = b_{HCl}$$

$$= E^\ominus - \frac{RT}{F}\ln\left(b/b^\ominus\right)^2 - \frac{RT}{F}\ln\left(\gamma_\pm\right)^2$$

$$= E^\ominus - \frac{2RT}{F}\ln\left(b/b^\ominus\right) - \frac{2RT}{F}\ln\left(\gamma_\pm\right).$$

From the Debye–Hückel limiting law [9.4] we know that

$$\log\gamma_\pm = -A\,|z_+z_-|\,I^{1/2} = -A\,|z_+z_-|\left\{\frac{1}{2}\left(z_+^2 b_+ + z_-^2 b_-\right)/b^\ominus\right\}^{1/2} \qquad [9.5];$$

$$(z_+^2 + z_-^2)b = (1+1)b = 2b,$$

$$\frac{\ln\gamma_\pm}{\ln(10)} = -A\left(b/b^\ominus\right)^{1/2},$$

$$\ln\gamma_\pm = -A\ln(10)\,\left(b/b^\ominus\right)^{1/2}.$$

Substitution into the working equation for E yields

$$E = E^\ominus - \frac{2RT}{F}\ln(b/b^\ominus) + \frac{2RTA\ln(10)}{F}(b/b^\ominus)^{1/2},$$

$$\boxed{E + \frac{2RT}{F}\ln(b/b^\ominus) = E^\ominus + \frac{2RTA\ln(10)}{F}(b/b^\ominus)^{1/2}}.$$

This equation indicates that a plot of $E + \frac{2RT}{F}\ln(b/b^\ominus)$ against $(b/b^\ominus)^{1/2}$ should be linear with an intercept equal to $E^\ominus$.

9.36 (a) E decreases, $E = E^\ominus - \frac{RT}{F}\ln\left(\frac{[\mathrm{Ag}^+]_\mathrm{L}}{[\mathrm{Ag}^+]_\mathrm{R}}\right)$.

(b) E increases, $E = E^\ominus - \frac{RT}{2F}\ln\left(\frac{p_\mathrm{R}}{p_\mathrm{L}}\right)$.

(c) E increases, $E = E^\ominus - \frac{RT}{2F}\ln\left(\frac{[\mathrm{Mn}^{2+}][\mathrm{Fe(CN)}_6^{3-}]^2}{[\mathrm{H}^+]^4[\mathrm{Fe(CN)}_6^{4-}]^2}\right)$.

(d) E increases, $E = E^\ominus - \frac{RT}{2F}\ln\left(\frac{[\mathrm{Br}^-]^2 p_{\mathrm{Cl}_3}}{[\mathrm{Cl}^-]^2}\right)$.

(e) E increases, $E = E^\ominus - \frac{RT}{2F}\ln\left(\frac{[\mathrm{Sn}^{2+}][\mathrm{Fe}^{3+}]^2}{[\mathrm{Sn}^{4+}][\mathrm{Fe}^{2+}]^2}\right)$.

(f) E increases, $E = E^\ominus - \frac{RT}{2F}\ln\left(\frac{[\mathrm{Fe}^{2+}][\mathrm{Mn}^{2+}]}{[\mathrm{H}^+]^4}\right)$.

9.37 (a) The cell reaction is $\mathrm{PbSO_4(s) + Fe(s) \rightarrow Pb(s) + Fe^{2+}(aq) + SO_4^{2-}(aq)}$.

The Nernst equation is

$$E = E^\ominus - \frac{RT}{2F}\ln[\mathrm{Fe}^{2+}][\mathrm{SO}_4^{2-}].$$

Increasing $[\mathrm{Fe}^{2+}]$ $\boxed{\text{decreases } E}$.

(b) The cell reaction is $\mathrm{Hg_2Cl_2(s) + H_2(g) \rightarrow 2Hg(l) + 2H^+(aq) + 2Cl^-(aq)}$.

The Nernst equation is

$$E = E^\ominus - \frac{RT}{2F}\ln\left(\frac{[\mathrm{H}^+]^2[\mathrm{Cl}^-]^2}{[p_{\mathrm{H}_2}]^4}\right).$$

Increasing $[\mathrm{H}^+]$ $\boxed{\text{decreases } E}$.

(c) The cell reaction is $2H_2(g) + O_2(g) \rightarrow 2H_2O(l)$.

The Nernst equation is

$$E = E^{\ominus} - \frac{RT}{4F} \ln\left(\frac{1}{p_{H_2}^2 p_{O_2}}\right).$$

Increasing $p(O_2)$ $\boxed{\text{increases } E}$.

(d) The cell reaction is $H_2(g) + O_2(g) \rightarrow H_2O_2(aq)$.

The Nernst equation is

$$E = E^{\ominus} - \frac{RT}{2F} \ln \frac{[H_2O_2]}{[p_{H_2} p_{O_2}]}.$$

Increasing $p(H_2)$ $\boxed{\text{increases } E}$.

(e) $H_2(g) + I_2(s) \rightarrow 2H^+(aq) + 2I^-(aq)$.

$$E = E^{\ominus} - \frac{RT}{2F} \ln \frac{[H^+]^2[I^-]^2}{p_{H_2}}.$$

(i) Increasing $[H^+]$ $\boxed{\text{decreases } E}$.

(ii) Increasing both $[H^+]$ and $[I^-]$ $\boxed{\text{decreases } E}$.

(f) $2Cu^+(aq) \rightarrow Cu(s) + Cu^{2+}(aq)$.

$$E = E^{\ominus} - \frac{RT}{2F} \ln \frac{[Cu^{2+}]}{[Cu^+]^2}.$$

Adding HCl(aq) has $\boxed{\text{no effect}}$ on E.

9.38 (a) The Nernst equation is

$$E = E^{\ominus} - \frac{RT}{2F} \ln \frac{[CH_3CHO][NADH][H^+]}{[CH_3CH_2OH][NAD^+]}.$$

Increasing pH implies decreasing $[H^+]$; therefore the cell potential $\boxed{\text{decreases}}$.

(b) The Nernst equation is

$$E = E^{\ominus} - \frac{RT}{2F} \ln \frac{[MgATP^{2-}]}{[ATP^{4-}][Mg^{2+}]}.$$

Increasing $[Mg^{2+}]$ from $MgSO_4$ $\boxed{\text{increases}}$ the cell potential.

(c) The Nernst equation is

$$E = E^{\ominus} - \frac{RT}{2F} \ln \frac{[\text{Cyt-c(ox)}][CH_3CH(OH)CO_2^-]}{[\text{Cyt-c(red)}][CH_3COCO_2^-][H^+]^2}.$$

Increasing $[CH_3CH(OH)CO_2^-]$ $\boxed{\text{decreases}}$ the cell potential.

9.39 R: $2Tl^+(aq) + 2e^- \rightarrow 2Tl(s)$ -0.34 V,

L: $Hg^{2+}(aq) + 2e^- \rightarrow Hg(l)$ $+0.83$ V.

(a) Combining for the cell potential $E^\ominus = E_R^\ominus - E_L^\ominus = -1.20$ V.

(b) Overall: $2Tl^+(aq) + Hg(l) \rightarrow 2Tl(s) + Hg^{2+}(aq)$.

Replacing activities by molar concentrations, we have $Q = \dfrac{[Hg^{2+}]}{[Tl^+]^2}$.

$$E = E^\ominus - \frac{RT}{\nu F}\ln Q = E^\ominus - \frac{25.7\,\text{mV}}{\nu}\ln Q$$

$$= -1.20\,\text{V} - \frac{25.7\,\text{mV}}{2} \times \ln\frac{0.150}{(0.93)^2}$$

$$= -1.20\,\text{V} + 0.023\,\text{V} = \boxed{-1.18\,\text{V}}.$$

9.40 Each reaction has to be broken down into the two half–reactions from which it is formed. The standard potential for the cell is calculated from the standard electrode potentials in Table D2.1. The standard Gibbs energies of the reaction are calculated from $\Delta_r G^\ominus = -\nu FE^\ominus$.

(a) (1) $Ca^{2+}(aq) + 2e^- \rightarrow Ca(s)$ $E_1^\ominus = -2.87$ V.

(2) $2H_2O(l) + 2e^- \rightarrow H_2(g) + 2OH^-(aq)$ $E_2^\ominus = -0.83$ V.

The overall reaction is then obtained from (2) – (1) and the $E^\ominus$ value for the cell reaction is calculated in the same manner, that is,

$$E^\ominus = E_2^\ominus - E_1^\ominus = -0.83\,\text{V} - (-2.87\,\text{V}) = +2.04\,\text{V}.$$

$$\Delta_r G^\ominus = -\nu FE^\ominus = -2 \times 96.485\,\text{kC mol}^{-1} \times 2.04\,\text{V} = \boxed{-394\,\text{kJ mol}^{-1}}.$$

(b) Same as above.

The standard Gibbs energies of reaction for reactions (c), (d), (e), and (f) are obtained by a procedure similar to that described for part (a). The results are:

(c) $E^\ominus = -0.39$ V.

$$\Delta_r G^\ominus = -\nu FE^\ominus = -2 \times 96.485\,\text{kC mol}^{-1} \times (-0.39\,\text{V}) = \boxed{+75\,\text{kJ mol}^{-1}}.$$

(d) $E^\ominus = +1.51$ V.

$$\Delta_r G^\ominus = -\nu FE^\ominus = -2 \times 96.485\,\text{kC mol}^{-1} \times 1.51\,\text{V} = \boxed{-291\,\text{kJ mol}^{-1}}.$$

(e) Same as above.

(f) $E^\ominus = -2.58$ V.

Therefore $\Delta_r G = -\nu FE^\ominus$.

$$\Delta_r G^\ominus = -\nu FE^\ominus = -2 \times 96.485\,\text{kC mol}^{-1} \times (-2.58\,\text{V}) = \boxed{-498\,\text{kJ mol}^{-1}}.$$

9.41 (a) See the solution to Exercise 9.22. $\Delta_r G^{\oplus} = \boxed{-440\,\text{kJ mol}^{-1}}$.

(b) $\Delta_r G^{\oplus} = -\nu F E^{\oplus}, \qquad \nu = 2$ [see the solution to Exercise 9.22]

$$= -2 \times 96.485\,\text{kC mol}^{-1} \times (-0.154\,\text{V})$$

$$= \boxed{29.7\,\text{kJ mol}^{-1}}.$$

(c) $\Delta_r G^{\oplus} = -\nu F E^{\oplus} \qquad \nu = 4$

$$= -4 \times 96.485\,\text{kC mol}^{-1} \times (+0.82\,\text{V})$$

$$= \boxed{-316\,\text{kJ mol}^{-1}}.$$

9.42 (a) $E^{\ominus} = \dfrac{-\Delta_r G^{\ominus}}{\nu F} = \dfrac{+62.5\,\text{kJ mol}^{-1}}{2 \times 96.485\,\text{kC mol}^{-1}} = \boxed{+0.324\,\text{V}}$.

(b) $E^{\ominus} = E^{\ominus}(Fe^{3+}, Fe^{2+}) - E^{\ominus}(Ag_2CrO_4, Ag, CrO_4^{2-})$.

Therefore,

$$E^{\ominus}(Ag_2CrO_4, Ag, CrO_4^{2-}) = E^{\ominus}(Fe^{3+}, Fe^{2+}) - E^{\ominus}$$

$$= +0.77 - 0.324\,\text{V} = \boxed{+0.45\,\text{V}}.$$

9.43 The cell reaction is $Ag^+(aq) + Cl^-(aq) \rightarrow AgCl(s)$.

$$E = E^{\ominus} - \frac{0.0592\,\text{V}}{\nu \log Q}.$$

$$\nu = 1, \quad Q = \frac{1}{b_{Ag^+} b_{Cl^-}} = \frac{1}{0.010 \times 0.025} = 4.0 \times 10^{-3}.$$

$$E^{\ominus} = E^{\ominus}(Ag^+, Ag) - E^{\ominus}(AgCl, Ag, Cl^-) = 0.80\,\text{V} - 0.22\,\text{V}$$

$$= 0.58\,\text{V}.$$

$$E = 0.58\,\text{V} - \frac{25.7\,\text{mV}}{1} \ln(4.0 \times 10^{-3}) = \boxed{0.72\,\text{V}}.$$

9.44 (a) $2Ag(s) + Cu^{2+}(aq) \rightarrow 2Ag^+(aq) + Cu(s)$.

$$E = E^{\ominus}(Cu^{2+}, Cu) - E^{\ominus}(Ag^+, Ag) = +0.34\,\text{V} - 0.80\,\text{V} = \boxed{-0046\,\text{V}}.$$

$$\Delta_r G^{\ominus} = 2\Delta_f G^{\ominus}(Ag^+, aq) - \Delta_f G^{\ominus}(Cu^{2+}, aq)$$

$$= [(2 \times 77.1) - (64.49)]\,\text{kJ mol}^{-1} = \boxed{+89.7\,\text{kJ mol}^{-1}}.$$

Alternatively, $\Delta_r G^{\ominus} = -\nu F E^{\ominus}$

$$= -(2) \times (96.485\,\text{kC mol}^{-1}) \times (-0.46\,\text{V})$$

$$= +88.8\,\text{kJ mol}^{-1}.$$

$$\Delta_r H^{\ominus} = 2\Delta_f H^{\ominus}(Ag^+, aq) - \Delta_f H^{\ominus}(Cu^{2+}, aq)$$

$$= [(2 \times 105.58) - (64.77)]\,\text{kJ mol}^{-1} = \boxed{146\,\text{kJ mol}^{-1}}.$$

$$\Delta_r S^\ominus = \frac{\Delta_r H^\ominus - \Delta_r G^\ominus}{T} \quad [\Delta_r G = \Delta_r H - T\Delta_r S]$$

$$\Delta_r S^\ominus = \boxed{188\,\mathrm{J\,mol^{-1}\,K^{-1}}}$$

(b) Therefore $\Delta_r G^\ominus(308\,\mathrm{K}) \approx [146 - 308 \times 0.188]\,\mathrm{kJ\,mol^{-1}}$

$$\approx \boxed{+88\,\mathrm{kJ\,mol^{-1}}}.$$

9.45 See the solution to Exercise 9.20 for the overall cell reaction. Note that $\nu = 4$ for this reaction.

$$E^\ominus_{\mathrm{cell}} = E^\ominus_{\mathrm{R}} - E^\ominus_{\mathrm{L}} = +1.23\,\mathrm{V} - (-0.34\,\mathrm{V}) = 1.57\,\mathrm{V}.$$

$$\Delta_r G^\ominus = -\nu F E^\ominus$$

$$= -4 \times 96.485\,\mathrm{kC\,mol^{-1}} \times 1.57\,\mathrm{V}$$

$$= \boxed{-606\,\mathrm{kJ\,mol^{-1}}}.$$

9.46 The couple is

$$\mathrm{CH_3COCOOH + 2H^+ + 2e^- \rightarrow CH_3CH(OH)COOH}.$$

The Nernst equation for the couple is

$$E = E^\ominus - \frac{RT}{2F}\ln\frac{a_{\mathrm{HL}}}{a_{\mathrm{HP}}a^2_{\mathrm{H^+}}}.$$

We assume $a_{\mathrm{HL}} = a_{\mathrm{HP}} = 1$; then

$$E = E^\ominus + \frac{RT}{F}\ln a_{\mathrm{H^+}} = E^\ominus - \frac{2.303RT}{F} \times \mathrm{pH}$$

$$= E^\ominus - 0.0615\,\mathrm{V} \times \mathrm{pH} \quad [\text{for } T = 310\,\mathrm{K}].$$

$$E = -0.19\,\mathrm{V} = E^\ominus - 0.0615\,\mathrm{V} \times 7.$$

$$E^\ominus = -0.19\,\mathrm{V} + 0.43\,\mathrm{V} = \boxed{+0.24\,\mathrm{V}}.$$

9.47 (a) $\mathrm{CO_3^{2-}(aq) + 3H^+(aq) + 2e^- \rightarrow HCO_3^-(aq) + H_2(g)}$.

$$\Delta_r G^\ominus = \Delta_f G^\ominus(\mathrm{HCO_3^-}) - \Delta_f G^\ominus(\mathrm{CO_3^{2-}})$$

$$= -586.77\,\mathrm{kJ\,mol^{-1}} - (-527.81\,\mathrm{kJ\,mol^{-1}})$$

$$= -58.96\,\mathrm{kJ\,mol^{-1}}.$$

$$E^\ominus = -\frac{\Delta_r G^\ominus}{\nu F} = \frac{58.96\,\mathrm{kJ\,mol^{-1}}}{2 \times 96.485\,\mathrm{kC\,mol^{-1}}}$$

$$= \boxed{+0.3108\,\mathrm{V}}.$$

(b) Combine the above half-reaction with

$2H_2O(l) + 2e^- \rightarrow H_2(g) + 2OH^-(aq),$ $\qquad E^\ominus = -0.83\ V,$

which yields (a) – (b):

$CO_3^{2-}(aq) + H^+(aq) \rightarrow HCO_3^-(aq)$

or $CO_3^{2-}(aq) + H_2O(l) \rightarrow HCO_3^-(aq) + OH^-(aq).$

$$E^\ominus = E_R^\ominus - E_L^\ominus = +0.3108\,V - (-0.83\,V) = \boxed{+1.14\,V}.$$

(c) $E = E^\ominus - \dfrac{RT}{2F} \ln \dfrac{a_{HCO_3^-} a_{OH^-}}{a_{CO_3^{2-}}}.$

(d) The change from the standard value is the change that results from

$a_{OH^-} = 1$ changing to $a_{OH^-} = 10^{-7}.$

That is,

$$E = E^\ominus = -\frac{RT}{2F} \ln(10^{-7}) = \frac{-59.160\,mV}{2} \log(10^{-7})$$
$$= \boxed{207\,mV}.$$

(e) $\Delta_r G^\ominus = -RT \ln K = -58.96\,kJ\,mol^{-1}$ [part (a)].

$$\ln K = \frac{-58.96\,kJ\,mol^{-1}}{-2.479\,kJ\,mol^{-1}} = 23.78.$$

$$K = 2.1 \times 10^{10}.$$

$$K_a = \frac{1}{K} \quad \text{[the reverse of reaction in (a)]}$$
$$= 4.7 \times 10^{-11}.$$

$$pK_a = \boxed{10.3}.$$

9.48 $Cu_3(PO_4)_2(s) \rightleftharpoons 3Cu^{2+}(aq) + 2PO_4^{3-}(aq).$

(a) $K_s = \left[Cu^{2+}\right]^3 \left[PO_4^{3-}\right]^2.$

The molar solubility of the salt is S; $\left[Cu^{2+}\right] = 3S$, $\left[PO_4^{3-}\right] = 2S$

Therefore,

$$K_s = 1.3 \times 10^{-37} = (3S)^3(2S)^2 = 108S^5,$$
$$S = 1.6 \times 10^{-8}\,mol\,dm^{-3}.$$

(b) The cell reaction is

R: $Cu^{2+}(aq) + 2e^- \rightarrow Cu(s)$ $\qquad +0.34\,V,$

L: $2H^+(aq) + 2e^- \rightarrow H_2(g)$ $\qquad 0.$

Overall: $Cu^{2+}(aq) + H_2(g) \rightarrow Cu(s) + 2H^+(aq)$ $\qquad +0.34\,V.$

Using the Nernst equation

$$E = E^\ominus - \frac{RT}{\nu F}\ln Q$$

$$= 0.34\,\text{V} - \frac{25.693\times 10^{-3}\,\text{V}}{2}\ln\frac{a^2_{\text{H}^+}}{a_{\text{Cu}^{2+}}}.$$

Because $a_{\text{Cu}^{2+}} \approx [\text{Cu}^{2+}] = 3S$, the following substitution can be made:

$$E = 0.34\,\text{V} - \frac{25.693\times 10^{-3}\,\text{V}}{2}\ln\frac{1}{3\times(1.6\times 10^{-8})}$$

$$= 0.34\,\text{V} - 0.22\,\text{V} = \boxed{0.12}.$$

9.49 In each case we use

$$\ln K = \frac{\nu FE^\ominus}{RT} \quad [9.16].$$

(a) $\text{Sn(s)} + \text{Sn}^{4+}\text{(aq)} \rightleftharpoons 2\text{Sn}^{2+}\text{(aq)}$.

R: $\text{Sn}^{4+}\text{(aq)} + 2\text{e}^- \rightarrow \text{Sn}^{2+}\text{(aq)}$ $+0.15\,\text{V}$,

L: $\text{Sn}^{2+}\text{(aq)} + 2\text{e}^- \rightarrow \text{Sn(s)}$ $-0.14\,\text{V}$,

$E^\ominus = +0.29\,\text{V}$.

$$\ln K = \frac{\nu FE^\ominus}{RT} = \frac{2\times 0.29\,\text{V}}{25.693\times 10^{-3}\,\text{V}} = +22.6, \qquad K = \boxed{6.5\times 10^9}.$$

(b) $\text{Sn(s)} + 2\text{AgBr(s)} \rightleftharpoons \text{SnBr}_2\text{(aq)} + 2\text{Ag(s)}$.

R: $\text{AgBr(s)} + \text{e}^- \rightarrow \text{Ag(s)} + \text{Br}^-\text{(aq)}$ $+0.07\,\text{V}$,

L: $\text{Sn}^{2+}\text{(aq)} + 2\text{e}^- \rightarrow \text{Sn(s)}$ $-0.14\,\text{V}$,

$E^\ominus = +0.21\,\text{V}$.

$$\ln K = \frac{2\times 0.21\,\text{V}}{25.693\times 10^{-3}\,\text{V}} = +16.3, \qquad K = \boxed{1.2\times 10^7}.$$

(c) $\text{Fe(s)} + \text{Hg(NO}_3)_2\text{(aq)} \rightleftharpoons \text{Hg(l)} + \text{Fe(NO}_3)_2\text{(aq)}$.

R: $\text{Hg}^{2+}\text{(aq)} + 2\text{e}^- \rightarrow \text{Hg(l)}$ $+1.62\,\text{V}$,

L: $\text{Fe}^{2+}\text{(aq)} + 2\text{e}^- \rightarrow \text{Fe(s)}$ $-0.44\,\text{V}$,

$E^\ominus = +2.06\,\text{V}$.

$$\ln K = \frac{2\times 2.06\,\text{V}}{25.693\times 10^{-3}\,\text{V}} = 160, \qquad \boxed{K = 4\times 10^{69}}.$$

(d) $Cd(s) + CuSO_4(aq) \rightleftharpoons Cu(s) + CdSO_4(aq)$.

$$R: Cu^{2+}(aq) + 2e^- \rightarrow Cu(s) \qquad +0.34\,V,$$

$$L: Cd^{2+}(aq) + 2e^- \rightarrow Cd(s) \qquad -0.40\,V,$$

$$E^\ominus = +0.74\,V.$$

$$\ln K = 57.6, \quad \boxed{K = 1.0 \times 10^{25}}.$$

(e) $Cu^{2+}(aq) + Cu(s) \rightleftharpoons 2Cu^+(aq)$.

$$R: Cu^{2+}(aq) + e^- \rightarrow Cu^+(aq) \qquad +0.16\,V,$$

$$L: Cu^+(aq) + e^- \rightarrow Cu(s) \qquad +0.52\,V,$$

$$E^\ominus = -0.36\,V.$$

$$\ln K = \frac{-0.36\,V}{25.693 \times 10^{-3}\,V} = -14.0, \qquad \boxed{K = 8.2 \times 10^{-7}}.$$

(f) $3Au^+(aq) \rightleftharpoons 2Au(s) + Au^{3+}(aq)$.

$$R: Au^+(aq) + e^- \rightarrow Au(s) \qquad 1.69\,V,$$

$$L: Au^{3+}(aq) + 3e^- \rightarrow Au(s) \qquad 1.40\,V,$$

$$E^\ominus = 0.29\,V.$$

$$\ln K = \frac{0.29\,V}{25.693 \times 10^{-3}\,V} = 11.3, \qquad \boxed{K = 8.0 \times 10^4}.$$

9.50 Assume that all activity coefficients are 1.

(1) $AgCl(s) \rightleftharpoons Ag^+(aq) + Cl^-(aq)$.

Since all stoichiometric coefficients are 1, $S(AgCl) = b(Ag^+) = b(Cl^-)$. Hence,

$$K_S = \frac{b(Ag^+) \times b(Cl^-)}{(b^\ominus)^2} = \frac{S^2}{(b^\ominus)^2} = (1.34 \times 10^{-5})^2 = \boxed{1.80 \times 10^{-10}}.$$

(2) $BaSO_4(s) \rightleftharpoons Ba^{2+}(aq) + SO_4^{2-}(aq)$.

$$S(BaSO_4) = b\left(Ba^{2+}\right) = b\left(SO_4^{2-}\right).$$

$$\text{As above, } K_S = \frac{S^2}{(b^\ominus)} = (9.51 \times 10^{-4})^2 = \boxed{9.04 \times 10^{-7}}.$$

9.51 The half-reaction is

$$Cr_2O_7^{2-}(aq) + 14H^+(aq) + 6e^- \rightarrow 2Cr^{3+}(aq) + 7H_2O(l).$$

The reaction quotient is

$$Q = \frac{a^2_{\mathrm{Cr}^3}}{a_{\mathrm{Cr_2O_7^{2-}}} a^{14}_{\mathrm{H}^+}} \qquad \nu = 6.$$

Hence,

$$E = \boxed{E^{\ominus} - \frac{RT}{6F} \ln\left(\frac{a^2_{\mathrm{Cr}^{3+}}}{a_{\mathrm{Cr_2O_7^{2-}}} a^{14}_{\mathrm{H}^+}}\right)}.$$

9.52 R: $2\mathrm{AgCl(s)} + 2\mathrm{e}^- \rightarrow 2\mathrm{Ag(s)} + 2\mathrm{Cl}^-(\mathrm{aq})$ +0.22 V,

L: $2\mathrm{H}^+(\mathrm{aq}) + 2\,\mathrm{e}^- \rightarrow \mathrm{H_2(g)}$ 0,

Overall: $2\,\mathrm{AgCl(s)} + \mathrm{H_2(g)} \rightarrow 2\,\mathrm{Ag(s)} + 2\,\mathrm{Cl}^-(\mathrm{aq}) + 2\,\mathrm{H}^+(\mathrm{aq})$.

$$Q = a^2_{\mathrm{H}^+} a^2_{\mathrm{Cl}^-} \qquad \nu = 2$$

$$= a^4_{\mathrm{H}^+} \qquad a_{\mathrm{H}^+} = a_{\mathrm{Cl}^-}.$$

Therefore, from the Nernst equation,

$$E = E^{\ominus} - \frac{RT}{2F} \ln a^4_{\mathrm{H}^+} = E^{\ominus} - \frac{2RT}{F} \ln a_{\mathrm{H}^+} = 0.312\,\mathrm{V}$$

$$= E^{\ominus} + 2.303\frac{2RT}{F}\,\mathrm{pH}.$$

Rearranging and substituting in:

$$\mathrm{pH} = \frac{F}{2 \times 2.303RT} \times (E - E^{\ominus}) = \frac{0.312\,\mathrm{V} - 0.22\,\mathrm{V}}{0.1183\,\mathrm{V}}$$

$$= \boxed{0.78}.$$

9.53 R: $\mathrm{AgBr(s)} + \mathrm{e}^- \rightarrow \mathrm{Ag(s)} + \mathrm{Br}^-(\mathrm{aq})$, $E^{\ominus} = +0.0713\,\mathrm{V}$,

L: $\mathrm{Ag}^+(\mathrm{aq}) + \mathrm{e}^- \rightarrow \mathrm{Ag(s)}$, $E^{\ominus} = +0.80\,\mathrm{V}$,

Overall: $\mathrm{AgBr(s)} \rightarrow \mathrm{Br}^-(\mathrm{aq}) + \mathrm{Ag}^+(\mathrm{aq})$, $E^{\ominus} = -0.73\,\mathrm{V}$.

Therefore, because the cell reaction is the solubility equilibrium, for a saturated solution there is no further tendency to dissolve and so $\boxed{E = 0}$.

We may check this common-sense result by a calculation:

$$E = E^{\ominus} - 0.025693\,\mathrm{V} \times \ln\,([\mathrm{Ag}^+][\mathrm{Br}^-]),$$

$$= -0.73\,\mathrm{V} - 0.025693\,\mathrm{V} \times \ln\,(2.6 \times 10^{-6})^2,$$

$$= (-0.73 + 0.66)\,\mathrm{V} = 0.07\,\mathrm{V}.$$

The discrepancy is probably a result of inaccuracy in the value of the molar solubility given.

9.54 R: $Ag^+(aq) + e^- \rightarrow Ag(s)$, $E^\ominus = +0.80\,V$,

L: $AgI(s) + e^- \rightarrow Ag(s) + I^-(aq)$, $E^\ominus = \boxed{-0.15\,V}$,

Overall: $Ag^+(aq) + I^-(aq) \rightarrow AgI(s)$, $\nu = 1$, $E^\ominus = +0.95\,V \approx 0.9509\,V$

(a) The solubility is obtained from $[Ag^+] = [I^-]$ and $S = [Ag^+]$.

$$K_s = [Ag^+]^2, \quad \text{so that}$$

$$S = (K_s)^{1/2}\,\text{mol dm}^{-3} = (8.45 \times 10^{-17})^{1/2}\,\text{mol dm}^{-3}$$

$$= \boxed{9.19 \times 10^{-9}\,\text{mol dm}^{-3}}.$$

(b) $\ln K = \dfrac{0.9509\,V}{25.693\,mV} = 37.010, \quad K = 1.184 \times 10^{16}.$

However, $K_s = K^{-1}$ because the solubility equilibrium is written as the reverse of the cell reaction. Therefore, $K_s = \boxed{8.5 \times 10^{17}}$.

9.55 We require two half-cell reactions, which, upon subtracting one (left) from the other (right), yields the given overall reaction. The half-reaction at the right electrode corresponds to reduction, that at the left electrode to oxidation; all half-reactions are listed in Table D2.1 as reduction reactions.

	$E^\ominus$
R: $Hg_2SO_4(s) + 2e^- \rightarrow 2Hg(l) + SO_4^{2-}(aq)$	$+0.62\,V$,
L: $PbSO_4(s) + 2e^- \rightarrow Pb(s) + SO_4^{2-}(aq)$	$-0.36\,V$,
R – L: $Pb(s) + Hg_2SO_4(s) \rightarrow PbSO_4(s) + 2Hg(l)$	$+0.98\,V$.

Hence, a suitable cell would be

$Pb(s)\,|\,PbSO_4(s)\,|\,H_2SO_4(aq)\,|\,Hg_2SO_4(s)\,|\,Hg(l)$

or, alternatively,

$Pb(s)\,|\,PbSO_4(s)\,|\,H_2SO_4(aq)\,||\,H_2SO_4(aq)\,|\,Hg_2SO_4(s)\,|\,Hg(l)$.

For the cell in which the only sources of electrolyte are the slightly soluble salts, $PbSO_4$ and Hg_2SO_4, the cell would be

$Pb(s)\,|\,PbSO_4(s)\,|\,H_2SO_4(aq)\,||\,H_2SO_4(aq)\,|\,Hg_2SO_4(s)\,|\,Hg(l)$.

The potential of this cell is given by the Nernst equation.

$$E = E^\ominus - \frac{RT}{\nu F}\,\text{In}Q, \quad \nu = 2.$$

$$Q = \frac{a_{Pb^{2+}} a_{SO_4^{2-}}}{a_{Hg_2^{2+}} a_{SO_4^{2-}}} = \frac{K_s(PbSO_4)}{K_s(Hg_2SO_4)}.$$

$$E = (0.98\ V) - \frac{RT}{2F} \ln \frac{K_s(PbSO_4)}{K_s(Hg_2SO_4)}.$$

$$= (0.98\ V) - \left(\frac{25.693 \times 10^{-3}\ V}{2}\right) \times \ln\left(\frac{1.6 \times 10^{-8}}{6.6 \times 10^{-7}}\right) \quad [\textit{CRC handbook}]$$

$$= (0.98\ V) + (0.05\ V) = \boxed{+1.03\ V}.$$

Chapter 10

The rates of reactions

Answers to discussion questions

10.1 The determination of a rate law is simplified by the isolation method in which the concentrations of all the reactants except one are in large excess. If B is in large excess, for example, then to a good approximation its concentration is constant throughout the reaction. Although the true rate law might be rate $= k[A][B]$, we can approximate [B] by $[B]_0$ and write

$$\text{Rate} = k'[A], \text{ where } k' = k[B]_0,$$

which has the form of a first-order rate law. Because the true rate law has been forced into first-order form by assuming that the concentration of B is constant, it is called a pseudo-first-order rate law. [A] has been isolated. The dependence of the rate on the concentration of each of the reactants may be found by isolating them in turn (by having all the other substances present in large excess), and so constructing a picture of the overall rate law.

In the method of initial rates, which is often used in conjunction with the isolation method, the rate is measured at the beginning of the reaction for several different initial concentrations of reactants. We shall suppose that the rate law for a reaction with A isolated is rate $= k[A]^a$; then its initial rate, rate_0 is given by the initial values of the concentration of A, and we write $\text{rate}_0 = k[A]_0^a$. Taking logarithms gives:

$$\log \text{rate}_0 = \log k + a \log[A]_0 \quad [10.11].$$

For a series of initial concentrations, a plot of the logarithms of the initial rates against the logarithms of the initial concentrations of A should be a straight lime with slope a.

The method of initial rates might not reveal the full rate law, for the products may participate in the reaction and affect the rate. For example, products participate in the synthesis of HBr, where the full rate law depends on the concentration of HBr. To avoid this difficulty, the rate law should be fitted to the data throughout the reaction. The fitting may be done, in simple cases at least, by using a proposed rate law to predict the concentration of any component at any time, and comparing it with the data.

Because rate laws are differential equations, we must integrate them if we want to find the concentrations as a function of time. Even the most complex rate laws may be integrated numerically. However, in a number of simple cases analytical solutions are easily obtained, and prove to be very useful. These are discussed in Example 10.2 and Figure 10.13 of the main text where it is shown that first-order rate data exhibit a linear plot when the logarithm of concentration is plotted against time, but a second-order rate law exhibits linearity with a plot of inverse concentration against time. The slope of the linear plot equals the rate constant in the latter case and the negative of the rate constant in the former case.

10.2 Consider a rate law of the form rate $= k\,[\mathrm{A}]^m\,[\mathrm{B}]^n$ where the concentration orders m and n equal either zero or a positive integer. If the sum $m + n$ equals zero, the rate is zeroth-order and the rate is independent of species concentration. If the sum equals either 1 or 2, the rate is first-order or second-order, respectively. In the case for which $m = 1$ and $n \neq 0$, the reaction order will appear to be 1 if the concentration of B is a large excess and [B] remains basically unchanged during the course of reaction. This is the pseudofirst-order reaction rate for which rate $= k\,[\mathrm{A}]\,[\mathrm{B}]^n = (k[\mathrm{B}]^n)[\mathrm{A}] = k'\,[\mathrm{A}]$ where $k' = k[\mathrm{B}]^n$ = pseudo-first-order rate constant.

10.3 The parameter A, which corresponds to the intercept of the line at $1/T = 0$ (at infinite temperature), is called the pre-exponential factor or the frequency factor. The parameter E_a, which is obtained from the slope of the line $(-E_\mathrm{a}/R)$, is called the activation energy. Collectively, the two quantities are called the Arrhenius parameters.

$$k = A\mathrm{e}^{-E_\mathrm{a}/RT} \;[10.19] \quad \text{or} \quad \ln k = \ln A - \frac{E_\mathrm{a}}{RT} \;[10.18].$$

The temperature dependence of some reactions is not Arrhenius-like, in the sense that a straight line is not obtained when ln k is plotted against $1/T$. However, it is still possible to define a general activation energy as

$$E_\mathrm{a} = RT^2\left(\frac{\mathrm{d}\ln k}{\mathrm{d}T}\right).$$

This definition reduces to the earlier one (as the slope of a straight line) for a temperature-independent activation energy. However, this latter definition is more general, because it allows E_a to be obtained from the slope (at the temperature of interest) of a plot of ln k against $1/T$ even if the Arrhenius plot is not a straight line. Non-Arrhenius behavior is sometimes a sign that quantum mechanical tunneling is playing a significant role in the reaction.

10.4 The Eyring equation [10.23] results from activated complex theory which is an attempt to account for the rate constants of bimolecular reactions of the form $\mathrm{A} + \mathrm{B} \rightleftharpoons \mathrm{C}^{\ddagger} \rightarrow \mathrm{P}$ in terms of the formation of an activated complex. In the formulation of the theory, it is assumed that the activated complex and the reactants are in equilibrium, and the concentration of activated complex is calculated in terms of an equilibrium constant, which in turn is calculated from the partition functions of the reactants and a postulated form of the activated complex. It is further supposed that one normal mode of the activated complex, the one corresponding to displacement along the reaction coordinate, has a very low force constant and displacement along this normal mode leads to products provided that the complex enters a certain configuration of its atoms, which is known as the transition state.

Solutions to exercises

10.5 $A = \log\left(\dfrac{I_0}{I}\right) = -\log\left(\dfrac{I}{I_0}\right) = -\log(0.398) = 0.400 \quad [10.1a] \text{ and } [10.2].$

$$[\text{cyt P450}] = \frac{A}{\varepsilon l} \quad [10.3]$$

$$= \frac{0.400}{\left(295\,\text{dm}^3\,\text{mol}^{-1}\,\text{cm}^{-1}\right)(0.65\,\text{cm})}$$

$$= 2.08 \times 10^{-3}\,\text{mol dm}^{-3} = \boxed{2.08\,\text{mmol dm}^{-3}}.$$

10.6 At the wavelength $\lambda_1 : \varepsilon_{\lambda_1,\text{A}} = \varepsilon_{\lambda_1,\text{B}} = \varepsilon_{\lambda_1}$ and $A_{\lambda_1} = \varepsilon_{\lambda_1,\text{A}}[\text{A}]\,l + \varepsilon_{\lambda_1,\text{B}}[\text{B}]\,l = ([\text{A}] + [\text{B}])\,\varepsilon_{\lambda_1} l.$

At the wavelength $\lambda_2 : A_{\lambda_2} = \varepsilon_{\lambda_2,\text{A}}[\text{A}]\,l + \varepsilon_{\lambda_2,\text{B}}[\text{B}]\,l.$

From the equation for wavelength λ_1, $[\text{B}] = \dfrac{A_{\lambda_1}}{\varepsilon_{\lambda_1} l} - [\text{A}]$. Substitution into the equation for wavelength λ_2 yields:

$$A_{\lambda_2} = \varepsilon_{\lambda_2,\text{A}}[\text{A}]\,l + \varepsilon_{\lambda_2,\text{B}}\left(\frac{A_{\lambda_1}}{\varepsilon_{\lambda_1} l} - [\text{A}]\right) l = \left(\varepsilon_{\lambda_2,\text{A}} - \varepsilon_{\lambda_2,\text{B}}\right)[\text{A}]\,l + \frac{\varepsilon_{\lambda_2,\text{B}} A_{\lambda_1}}{\varepsilon_{\lambda_1}},$$

$$\boxed{[\text{A}] = \frac{\varepsilon_{\lambda_1} A_{\lambda_2} - \varepsilon_{\lambda_2,\text{B}} A_{\lambda_1}}{\varepsilon_{\lambda_1}\left(\varepsilon_{\lambda_2,\text{A}} - \varepsilon_{\lambda_2,\text{B}}\right) l}}.$$

Substitution into the equation for wavelength λ_1 yields

$$[\text{B}] = \frac{A_{\lambda_1}}{\varepsilon_{\lambda_1} l} - \frac{\varepsilon_{\lambda_1} A_{\lambda_2} - \varepsilon_{\lambda_2,\text{B}} A_{\lambda_1}}{\varepsilon_{\lambda_1}\left(\varepsilon_{\lambda_2,\text{A}} - \varepsilon_{\lambda_2,\text{B}}\right) l} = \frac{\left(\varepsilon_{\lambda_2,\text{A}} - \varepsilon_{\lambda_2,\text{B}}\right) A_{\lambda_1} - \varepsilon_{\lambda_1} A_{\lambda_2} + \varepsilon_{\lambda_2,\text{B}} A_{\lambda_1}}{\varepsilon_{\lambda_1}\left(\varepsilon_{\lambda_2,\text{A}} - \varepsilon_{\lambda_2,\text{B}}\right) l},$$

$$\boxed{[\text{B}] = \frac{\varepsilon_{\lambda_2,\text{A}} A_{\lambda_1} - \varepsilon_{\lambda_1} A_{\lambda_2}}{\varepsilon_{\lambda_1}\left(\varepsilon_{\lambda_2,\text{A}} - \varepsilon_{\lambda_2,\text{B}}\right) l}}.$$

NOTE A wavelength at which two components have the same molar absorption coefficient is known as an **isosbestic point or wavelength**.

10.7 $$[\text{try}] = \frac{\varepsilon_{\text{tyr}2} A_1 - \varepsilon_{\text{tyr}1} A_2}{\left(\varepsilon_{\text{try}1}\varepsilon_{\text{tyr}2} - \varepsilon_{\text{try}2}\varepsilon_{\text{tyr}1}\right) l} \quad [10.4]$$

$$= \frac{\left(1.50 \times 10^3\right)(0.660) - \left(1.12 \times 10^4\right)(0.221)}{\left\{\left(2.00 \times 10^3\right)\left(1.50 \times 10^3\right) - \left(5.40 \times 10^3\right)\left(1.12 \times 10^4\right)\right\}\left(\dfrac{\text{dm}^3}{\text{mol cm}}\right)(1.00\,\text{cm})},$$

$$\boxed{[\text{try}] = 2.58 \times 10^{-5}\,\text{mol dm}^{-3}}.$$

$$[\text{tyr}] = \frac{\varepsilon_{\text{try1}}A_2 - \varepsilon_{\text{try2}}A_1}{(\varepsilon_{\text{try1}}\varepsilon_{\text{tyr2}} - \varepsilon_{\text{try2}}\varepsilon_{\text{tyr1}})\,l} \quad [10.4]$$

$$= \frac{(2.00 \times 10^3)(0.221) - (5.40 \times 10^3)(0.660)}{\{(2.00 \times 10^3)(1.50 \times 10^3) - (5.40 \times 10^3)(1.12 \times 10^4)\}\left(\frac{\text{dm}^3}{\text{mol cm}}\right)(1.00\,\text{cm})},$$

$$\boxed{[\text{tyr}] = 5.43 \times 10^{-5}\,\text{mol dm}^{-3}}.$$

10.8 The isosbestic wavelength $\lambda_1 = 294$ nm simplifies equations[10.4] (see Exercise 10.6) to

$$[\text{tyr}] = \frac{\varepsilon_{\lambda_1}A_{\lambda_2} - \varepsilon_{\lambda_2,\text{tyr}}A_{\lambda_1}}{\varepsilon_{\lambda_1}(\varepsilon_{\lambda_2,\text{try}} - \varepsilon_{\lambda_2,\text{tyr}})\,l} \quad \text{and} \quad [\text{tyr}] = \frac{\varepsilon_{\lambda_2,\text{try}}A_{\lambda_1} - \varepsilon_{\lambda_1}A_{\lambda_2}}{\varepsilon_{\lambda_1}(\varepsilon_{\lambda_2,\text{try}} - \varepsilon_{\lambda_2,\text{tyr}})\,l}.$$

$$[\text{try}] = \frac{(2.38 \times 10^3)(0.676) - (1.58 \times 10^3)(0.468)}{(2.38 \times 10^3)(5.23 \times 10^3 - 1.58 \times 10^3)\left(\frac{\text{dm}^3}{\text{mol cm}}\right)(1.00\,\text{cm})},$$

$$\boxed{[\text{try}] = 1.00 \times 10^{-4}\,\text{mol dm}^{-3}}.$$

$$[\text{tyr}] = \frac{(5.23 \times 10^3)(0.468) - (2.38 \times 10^3)(0.676)}{(2.38 \times 10^3)(5.23 \times 10^3 - 1.58 \times 10^3)\left(\frac{\text{dm}^3}{\text{mol cm}}\right)(1.00\,\text{cm})},$$

$$\boxed{[\text{tyr}] = 9.66 \times 10^{-5}\,\text{mol dm}^{-3}}.$$

10.9 Because the rate of formation of C is known, the reaction stoichiometry can be used to determine the rates of consumption and formation of the other participants in the reaction.

$$\text{rate of consumption of A} = \frac{2}{3} \times \text{rate of formation of C} = \frac{2}{3} \times (2.2\,\text{mol dm}^{-3}\,\text{s}^{-1})$$

$$= \boxed{1.5\,\text{mol dm}^{-3}\,\text{s}^{-1}}.$$

$$\text{rate of consumption of B} = \frac{1}{3} \times \text{rate of formation of C}$$

$$= \boxed{0.73\,\text{mol dm}^{-3}\,\text{s}^{-1}}.$$

$$\text{rate of consumption of D} = \frac{2}{3} \times \text{rate of formation of C}$$

$$= \boxed{1.5\,\text{mol dm}^{-3}\,\text{s}^{-1}}.$$

10.10 The rate has units of concentration per unit time (mol dm^{-3} s^{-1}); hence

$$\text{mol dm}^{-3}\text{s}^{-1} = [\text{units of } k] \times (\text{mol dm}^{-3})^3.$$

$$\text{Therefore, [units of } k] = \frac{\text{mol dm}^{-3}\text{s}^{-1}}{(\text{mol dm}^{-3})^3} = \boxed{\text{mol}^{-2}\,\text{dm}^{-6}\,\text{s}^{-1}}.$$

10.11 (a) Concentrations in (molecules m^{-3}) = $N\,\text{m}^{-3}$

(i) Second-order:

$$\text{rate} = N\,\text{m}^{-3}\,\text{s}^{-1} = [k] \times (N\,\text{m}^{-3})^2$$

where $[k]$ = units of k and N = number of molecules.

$$\text{Then, } [k] = \frac{N\,\text{m}^{-3}\,\text{s}^{-1}}{(N\,\text{m}^{-3})^2} = N^{-1}\,\text{m}^3\,\text{s}^{-1}$$

Because N is unitless, $[k] = \boxed{\text{m}^3\,\text{s}^{-1}}$,

though loosely we may say molecules^{-1} m^3 s^{-1}.

(ii) Third-order $[k] = \dfrac{N\text{m}^{-3}\,\text{s}^{-1}}{(N\,\text{m}^{-3})^3} = N^{-2}\text{m}^6\,\text{s}^{-1}$

or $[k] = \boxed{\text{m}^6\,\text{s}^{-1}}$.

(b) Pressures in kilopascals

(i) Second-order:

$$\text{rate} = \text{kPa s}^{-1} = [k] \times (\text{kPa})^2.$$

$$[k] = \frac{\text{kPa s}^{-1}}{(\text{kPa})^2} = \boxed{\text{kPa}^{-1}\,\text{s}^{-1}}.$$

(ii) Third-order:

$$[k] = \frac{\text{kPa s}^{-1}}{(\text{kPa})^3} = \boxed{\text{kPa}^{-2}\,\text{s}^{-1}}.$$

10.12 (a) We fit the transformed data to equation (10.11)

$$\log \text{rate}_0 = \log k' + a \log[\text{A}]_0 \qquad \text{A} = \text{C}_6\text{H}_{12}\text{O}_6.$$

log(rate$_0$/mol dm^{-3} s^{-1})	0.699	0.881	1.19	1.30
log([A])$_0$/mmol dm^{-3})	0.00	0.188	0.494	0.604

The plot is shown in Figure 10.1. The slope $a = 0.998 \approx 1$.

Therefore order = $\boxed{1}$.

(b) From the fit, $\log k' = 3.689$, $k' = \boxed{4.89 \times 10^3\ \text{s}^{-1}}$.

COMMENT In the plot, the concentrations used are mmol dm^{-3}, rather than mol dm^{-3}. This choice does not affect the slope, but does affect the value of k'. The k' above is obtained after converting concentrations to mol dm^{-3}. From the data of this exercise, we can only determine k', not the true rate constant k, which probably corresponds to an overall second-order rate, and therefore would not have the units s^{-1}.

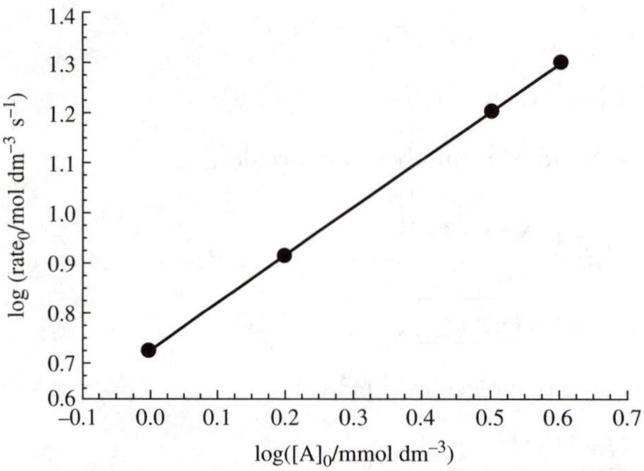

Fig 10.1

10.13 (a) We convert the concentration of the complex to mol dm^{-3} and fit the data first to

(1) $\log \text{rate}_0 = \log k' + a \log [\text{complex}]$

where $k' = k[\text{Y}]_0^b$, then to

(2) $\log k' = \log k + b \log[\text{Y}]_0$

log [complex]	−2.096	−2.035	−1.917
$\log \text{rate}_0$ (a)	2.097	2.158	2.279
$\log \text{rate}_0$ (b)	2.806	2.863	2.982
$\log[\text{Y}]_0$:	(a)−2.569,	(b)−2.215.	

From the fit of the data to equation (1) we obtain

$a = 1.018 \approx 1$ for data (a), $\log k' = 4.230$.

$a = 0.987 \approx 1$ for data (b), $\log k' = 4.873$.

The order of the reaction with respect to [complex] is seen to be $\boxed{1}$.

From the fit of the data to equation (2) we obtain $b = 1.816$.

This value is closest to 2, so we suggest that the order with respect to [Y] is $\boxed{2}$. But this is not a very satisfying result.

(b) The value of log k obtained from the fit is

$\log k = 8.896$

and then $k = 7.87 \times 10^8\ \text{dm}^6\ \text{mol}^{-2}\ \text{s}^{-1}$. Solving for k from equation (2) after setting $b = 2$,

$$\log k = \log k' - 2\log[\text{Y}]_0$$

$$= 4.230 - 2(-2.569)\ \text{[data from (a)]}$$

$$= 9.368.$$

$$k = \boxed{2.33 \times 10^9\ \text{dm}^6\ \text{mol}^{-2}\ \text{s}^{-1}}.$$

Using data from part (b), the result is $k = 2.09 \times 10^9\ \text{dm}^6\ \text{mol}^{-2}\ \text{s}^{-1}$. This can be considered satisfactory agreement in view of the fact that b was found to be 1.816 rather than 2.

10.14 Because partial pressures and concentrations are proportional we may write

$$\text{rate} = k[\text{N}_2\text{O}_5] = kp_{\text{N}_2\text{O}_5}.$$

$$k\,t_{1/2} = \ln 2, \qquad t_{1/2} = \frac{\ln 2}{(3.38 \times 10^{-5}\text{s}^{-1})}.$$

$$t_{1/2} = \boxed{2.05 \times 10^4\ \text{s}}.$$

(a) $p_{\text{N}_2\text{O}_5} = p^0_{\text{N}_2\text{O}_5}\text{e}^{-kt}$.

After 10 s very little N_2O_5 will have decomposed (note the half-life).

$$p_{\text{N}_2\text{O}_5} = 88.3\ \text{kPa} \times \text{e}^{-(0.0000338)\times(10)} = \boxed{88.3\ \text{kPa}}.$$

$$p(\text{total}) \approx \boxed{88.3\ \text{kPa}}.$$

(b) $p_{\text{N}_2\text{O}_5} = 88.3\ \text{kPa} \times \text{e}^{-(0.0000338)\times(600)} = \boxed{86.5\ \text{kPa}}$.

$$p(\text{total}) \approx 86.5\ \text{kPa} + \left(\frac{4}{2} \times 1.8\ \text{kPa}\right) + \left(\frac{1}{2} \times 1.8\ \text{kPa}\right)$$

$$= \boxed{91.0\ \text{kPa}}.$$

10.15 Use $\ln \dfrac{[\text{A}]_0}{[\text{A}]} = kt$ [10.13a].

Solve for k.

$$k = \frac{\ln\left(\dfrac{220}{56.0}\right)}{1.22 \times 10^4\ \text{s}} = \boxed{1.12 \times 10^{-4}\ \text{s}^{-1}}.$$

10.16 The reaction is

$$\text{CH}_3\text{COCO}_2^-(\text{aq}) + \text{H}^+(\text{aq}) \rightarrow \text{CH}_3\text{CHO}(\text{aq}) + \text{CO}_2(\text{g}).$$

We can assume that the concentration of pyruvate (P) decreases in proportion to the increase in p_{CO_2}. For the $CO_2(g)$ formed, we may write

$$pV = nRT.$$

$$n_{CO_2} = \frac{pV}{RT} = \frac{V}{RT}p_{CO_2} = \Delta n_P.$$

$$\Delta n_P = \frac{250 \times 10^{-6}\,\text{m}^3 \times 100\,\text{Pa}}{8.3145\,\text{J atm K}^{-1}\,\text{mol}^{-1} \times 293\,\text{K}} = 0.0103\,\text{mmol}.$$

$$[P] = [P]_0 - \frac{\Delta n_P}{0.100\,\text{dm}^3} = 3.23\,\text{mmol dm}^{-3} - 0.1026\,\text{mmol dm}^{-3}$$
$$= 3.13\,\text{mmol dm}^{-3}.$$

For a first-order reaction,

$$[P] = [P]_0\,e^{-kt}.$$

$$k = \frac{\ln\left(\frac{[P]_0}{[P]}\right)}{t}$$
$$= \frac{\ln\left(\frac{3.23\,\text{mmol dm}^{-3}}{3.13\,\text{mmol dm}^{-3}}\right)}{522\,\text{s}}$$
$$= \boxed{4.2 \times 10^{-5}\,\text{s}^{-1}}.$$

10.17 $$\frac{1}{[A]_t} - \frac{1}{[A]_0} = kt \quad [10.15a].$$

$$\frac{1}{56.0\,\text{mmol dm}^{-3}} - \frac{1}{220\,\text{mmol dm}^{-3}} = k \times 1.22 \times 10^4\,\text{s}.$$

$$k = 1.09 \times 10^{-6}\,\text{dm}^3\,\text{mmol}^{-1}\,\text{s}^{-1} = \boxed{1.09 \times 10^{-3}\,\text{dm}^3\,\text{mol}^{-1}\,\text{s}^{-1}}.$$

10.18 For a first-order reaction, we have

$$\ln\frac{[A]_0}{[A]} = kt \quad [10.13a].$$

Solving for k,

$$k = \frac{\ln\frac{[A]_0}{[A]}}{t} = \frac{\ln\left(\frac{220\,\text{mmol, dm}^{-3}}{56.0\,\text{mmol dm}^{-3}}\right)}{1.22 \times 10^4\,\text{s}}$$
$$= \boxed{1.12 \times 10^{-4}\,\text{s}^{-1}}.$$

10.19 We have the reaction

$$2NO(g) + Cl_2(g) \rightarrow 2NOCl(g).$$

This is stated to be a pseudo-second-order reaction in NO, which implies that it is really third-order, second-order in NO and first-order in Cl_2.

$$\text{rate} = kp_{NO}^2 p_{Cl_2} = kp_{Cl_2} \times p_{NO}^2 = k' p_{NO}^2$$

where k' is the pseudo-second-order rate constant. Without knowledge of p_{Cl_2}, we cannot calculate k, but we can calculate k'.

At $t = 0$, $p_{NOCl} = 0$, $p_{NO} = 300\,\text{Pa}$.

At $t = 522\,\text{s}$, $p_{NOCl} = 100\,\text{Pa}$, $p_{NO} = (300 - 100)\,\text{Pa} = 200\,\text{Pa}$.

$$\frac{1}{p_{NO}} - \frac{1}{p_{NO}(O)} = k't.$$

$$\frac{1}{200\,\text{Pa}} - \frac{1}{300\,\text{Pa}} = k' \times 522\,\text{s}.$$

$$k' = \boxed{3.19 \times 10^{-6}\,\text{Pa}^{-1}\,\text{s}^{-1}}.$$

10.20 $-\dfrac{d[A]}{dt} = k$ [zero-order in [A]].

$$\int_{[A]_0}^{[A]} d[A] = -\int_0^t k dt.$$

$$[A] - [A]_0 = -kt.$$

(a) Because pressures and concentrations are proportional, we may write

$$p - p_0 = -kt.$$

$$k = \frac{p_0 - p}{t} = \frac{(21 - 10)\,\text{kPa}}{770\,\text{s}} = \boxed{0.014\,\text{kPa}\,\text{s}^{-1}}.$$

(b) $t = \dfrac{p_0}{k} = \dfrac{21\,\text{kPa}}{0.014\,\text{kPa}\,\text{s}^{-1}} = \boxed{1.5 \times 10^3\,\text{s}}$.

10.21 (a) $\boxed{\text{rate} = k[\text{ICl}][H_2]}$.

In order to deduce this rate law, compare experiments that have an identical initial H_2 concentration then compare experiments that have an identical initial ICl concentration. Experiments 1 and 2 have identical $[H_2]_0$ values but the second has twice the $[\text{ICl}]_0$ value and an initial rate that is twice as large. The rate must be first-order in [ICl]. Similarly, experiments 2 and 3 have identical $[\text{ICl}]_0$ values, but the third has thrice the $[H_2]_0$ value and an initial rate that is three times as large. Once again, the rate is proportional to the concentrations so it must be first-order in $[H_2]$.

(b) $k = \text{rate}/([\text{ICl}][H_2]) = 3.7 \times 10^{-7}\,\text{mol}\,\text{dm}^{-3}\,\text{s}^{-1}/(1.5 \times 10^{-3}\,\text{mol}\,\text{dm}^{-3})^2$

$= \boxed{0.16\,\text{dm}^3\,\text{mol}^{-1}\,\text{s}^{-1}}$.

(c) $\text{rate} = (0.16\,\text{dm}^3\,\text{mol}^{-1}\,\text{s}^{-1} \times (4.7 \times 10^{-3}\,\text{mol}\,\text{dm}^{-3}) \times (2.7 \times 10^{-3}\,\text{mol}\,\text{dm}^{-3})$

$= \boxed{2.0 \times 10^{-6}\,\text{mol}\,\text{dm}^{-3}\,\text{s}^{-1}}$.

10.22 (a) Figure 10.2 is a plot of ln [HI] against t while Figure 10.3 is a plot of 1/[HI] against t. Since the plot of Figure 10.2 is non-linear, we conclude that the reaction rate is not first-order in [HI]. The plot of Figure 10.3 is linear so we conclude that the reaction rate is $\boxed{\text{second-order}}$ in [HI].

(b) The linear regression fit of the second-order rate is displayed in Figure 10.3. The rate constant equals the regression slope. $k = \text{slope} = \boxed{7.80 \times 10^{-3}\ \text{dm}^3\ \text{mol}^{-1}\ \text{s}^{-1}}$.

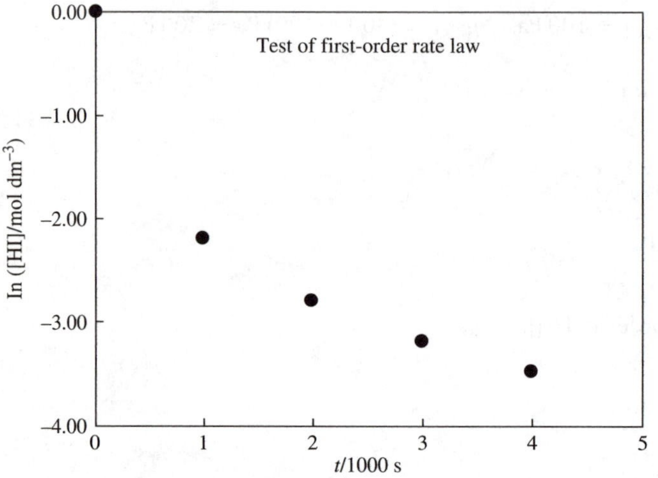

Fig 10.2

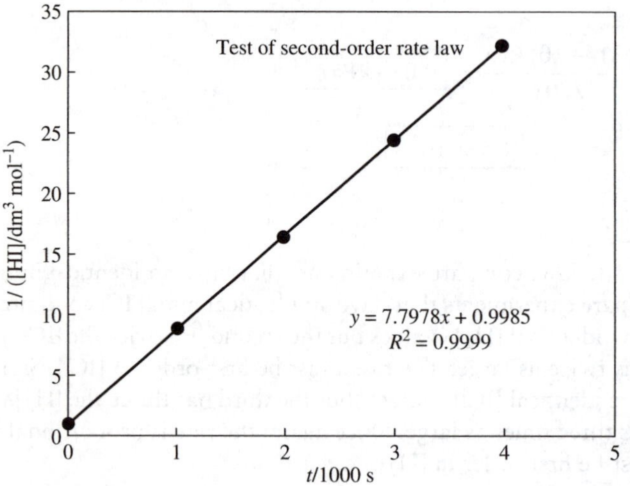

Fig 10.3

10.23 (a) Figure 10.4 is a plot of ln [HI] against t, while Figure 10.5 is a plot of 1/[HI] against t. Since the plot of Figure 10.4 is non-linear, we conclude that the reaction rate is not first-order in [HI]. The plot of Figure 10.5 is linear so we conclude that the reaction rate is $\boxed{\text{second-order}}$ in [HI].

(b) The linear regression fit of the second-order rate is displayed in Figure 10.5. The rate constant equals the regression slope. $k = \text{slope} = \boxed{1.32\ \text{dm}^3\ \text{mol}^{-1}\ \text{s}^{-1}}$.

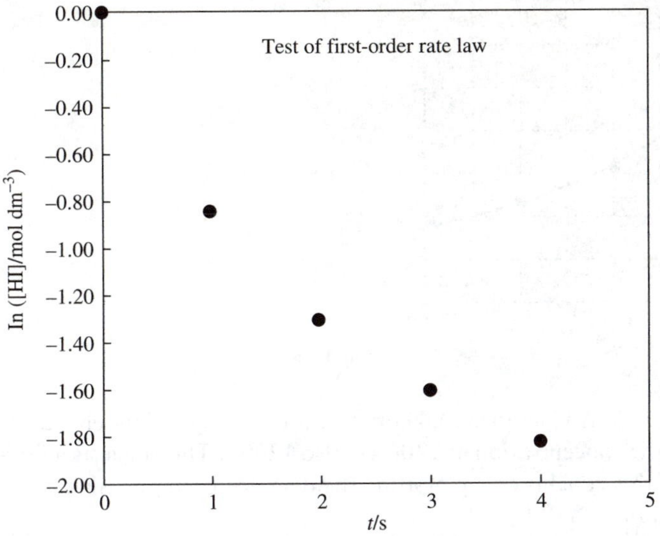

Fig 10.4

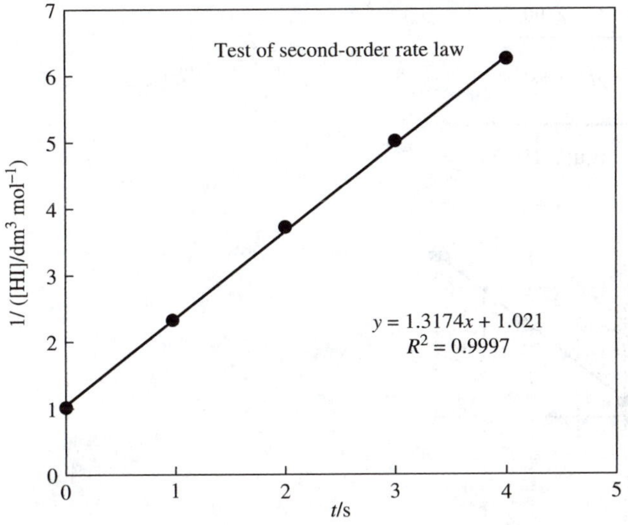

Fig 10.5

10.24 $[B]_\infty = \frac{1}{2}[A]_0$; hence $[A]_0 = 0.624\,\text{mol dm}^{-3}$. For the reaction $2A \rightarrow B$, $[A] = [A]_0 - 2[B]$. We can therefore draw up the following table.

t/s	0	600	1200	1800	2400
$[B]/(\text{mol dm}^{-3})$	0	0.089	0.153	0.200	0.230
$[A]/(\text{mol dm}^{-3})$	0.624	0.446	0.318	0.224	0.164

The data are plotted in Figure 10.6.

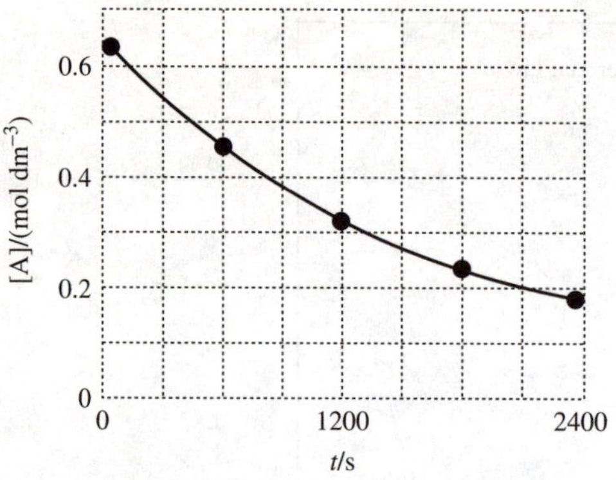

Fig 10.6

We see that the half-life of A from its initial concentration is approximately 1200 s, and that its half-life from the concentration at 1200 s is also 1200 s. This suggests a first-order reaction. We confirm this conclusion by plotting the data accordingly, using

$$\ln\frac{[A]_0}{[A]} = k_A t \quad \text{if} \quad \frac{d[A]}{dt} = -k_A[A].$$

First, draw up the table.

t/s	0	600	1200	1800	2400
$\ln\frac{[A]_0}{[A]}$	0	0.34	0.67	1.02	1.34

and plot the points. (See Figure 10.7.)

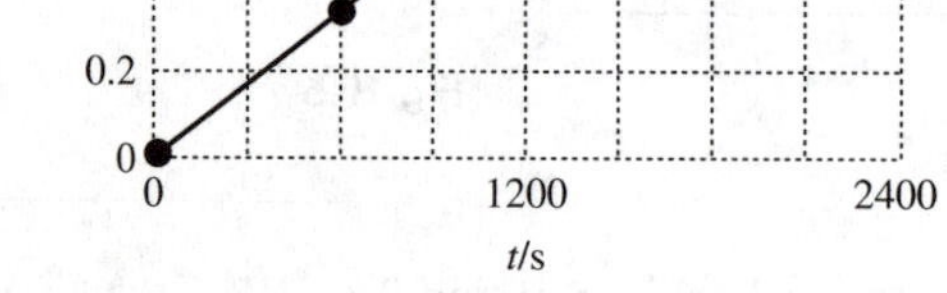

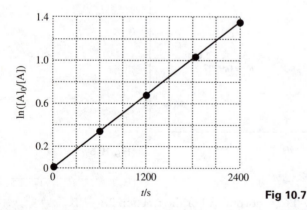

Fig 10.7

The points lie on a straight line, which confirms first-order kinetics. Because the slope of the line is 5.6×10^{-4}, we conclude that $k_A = 5.6 \times 10^{-4}\,\text{s}^{-1}$.

10.25 $\text{rate} = -\dfrac{d[A]}{dt} = k[A]^3.$

$$\frac{d[A]}{[A]^3} = -k\,dt.$$

$$\int_{[A]_0}^{[A]} \frac{d[A]}{[A]^3} = -k \int_0^t dt.$$

We now use the standard integral $\int x^n dy = \frac{x^{n+1}}{n+1}$, $n \neq -1$, which implies that

$$\int_a^b x^{-3} dx = \left.\frac{x^{-2}}{-2}\right|_a^b = -\left.\frac{1}{2x^2}\right|_a^b = \frac{1}{2a^2} - \frac{1}{2b^2}$$

Therefore,

$$\frac{1}{2[A]_0^2} - \frac{1}{2[A]^2} = -kt, \text{ or}$$

$$\frac{1}{[A]^2} = \frac{1}{[A]_0^2} + 2kt$$

$$\boxed{\frac{1}{[A]^2} = \frac{1}{[A]_0^2} + 2kt}.$$

Confirmation of a third-order reaction rate is provided by a linear plot of $\frac{1}{[A]^2}$ against t.

The third-order rate constant equals the linear regression slope divided by 2.

10.26 A + B → products.

Let $[A] = [A]_0 - x$; then $d[A] = -dx$ and $[B] = [B]_0 - x$ because of the 1:1 reaction stoichiometry.

$$\text{rate} = -\frac{d[A]}{dt} = k\,[A]\,[B] \quad \text{[Differential form of rate expression].}$$

$$\frac{dx}{dt} = k\left([A]_0 - x\right)\left([B]_0 - x\right) \quad \text{or} \quad \frac{dx}{\left([A]_0 - x\right)\left([B]_0 - x\right)} = kdt.$$

$$\int_0^x \frac{dx}{\left([A]_0 - x\right)\left([B]_0 - x\right)} = \int_0^t kdt.$$

$$\int_0^x \frac{dx}{\left([A]_0 - x\right)\left([B]_0 - x\right)} = kt.$$

(a) $[A]_0 \neq [B]_0$ case.

We use the standard integral

$$\int \frac{dx}{(a-x)(b-x)} = \frac{1}{b-a}\left(\ln\frac{1}{a-x} - \ln\frac{1}{b-x}\right) + \text{constant},$$

which implies that

$$\int_0^x \frac{dx}{(a-x)(b-x)} = \left.\left\{\frac{1}{b-a}\left(\ln\frac{1}{a-x} - \ln\frac{1}{b-x}\right) + \text{constant}\right\}\right|_x - \left.\left\{\frac{1}{b-a}\left(\ln\frac{1}{a-x} - \ln\frac{1}{b-x}\right) + \text{constant}\right\}\right|_0$$

$$= \frac{1}{b-a}\left(\ln\frac{1}{a-x} - \ln\frac{1}{b-x}\right) - \frac{1}{b-a}\left(\ln\frac{1}{a} - \ln\frac{1}{b}\right)$$

$$= \frac{1}{b-a}\ln\left\{\left(\frac{a}{a-x}\right)\left(\frac{b-x}{b}\right)\right\}$$

and the general form of the integrated rate expression becomes

$$\boxed{\frac{1}{[\mathrm{B}]_0 - [\mathrm{A}]_0}\ln\left\{\left(\frac{[\mathrm{A}]_0}{[\mathrm{A}]_0 - x}\right)\left(\frac{[\mathrm{B}]_0 - x}{[\mathrm{B}]_0}\right)\right\} = kt}$$

(b) $[\mathrm{A}]_0 = [\mathrm{B}]_0$ case.

We use the standard integral

$$\int \frac{\mathrm{d}x}{(a-x)^2} = \frac{1}{a-x} + \text{constant},$$

which implies that

$$\int_0^x \frac{\mathrm{d}x}{(a-x)^2} = \left\{\frac{1}{a-x} + \text{constant}\right\}\Bigg|_x - \left\{\frac{1}{a-x} + \text{constant}\right\}\Bigg|_0$$

$$= \frac{1}{a-x} - \frac{1}{a}$$

and the general form of the integrated rate expression becomes

$$\boxed{\frac{1}{[\mathrm{A}]_0 - x} - \frac{1}{[\mathrm{A}]_0} = kt} \quad \text{or} \quad \boxed{x = \frac{[\mathrm{A}]_0^2 kt}{1 + [\mathrm{A}]_0\, kt}}.$$

10.27 At $t_{1/2}[\mathrm{A}] = \frac{1}{2}[\mathrm{A}]_0$.

$(1/2)^n = 1/64, \qquad n = 6$ half-lives needed.

$t = t_{1/2} \times 6 = 221\text{s} \times 6 = \boxed{1.33 \times 10^3\ \text{s}}$.

10.28 $k = \frac{\ln 2}{t_{1/2}} = \frac{0.693}{5730\ \text{y}} = 1.209 \times 10^{-4}\text{y}^{-1}$.

$\ln\frac{[\mathrm{A}]_0}{[\mathrm{A}]} = kt$.

$\ln\frac{1.00}{0.69} = (1.21 \times 10^{-4}\text{y}^{-1})\, t$.

$t = \boxed{3.1 \times 10^3\ \text{y}}$.

10.29 $[^{90}\mathrm{Sr}] = [^{90}\mathrm{Sr}]_0\, \mathrm{e}^{-kt}, \quad k = \frac{\ln 2}{t_{1/2}}$.

$k = \frac{\ln 2}{28.1\ \text{y}} = 0.0247\ \text{y}^{-1}$.

Concentrations and masses are proportional to each other, so we may write $m = m_0\,e^{-kt}$.

(a) $m = 1.00\mu g \times e^{-(0.0247\,y^{-1} \times 19\,y)} = \boxed{0.63\,\mu g}$.

(b) $m = 1.00\mu g \times e^{-(0.0247\,y^{-1} \times 75\,y)} = \boxed{0.16\,\mu g}$.

10.30 This is a reaction of the type A + B → P.

This integrated rate law is given in Table 10.4 and is

$$[P] = \frac{[A]_0\,[B]_0\left[1 - \exp\left[\left([B]_0 - [A]_0\right)kt\right]\right]}{[A]_0 - [B]_0 \exp\left[\left([B]_0 - [A]_0\right)kt\right]}.$$

(a) $$[P] = \frac{(0.150)\,(0.055)\,[1 - \exp[(0.055 - 0.150) \times 0.11 \times 15]]}{0.150 - 0.055\,\exp\,[(0.055 - 0.150) \times 0.11 \times 15]}\ \mathrm{mol\,dm^{-3}}$$

$$= 1.16 \times 10^{-2}\ \mathrm{mol\,dm^{-3}}.$$

$$[CH_3COOC_2H_5] = 0.150\ \mathrm{mol\,dm^{-3}} - 0.012\ \mathrm{mol\,dm^{-3}} = \boxed{0.138\ \mathrm{mol\,dm^{-3}}}.$$

(b) Perform the above calculation of [P] with $t = 900\,\mathrm{s}$.

$$[P] = 5.50 \times 10^{-2}\ \mathrm{mol\,dm^{-3}}.$$

$$[CH_3COOC_2H_5] = 0.150\ \mathrm{mol\,dm^{-3}} - 0.055\ \mathrm{mol\,dm^{-3}} = \boxed{0.095\ \mathrm{mol\,dm^{-3}}}.$$

10.31 The amount of A in a second-order reaction is

$$\frac{1}{[A]} - \frac{1}{[A]_0} = kt.$$

Therefore,

$$t = \frac{1}{k}\left\{\frac{1}{[A]_t} - \frac{1}{[A]_0}\right\}$$

$$= \frac{1}{1.24 \times 10^{-3}\ \mathrm{mol^{-1}\,dm^3\,s^{-1}}} \times \left\{\frac{1}{0.026\ \mathrm{mol\,dm^{-3}}} - \frac{1}{0.260\ \mathrm{mol\,dm^{-3}}}\right\}$$

$$= \boxed{2.8 \times 10^4\ \mathrm{s}}.$$

10.32 $$\ln\left(\frac{k'}{k}\right) = \frac{E_a}{R}\left(\frac{1}{T} - \frac{1}{T'}\right)\quad [10.20].$$

Solve the above equation for E_a.

$$E_a = \frac{R\ln(k'/k)}{\left(\frac{1}{T} - \frac{1}{T'}\right)} = \frac{8.3145\ \mathrm{J\,K^{-1}\,mol^{-1}} \times \ln\left(\frac{1.38 \times 10^{-3}}{1.78 \times 10^{-4}}\right)}{\frac{1}{292\ \mathrm{K}} - \frac{1}{310\ \mathrm{K}}}$$

$$= \boxed{85.6\ \mathrm{kJ\,mol^{-1}}}.$$

For A, use

$$A = k \times e^{E_a/RT}$$

$$= 1.78 \times 10^{-4}\ \text{mol dm}^{-3}\ \text{s}^{-1} \times e^{85600/8.3145\times 292}$$

$$= \boxed{3.66 \times 10^{11}\ \text{mol dm}^{-3}\ \text{s}^{-1}}.$$

10.33 Solve equation 10.20 for T'

$$\frac{1}{T'} = \frac{1}{T} - \frac{R}{E_a}\ln\left(\frac{k'}{k}\right).$$

$$\frac{1}{T'} = \frac{1}{298\ \text{K}} - \frac{8.3145\ \text{J K}^{-1}\ \text{mol}^{-1}}{9.91 \times 10^{4}\ \text{J mol}^{-1}} \times \ln(1.10)$$

$$= 3.348 \times 10^{-3}\ \text{K}^{-1}.$$

$$T' = 298.7\ \text{K}$$

which becomes $\boxed{299\ \text{K}}$ (3 sig. figs).

10.34 In a relative sense, namely, the ratio k'/k, the reaction with the greater activation energy is more temperature-dependent, that is, $\boxed{E_a = 52\ \text{kJ mol}^{-1}}$.

In an absolute sense ($k' - k$), the reaction with the smaller activation energy is more temperature-dependent, here $\boxed{E_a = 25\ \text{kJ mol}^{-1}}$. Rate constants are greater for reactions with smaller activation energies.

10.35 We use the expression for E_a derived in the solution to Exercise 10.32.

$$E_a = \frac{R\ln(k'/k)}{\left(\frac{1}{T} - \frac{1}{T'}\right)}.$$

$$k' = 1.23\,\text{k}.$$

$$E_a = \frac{8.3145\ \text{J K}^{-1}\ \text{mol}^{-1} \times \ln(1.23)}{\frac{1}{293\ \text{K}} - \frac{1}{300\ \text{K}}}$$

$$= \boxed{21.6\ \text{kJ mol}^{-1}}.$$

10.36 According to [10.18], a plot of $\ln k$ against $1/T$ will be linear with a slope equal to $-E_a/R$. The plot and linear regression fit of the data are shown in Figure 10.8.

$$E_a = -\text{slope} \times R = -(33 \times 10^{3}\ \text{K}) \times \left(8.3145\ \text{J K}^{-1}\ \text{mol}^{-1}\right) = \boxed{27\bar{4}\ \text{kJ mol}^{-1}}.$$

10.37 We use the expression for E_a derived in the solution to Exercise 10.32 and assume that the ratio of the rates of spoilage is the ratio of the rate constants.

$$E_a = \frac{R\ln(k'/k)}{\left(\frac{1}{T} - \frac{1}{T'}\right)}.$$

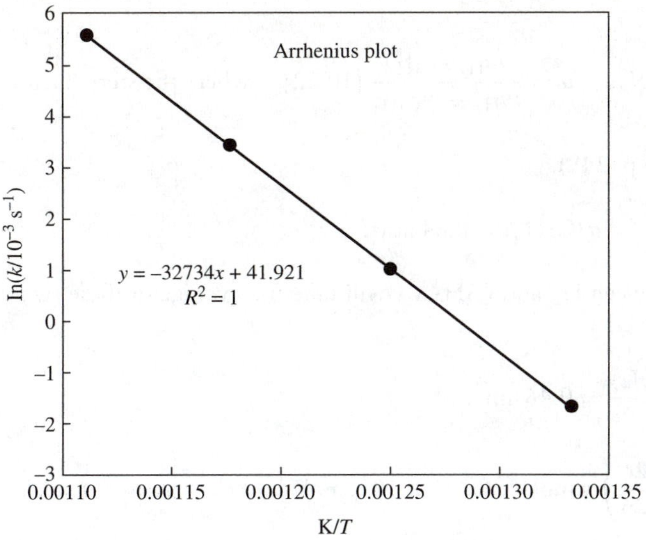

Fig 10.8

$$E_a = \frac{8.3145\,\mathrm{J\,K^{-1}\,mol^{-1}} \times \ln(40)}{\dfrac{1}{277\,\mathrm{K}} - \dfrac{1}{298\,\mathrm{K}}} = \boxed{120\,\mathrm{kJ\,mol^{-1}}}.$$

10.38 This is the inverse of the result in Exercise 10.35. Hence, $E_a = \boxed{-21.6\,\mathrm{kJ\,mol^{-1}}}$ but this can hardly be interpreted as an activation energy.

10.39 $$E_a = \frac{R\ln(k'/k)}{\left(\dfrac{1}{T} - \dfrac{1}{T'}\right)}.$$

$$k = \frac{\ln 2}{t_{1/2}} \quad \text{and} \quad k' = \frac{\ln 2}{2\,t_{1/2}}.$$

$$k'/k = 1/2.$$

$$E_a = \frac{8.3145\,\mathrm{J\,K^{-1}\,mol^{-1}} \times \ln(1/2)}{\dfrac{1}{283\,\mathrm{K}} - \dfrac{1}{293\,\mathrm{K}}} = \boxed{48\,\mathrm{kJ\,mol^{-1}}}.$$

10.40 $$\ln\left(\frac{k'}{k}\right) = \frac{E_a}{R}\left(\frac{1}{T} - \frac{1}{T'}\right)$$

$$= \frac{2.51 \times 10^5\,\mathrm{J\,mol^{-1}}}{8.315\,\mathrm{J\,K^{-1}\,mol^{-1}}}\left(\frac{1}{728\,\mathrm{K}} - \frac{1}{823\,\mathrm{K}}\right).$$

$$= 4.797$$

$$\frac{k'}{k} = 120 = \frac{t_{1/2}}{t'_{1/2}}.$$

$$t'_{1/2} = \frac{6.5 \times 10^6\,\mathrm{s}}{120} = \boxed{5.4 \times 10^4\,\mathrm{s}}.$$

10.41 We use

$$A = P\sigma\left(\frac{8kT}{\pi\mu}\right)^{1/2} N_A, \quad \mu = \frac{m_{H_2}m_{C_2H_4}}{m_{H_2}+m_{C_2H_4}}\ [10.22] \quad \text{where the steric factor,}$$

$P = 1.7 \times 10^{-6}$ (see p. 249).

$\sigma(H_2) = 0.27\,\text{nm}^2, \qquad \sigma(C_2H_4) = 0.64\,\text{nm}^2.$

For the effective σ between H_2 and C_2H_4 we will take the average of these two values, that is,

$$\sigma = \frac{\sigma(H_2) + \sigma(C_2H_4)}{2} = 0.46\,\text{nm}^2.$$

$$\mu = \left(\frac{2.016 \times 26.026}{2.016 + 26.026}\right) \text{g mol}^{-1}/6.022 \times 10^{23}\,\text{mol}^{-1}.$$

$$= 3.11 \times 10^{-24}\,\text{g} = 3.15 \times 10^{-27}\,\text{kg}.$$

$$A = 1.7 \times 10^{-6} \times 0.46 \times 10^{-18}\,\text{m}^2$$

$$\times \left(\frac{8 \times 1.381 \times 10^{-23}\,\text{J K}^{-1} \times 673\,\text{K}}{\pi \times 3.11 \times 10^{-27}\,\text{kg}}\right)^{1/2} \times 6.022 \times 10^{23}\,\text{mol}^{-1}$$

$$= 1.30 \times 10^3\,\text{m}^3\,\text{mol}^{-1}\,\text{s}^{-1}$$

$$= \boxed{1.30 \times 10^6\,\text{dm}^3\,\text{mol}^{-1}\,\text{s}^{-1}}.$$

10.42 Consider the example discussed in the text.

$K + Br_2 \rightarrow KBr + Br$

We will estimate the value of P based on a harpoon mechanism for this reaction by calculating the distance at which it is energetically favorable for the electron to leap from K to Br_2.

We should begin by identifying all the contributions to the energy of interaction between the colliding species. There are three contributions to the energy of the process $K + Br_2 \rightarrow K^+ + Br_2^-$. The first is the ionization energy, I, of K. The second is the electron affinity, E_{ea}, of Br_2. The third is the Coulombic interaction energy between the ions when they have been formed: when their separation is R this energy is $-e^2/4\pi\varepsilon_0 R$. The electron flips across when the sum of these three contributions changes from positive to negative (that is, when the sum is zero).

The net change in energy when the transfer occurs at a separation R is

$$E = I - E_{ea} - \frac{e^2}{4\pi\varepsilon_0 R}$$

The ionization energy I is larger than E_{ea}, so E becomes negative only when R has decreased to less than some critical value R^* given by

$$\frac{e^2}{4\pi\varepsilon_0 R^*} = I - E_{ea}$$

When the particles are at this separation, the harpoon shoots across from K to Br_2, so we can identify the reactive cross-section as $\sigma^* = \pi R^{*^2}$. This value of σ^*

$$P = \sigma^*/\sigma = \frac{R^{*2}}{d^2} = \left\{\frac{e^2}{4\pi\varepsilon_0 d(I - E_{ea})}\right\}^2$$

where $d = R(\mathrm{K}) + R(\mathrm{Br_2})$. With $I = 420\,\mathrm{kJ\,mol^{-1}}$ (corresponding to $7.0 \times 10^{-19}\,\mathrm{J}$), $E_{ea} \approx 250\,\mathrm{kJ\,mol^{-1}}$ (corresponding to $4.2 \times 10^{-19}\,\mathrm{J}$), and $d = 500\,\mathrm{pm}$, we find $P = 5.2$, in fair agreement with the experimental value (4.1).

We may use these calculated values to calculate σ^*, the reaction cross-section.

$$\sigma^* = \pi R^{*^2} = \pi P d^2$$

$$= \pi \times 2.7 \times (5.0 \times 10^{-10}\,\mathrm{m})^2$$

$$= 2.1 \times 10^{-18}\,\mathrm{m^2} = \boxed{2.1\,\mathrm{nm^2}}.$$

10.43 We use

$$k_{rate} = \left(\frac{kT}{h}\,e^{\Delta^\ddagger S/R}\right)\left(e^{-\Delta^\ddagger H/RT}\right) = Ae^{-\Delta^\ddagger H/RT}.$$

Let us assume that the factor $A = kT/h\,e^{-\Delta^\ddagger S/R}$ is essentially a constant over this small temperature range (10 K). Then we can determine $\Delta^\ddagger H$ from fitting the data to $\ln k_{rate} = \ln A - \Delta^\ddagger H/RT$.

k_{rate}	$1.2 \times 10^{-7}\,\mathrm{s^{-1}}$	$4.6 \times 10^{-7}\,\mathrm{s^{-1}}$
$\ln k_{rate}$	-15.94	-14.59
T	333 K	343 K
$\dfrac{1/T}{\mathrm{K^{-1}}}$	3.003×10^{-3}	2.915×10^{-3}

We find the slope of $\ln k_{rate}$ against $\left(\frac{1}{T}\right)$ to be

$-1.53 \times 10^4\,\mathrm{K} = -\Delta^\ddagger H/R$. Then

$$\Delta^\ddagger H = R \times 1.53 \times 10^4\,\mathrm{K} = 1.27 \times 10^5\,\mathrm{J\,mol^{-1}} = \boxed{127\,\mathrm{kJ\,mol^{-1}}}.$$

We also find $\ln A = 29.9$ and

$$\ln A = \ln\left(\frac{kT}{h}\right) + \Delta^\ddagger S/R.$$

Solving for $\Delta^{\ddagger}S$,

$$\Delta^{\ddagger}S = R\ln A + R\ln\left(\frac{h}{kt}\right) \qquad \text{[use average } T = 338\,\text{K]}$$

$$= (8.3145\,\text{J}\,\text{K}^{-1}\,\text{mol}^{-1} \times 29.9) + R\ln\left(\frac{6.626\times10^{-34}}{1.382\times10^{-23}\times 338}\right)$$

$$= 248.6\,\text{J}\,\text{K}^{-1}\,\text{mol}^{-1} - 245.96\,\text{J}\,\text{K}^{-1}\,\text{mol}^{-1}$$

$$= 2.6\,\text{J}\,\text{K}^{-1}\,\text{mol}^{-1}.$$

Then $\Delta^{\ddagger}G = \Delta^{\ddagger}H - T\Delta^{\ddagger}S$

$$= 127\,\text{kJ}\,\text{mol}^{-1} - 338\,\text{K}\times 2.6\,\text{J}\,\text{K}^{-1}\,\text{mol}^{-1}$$

$$= \boxed{126\,\text{kJ}\,\text{mol}^{-1}}.$$

10.44 We will assume that A does not change over the temperature range involved; then

$$\Delta^{\ddagger}S(60°\text{C}) = R\ln A + R\ln\left(\frac{h}{kT}\right)$$

$$= 248.6\,\text{J}\,\text{K}^{-1}\,\text{mol}^{-1} + R\ln\left(\frac{6.626\times10^{-34}}{1.382\times10^{-23}\times 333}\right)$$

$$= \boxed{2.7\,\text{J}\,\text{K}^{-1}\,\text{mol}^{-1}}.$$

$$\Delta^{\ddagger}S(70°\text{C}) = \boxed{2.5\,\text{J}\,\text{K}^{-1}\,\text{mol}^{-1}}.$$

There is little change between the two temperatures; $\Delta^{\ddagger}S$ remains constant at the level of the approximations employed.

Chapter 11

Accounting for the rate laws

Answers to discussion questions

11.1 Figure 11.1 sketches the concentration variations for second-order reversible steps with an assumed equilibrium constant equal to 1. This figure and Figure 11.2 of the text are qualitatively comparable.

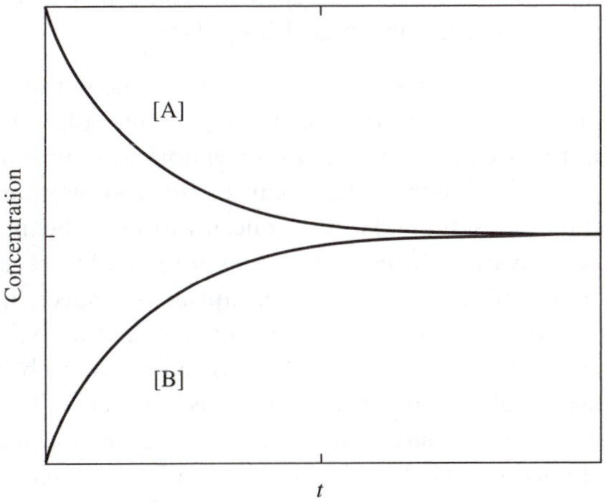

Fig 11.1

11.2 The rate-determining step is not just the slowest step: it must be slow *and* be a crucial gateway for the formation of products. If a faster reaction can also lead to products, then the slowest step is irrelevant because the slow reaction can then be side-stepped. The rate-determining step is like a slow ferry crossing between two fast highways: the overall rate at which traffic can reach its destination is determined by the rate at which it can make the ferry crossing.

If the first step in a mechanism is the slowest step with the highest activation energy, then it is rate-determining, and the overall reaction rate is equal to the rate of the first step because all subsequent steps are so fast that once the first intermediate is formed it results immediately in the formation of products. Once over the initial barrier, the intermediates cascade into products. However, a rate-determining step may also stem from the low concentration of a crucial reactant or catalyst and need not correspond to the step with highest activation barrier. A rate-determining step arising from the low activity of a crucial enzyme can sometimes be identified by determining whether or not the reactants and products for that step are in equilibrium: if the reaction is not at equilibrium it suggests that the step may be slow enough to be rate-determining.

11.3 The expression $k = k_a k_b[\mathrm{A}]/(k_b + k'_a[\mathrm{A}])$ for the effective rate constant of a unimolecular reaction A→P is based on the validity of the assumption of the existence of the pre-equilibrium $\mathrm{A}+\mathrm{A} \rightleftarrows \mathrm{A}^* + \mathrm{A}(k_a, k'_a)$. This can be a good assumption if both k_a and k'_a are much larger than k_b. The expression for the effective rate constant, k, can be rearranged to

$$\frac{1}{k} = \frac{k'_a}{k_a k_b} + \frac{1}{k_a[\mathrm{A}]}.$$

Hence, a test of the theory is to plot $1/k$ against $1/[\mathrm{A}]$, and to expect a straight line. Another test is based on the prediction from the Lindemann–Hinshelwood mechanism that, as the concentration (and therefore the partial pressure) of A is reduced, the reaction should switch to overall second-order kinetics. Whereas the mechanism agrees in general with the switch in order of unimolecular reactions, it does not agree in detail. A typical graph of $1/k$ against $1/[\mathrm{A}]$ has a pronounced curvature, corresponding to a larger value of k (a smaller value of $1/k$) at high pressures (low $1/[\mathrm{A}]$) than would be expected by extrapolation of the reasonably linear low pressure (high $1/[\mathrm{A}]$) data.

11.4 The Michaelis–Menten mechanism of enzyme activity models the enzyme with one active site that, weakly and reversibly, binds a substrate in homogeneous solution. It is a three-step mechanism. The first and second steps are the reversible formation of the enzyme–substrate complex (ES). The third step is the decay of the complex into the product. The steady-state approximation is applied to the concentration of the intermediate (ES) and its use simplifies the derivation of the final rate expression. However, the justification for the use of the approximation with this mechanism is suspect, in that both rate constants for the reversible steps may not be as large, in comparison to the rate constant for the decay to products, as they need to be for the approximation to be valid. The simplest form of the mechanism applies only when $k_b \gg k'_a$. Nevertheless, the form of the rate equation obtained does seem to match the principal experimental features of enzyme-catalyzed reactions; it explains why there is a maximum in the reaction rate and provides a mechanistic understanding of the turnover number. The model may be expanded to include multisubstrate reactions and inhibition.

Solutions to exercises

11.5 $K = \dfrac{k}{k'}$ [11.1].

This relation applies to a second-order reaction as well as to a first-order reaction.

$k = 7.4 \times 10^7\ \mathrm{dm^3\ mol^{-1}\ s^{-1}}$.

$$k' = \frac{k}{K} = \frac{7.4 \times 10^7\ \mathrm{dm^3\ mol^{-1}\ s^{-1}}}{235} = \boxed{3.1 \times 10^5\ \mathrm{dm^3\ mol^{-1}\ s^{-1}}}.$$

11.6 $\mathrm{A} \underset{k'}{\overset{k}{\rightleftharpoons}} \mathrm{B}$.

The rate expressions for [B], [A] are:

$$\frac{\mathrm{d[B]}}{\mathrm{d}t} = k[\mathrm{A}] - k'[\mathrm{B}],$$

$$\frac{\mathrm{d[A]}}{\mathrm{d}t} = -k[\mathrm{A}] + k'[\mathrm{B}],$$

and the solutions to be tested are given in eqn 11.2.

$$[\mathrm{B}] = \frac{k\left(1 - \mathrm{e}^{-(k+k')t}[\mathrm{A}]_0\right)}{k+k'},$$

$$[\mathrm{A}] = \frac{\left(k' + k\mathrm{e}^{-(k+k')t}\right)[\mathrm{A}]_0}{k+k'}.$$

To test the solutions, we substitute the expressions for [B], [A] into the rate expressions. The left-side substitution requires that we differentiate

$$\frac{k}{k+k'}\,(k+k')[\mathrm{A}]_0\mathrm{e}^{-\left(k+k'\right)t} = \frac{k\left(k' + k\mathrm{e}^{-(k+k')t}\right)[\mathrm{A}]_0}{k+k'} - \frac{k'k\left(1 - \mathrm{e}^{-(k+k')t}\right)[\mathrm{A}]_0}{k+k'}$$

which simplifies to

$$k[\mathrm{A}]_0\mathrm{e}^{-\left(k+k'\right)t} = k[\mathrm{A}]_0\mathrm{e}^{-\left(k+k'\right)t}.$$

Because both sides are equal, the expression for [B] is a solution to the rate expression for [B]. Similar substitutions for [A] into the $\frac{\mathrm{d[A]}}{\mathrm{d}t}$ rate expression prove it to be a solution.

11.7 $\mathrm{A} \rightleftharpoons \mathrm{B}$.

$$\frac{\mathrm{d[A]}}{\mathrm{d}t} = -k[\mathrm{A}] + k'[\mathrm{B}], \quad \frac{\mathrm{d[B]}}{\mathrm{d}t} = -k'[\mathrm{B}] + k[\mathrm{A}].$$

$[\mathrm{A}] + [\mathrm{B}] = [\mathrm{A}]_0 + [\mathrm{B}]_0$ at all times.

Therefore, $[\mathrm{B}] = [\mathrm{A}]_0 + [\mathrm{B}]_0 - [\mathrm{A}]$.

$$\frac{\mathrm{d[A]}}{\mathrm{d}t} = -k[\mathrm{A}] + k'\{[\mathrm{A}]_0 + [\mathrm{B}]_0 - [\mathrm{A}]\} = -(k+k')[\mathrm{A}] + k'([\mathrm{A}]_0 + [\mathrm{B}]_0)$$

and $\displaystyle\int_{[\mathrm{A}]_0}^{[\mathrm{A}]_0} \frac{\mathrm{d[A]}}{(k+k')[\mathrm{A}] + k'([\mathrm{A}]_0 + [\mathrm{B}]_0)} = -\int_0^t \mathrm{d}t.$

The solution is $[\mathrm{A}] = \dfrac{k'([\mathrm{A}]_0 + [\mathrm{B}]_0) + (k[\mathrm{A}]_0 - k'[\mathrm{B}]_0)\mathrm{e}^{-\left(k+k'\right)t}}{k+k'}$.

Setting $[\mathrm{B}]_0 = 0$,

$$[\mathrm{A}] = \frac{k'[\mathrm{A}]_0 + k[\mathrm{A}]_0\mathrm{e}^{-\left(k+k'\right)t}}{k+k'} = \frac{\left(k' + k\mathrm{e}^{-\left(k+k'\right)t}\right)[\mathrm{A}]_0}{k+k'},$$

which is the expression for [A] in eqn 11.2. Then

$$[\mathrm{B}] = [\mathrm{A}]_0 - [\mathrm{A}] = \frac{k(1 - \mathrm{e}^{-(k+k')t})[\mathrm{A}]_0}{k + k'}$$

which agrees with [B] in eqn 11.2.

11.8 The rate laws are

$$(1)\ \frac{\mathrm{d}[\mathrm{A}]}{\mathrm{d}t} = -k_1[\mathrm{A}],$$

$$(2)\ \frac{\mathrm{d}[\mathrm{I}]}{\mathrm{d}t} = -k_1[\mathrm{A}] - k_2[\mathrm{I}],$$

$$(3)\ \frac{\mathrm{d}[\mathrm{P}]}{\mathrm{d}t} = k_2[\mathrm{I}].$$

Differentiating [11.5a] yields (1) directly. Differentiating [11.5c] yields (3) almost as directly.

$$\frac{\mathrm{d}[\mathrm{P}]}{\mathrm{d}t} = \frac{1}{k_2 - k_1}(-k_1k_2\mathrm{e}^{-k_2t} + k_1k_2\mathrm{e}^{-k_1t})[\mathrm{A}]_0$$

$$= \frac{k_1k_2}{k_2 - k_1}(\mathrm{e}^{-k_1t} - \mathrm{e}^{-k_2t})[\mathrm{A}]_0$$

$$= k_2[\mathrm{I}], \quad \text{which is (3).}$$

From [11.5b], after differentiating, we have

$$(4)\ \frac{\mathrm{d}[\mathrm{I}]}{\mathrm{d}t} = \frac{k_1}{k_2 - k_1}(-k_1\mathrm{e}^{-k_1t} + k_2\mathrm{e}^{-k_2t})[\mathrm{A}]_0.$$

The reduction of (4) to (2) requires a little more algebraic manipulation. In this case, it seems easier to work backwards. Thus, substitute [11.5a] and [11.5b] into (2). We obtain

$$\frac{\mathrm{d}[\mathrm{I}]}{\mathrm{d}t} = k_1\mathrm{e}^{-k_1t}[\mathrm{A}]_0 - k_2\left(\frac{k_1}{k_2 - k_1}\right)(\mathrm{e}^{-k_1t} - \mathrm{e}^{-k_2t})[\mathrm{A}]_0$$

$$= \left(\frac{k_1}{k_2 - k_1}\right)(k_2 - k_1)\mathrm{e}^{-k_1t}[\mathrm{A}]_0 - \left(\frac{k_1}{k_2 - k_1}\right)k_2(\mathrm{e}^{-k_1t} - \mathrm{e}^{-k_2t})[\mathrm{A}]_0$$

$$= \left(\frac{k_1}{k_2 - k_1}\right)[(k_2 - k_1)\mathrm{e}^{-k_1t}[\mathrm{A}]_0 - k_2(\mathrm{e}^{-k_1t} - \mathrm{e}^{-k_2t})[\mathrm{A}]_0]$$

$$= \left(\frac{k_1}{k_2 - k_1}\right)[k_2\mathrm{e}^{-k_1t}[\mathrm{A}]_0 - k_1\mathrm{e}^{-k_1t}[\mathrm{A}]_0 - k_2\mathrm{e}^{-k_1t}[\mathrm{A}]_0 + k_2\mathrm{e}^{-k_1t}[\mathrm{A}]_0].$$

The first and third terms in the brackets cancel leaving

$$\frac{\mathrm{d}[\mathrm{I}]}{\mathrm{d}t} = \left(\frac{k_1}{k_2 - k_1}\right)(-k_1\mathrm{e}^{-k_1t} + k_2\mathrm{e}^{-k_2t})[\mathrm{A}]_0$$

which is (4) above.

11.9 Differentiating [11.5b] yields eqn (4) in the solution to Exercise 11.8. Setting (4) to zero, we have

$$k_1 e^{-k_1 t} = k_2 e^{-k_2 t}$$

$$\text{or} \quad \frac{k_1}{k_2} = e^{(k_1 - k_2)t}$$

$$\text{or} \quad \ln\left(\frac{k_1}{k_2}\right) = (k_1 - k_2)t.$$

$$t = \frac{1}{k_1 - k_2} \ln\left(\frac{k_1}{k_2}\right)$$

which is eqn 11.6 of the text.

11.10 We use eqn 11.6 after solving for k_1 and k_2 from the half-lives.

$$k_1 = \frac{\ln 2}{22.5\,\text{d}} = 3.08 \times 10^{-2}\,\text{d}^{-1}.$$

$$k_2 = \frac{\ln 2}{33.0\,\text{d}} = 2.10 \times 10^{-2}\,\text{d}^{-1}.$$

$$t = \frac{1}{k_1 - k_2} \ln \frac{k_1}{k_2}$$

$$= \frac{1}{(3.08 - 2.10) \times 10^{-2}\,\text{d}^{-1}} \ln(3.08/2.10)$$

$$= \boxed{39.1\,\text{d}}\,.$$

11.11 For the simple irreversible, parallel reactions $R \to P_1$ and $R \to P_2$, shown in Figure 11.2, the product P_1 is thermodynamically favored. However, the rate at which each product appears does not depend upon thermodynamic favorability. Rate constants depend upon activation energy according to the expression $k = Ae^{-E_a/RT}$ [10.19]. The activation energy for the formation of P_1 is much larger than that for formation of P_2. At low and moderate temperatures, the large activation energy may not be readily available in the environment of the reactant R and P_1 either cannot form or forms at a slow rate. The much smaller activation energy for P_2 formation is readily available and, consequently, P_2 is produced even though it is not the thermodynamically favored product. This kinetic control yields the ratio $[P_1]/[P_2] = k_1/k_2 < 1$. Details of the temperature dependence can be mathematically examined.

$$\frac{[P_1]}{[P_2]} = \frac{k_1}{k_2} = \frac{A_1 e^{-E_{a1}/RT}\ [10.19]}{A_2 e^{-E_{a2}/RT}} = \left(\frac{A_1}{A_2}\right) e^{-(E_{a1} - E_{a2})/RT} = \left(\frac{A_1}{A_2}\right) e^{-\Delta E_a/RT}$$

where $\Delta E_a = E_{a1} - E_{a2} > 0$.

Since, as $T \to$ larger, $\Delta E_a/RT \to$ smaller and $e^{-\Delta E_a/RT} \to$ larger (but is always less than 1). Consequently, the above equation indicates that, as $T \to$ larger, the $\boxed{\text{concentration ratio } [P_1]/[P_2] \to \text{larger}}$. This is expected because the high requisite activation energy for the formation of P_1 is available in the high temperature environment of R.

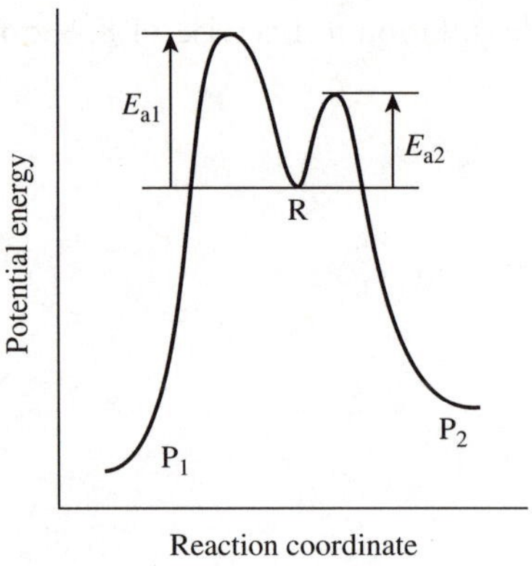

Fig 11.2

11.12 $k_d = \frac{8RT}{3\eta}\ [11.17] = \frac{8 \times (8.3145\ \mathrm{J\,K^{-1}\,mol^{-1}}) \times (298\ \mathrm{K})}{3\eta}$

$$= \frac{6.61 \times 10^3\ \mathrm{J\,mol^{-1}}}{\eta} = \frac{6.61 \times 10^3\ \mathrm{kg\,m^2\,s^{-2}\,mol^{-1}}}{(\eta/\mathrm{kg\,m^{-1}\,s^{-1}}) \times \mathrm{kg\,m^{-1}\,s^{-1}}} = \frac{6.61 \times 10^3\ \mathrm{m^3\,mol^{-1}\,s^{-1}}}{(\eta/\mathrm{kg\,m^{-1}\,s^{-1}})}$$

$$= \frac{6.61 \times 10^6\ \mathrm{dm^3\,mol^{-1}\,s^{-1}}}{(\eta/\mathrm{kg\,m^{-1}\,s^{-1}})} = \frac{6.61 \times 10^9\ \mathrm{dm^3\,mol^{-1}\,s^{-1}}}{(\eta/10^{-3}\ \mathrm{kg\,m^{-1}\,s^{-1}})}.$$

(a) Water, $\eta = 1.00 \times 10^{-3}\ \mathrm{kg\,m^{-1}\,s^{-1}}$.

$$k_d = \frac{6.61 \times 10^9}{1.00}\ \mathrm{dm^3\,mol^{-1}\,s^{-1}} = 6.61 \times 10^9\ \mathrm{dm^3\,mol^{-1}\,s^{-1}}$$

$$= \boxed{6.61 \times 10^6\ \mathrm{m^3\,mol^{-1}\,s^{-1}}}.$$

(b) Pentane, $\eta = 0.22 \times 10^{-3}\ \mathrm{kg\,m^{-1}\,s^{-1}}$.

$$k_d = \frac{6.61 \times 10^9}{0.22}\ \mathrm{dm^3\,mol^{-1}\,s^{-1}} = 3.0 \times 10^{10}\ \mathrm{dm^3\,mol^{-1}\,s^{-1}}$$

$$= \boxed{3.0 \times 10^7\ \mathrm{m^3\,mol^{-1}\,s^{-1}}}.$$

11.13 (a) $J = -D \times (\text{concentration gradient})$ [Fick's first law [11.18b]]

$$= -(5.22 \times 10^{-10}\ \mathrm{m^2\,s^{-1}}) \times (-0.10\ \mathrm{mol\,dm^{-3}\,m^{-1}}) \times \left(\frac{1\ \mathrm{dm^3}}{10^{-3}\ \mathrm{m^3}}\right)$$

$$= \boxed{5.2 \times 10^{-8}\ \mathrm{mol\,m^{-2}\,s^{-1}}}.$$

(b) $n = JA\Delta t$ [See Illustration 11.3].

$$= (5.2\times10^{-8}\,\mathrm{mol\,m^{-2}\,s^{-1}})\times(5.0\,\mathrm{mm^2})\times(60\,\mathrm{s})\times\left(\frac{10^{-6}\,\mathrm{m^2}}{1\,\mathrm{mm^2}}\right)$$

$$= \boxed{1.6\times10^{-11}\,\mathrm{mol}}.$$

11.14 $d = (2Dt)^{1/2}$ or $t = \dfrac{d^2}{2D}$ [11.20].

(a) $$t = \frac{(1.00\times10^{-3}\,\mathrm{m})^2}{2(0.522\times10^{-9}\,\mathrm{m^2\,s^{-1}})\ [\text{Table 11.1}]} = 958\,\mathrm{s} = \boxed{16.0\,\mathrm{min}}.$$

(b) $$t = \frac{(1.00\times10^{-2}\,\mathrm{m})^2}{2(0.522\times10^{-9}\,\mathrm{m^2\,s^{-1}})\ [\text{Table 11.1}]} = 9.58\times10^{4}\,\mathrm{s} = \boxed{26.7\,\mathrm{h}}.$$

(c) $$t = \frac{(1.00\,\mathrm{m})^2}{2(0.522\times10^{-9}\,\mathrm{m^2\,s^{-1}})\ [\text{Table 11.1}]} = 9.58\times10^{8}\,\mathrm{s} = \boxed{30.4\,\mathrm{y}}.$$

11.15 $D = \dfrac{\lambda^2}{2\tau}$ Einstein–Smoluchowski equation [11.21].

(a) $$D = \frac{(150\times10^{-12}\,\mathrm{m})^2}{2(1.8\times10^{-12}\,\mathrm{s})} = \boxed{6.3\times10^{-9}\,\mathrm{m^2\,s^{-1}}}.$$

(b) $$D = \frac{(75\times10^{-12}\,\mathrm{m})^2}{2(1.8\times10^{-12}\,\mathrm{s})} = \boxed{1.6\times10^{-9}\,\mathrm{m^2\,s^{-1}}}.$$

11.16 Molecular oxygen, O_2, has a radius of about 180 pm and it provides an interesting example of the time required for a small molecule to traverse a lipid bilayer membrane.

$$D_{\mathrm{O_2/lipid\ bilayer}} \simeq \frac{kT}{6\pi\eta a}\ [11.23] = \frac{(1.381\times10^{-23}\,\mathrm{J\,K^{-1}})\times(310.\,\mathrm{K})}{6\pi(0.010\,\mathrm{kg\,m^{-1}\,s^{-1}})\times(180\times10^{-12}\,\mathrm{m})}$$

$$= 1.3\times10^{-10}\,\mathrm{m^2\,s^{-1}}.$$

$$t_{\mathrm{O_2/lipid\ bilayer}} \simeq \frac{d^2}{2D}\ [11.20] = \frac{(0.50\times10^{-9}\,\mathrm{m})^2}{2(1.3\times10^{-10}\,\mathrm{m^2\,s^{-1}})} = 9.6\times10^{-10}\,\mathrm{s} = \boxed{0.96\,\mathrm{ns}}.$$

COMMENT A typical bilayer thickness is closer to 5.0 nm rather than 0.50 nm, so the time calculated here is abnormally short.

11.17 Diffusion is only important in the absence of macroscopic fluid flow or convection or turbulence as may be the case when microscopic spatial constraints exist such as passage of gases through lung alveoli or the passage of neurotransmitter molecules across a synaptic gap (see Exercise 11.16). An extraordinarily large time is required for a molecule to move across a lake by diffusion alone.

$$t_{\mathrm{H_2O}} \simeq \frac{d^2}{2D}\ [11.20] = \frac{(1.0\times10^{2}\,\mathrm{m})^2}{2(2.26\times10^{-9}\,\mathrm{m^2\,s^{-1}})} = 2.2\times10^{12}\,\mathrm{s}$$

$$= \boxed{7.0\times10^{4}\,\mathrm{y}}\quad [\text{see Self-test 11.4}].$$

11.18 Let N be the number of steps that a molecule takes in time t and let τ be the time each step takes; $\tau = t/N$.

$$D = \frac{\lambda^2}{2\tau}\ [11.21] = \frac{\lambda^2 N}{2t} \quad \text{or} \quad t = \frac{\lambda^2 N}{2D}.$$

Substitution into [11.20] gives:

$$d = (2Dt)^{1/2} = \left(\frac{2D\lambda^2 N}{2D}\right)^{1/2} = (\lambda^2 N)^{1/2} \quad \text{or} \quad N = \left(\frac{d}{\lambda}\right)^2.$$

For $d = 1000\lambda$, $N = \left(\frac{1000\lambda}{\lambda}\right)^2 = \boxed{1\times10^6 \text{ steps}}$.

11.19 $\eta = \eta_0 e^{E_a/RT}$ [11.24].

$$\frac{\eta_{T_1}}{\eta_{T_2}} = \frac{e^{E_a/RT_1}}{e^{E_a/RT_2}} = e^{\frac{E_a}{R}\left(\frac{1}{T_1}-\frac{1}{T_2}\right)}.$$

$$\ln\left(\frac{\eta_{T_1}}{\eta_{T_2}}\right) = \frac{E_a}{R}\left(\frac{1}{T_1}-\frac{1}{T_2}\right) \quad \text{or} \quad E_a = R\left(\frac{1}{T_1}-\frac{1}{T_2}\right)^{-1}\ln\left(\frac{\eta_{T_1}}{\eta_{T_2}}\right).$$

$$E_a = (8.3145\,\mathrm{J\,mol^{-1}\,K^{-1}}) \times \left(\frac{1}{293\,\mathrm{K}} - \frac{1}{303\,\mathrm{K}}\right)^{-1} \ln\left(\frac{1.0019}{0.7982}\right) = \boxed{16.8\,\mathrm{kJ\,mol^{-1}}}.$$

11.20 $k_{\text{rate}} = \frac{kT}{h} e^{-\Delta^{\ddagger}G/RT}$ [10.23].

The ratio of the rates is then given by

$$\frac{k(\text{cat})}{k(\text{uncat})} = e^{\left[-(10-100)\,\mathrm{kJ\,mol^{-1}}/\left(8.3145\times10^{-3}\,\mathrm{kJ\,mol^{-1}\,K^{-1}}\times 310\,\mathrm{K}\right)\right]}$$

$$= \boxed{1.5\times10^{15}}.$$

11.21 The first step is rate-determining; hence rate $= k[H_2O_2][Br^-]$.

The reaction is $\boxed{\text{first-order in } H_2O_2 \text{ and in } Br^-}$ and $\boxed{\text{second-order overall}}$.

11.22 We assume a pre-equilibrium (that step is fast), and write

$$K = \frac{[A]^2}{[A_2]} \quad \text{so that } [A] = K^{1/2}[A_2]^{1/2}.$$

The rate-determining step then gives

$$\text{rate} = k_2[A][B] = k_2K^{1/2}[A_2]^{1/2}[B] = k_{\text{eff}}[A_2]^{1/2}[B]$$

where $k_{\text{eff}} = k_2K^{1/2}$.

11.23 We assume a pre-equilibrium (as the initial step is fast), and write

$$K = \frac{[\text{unstable helix}]}{[\text{A}][\text{B}]}, \text{ indicating } [\text{unstable helix}] = K[\text{A}][\text{B}].$$

The rate-determining step then gives

$$\text{rate} = \frac{d[\text{double helix}]}{dt} = k_2[\text{unstable helix}] = \boxed{k_2K[\text{A}][\text{B}]} = k[\text{A}][\text{B}]\ [k = k_2K].$$

The equilibrium constant is the outcome of the two processes:

$$\text{A} + \text{B} \underset{k_2}{\overset{k_1}{\rightleftharpoons}} \text{B unstable helix}, K = \frac{k_1}{k_1'}.$$

Therefore, with rate $= k[\text{A}][\text{B}]$, $\boxed{k = \frac{k_1k_2}{k_1'}}$.

11.24 We assume that the steady-state approximation applies to [O] (but see the question below). Then

$$\frac{d[\text{O}]}{dt} = 0 = k_1[\text{O}_3] - k_1'[\text{O}][\text{O}_2] - k_2[\text{O}][\text{O}_3].$$

Solving for [O],

$$[\text{O}] = \frac{k_1[\text{O}_3]}{k_1'[\text{O}_2] + k_2[\text{O}_3]}.$$

$$\text{rate} = -\frac{1}{2}\frac{d[\text{O}_3]}{dt}.$$

$$\frac{d[\text{O}_3]}{dt} = -k_1[\text{O}_3] + k_1'[\text{O}][\text{O}_2] - k_2[\text{O}][\text{O}_3].$$

Substituting for [O] from above,

$$\frac{d[\text{O}_3]}{dt} = -k_1[\text{O}_3] + \frac{k_1[\text{O}_3](k_1'[\text{O}_2] - k_2[\text{O}_3])}{k_1'[\text{O}_2] + k_2[\text{O}_3]}$$

$$= \frac{-k_1[\text{O}_3](k_1'[\text{O}_2] + k_2[\text{O}_3]) + k_1[\text{O}_3](k_1'[\text{O}_2] - k_2[\text{O}_3])}{k_1'[\text{O}_2] + k_2[\text{O}_3]}$$

$$= \frac{-2k_1k_2[\text{O}_3]^2}{k_1'[\text{O}_2] + k_2[\text{O}_3]}.$$

$$\boxed{\text{rate} = \frac{k_1k_2[\text{O}_3]^2}{k_1'[\text{O}_2] + k_2[\text{O}_3]}}.$$

If the second step is slow, then $k_2[O_3] \ll k_1'[O_2]$ and the rate reduces to

$$\text{rate} = \frac{k_1 k_2 [O_3]^2}{k_1'[O_2]}.$$

which is second-order in $[O_3]$ and -1 order in $[O_2]$.

QUESTION Can you determine the rate law expression if the first step of the proposed mechanism is a rapid pre-equilibrium? Under what conditions does the rate expression above reduce to the case of rapid pre-equilibrium?

11.25 $$\frac{d[A^-]}{dt} = k_1[AH][B] - k_2[A^-][BH^+] - k_3[A^-][AH] = 0.$$

Therefore, $$[A^-] = \boxed{\frac{k_1[AH][B]}{k_2[BH^+] + k_3[AH]}}$$

and the rate of formation of product is

$$\frac{d[P]}{dt} = k_3[AH][A^-] = \boxed{\frac{k_1 k_3 [AH]^2[B]}{k_2[BH^+] + k_3[AH]}}.$$

11.26 The rate of production of the product is

$$\frac{d[BH^+]}{dt} = k_2[HAH^+][B].$$

HAH^+ is an intermediate involved in a rapid pre-equilibrium.

$$\frac{[HAH^+]}{[HA][H^+]} = \frac{k_1}{k_1'} \quad \text{so} \quad [HAH^+] = \frac{k_1[HA][H^+]}{k_1'}$$

and $$\frac{d[BH^+]}{dt} = \boxed{\frac{k_1 k_2}{k_1'}[HA][H^+][B]}.$$

This rate law can be made independent of $[H^+]$, if the source of H^+ is the acid HA, for then H^+ is given by another equilibrium.

$$\frac{[H^+][A^-]}{[HA]} = K_a = \frac{[H^+]^2}{[HA]} \quad \text{so} \quad [H^+] = (K_a[HA])^{1/2}$$

and $$\frac{d[BH^+]}{dt} = \boxed{\frac{k_1 k_2 K_a^{1/2}}{k_1'}[HA]^{3/2}[B]}.$$

11.27 $E + S \underset{k_a'}{\overset{k_a}{\rightleftharpoons}} ES \xrightarrow{k_b} E + P.$

In the pre-equilibrium approximation, the intermediate ES is in equilibrium with the reactants E and S. This requires equality between the rate of formation of ES and the rate of ES dissociation to reactants.

rate of formation of ES = rate of ES dissociation to reactants.

$$k_a[E][S] = k_a'[ES],$$

$$\frac{[ES]}{[E][S]} = \frac{k_a}{k_a'} = K \quad \text{or} \quad [ES] = K[E][S].$$

If $[E]_0$ is the total enzyme concentration, then $[E] = [E]_0 - [ES]$ by conservation of mass.

$$[ES] = K[E][S] = K\big([E]_0 - [ES]\big)[S],$$

$$[ES] + \frac{[ES]}{K[S]} = [E]_0,$$

$$[ES] = \frac{[E]_0}{1 + \dfrac{1}{K}} = \frac{[E]_0[S]}{[S] + \dfrac{1}{K}}.$$

Substitution in the expression for the rate of formation of P gives:

$$\boxed{\text{rate of formation of P} = k_b[ES] = \frac{k_b[E]_0[S]}{[S] + \dfrac{1}{K}}} \quad \text{pre-equilibrium approximation, } K = \frac{k_a}{k_a'};$$

$$\text{rate of formation of P} = k_b[ES] = \frac{k_b[E]_0[S]}{[S] + K_M} \quad \text{steady-state approximation [11.25].}$$

Comparison of the pre-equilibrium approximation with the steady-state approximation shows that the two approximations are the same when $\dfrac{1}{K} = K_M$.

Since $\dfrac{1}{K} = \dfrac{k_a'}{k_a}$ and $K_M = \dfrac{k_a' + k_b}{k_a}$ [11.26], the two approximations are identical when $\boxed{k_a' \gg k_b}$.

11.28 Maximum velocity $= k_b[E]_0$ [11.28].

Also, rate $= k[E]_0$ with $k = \dfrac{k_b[S]}{K_M + [S]}$ [11.25].

Therefore,

$$\text{rate} = v = \frac{k_b[S][E]_0}{K_M + [S]}. \text{ Rearranging,}$$

$$k_b[\mathrm{E}]_0 = \left\{\frac{K_\mathrm{M} + [\mathrm{S}]}{[\mathrm{S}]}\right\} v$$

$$= \left\{\frac{0.045\,\mathrm{mol\,dm^{-3}} + 0.110\,\mathrm{mol\,dm^{-3}}}{0.110\,\mathrm{mol\,dm^{-3}}}\right\} \times (1.15 \times 10^{-3}\,\mathrm{mol\,dm^{-3}\,s^{-1}})$$

$$= \boxed{1.62 \times 10^{-3}\,\mathrm{mol\,dm^{-3}\,s^{-1}}}.$$

11.29 We have rate $= \dfrac{k_b[\mathrm{E}]_0[\mathrm{S}]}{K_\mathrm{M} + [\mathrm{S}]}$ [11.29] with $k_b[\mathrm{E}]_0$ being the maximum rate.

So the condition is

$$\frac{[\mathrm{S}]}{K_\mathrm{M} + [\mathrm{S}]} = \frac{1}{2} \quad \text{which is satisfied when } \boxed{[\mathrm{S}] = K_\mathrm{M}}.$$

11.30 We start with the Lineweaver–Burk expression, eqn 11.30.

$$\frac{1}{v} = \frac{1}{v_\mathrm{max}} + \left(\frac{K_\mathrm{M}}{v_\mathrm{max}}\right)\frac{1}{[\mathrm{S}]}.$$

Multiply both sides of this equation by v/v_max.

$$v_\mathrm{max} = v + K_\mathrm{M}\left(\frac{v}{[\mathrm{S}]}\right)$$

or

$$v = v_\mathrm{max} - K_\mathrm{M}\left(\frac{v}{[\mathrm{S}]}\right).$$

Thus, a plot of v against $v/[\mathrm{S}]$ yields a straight line with slope equal to $-K_\mathrm{M}$ and intercept v_max.

11.31 We fit the data to the Lineweaver–Burk equation [11.30]. Hence, draw up the following table after converting to a consistent set of concentration units ($\mathrm{mmol\,dm^{-3}}$).

$(1/[\mathrm{S}])/(\mathrm{dm^3\,mmol^{-1}})$	1.0	0.50	0.33	0.25	0.20
$(1/v)/(\mathrm{dm^3\,s\,mmol^{-1}})$	9.1×10^2	5.6×10^2	4.3×10^2	3.8×10^2	3.4×10^2

A plot of these values is shown in Figure 11.3. The curve is linear with slope 7.0×10^2 s and intercept $2.1 \times 10^2\,\mathrm{dm^3\,s\,mmol^{-1}}$. We calculate v_max with the intercept.

$$v_\mathrm{max} = \frac{1}{\text{intercept}} = \frac{1}{2.1 \times 10^2\,\mathrm{dm^3\,s\,mmol^{-1}}} = \boxed{4.8 \times 10^{-3}\,\mathrm{mmol\,dm^{-3}\,s^{-1}}}.$$

The slope is $K_\mathrm{M}/v_\mathrm{max}$; hence

$$K_\mathrm{M} = v_\mathrm{max} \times \text{slope}$$

$$= 4.8 \times 10^{-3}\,\mathrm{mmol\,dm^{-3}\,s^{-1}} \times 7.0 \times 10^2\,\mathrm{s}$$

$$= \boxed{3.4\,\mathrm{mmol\,dm^{-3}}}.$$

The maximum turnover number k_b is given by

$$k_b = \frac{v_{max}}{[E]_0} = \frac{4.8 \times 10^{-3}\,\text{mmol dm}^{-3}\,\text{s}^{-1}}{12.5 \times 10^{-3}\,\text{mmol dm}^{-3}} = \boxed{0.38\,\text{s}^{-1}}.$$

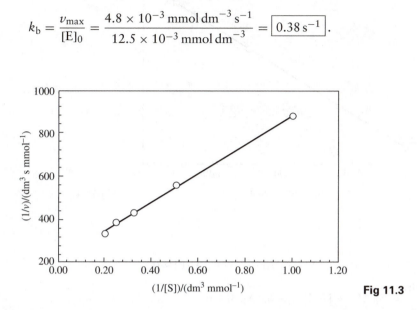

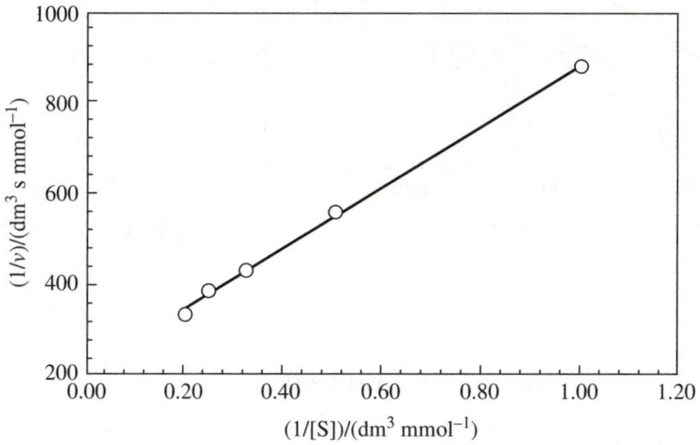

Fig 11.3

11.32 We follow the same procedure as in the solution to Exercise 11.31. We draw up the following table.

$(1/[S])/(\text{dm}^3\,\mu\text{mol}^{-1})$	1.67	1.25	0.71	0.50	0.33
$(1/v)/(\text{dm}^3\,\text{s}\,\mu\text{mol}^{-1})$	1.23	1.03	0.769	0.680	0.592

A plot of $1/v$ against $1/[S]$ is shown in the Figure 11.4.

The intercept is $1/v_{max} = 0.437\,\text{dm}^3\,\text{s}\,\mu\text{mol}^{-1}$. Therefore, $v_{max} = \boxed{2.29\,\mu\text{mol dm}^{-3}\,\text{s}^{-1}}$.

The slope is $K_M/v_{max} = 0.474\,\text{s}$. Therefore,

$$K_M = 0.0474\,\text{s} \times 2.29\,\mu\text{mol dm}^{-3} = \boxed{1.08\,\mu\text{mol dm}^{-3}}.$$

The maximum turnover number k_b is

$$k_b = \frac{v_{max}}{[E]_0} = \frac{2.29\,\mu\text{mol dm}^{-3}\,\text{s}^{-1}}{0.020\,\mu\text{mol dm}^{-3}} = \boxed{1.1 \times 10^2\,\text{s}^{-1}}.$$

11.33 $\boxed{\text{Step 1: initiation}}$ [radicals formed]; $\boxed{\text{Steps 2 and 3: propagation}}$ [new radicals formed]; $\boxed{\text{Step 4: termination}}$ [non-radical product formed].

$$\frac{d[AH]}{dt} = -k_a[AH] - k_c[AH][B].$$

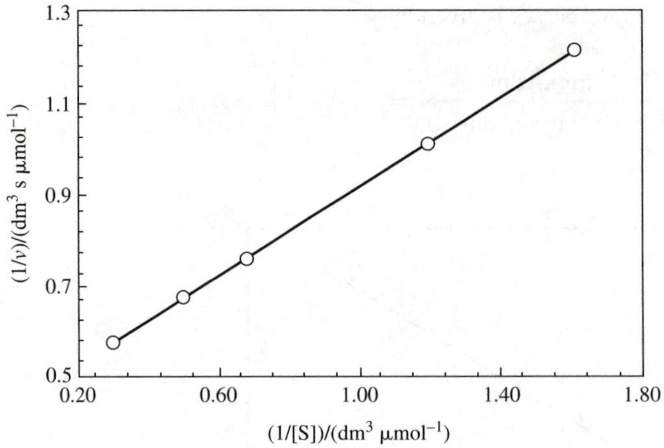

Fig 11.4

(i) $\dfrac{d[A]}{dt} = k_a[AH] - k_b[A] + k_c[AH][B] - k_d[A][B] \approx 0.$

(ii) $\dfrac{d[B]}{dt} = k_b[A] - k_c[AH][B] - k_d[A][B] \approx 0.$

(i + ii) $[A][B] = \left(\dfrac{k_a}{2k_d}\right)[AH].$

(i − ii) $[A] = \left(\dfrac{k_a + 2k_c[B]}{2k_b}\right)[AH].$

Then, solving for [A],

$$[A] = k[AH], \quad k = \left(\frac{k_a}{4k_b}\right) \times \left[1 + \left(1 + \frac{8k_bk_c}{k_ak_d}\right)^{1/2}\right]$$

from which it follows that

$$[B] = \frac{k_a[AH]}{2k_d[A]} = \frac{k_a}{2kk_d}$$

and hence that $\dfrac{d[AH]}{dt} = -k_a[AH] - \left(\dfrac{k_ak_c}{2kk_d}\right)[AH] = -k_{\text{eff}}[AH]$

with $\boxed{k_{\text{eff}} = k_a + \dfrac{k_ak_c}{2kk_d}}$.

Therefore, the decomposition is first-order in AH.

11.34 $\dfrac{d[R]}{dt} = 2k_1[R_2] - k_2[R][R_2] + k_3[R'] - 2k_4[R]^2.$

$$\frac{d[R']}{dt} = k_2[R][R_2] - k_3[R'].$$

Apply the steady-state approximation to both equations.

$$2k_1[R_2]-k_2[R][R_2]+k_3[R']-2k_4[R]^2=0,$$

$$k_2[R][R_2]-k_3[R']=0.$$

The second solves to $[R']=\frac{k_2}{k_3}[R][R_2]$

and then the first solves to $[R]=\left(\frac{k_1}{k_4}[R_2]\right)^{1/2}$.

Therefore, $\frac{d[R_2]}{dt}=-k_1[R_2]-k_2[R_2][R]=\boxed{-k_1[R_2]-k_2\left(\frac{k_1}{k_4}\right)^{1/2}[R_2]^{3/2}}$.

11.35 (a) The radical chain mechanism for the formation of HBr(g) from the elements is discussed in detail in Section 11.15. Using the steady-state approximation for the intermediate of atomic hydrogen and atomic bromine, the mechanism indicates that

$$\text{rate of formation of HBr}=\frac{2k_b(k_a/k_d)^{1/2}[H_2][Br_2]^{3/2}}{[Br_2]+(k_c/k'_b)[HBr]}\quad[11.33]$$

$$=\frac{k[H_2][Br_2]^{3/2}}{[Br_2]+k'[HBr]}$$

where $k=2k_b(k_a/k_d)^{1/2}$ and $k'=k_c/k'_b$.

This rate has the same mathematical form as the experimental rate law and, although the form does not prove that the proposed mechanism is correct, the match is consistent with the proposed mechanism.

(b) (i) When the concentration of HBr is very low so that $k'[HBr]\ll[Br_2]$, the formation rate simplifies.

$$\text{rate of formation of HBr}=\frac{k[H_2][Br_2]^{3/2}}{[Br_2]+k'[HBr]}=\frac{k[H_2][Br_2]^{3/2}}{[Br_2]}=k[H_2][Br_2]^{1/2}.$$

Under this condition, the rate is $\boxed{\text{first-order in }[H_2]}$ and $\boxed{\text{half-order in }[Br_2]}$. The proposed mechanism suggests that the condition is expected when the rate of the propagation step for atomic bromine formation is much greater than the rate of the retardation step for atomic bromine formation, a step that consumes HBr (i.e. $k_c[H][HBr]\ll k'_b[H][Br_2]$).

(ii) When the concentration of HBr is very high so that $k'[HBr]\gg[Br_2]$, the formation rate simplifies.

$$\text{rate of formation of HBr}=\frac{k[H_2][Br_2]^{3/2}}{[Br_2]+k'[HBr]}=\frac{k[H_2][Br_2]^{3/2}}{k'[HBr]}$$

$$=k''\frac{[H_2][Br_2]^{3/2}}{[HBr]}.$$

Under this condition, the rate is first-order in $[H_2]$, three-halves-order in $[Br_2]$, and negative-first-order in [HBr]. The proposed mechanism suggests that the condition is expected when the rate of the propagation step for atomic bromine formation is much less than the rate of the retardation step for atomic bromine formation, a step that consumes HBr (i.e. $k_c[H][HBr] \gg k'_b[H][Br_2]$).

Chapter 12

Quantum theory

Answers to discussion questions

12.1 At the end of the nineteenth century and the beginning of the twentieth, there were many experimental results on the properties of matter and radiation that could not be explained on the basis of established physical principles and theories. Here we list only some of the most significant.

(1) The energy density distribution of black-body radiation as a function of wavelength.
(2) The heat capacities of monatomic solids such as copper metal.
(3) The absorption and emission spectra of atoms and molecules, especially the line spectra of atoms.
(4) The frequency dependence of the kinetic energy of emitted electrons in the photoelectric effect.
(5) The diffraction of electrons by crystals in a manner similar to that observed for X-rays.

12.2 In quantum mechanics, particles are said to have wave characteristics. The fact of the existence of the particles then requires that the wavelengths of the waves representing them be such that the wave does not experience destructive interference upon reflection by a barrier or in its motion around a closed loop. This requirement restricts the wavelength to values $\lambda = 2/n \times L$, where L is the length of the path and n is a positive integer. Then, using the relations $\lambda = h/p$ and $E = p^2/2m$, the energy is quantized at $E = n^2h^2/8mL^2$. This derivation applies specifically to the particle in a box but the derivation is similar for the particle on a ring (see Section 12.10).

12.3 The lowest energy level possible for a confined quantum mechanical system is the zero-point energy, and zero-point energy is not zero energy. The system must have at least that minimum amount of energy even at absolute zero. The physical reason is that, if the particle is confined, its position is not completely uncertain, and therefore its momentum, and hence its kinetic energy, cannot be exactly zero. The particle in a box, the harmonic oscillator, the particle on a ring or on a sphere, the hydrogen atom, and many other systems we will encounter, all have zero-point energy.

12.4 The physical origin of tunnelling is related to the probability density of the particle, which according to the Born interpretation is the square of the wavefunction that represents the particle. This interpretation requires that the wavefunction of the system be everywhere continuous, even at barriers. Therefore, if the wavefunction is non-zero on one side of a barrier it must be non-zero on the other side of the barrier, and this implies that the particle has tunnelled through the barrier. The transmission probability depends upon the mass of the particle (specifically $m^{1/2}$): the greater the mass the smaller the probability of tunnelling. Electrons and protons have small masses, molecular groups large masses; therefore, tunnelling effects are more observable in processes involving electrons and protons.

Solutions to exercises

12.5 We use the Wien displacement law, $T\lambda_{\max} = 2.9\,\text{mm K}$.

$$\lambda_{\max} = \frac{2.9\,\text{mm K}}{2773\,\text{K}} = \boxed{1.0 \times 10^{-6}\,\text{m}}.$$

This radiation is in the near infrared region of the spectrum.

12.6 $E = P \times t, \quad E = Nh\nu = Nh\dfrac{c}{\lambda} = P \times t.$

$$N = \frac{P \times t}{h \times \dfrac{c}{\lambda}} = \frac{0.68 \times 10^{-6}\,\text{W} \times 1.0\,\text{s}}{6.626 \times 10^{-34}\,\text{J s} \times \left(\dfrac{3.00 \times 10^{8}\,\text{m s}^{-1}}{2.45 \times 10^{-7}\,\text{m}}\right)}.$$

$$\boxed{N = 8.4 \times 10^{11}}.$$

12.7 Assume that the wavelength quoted corresponds roughly to $\lambda_{\max}$. Then

$$T = \frac{2.9\,\text{mm K}}{\lambda_{\max}} = \frac{2.9 \times 10^{-3}\,\text{m K}}{6.5 \times 10^{-7}\,\text{m}} = \boxed{4.5 \times 10^{3}\,\text{K}}.$$

$\boxed{\text{This radiation is not thermal radiation.}}$

12.8 $P = M \times A, \quad M = aT^4\,[12.3].$

$$P = aT^4 \times A = (5.67 \times 10^{-8}\,\text{W m}^{-2}\,\text{K}^{-4}) \times (3.00 \times 10^{3}\,\text{K})^4$$
$$\times (5.0\,\text{cm} \times 2.0\,\text{cm} \times 10^{-4}\,\text{m}^2/\text{cm}^2)$$
$$= \boxed{4.6 \times 10^{3}\,\text{W}}.$$

12.9 The Planck energy density distribution reveals that the constant of Wien's displacement law equals $hc/5k$ (see Numerical problem 8.10 in Atkins and de Paula, *Physical Chemistry*, 8th edition, 2006).

$$\lambda_{\max} T = \frac{c_2}{5} \quad \text{where } c_2 = \frac{hc}{k}.$$

Therefore, $\lambda_{\max}T = \dfrac{hc}{5k}$ and, if we find the mean of the $\lambda_{\max}T$ values, we can obtain h from the equation $h = \dfrac{5k}{c}(\lambda_{\max}T)_{\text{mean}}$. We draw up the following table.

$\theta/°\text{C}$	1000	1500	2000	2500	3000	3500
T/K	1273	1773	2273	2773	3273	3773
$\lambda_{\max}/\text{nm}$	2181	1600	1240	1035	878	763
$\lambda_{\max}T/(10^6\ \text{nm K})$	2.776	2.837	2.819	2.870	2.874	2.879

The mean is 2.84×10^6 nm K with a standard deviation of 0.04×10^6 nm K

$$\text{and } h = \frac{(5) \times (1.38066 \times 10^{-23}\ \text{J K}^{-1}) \times (2.84 \times 10^{-3}\ \text{m K})}{2.99792 \times 10^8\ \text{m s}^{-1}} = \boxed{6.54 \times 10^{-34}\ \text{J s}^{-1}}.$$

COMMENT Planck's estimate of the constant h in his first paper of 1900 on black-body radiation was 6.55×10^{-27} erg s(1 erg $= 10^{-7}$ J) which is remarkably close to the current value of 6.626×10^{-34} J s and is essentially the same as the value obtained above. Also from his analysis of the experimental data he obtained values of k (the Boltzmann constant), N_A (the Avogadro constant), and e (the fundamental charge). His values of these constants remained the most accurate for almost 20 years.

12.10 $\Delta E = h\omega = h\nu = \dfrac{h}{t}\left[T = \text{period} = \dfrac{1}{\nu} = \dfrac{2\pi}{\omega}\right].$

(a) $\Delta E = 6.626 \times 10^{-34}\ \text{J s} \times 1.0 \times 10^{15}\ \text{s}^{-1} = 6.6 \times 10^{-19}\ \text{J}$ corresponding to $N_A \times 6.6 \times 10^{-19}\ \text{J} = 4.0 \times 102\ \text{kJ mol}^{-1}$.

(b) $\Delta E = \dfrac{6.626 \times 10^{-34}\ \text{J s}}{2.0 \times 10^{-14}\ \text{s}} = 3.3 \times 10^{-20}\ \text{J} = \boxed{20\ \text{kJ mol}^{-1}}.$

(c) $\Delta E = \dfrac{6.626 \times 10^{-34}\ \text{J s}}{0.50\ \text{s}} = 1.3 \times 10^{-33}\ \text{J} = \boxed{8.0 \times 10^{-3}\ \text{kJ mol}^{-1}}.$

12.11 $P = \dfrac{E}{t}.$

$$N = \frac{P}{h\nu} = \frac{P\lambda}{hc} \quad [P = \text{power in J s}^{-1}]$$

$$= \frac{P\lambda}{(6.626 \times 10^{-34}\ \text{J Hz}^{-1}) \times (2.998 \times 10^8\ \text{m s}^{-1})}$$

$$= \frac{(P/\text{W}) \times (\lambda/\text{nm})\ \text{s}^{-1}}{1.99 \times 10^{-16}} = 5.03 \times 10^{15}\ (P/\text{W}) \times (\lambda/\text{nm})\ \text{s}^{-1}.$$

(a) $N = (5.03 \times 10^{15}) \times 1.0 \times 350\ \text{s}^{-1} = \boxed{1.8 \times 10^{18}\ \text{s}^{-1}}.$

(b) $N = (5.03 \times 10^{15}) \times 100 \times 350\ \text{s}^{-1} = \boxed{1.8 \times 10^{20}\ \text{s}^{-1}}.$

12.12 $$N = \frac{P}{h\nu} = \frac{45 \times 10^3\,\mathrm{W}}{6.626 \times 10^{-34}\,\mathrm{J\,s} \times 98.4 \times 10^6\,\mathrm{s^{-1}}}\,[\mathrm{W} = \mathrm{J\,s^{-1}}],$$

$$N = \boxed{6.90 \times 10^{29}\,\mathrm{s^{-1}}}.$$

12.13 $$\Phi = 2.14\,\mathrm{eV} = 2.14\,\mathrm{eV} \times 1.602 \times 10^{-19}\,\mathrm{J\,eV^{-1}}$$

$$= 3.43 \times 10^{-19}\,\mathrm{J}.$$

$$\frac{1}{2}m_e v^2 = h\nu - \Phi = \frac{hc}{\lambda} - \Phi\,[12.6].$$

(a) $$\frac{hc}{\lambda} = \frac{(6.626 \times 10^{-34}\,\mathrm{J\,s}) \times 2.998 \times 10^8\,\mathrm{m\,s^{-1}}}{(750 \times 10^{-9}\,\mathrm{m})}$$

$$= 2.65 \times 10^{-19}\,\mathrm{J} < \Phi. \boxed{\text{Therefore, no ejection occurs}}.$$

(b) $$\frac{hc}{\lambda} = \frac{(6.626 \times 10^{-34}\,\mathrm{J\,s}) \times 2.998 \times 10^8\,\mathrm{m\,s^{-1}}}{(250 \times 10^{-9}\,\mathrm{m})}$$

$$= 7.95 \times 10^{-19}\,\mathrm{J}$$

Hence $1/2\,mv^2 = (7.95 - 3.34) \times 10^{-19}\,\mathrm{J} = \boxed{4.52 \times 10^{-19}\,\mathrm{J}}$.

$$v = \left(\frac{2 \times 4.52 \times 10^{-19}\,\mathrm{J}}{9.109 \times 10^{-13}\,\mathrm{kg}}\right)^{1/2} = \boxed{996\,\mathrm{km\,s^{-1}}}.$$

12.14 $p = mv$ and $p = \dfrac{h}{\lambda}$.

Therefore,

$$v = \frac{h}{m\lambda} = \frac{(6.626 \times 10^{-34}\,\mathrm{J\,s})}{(9.109 \times 10^{-31}\,\mathrm{kg}) \times (0.55 \times 10^{-9}\,\mathrm{m})} = \boxed{1.32 \times 10^6\,\mathrm{m\,s^{-1}}}.$$

12.15 $$\lambda = \frac{h}{p} = \frac{h}{mv}.$$

(a) $$\lambda = \frac{(6.626 \times 10^{-34}\,\mathrm{J\,s})}{(1.00\,\mathrm{m\,s^{-1}}) \times 1.0 \times 10^{-3}\,\mathrm{kg}} = \boxed{6.6 \times 10^{-31}\,\mathrm{m}}.$$

(b) $$\lambda = \frac{(6.626 \times 10^{-34}\,\mathrm{J\,s})}{(1.0 \times 10^8\,\mathrm{m\,s^{-1}}) \times 1.0 \times 10^{-3}\,\mathrm{kg}} = \boxed{6.6 \times 10^{-39}\,\mathrm{m}}.$$

(c) $$\lambda = \frac{(6.626 \times 10^{-34}\mathrm{J\,s})}{4.003 \times (1.6605 \times 10^{-27}\,\mathrm{kg}) \times (1.0 \times 10^3\,\mathrm{m\,s^{-1}})} = \boxed{99.7\,\mathrm{pm}}.$$

12.16 In order to avoid confusion between the potential difference V and the unit volt, we represent the potential difference below as $\Delta\phi$. Then

$$\frac{1}{2}mv^2 = e\Delta\phi, \quad \text{implying that} \quad v = \left(\frac{2e\Delta\phi}{m}\right)^{1/2} \quad \text{and}$$

$p = mv = (2me\Delta\phi)^{1/2}$. Therefore,

$$\lambda = \frac{h}{p} = \left(\frac{h}{(2me\Delta\phi)^{1/2}}\right)$$

$$= \frac{(6.626 \times 10^{-34}\,\text{J s})}{\{2 \times (9.109 \times 10^{-31}\,\text{kg}) \times (1.602 \times 10^{-19}\,\text{C} \times \Delta\phi)\}^{1/2}}$$

$$= \frac{1.226\,\text{nm}}{(\Delta\phi/\text{V})^{1/2}} \; [1\,\text{J} = 1\,\text{C V}].$$

(a) $\Delta\phi = 1.00\,\text{V}, \lambda = \boxed{1.23\,\text{nm}}$.

(b) $\Delta\phi = 1.0\,\text{kV}, \lambda = \dfrac{1.226\,\text{nm}}{31.6} = \boxed{39\,\text{pm}}$.

(c) $\Delta\phi = 100\,\text{kV}, \lambda = \dfrac{1.226\,\text{nm}}{316.2} = \boxed{3.88\,\text{pm}}$.

12.17 $m = 85\,\text{kg}, \quad v = 8.0\,\text{km h}^{-1}$.

$$\lambda = \frac{h}{p} = \frac{h}{mv} \quad \text{[de Broglie relation [12.7]]}$$

$$= \frac{6.63 \times 10^{-34}\,\text{J s}}{(85\,\text{kg}) \times (8.0 \times 10^{3}\,\text{m h}^{-1})}\left(\frac{3600\,\text{s}}{1\,\text{h}}\right) = \boxed{3.5 \times 10^{-36}\,\text{m}}.$$

This extraordinarily small wavelength is much, much smaller than the diameter of a hydrogen nucleus and the calculation illustrates the hopelessness of measuring the de Broglie wavelength of a macroscopic object. The de Broglie wavelength does increase as the speed of an object decreases and, according to the quantum behavior of a particle in a one-dimensional box of length L, the de Broglie wavelength may be as long as $2L$.

12.18 $p = \dfrac{h}{\lambda}$.

(a) $p = \dfrac{(6.626 \times 10^{-34}\,\text{J s})}{(725 \times 10^{-9}\,\text{m})} = \boxed{9.14 \times 10^{-28}\,\text{kg m s}^{-1}}$.

(b) $p = \dfrac{(6.626 \times 10^{-34}\,\text{J s})}{(75 \times 10^{-12}\,\text{m})} = \boxed{8.8 \times 10^{-24}\,\text{kg m s}^{-1}}$.

(c) $p = \dfrac{(6.626 \times 10^{-34}\,\text{J s})}{(20\,\text{m})} = \boxed{3.3 \times 10^{-35}\,\text{kg m s}^{-1}}$.

12.19 $E = h\nu = \dfrac{hc}{\lambda}$.

$hc = (6.6208 \times 10^{-34}\,\text{J s}) \times (2.99792 \times 10^{8}\,\text{m s}^{-1}) = 1.986 \times 10^{-25}\,\text{J m}$.

$$N_A hc = \left(6.02214 \times 10^{23}\,\text{mol}^{-1}\right) \times \left(1.986 \times 10^{-25}\,\text{J m}\right)$$

$$= 0.1196\,\text{J m mol}^{-1}.$$

We can therefore draw up the following table

λ	E/J	E/(kJ mol^{-1})
(a) 600 nm	3.31×10^{-19}	199
(b) 550 nm	3.61×10^{-19}	218
(c) 400 nm	4.97×10^{-19}	299
(d) 200 nm	9.93×10^{-15}	598
(e) 150 pm	1.32×10^{-15}	7.98×10^5
(f) 1.00 cm	1.99×10^{-23}	0.012

12.20 $p = \dfrac{h}{\lambda} = \dfrac{6.626 \times 10^{-34}\,\text{J s}}{\lambda} = mv.$

$$v = \frac{(6.626 \times 10^{-34}\,\text{J s})}{(300 \times 10^{-9}\,\text{m}) \times 1.0 \times 10^{-3}\,\text{kg}}$$

$$= \boxed{2.2 \times 10^{-24}\,\text{m s}^{-1}}.$$

12.21 The momentum per photon of wavelength 650 nm is

$$p = \frac{h}{\lambda} = \frac{6.626 \times 10^{-34}\,\text{J s}}{650 \times 10^{-9}\,\text{m}} = 1.02 \times 10^{-27}\,\text{kg m s}^{-2}$$

and this is also the change of momentum per photon absorbed by the fabric. The rate of photon bombardment is the power (rate of energy bombardment) divided by the energy per photon (hc/λ). So, the number of photons per second is given by

$$\frac{N}{t} = \frac{P\lambda}{hc} = \frac{1 \times 10^{3}\,\text{J s}^{-1} \times 6.50 \times 10^{-7}\,\text{m}}{6.626 \times 10^{-34}\,\text{J s} \times 2.998 \times 10^{8}\,\text{m s}^{-1}}$$

$$= 3.27 \times 10^{21}\,\text{s}^{-1}.$$

(a) The force is rate of change of momentum, so

force = number of photons per second × momentum per photon

$$= 3.27 \times 10^{21}\text{s}^{-1} \times 1.02 \times 10^{-27}\,\text{kg m s}^{-1}$$

$$= \boxed{3.34 \times 10^{-6}\,\text{kg m s}^{-2}} = \boxed{3.34 \times 10^{-6}\,\text{N}}.$$

(b) $p = \dfrac{F}{A} = \dfrac{3.34 \times 10^{-6}\,\text{kg m s}^{-2}}{1.0 \times 10^{6}\,\text{m}^2} = 3.34 \times 10^{-12}\,\text{kg m}^{-1}\,\text{s}^{-2}$

$$= \boxed{3.34 \times 10^{-12}\,\text{Pa}}.$$

(c) speed $= \dfrac{\text{force}}{\text{mass}} \times \text{time}$ or $v = \dfrac{F}{M} \times t.$

$$t = \frac{mv}{F} = \frac{1.0\,\text{kg} \times 1.0\,\text{m s}^{-1}}{3.34 \times 10^{-6}\,\text{kg m s}^{-2}}$$

$$= \boxed{3.0 \times 10^{5}\,\text{s}}\ (\text{about } 83\ \text{h})\,.$$

Not a very practical method!

12.22 This is essentially the photoelectric effect with the work function Φ being the ionization energy I. Hence,

$$\frac{1}{2}m_e v^2 = h\nu - I = \frac{hc}{\lambda} - I.$$

Solving for λ,

$$\lambda = \frac{hc}{I + \frac{1}{2}mv^2} = \frac{(6.626 \times 10^{-34}\,\text{J s}) \times (2.998 \times 10^{8}\,\text{m s}^{-1})}{(3.44 \times 10^{-18}\,\text{J}) + \frac{1}{2}(9.109 \times 10^{-31}\,\text{kg}) \times (1.03 \times 10^{6}\,\text{m s}^{-1})^2}$$

$$= 5.06 \times 10^{-8}\,\text{m} = \boxed{50.6\,\text{nm}}\,.$$

QUESTION What is the energy of the photon?

12.23 $$\frac{1}{2}mv^2 = h\nu - I, \quad \nu = \frac{c}{\lambda}.$$

$$I = \frac{hc}{\lambda} - \frac{1}{2}mv^2 = \frac{(6.626 \times 10^{-34}\,\text{J s}) \times 2.998 \times 10^{8}\,\text{m s}^{-1}}{150 \times 10^{-12}\,\text{m}} - \frac{1}{2}\left(9.109 \times 10^{-31}\,\text{kg}\right)$$

$$\times \left(2.24 \times 10^{7}\,\text{m s}^{-1}\right)^2$$

$$= \boxed{1.11 \times 10^{-15}\,\text{J}}\,.$$

12.24 The Born interpretation (Section 12.7) of a normalized wavefunction states that the probability, P, of finding a particle in a very small region equals $\psi^2\delta V$. When consideration is given to a particle in a one-dimensional box, this becomes $P = \psi^2\delta x$ where ψ is evaluated at the mid-point of the δx range. This method is an estimate for which improvements require the use of calculus (see Exercise 12.25).

$$P = \psi^2\delta x = \left\{\left(\frac{2}{L}\right)^{1/2} \sin^2\left(\frac{2\pi x}{L}\right)\right\}^2 \delta x = \left(\frac{2}{L}\right) \sin^2\left(\frac{2\pi x}{L}\right)\ \delta x.$$

(a) $$P = \left(\frac{2}{10\,\text{nm}}\right) \times \sin^2\left(\frac{2\pi \times 0.15\,\text{nm}}{10\,\text{nm}}\right) \times (0.2\,\text{nm} - 0.1\,\text{nm}) = \boxed{1.\overline{77} \times 10^{-4}}\,.$$

(b) $$P = \left(\frac{2}{10\,\text{nm}}\right) \sin^2\left(\frac{2\pi \times 5.05\,\text{nm}}{10\,\text{nm}}\right) (5.2\,\text{nm} - 4.9\,\text{nm}) = \boxed{5.\overline{92} \times 10^{-5}}\,.$$

12.25 When the Born interpretation (Section 12.7) describes the infinitesimally small probability, dP, of finding a particle in an infinitely small region dV, it is written as the differential equation $dP = \psi^2 dV$ where the coordinates of ψ are those of the infinitesimal volume

dV. To find the probability that the particle will be found in the region of V, integration must be performed over the region.

$$P = \int_{\text{region}} dV = \int_{\text{region}} \psi^2 \, dV.$$

When consideration is given to a particle in a one-dimensional box, where the region is between x_1 and x_2, this becomes

$$P = \int_{x_1}^{x_2} \psi^2 \, dx, \quad \text{where } \psi = \left(\frac{2}{L}\right)^{1/2} \sin\left(\frac{2\pi x}{L}\right).$$

$$P = \int_{x_1}^{x_2} \left\{\left(\frac{2}{L}\right)^{1/2} \sin\left(\frac{2\pi x}{L}\right)\right\}^2 dx = \left(\frac{2}{L}\right) \int_{x_1}^{x_2} \sin^2\left(\frac{2\pi x}{L}\right) dx.$$

Using the standard integral $\int \sin^2(ax)\, dx = \frac{x}{2} - \frac{\sin(2ax)}{4a}$, the working equation becomes

$$P = \left(\frac{2}{L}\right)\left[\frac{x}{2} - \frac{\sin\left(2\left(\frac{2\pi}{L}\right)x\right)}{4\left(\frac{2\pi}{L}\right)}\right]_{x_1}^{x_2} = \left[\frac{x}{L} - \frac{1}{4\pi}\sin\left(\frac{4\pi x}{L}\right)\right]_{x_1}^{x_2}.$$

(a) $$P = \left[\frac{x}{10\,\text{nm}} - \frac{1}{4\pi}\sin\left(\frac{4\pi x}{10\,\text{nm}}\right)\right]_{0.1\,\text{nm}}^{0.2\,\text{nm}} = 1.\overline{84} \times 10^{-4}.$$

Error of the Exercise 12.24 approximation:

$$\frac{1.\overline{84} \times 10^{-4} - 1.\overline{77} \times 10^{-4}}{1.\overline{84} \times 10^{-4}} \times 100 = 12.9\%.$$

(b) $$P = \left[\frac{x}{10\,\text{nm}} - \frac{1}{4\pi}\sin\left(\frac{4\pi x}{10\,\text{nm}}\right)\right]_{4.9\,\text{nm}}^{5.2\,\text{nm}} = 2.\overline{36} \times 10^{-4}.$$

Error of the Exercise 12.24 approximation:

$$\left|\frac{2.\overline{36} \times 10^{-4} - 5.\overline{92} \times 10^{-5}}{2.\overline{36} \times 10^{-4}}\right| \times 100 = 74.9\%.$$

12.26 $$P = \int_{x_1}^{x_2} \psi^2 \, dx, \quad \text{where } \psi = \left(\frac{2}{L}\right)^{1/2} \sin\left(\frac{\pi x}{L}\right).$$

$$P = \int_{x_1}^{x_2} \left\{\left(\frac{2}{L}\right)^{1/2} \sin\left(\frac{\pi x}{L}\right)\right\}^2 dx = \left(\frac{2}{L}\right) \int_{x_1}^{x_2} \sin^2\left(\frac{\pi x}{L}\right) dx.$$

Using the standard integral $\int \sin^2(ax)\, dx = \frac{x}{2} - \frac{\sin(2ax)}{4a}$, the working equation becomes

$$P = \left(\frac{2}{L}\right)\left[\frac{x}{2} - \frac{\sin\left(2\left(\frac{\pi}{L}\right)x\right)}{4\left(\frac{\pi}{L}\right)}\right]_{x_1}^{x_2} = \left[\frac{x}{L} - \frac{1}{2\pi}\sin\left(\frac{2\pi x}{L}\right)\right]_{x_1}^{x_2}.$$

(a) $P = \left[\frac{x}{L} - \frac{1}{2\pi}\sin\left(\frac{2\pi x}{L}\right)\right]_0^{L/3} = \left[x - \frac{1}{2\pi}\sin(2\pi x)\right]_0^{1/3} = \boxed{0.196}$.

(b) $P = \left[\frac{x}{L} - \frac{1}{2\pi}\sin\left(\frac{2\pi x}{L}\right)\right]_{L/3}^{2L/3} = \left[x - \frac{1}{2\pi}\sin(2\pi x)\right]_{1/3}^{2/3} = \boxed{0.609}$.

(c) $P = \left[\frac{x}{L} - \frac{1}{2\pi}\sin\left(\frac{2\pi x}{L}\right)\right]_{2L/3}^{L} = \left[x - \frac{1}{2\pi}\sin(2\pi x)\right]_{2/3}^{1} = \boxed{0.196}$.

Note that the probabilities sum to 1.

12.27 $\Delta p \approx 0.0100$ per cent of p, and $p = m_p v$

$$= p \times 1.00 \times 10^{-4}.$$

$$\Delta x = \frac{\hbar}{2\Delta p} \quad [12.9]$$

$$\approx \frac{(1.055 \times 10^{-34}\,\mathrm{J\,s})}{2 \times (1.673 \times 10^{-27}\,\mathrm{kg}) \times (3.5 \times 10^5\,\mathrm{m\,s^{-1}}) \times (1.00 \times 10^{-4})}$$

$$= 9.0 \times 10^{-10}\,\mathrm{m}, \text{ or } \boxed{90\,\mathrm{nm}}.$$

12.28 $\Delta p \Delta x \geq \frac{1}{2}\hbar, \quad \Delta p = m\Delta v.$

$$\Delta v_{\min} = \frac{\hbar}{2m\Delta x} = \frac{1.055 \times 10^{-34}\,\mathrm{J\,s}}{2 \times 0.500\,\mathrm{kg} \times 5.0 \times 10^{-6}\,\mathrm{m}} = \boxed{2.1 \times 10^{-29}\,\mathrm{m\,s^{-1}}}.$$

12.29 $\Delta p \Delta x \geq \frac{1}{2}\hbar, \quad \Delta p = m\Delta v.$

$$\Delta x_{\min} = \frac{\hbar}{2m\Delta x} = \frac{1.055 \times 10^{-34}\,\mathrm{J\,s}}{2 \times 0.005\,\mathrm{kg} \times 5.0 \times 10^{-6}\,\mathrm{m}} = \boxed{1 \times 10^{-26}\,\mathrm{m}}.$$

12.30 The minimum uncertainty in position is $\boxed{100\,\mathrm{pm}}$. Therefore, because $\Delta x \Delta p \geq \frac{1}{2}\hbar$,

$$\Delta p \geq \frac{\hbar}{2\Delta x} = \frac{1.0546 \times 10^{-34}\,\mathrm{J\,s}}{2\,(100 \times 10^{-12}\,\mathrm{m})} = 5.3 \times 10^{-25}\,\mathrm{kg\,m\,s^{-1}}.$$

$$\Delta v = \frac{\Delta p}{m} = \frac{5.3 \times 10^{-25}\,\mathrm{kg\,m\,s^{-1}}}{9.11 \times 10^{-31}\,\mathrm{kg}} = \boxed{5.8 \times 10^{-5}\,\mathrm{m\,s^{-1}}}.$$

12.31 $E_n = \frac{n^2 h^2}{8\,mL^2}.$

$$E_2 - E_1 = \frac{(2)^2 h^2}{8\,mL^2} - \frac{(1)^2 h^2}{8\,mL^2}$$

$$= \frac{3 \times (6.626 \times 10^{-34}\,\mathrm{J\,s})^2}{8\,(1.672 \times 10^{-27}\,\mathrm{kg})\,(1.0 \times 10^{-9}\,\mathrm{m})^2}$$

$$= \boxed{9.85 \times 10^{-23}\,\mathrm{J}}.$$

12.32 The maximum probability occurs at $\frac{\pi x}{L} = \frac{\pi}{2}$, corresponding to $x = \frac{1}{2L}$.

At that location

$$P_{\max} \propto \psi^2 = \left(\frac{2}{L}\right)\sin^2\frac{\pi}{2} = \frac{2}{L}$$

$$50\%\ \max = 0.5\frac{2}{L} = \frac{1}{L}$$

We need to find the location at which $P_{1/2} = \frac{1}{L}$. This corresponds to the angle at which $\sin^2\theta = \frac{1}{2}$. This is $\theta = \frac{\pi}{4}$, corresponding to, $\boxed{x = \frac{L}{4}}$ or $\boxed{x = \frac{3L}{4}}$

12.33 (a)

$$\Delta E = (5^2 - 4^2)\frac{h^2}{8mL^2} = \frac{9h^2}{8mL^2}$$

$$= \frac{9 \times (6.626 \times 10^{-34}\,\mathrm{J\,s})^2}{8 \times (9.109 \times 10^{-31}\,\mathrm{kg})(5 \times 10^{-9}\,\mathrm{m})^2}$$

$$= \boxed{2.17 \times 10^{-20}\,\mathrm{J}}.$$

(b) $\Delta E = h\nu = \frac{hc}{\lambda}$, $\quad \lambda = \frac{hc}{\Delta E}$.

$$\lambda = (6.62608 \times 10^{-34}\,\mathrm{J\,s}) \times (2.99792 \times 10^{8}\,\mathrm{m\,s^{-1}})/\left(2.17 \times 10^{-20}\,\mathrm{J}\right)$$

$$= \boxed{9.16 \times 10^{-6}\,\mathrm{m}}.$$

12.34

$$\int_0^\infty \psi^2 dx \int_0^L \psi^2 dx = \int_0^L A^2 dx = 1$$

$$= A^2 x\,\big|_0^L = A^2 L = 1.$$

Therefore $A^2 = \frac{1}{L}, A = \left(\frac{1}{L}\right)^{1/2}$.

The normalized wave function is $\boxed{\psi = \left(\frac{1}{L}\right)^{1/2}}$.

12.35 Carotene is a chain of 22 carbon atoms; therefore there are 21 alternating single and double bonds. There are 11 double bonds, hence 11 π-bonds, each of which has two electrons.

$$L = 21 \times (1.40 \times 10^{-10}\,\mathrm{m}) = 2.94 \times 10^{-9}\,\mathrm{m}.$$

$$\Delta E = E_{12} - E_{11} = (2n+1) \times \frac{h^2}{8mL^2} = (2 \times 11 + 1) \times \frac{h^2}{8mL^2}$$

$$= \frac{(23) \times (6.626 \times 10^{-34}\,\mathrm{J\,s})^2}{(8) \times (9.11 \times 10^{-31}\,\mathrm{kg}) \times (2.94 \times 10^{-9})^2}$$

$$= 1.60\overline{3} \times 10^{-19}\,\mathrm{J} = 1.60 \times 10^{-19}\,\mathrm{J}.$$

$$\nu = \frac{\Delta E}{h} = \frac{1.60\overline{3} \times 10^{-19}\,\text{J}}{6.626 \times 10^{-34}\,\text{J s}} = 2.42 \times 10^{14}\,\text{s}^{-1} = 2.42 \times 10^{14}\,\text{Hz}.$$

$$\lambda = \frac{c}{\nu} = \frac{2.998 \times 10^{8}\,\text{m s}^{-1}}{2.42 \times 10^{14}\,\text{s}^{-1}} = \boxed{1.24 \times 10^{-6}\,\text{m}}.$$

COMMENT The observed wavelength for this transition is 4.5×10^{-7} m, so our crude model is off by more than a factor of 2. But it is the right order of magnitude.

12.36 (a)

$$\begin{aligned} I &= mr^2 \\ &= 1.008\,\text{u} \times 1.6605 \times 10^{-27}\,\text{kg/u} \times (1.61 \times 10^{-10}\,\text{m})^2 \\ &= \boxed{4.34 \times 10^{-47}\,\text{kg m}^2}. \end{aligned}$$

(b)

$$E_{m_l} = \frac{m_l^2 \hbar^2}{2I} \quad [12.15].$$

$$E_0 = 0\ [m_l = 0].$$

$$E_1 = \frac{\hbar^2}{2I}\ [m_l = 1].$$

$$\Delta E = E_1 - E_0 = h\nu = \frac{hc}{\lambda} = \frac{\hbar^2}{2I} = \frac{h^2}{8\pi^2 I}.$$

$$\lambda = \frac{8\pi^2 cI}{h} = \frac{8\pi^2 \times 2.998 \times 10^{8}\,\text{m s}^{-1} \times 4.34 \times 10^{-47}\,\text{kg m}^2}{6.626 \times 10^{-34}\,\text{J s}},$$

$$\lambda = 1.55 \times 10^{-3}\,\text{m} = \boxed{1.55\,\text{nm}}.$$

This wavelength is in the microwave region of the electromagnetic spectrum.

12.37 A period T is the reciprocal of the frequency ν. Therefore $\nu = 1\text{s}^{-1}$.

$$\nu = \frac{1}{2\pi}\left(\frac{k}{m}\right)^{1/2} \quad [12.18].$$

Solve for the force constant, k.

$$\begin{aligned} k &= (2\pi\nu)^2 m \\ &= (2\pi \times 1\,\text{s}^{-1})^2 \times 1 \times 10^{-3}\,\text{kg} \\ &= 0.04\,\text{kg s}^{-2} = \boxed{0.04\,\text{N m}^{-1}}. \end{aligned}$$

12.38 (a)

$$\begin{aligned} \nu &= \frac{1}{2\pi}\left(\frac{k}{m}\right)^{1/2} \quad [12.18] \\ &= \frac{1}{2\pi}\left(\frac{314\,\text{N m}^{-1}}{1.008\,\text{u} \times 1.6605 \times 10^{-27}\,\text{kg/u}}\right)^{1/2} = \boxed{6.89 \times 10^{13}\,\text{s}^{-1}}. \end{aligned}$$

(b) $\lambda = \dfrac{c}{\nu} = \dfrac{2.998 \times 10^{8}\,\text{m}\,\text{s}^{-1}}{2.42 \times 10^{13}\,\text{s}^{-1}} = 4.35 \times 10^{-6}\,\text{m} = \boxed{4.35\,\mu\text{m}}$.

12.39 The D—I bond and the H—I bond are expected to have almost identical bond strengths and identical bonding force constants, k, because bonding is an electronic, not a mass/isotopic, property. However, the vibrational frequency does have a mass dependence.

$$\frac{\nu_{\text{DI}}}{\nu_{\text{HI}}} = \frac{\dfrac{1}{2\pi}\left(\dfrac{k}{m_{\text{D}}}\right)^{1/2}}{\dfrac{1}{2\pi}\left(\dfrac{k}{m_{\text{H}}}\right)^{1/2}} \;[12.18] = \left(\frac{m_{\text{H}}}{m_{\text{D}}}\right)^{1/2} = \left(\frac{1}{2}\right)^{1/2} = 0.707$$

When hydrogen-1 is replaced by hydrogen-2 (deuterium) in H—I, the vibrational frequency is reduced by a factor of 0.707.

12.40 (a) $N^2 \displaystyle\int_{-\infty}^{\infty} \left(\text{e}^{-ax^2/2}\right)^2 \text{d}x = 1$ (See Derivation 12.2),

$$N^2 \int_{-\infty}^{\infty} \text{e}^{-ax^2}\,\text{d}x = 1.$$

Using the standard, definite integral $\displaystyle\int_{-\infty}^{\infty} \text{e}^{-ax^2}\,\text{d}x = \left(\frac{\pi}{a}\right)^{1/2}$, we find that

$$N^2 \left(\frac{\pi}{a}\right)^{1/2} = 1 \quad \text{or} \quad \boxed{N = \left(\frac{a}{\pi}\right)^{1/4}}.$$

The normalized wavefunction is $\psi = \left(\dfrac{a}{\pi}\right)^{1/4} \text{e}^{-ax^2/2}$.

(b) The function $\psi = \left(\dfrac{a}{\pi}\right)^{1/4} \text{e}^{-ax^2/2}$ is a 'bell' or 'Gaussian' curve with a maximum $x = 0$. Consequently, ψ^2 has a maximum value at the displacement $x = 0$ and the Born interpretation of ψ^2 (see Section 12.7) indicates that the displacement $x = 0$ is the most probable displacement. It is instructive to use calculas to find the maximum. This requires identification of the displacement for which $\text{d}\psi/\text{d}x = 0$ and showing that $\text{d}\psi/\text{d}x > 0$ before the maximum and $\text{d}\psi/\text{d}x < 0$ after the maximum

$$\frac{\text{d}\psi}{\text{d}x} = \frac{\text{d}}{\text{d}x}\left\{\left(\frac{a}{\pi}\right)^{1/4} \text{e}^{-ax^2/2}\right\} = \left(\frac{a}{\pi}\right)^{1/4} \frac{\text{d}}{\text{d}x}\,\text{e}^{-ax^2/2} = -a\,x\,\text{e}^{-ax^2/2}.$$

The factor of x in the first derivative indicates that the derivative equals zero when $x = 0$. Furthermore, the formula for the derivative clearly shows that the derivative is positive when $x < 0$ and the derivative is negative when $x > 0$. The function is a maximum at $x = 0$.

Chapter 13

Atomic structure

Answers to discussion questions

13.1 (1) The principal quantum number, n, determines the energy of a hydrogenic atomic orbital through eqn 13.4a.

(2) The azimuthal quantum number, l, determines the magnitude of the angular momentum of a hydrogenic atomic orbital through the formula $\{l(l+1)\}^{1/2}\hbar$.

(3) The magnetic quantum number, m_l, determines the z-component of the angular momentum of a hydrogenic orbital through the formula $m_l\hbar$.

(4) The spin quantum number, s, determines the magnitude of the spin angular momentum through the formula $\{s(s+1)\}^{1/2}\hbar$. For hydrogenic atomic orbitals, s can only be $1/2$.

(5) The spin quantum number, m_s, determines the z-component of the spin angular momentum through the formula $m_s\hbar$. For hydrogenic atomic orbitals, m_s can only be $\pm 1/2$.

13.2 (a) A boundary surface for a hydrogenic orbital is drawn to contain most (say, 90%) of the probability density of an electron in that orbital. Its shape varies from orbital to orbital because the electron density distribution is different for different orbitals.

(b) The radial distribution function gives the probability that the electron will be found anywhere within a shell of radius r around the nucleus. It gives a better picture of where the electron is likely to be found with respect to the nucleus than the probability density, which is the square of the wavefunction.

13.3 In the simplest orbital approximation, many-electron atomic wavefunctions are represented as a simple product of one-electron wavefunctions. That is, each electron is in its 'own' orbital and the approximate atomic wavefunction is $\psi = \psi(1)\psi(2)\psi(3)\cdots$ where $\psi(i)$ is the orbital of the ith electron. Orbitals in heavier atoms resemble hydrogenic orbitals but nuclear charges are modified to account for screening. Parameters are used to minimize the energy of the approximate wavefunction. The orbital approximation allows us to express the electronic structure of the atom by reporting the electronic configuration as a list of occupied orbitals. This simple model does not yield accurate spectral and reaction information but it is very useful for the conceptual framework of organic and inorganic chemistry. The orbital approximation is based on the disregard of significant portions of the electron-electron interaction terms in the many-electron Hamiltonian, so we cannot expect that it will be quantitatively accurate. By abandoning the orbital approximation, we could in principle obtain essentially exact energies; however, its conceptual advantages are very significant. The everyday vocabulary of chemistry relies heavily upon the description of atomic structure in terms of orbitals.

At the somewhat more sophisticated level, the many-electron wavefunctions are written as select linear combinations of product orbitals that explicitly satisfy the Pauli exclusion principle as described in the section *Further information* 13.1. Sections 14.16 to 14.19 present a discussion of more advanced methods of computational chemistry.

13.4 (a) The selection rules for hydrogenic atoms are

$$\Delta n = \pm 1, \pm 2, \cdots \quad \Delta l = \pm 1 \quad \Delta m_l = 0, \pm 1.$$

In a spectroscopic transition, the atom emits or absorbs a photon. Photons have a spin angular momentum of 1. Therefore, because of the transition, the angular momentum of the electromagnetic field has changed by $\pm 1\hbar$. The principle of the conservation of angular momentum then requires that the angular momentum of the atom has undergone an equal and opposite change in angular momentum. Hence, the selection rule on $\Delta l = \pm 1$. The principal quantum number n can change by any amount since n does not directly relate to angular momentum. The selection rule on Δm_l is harder to account for on the basis of these simple considerations alone. One has to evaluate the transition dipole moment between the wavefunctions representing the initial and final states involved in the transition. See Section 19.3 for additional information.

(b) The selection rules for relatively light many-electron atoms are

$$\Delta S = 0, \quad \Delta L = 0, \pm 1, \quad \Delta l = \pm 1$$

$$\Delta J = 0, \pm 1, \text{ but } J = 0 \leftrightarrow J = 0 \text{ is forbidden.}$$

A change in the total spin angular momentum is forbidden in an electronic transition because light does not directly affect spin. This important selection rule applies to both atoms and molecules. Additional rules arise from the conservation of total angular momentum in the atom-radiation system. As discussed in part (a) the orbital angular momentum of an individual electron must change. This may, or may not, affect the total orbital angular momentum.

Solutions to exercises

13.5 $$\tilde{\nu} = \frac{1}{\lambda} = R_H\left(\frac{1}{n_1^2} - \frac{1}{n_2^2}\right) \quad [13.1]$$

and $n_1 = 2$, $n_2 = 5$.

$$R_H = 109677\,\text{cm}^{-1} = 1.09677 \times 10^7\,\text{m}^{-1}.$$

Solving for λ,

$$\lambda = (1.09677 \times 10^7\,\mathrm{m}^{-1})^{-1}\left(\frac{1}{4} - \frac{1}{25}\right)^{-1}$$

$$= \boxed{434\,\mathrm{nm}}.$$

13.6 $$\frac{1}{\lambda} = R_H\left(\frac{1}{9} - \frac{1}{n^2}\right)$$

and therefore

$$n = \left(\frac{1}{9} - \frac{1}{\lambda R_H}\right)^{-1/2} = \left(\frac{1}{9} - \frac{\nu}{cR_H}\right)^{-1/2}$$

$$= \left(\frac{1}{9} - \frac{2.7415 \times 10^{14}\,\mathrm{s}^{-1}}{(2.9979 \times 10^8\,\mathrm{m\,s}^{-1}) \times (1.0968 \times 10^7\,\mathrm{m}^{-1})}\right)^{-1/2} = 6.005$$

Therefore $\boxed{n = 6}$.

13.7 $$\frac{1}{\lambda} = \frac{1}{486.1 \times 10^{-7}\,\mathrm{cm}} = 20572\,\mathrm{cm}^{-1}.$$

(a) Hence, the term lies at $27414\,\mathrm{cm}^{-1} - 20572\,\mathrm{cm}^{-1} = \boxed{6842\,\mathrm{cm}^{-1}}$.

(b) $E = \frac{hc}{\lambda} = (6.626 \times 10^{-34}\,\mathrm{J\,s}) \times 2.998 \times 10^8\,\mathrm{m\,s}^{-1} \times 6842\,\mathrm{cm}^{-1} \times 100\,\mathrm{m}^{-1}/\mathrm{cm}^{-1}$

$$E = \boxed{1.36 \times 10^{-19}\,\mathrm{J}}.$$

13.8 $$\tilde{\nu}(\mathrm{H}) = R_{\mathrm{H}}\left(\frac{1}{n_1^2} - \frac{1}{n_2^2}\right)$$

$$= 1.09677 \times 10^7\,\mathrm{cm}^{-1} \times \left(\frac{1}{1^2} - \frac{1}{2^2}\right)$$

$$= 8.22578 \times 10^6\,\mathrm{cm}^{-1}.$$

$$\frac{R_{\mathrm{D}}}{R_{\mathrm{H}}} = \frac{\mu_{\mathrm{D}}}{\mu_{\mathrm{H}}} = \frac{m_{\mathrm{e}}m_{\mathrm{D}}/(m_{\mathrm{e}} + m_{\mathrm{D}})}{m_{\mathrm{e}}m_{\mathrm{H}}/(m_{\mathrm{e}} + m_{\mathrm{H}})} = \frac{m_{\mathrm{D}}/(m_{\mathrm{e}} + m_{\mathrm{H}})}{m_{\mathrm{H}}/(m_{\mathrm{e}} + m_{\mathrm{D}})}.$$

$$m_{\mathrm{D}} = 2.014\,\mathrm{u}, \quad m_{\mathrm{H}} = 1.008\,\mathrm{u}, \quad m_{\mathrm{e}} = 5.486 \times 10^{-4}\,\mathrm{u}.$$

$$\frac{R_{\mathrm{D}}}{R_{\mathrm{H}}} = \frac{2.014(5.486 \times 10^{-4} + 1.008)}{1.008(5.486 \times 10^{-4} + 2.014)} = 1.000272.$$

The difference in wavenumber is then

$$0.000272 \times 8.22578 \times 10^6\,\mathrm{cm}^{-1} = \boxed{2.24 \times 10^3\,\mathrm{cm}^{-1}}.$$

13.9 Examine eqn 13.4a and see that

$$\Delta E \propto Z^2\left(\frac{1}{n_1^2}-\frac{1}{n_2^2}\right)\propto \nu.$$

We look for n_1 (He^+) and n_2 (He^+) that satisfies

$$\left(\frac{1}{1^2}-\frac{1}{2^2}\right)=4\left(\frac{1}{n_1^2(\mathrm{He}^+)}-\frac{1}{n_2^2(\mathrm{He}^+)}\right).$$

Clearly, $n_1\,(\mathrm{He}^+) = 2n_1(\mathrm{H}) = 2,$

$$n_2\,(\mathrm{He}^+) = 2n_2(\mathrm{H}) = 4.$$

An allowed transition would be $\boxed{4p \to 2s}$ in He^+.

13.10 We use $I = Z^2hcR_H$.

For He^+, $I = 4hcR_H = 54.36\,\mathrm{eV}$.

For Li^{2+}, $I = 9hcR_H = \frac{9}{4}\times 54.36\,\mathrm{eV}$

$$= \boxed{122.31\,\mathrm{eV}}.$$

13.11 $n = 4$ for the N shell.

$n^2 = 4^2 = \boxed{16 \text{ orbitals}}$.

13.12 All lines in the hydrogen spectrum fit the Rydberg formula

$$\frac{1}{\lambda} = R_H\left(\frac{1}{n_1^2}-\frac{1}{n_2^2}\right)\left[1, \text{ with } \tilde{\nu}=\frac{1}{\lambda}\right] R_H = 109677\,\mathrm{cm^{-1}}.$$

(a) Find n_1 from the value of λ_{max}, which arises from the transition $n_1+1 \to n_1$

$$\frac{1}{\lambda_{max}R_H} = \frac{1}{n_1^2}-\frac{1}{(n_1+1)^2} = \frac{2n_1+1}{n_1^2(n_1+1)^2}.$$

$$\lambda_{max}R_H = \frac{n_1^2(n_1+1)^2}{2n_1+1} = (12368\times10^{-9}\,\mathrm{m})\times(109677\times10^2\,\mathrm{m^{-1}}) = 135.65.$$

Because $n_1 = 1, 2, 3$, and 4 have already been accounted for, try $n_1 = 5, 6, \ldots$. With $n_1 = 6$ we get $\dfrac{n_1^2(n_1+1)^2}{2n_1+1} = 136$. Hence, the Humphreys series is $\boxed{n_2 \to 6}$ and the transitions are given by

(b) $\dfrac{1}{\lambda} = (109\,677\,\mathrm{cm^{-1}})\times\left(\dfrac{1}{36}-\dfrac{1}{n_2^2}\right), n_2 = 7, 8, \ldots$

and occur at $\boxed{\text{12372 nm, 7503 nm, 5908 nm, 5908 nm, 5129 nm, …3908 nm (at } n_2 = 15\text{), converging to 3282 nm as } n_2 \to \infty}$, in agreement with the quoted experimental result.

13.13 $\lambda_{max}(H) = 12368\,nm.$

$$\Delta E = h\nu = \frac{hc}{\lambda}.$$

$\Delta E\,(He^+) \approx Z^2\Delta E(H) = 4\Delta E_H$; therefore

$$\lambda_{max}(He^+) \approx \frac{1}{4}\lambda_{max}(H) = \frac{12368\,nm}{4} = \boxed{3092\,nm}.$$

13.14 (a) All lines in the hydrogen spectrum fit the Rydberg formula

$$\frac{1}{\lambda} = R_H\left(\frac{1}{n_1^2} - \frac{1}{n_2^2}\right)\left[1, \text{ with } \tilde{\nu} = \frac{1}{\lambda}\right] R_H = 109677\,cm^{-1}.$$

Find n_1 from the value of λ_{max}, which arises from the transition $n_1 + 1 \rightarrow n_1$

$$\frac{1}{\lambda_{max}R_H} = \frac{1}{n_1^2} - \frac{1}{(n_1+1)^2} = \frac{2n_1+1}{n_1^2(n_1+1)^2},$$

$$\lambda_{max}R_H = \frac{n_1^2(n_1+1)^2}{2n_1+1} = (656.46 \times 10^{-9}\,m) \times (109677 \times 10^2\,m^{-1}) = 7.20,$$

and hence $n_1 = 2$, as determined by trial and error substitution. Therefore, the transitions are given by

$$\tilde{\nu} = \frac{1}{\lambda} = (109677\,cm^{-1}) \times \left(\frac{1}{4} - \frac{1}{n_2^2}\right), \quad n_2 = 3, 4, 5, 6.$$

The next line has $n_2 = 7$, and occurs at

$$\tilde{\nu} = \frac{1}{\lambda} = (109677\,cm^{-1})\left(\frac{1}{4} - \frac{1}{49}\right) = 2.5181 \times 10^4\,cm^{-1}, \text{ which}$$

corresponds to $\boxed{397.13\,nm}$.

(b) The energy required to ionize the atom is obtained by letting $n_2 \rightarrow \infty$. Then

$$\tilde{\nu}_\infty = \frac{1}{\lambda_\infty} = (109\,677\,cm^{-1}) \times \left(\frac{1}{4} - 0\right) = \boxed{274\,19\,cm^{-1}}, \text{ or } \boxed{3.40\,eV}.$$

(The answer, 3.40 eV, is the ionization energy of an H atom that is already in an excited state, with $n = 2$.)

COMMENT The series with $n_1 = 2$ is the Balmer series.

13.15 A Lyman series corresponds to $n_1 = 1$; hence

$$\tilde{\nu} = R_{Li^{2+}}\left(1 - \frac{1}{n_2}\right), \quad n = 2, 3, \ldots \left[\tilde{\nu} = \frac{1}{\lambda}\right].$$

Therefore, if the formula is appropriate, we expect to find that

$\tilde{\nu}\left(1-\frac{1}{n_2}\right)^{-1}$ is a constant ($R_{Li^{2+}}$). We therefore draw up the following table.

n	2	3	4
$\tilde{\nu}/cm^{-1}$	740 747	877 924	925 933
$\tilde{\nu}\left(1-\frac{1}{n_2}\right)^{-1} / cm^{-1}$	987 663	987 665	987 662

Hence, the formula does describe the transitions, and $\boxed{R_{Li^{2+}} = 987663\,cm^{-1}}$.

The Balmer transitions lie at

$$\tilde{\nu} = R_{Li^{2+}}\left(\frac{1}{4}-\frac{1}{n^2}\right) \qquad n = 3, 4, \ldots$$

$$= (987663\,cm^{-1}) \times \left(\frac{1}{4}-\frac{1}{n^2}\right) = \boxed{137175\,cm^{-1},\ 185187\,cm^{-1}, \ldots}.$$

The ionization energy of the ground-state ion is given by

$$\tilde{\nu} = R_{Li^{2+}}\left(1-\frac{1}{n_2}\right) \qquad n \to \infty$$

and hence corresponds to

$$\tilde{\nu} = 987\,663\,cm^{-1}, \quad \text{or } \boxed{122.5\,eV}.$$

13.16 The probability density varies as

$$\psi^2 = \frac{1}{\pi a_0^3}\,e^{-2r/a_0}.$$

The maximum value is at $r = 0$ and ψ^2 is 25 per cent of the maximum when $e^{-2r/a_0} = 0.25$, so that $r = 1/2\,a_0 \ln(0.25)$, which is at $\boxed{r = 0.693\,a_0}$, which corresponds to 36.7 pm.

13.17 The radial distribution function varies as

$$P = 4\pi r^2\psi^2 = \frac{4r^2}{a_0^3}\,e^{-2r/a_0}.$$

The maximum value of P occurs at $r = a_0$ because

$$\frac{dP}{dr} \propto \left(2r - \frac{2r^2}{a_0}\right) e^{-2r/a_0} = \text{ at } r = a_0 \text{ and } P_{max} = \frac{4}{a_0}\,e^{-2}.$$

P falls to a fraction f of its maximum when

$$f = \frac{\dfrac{4r^2}{a_0^3}\,e^{-2r/a_0}}{\dfrac{4}{a_0}\,e^{-2}} = \frac{r^2}{a_0^2}\,e^2\,e^{-2r/a_0}.$$

Therefore solve

$$\frac{f^{1/2}}{e} = \left(\frac{r}{a_0}\right) e^{-r/a_0}.$$

(a) $f = 0.25$

solves to $r = 0.7569\,a_0$ or $0.2431\,a_0 = \boxed{40\text{ pm or }13\text{ pm}}$.

(b) $f = 0.10$

solves to $r = 0.554\,a_0$ or $0.446\,a_0 = \boxed{29\text{ or }24\text{ pm}}$.

13.18 There are 3 degenerate p-orbitals for any principal quantum number, n.

Each p-orbital has 2 lobes; therefore, for any given lobe the probability is $\boxed{1/6}$.

13.19 $V = 5.0\,\text{pm}^3 = \frac{4}{3}\pi r^3$ [assume a spherical volume].

$$r = \left(\frac{3V}{4\pi}\right)^{1/3} = \left(\frac{3 \times 5.0 \times 10^{-36}\,\text{m}^3}{4\pi}\right)^{1/3} = 1.06 \times 10^{-12}\,\text{m} = 1.06\,\text{pm}.$$

$r/a_0 = (1.06\,\text{pm})/(52.9\,\text{pm}) = 0.0200.$

As the mean radius of the electron in the ground state ($n = 1$ and $l = 0$) of a hydrogen like-atom is $3a_0/2Z$ [13.5] $= 79.4\,\text{pm}/Z$, it is probably safe to assume that $\psi^2(r)$ is a constant within the spherical volume with $r = 1.06$ pm, Therefore,

$$\psi^2(r) = \frac{Z^2}{\pi a_0^3} e^{-2Zr/a_0} = \text{constant}.$$

(a) For the hydrogen atom $Z = 1$.

Evaluate $\psi^2(r)$ at $r = 0.53$ pm, which is half the radius of V.

$$\psi^2 = \frac{1}{\pi\,(53\,\text{pm})^3} e^{-0.020} = \frac{0.98}{\pi(53\,\text{pm})^3}.$$

Then,

$$\text{Probability} = \int \psi^2(r)\delta V \approx \psi^2 \delta V$$

$$= \frac{0.98}{\pi(53\,\text{pm})^3} \times 5.0\,\text{pm}^3 = \boxed{1.1 \times 10^{-5}}.$$

(b) For the He^+ ion $Z = 2$.

Evaluate $\psi^2(r)$ at $r = 0.53$ pm, which is half the radius of V.

$$\psi^2 = \frac{2^2}{\pi(53\,\text{pm})^3} e^{-0.040} = \frac{4 \times (0.96)}{\pi(53\,\text{pm})^3}.$$

Then,

$$\text{Probability} = \int \psi^2(r)\delta V \approx \psi^2 \delta V$$

$$= \frac{4 \times (0.96)}{\pi(53\,\text{pm})^3} \times 5.0\,\text{pm}^3 = \boxed{4.1 \times 10^{-5}}.$$

13.20 The most probable distance of a 1*s* electron from the nucleus occurs when the first derivative of the radial distribution function equals zero.

$$P_{1s} = 4\pi r^2 \psi_{1s}^2 \ [13.8a] = 4\pi r^2 \left(Ne^{-r/a_0}\right)^2 \ [13.7] = 4\pi N^2 \left(r^2 e^{-2r/a_0}\right),$$

$$\frac{dP_{1s}}{dr} = 4\pi N^2 \frac{d\left(r^2 e^{-2r/a_0}\right)}{dr} = 4\pi N^2 \left\{2re^{-2r/a_0} + r^2\left(-\frac{2}{a_0}e^{-2r/a_0}\right)\right\}$$

$$= 8\pi N^2 \left\{1 - \frac{r}{a_0}\right\} re^{-2r/a_0}.$$

The derivative equals zero when the factor $1 - r/a_0$ equals zero. Therefore, $\boxed{r_{max} = a_0}$.

13.21 We will make the assumption that ψ^2 is a constant within this very small volume. Then

$$\text{Probability} = \int \psi^2(r)\delta V \approx \psi^2 \delta V \ \text{ with } \ \delta V = 1.0\,\text{pm}^3.$$

$$\psi^2 = \frac{1}{32\pi a_0^3}\left(2 - \frac{r}{a_0}\right)^2 e^{-r/a_0} = 6.72 \times 10^{-8}\,\text{pm}^{-3}\left(2 - \frac{r}{a_0}\right)^2 e^{-r/a_0}.$$

(a) $\psi^2 = 6.72 \times 10^{-8}\,\text{pm}^{-3} \times 2^2 \times 1 = 2.7 \times 10^{-7}\,\text{pm}^{-3}$

$\psi^2\delta V = \boxed{2.7 \times 10^{-7}}$.

(b) $\psi^2 = 6.72 \times 10^{-8}\,\text{pm}^{-3} \times 1 \times e^{-1} = 9.9 \times 10^{-8}\,\text{pm}^3$.

$\psi^2\delta V = \boxed{9.9 \times 10^{-8}}$.

(c) $\psi^2 = 0, \quad \psi^2\delta V = \boxed{0}$.

13.22 The most probable distance of a 2*s* electron from the nucleus may be determined by plotting the radial distribution function against r/a_0 and using the trace function of the plotting software to evaluate the coordinates of the maximum. The following function is plotted in Figure 13.1. The plot reveals that $\boxed{r_{max} = 5.235a_0}$.

$$P_{2s} = 4\pi r^2 \psi_{2s}^2 \ [13.8a] = 4\pi r^2 \left\{\left(\frac{1}{32\pi a_0^3}\right)^{1/2}\left(2 - \frac{r}{a_0}\right)e^{-r/2a_0}\right\}^2 \quad [\text{Exercise 13.21}]$$

$$= \frac{1}{8a_0^3}\left(2 - \frac{r}{a_0}\right)^2 r^2 e^{-r/a_0} = \boxed{\frac{1}{8a_0}(2 - x)^2 x^2 e^{-x} \quad \text{where } x = r/a_0}.$$

13.23 $$P_{2s} = 4\pi r^2 \psi_{2s}^2 \ [13.8a] = 4\pi r^2 \left\{\left(\frac{1}{32\pi a_0^3}\right)^{1/2}\left(2 - \frac{r}{a_0}\right)e^{-r/2a_0}\right\}^2 \quad [\text{Exercise 13.21}].$$

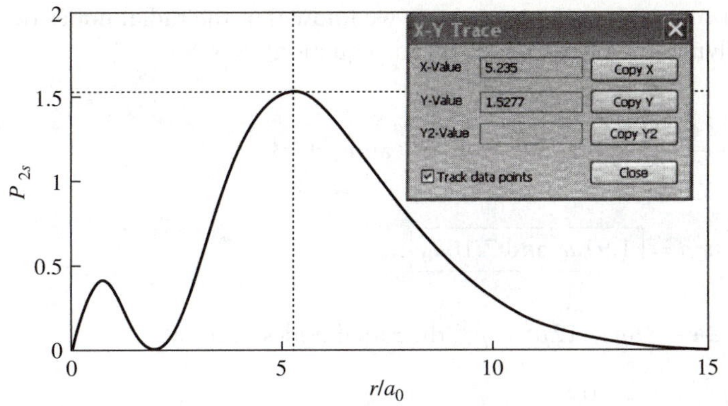

Fig 13.1

Let $x = r/a_0$; then

$$P_{2s} = \frac{1}{8a_0}\left\{x^2(2-x)^2\,e^{-x}\right\} = \frac{1}{8a_0}\left(4x^2 - 4x^3 + x^4\right)e^{-x}.$$

$$\frac{dP_{2s}}{dr} = \frac{dx}{dr}\frac{dP_{2s}}{dx} = \frac{1}{8a_0^2}\frac{d\left\{\left(4x^2-4x^3+x^4\right)e^{-x}\right\}}{dx}$$

$$= \frac{1}{8a_0^2}\left\{\left(8x - 12x^2 + 4x^3\right)e^{-x} + \left(4x^2 - 4x^3 + x^4\right)\left(-e^{-x}\right)\right\}$$

$$= \frac{1}{8a_0^2}\left\{8 - 16x + 8x^2 - x^3\right\}x\,e^{-x}$$

$$= -\frac{1}{8a_0^2}\times(x-2)\times\left(x^2 - 6x - 4\right)x\,e^{-x}.$$

The derivative equals zero when $x = r/a_0 = 0,\ 3-5^{1/2},\ 2,\ 3+5^{1/2}$, and ∞. These correspond to the radial distribution function being a minimum, a maximum, a minimum, a maximum, and a minimum, respectively. The following ratio identifies the maximum that is most probable.

$$\frac{P_{2s}\left(x = 3+5^{1/2}\right)}{P_{2s}\left(x = 3-5^{1/2}\right)} = \frac{\left[\left(4x^2-4x^3+x^4\right)e^{-x}\right]_{x=3+5^{1/2}}}{\left[\left(4x^2-4x^3+x^4\right)e^{-x}\right]_{x=3-5^{1/2}}} = \frac{1.528}{0.415} = 3.68.$$

The most probable distance is $\boxed{(3+5^{1/2})a_0}$.

13.24 Let $\rho = 2Zr/na_0$ where $a_0 = 52.92$ pm and $Z = 1$ for the hydrogen atom.

(a) The values of Zr/a_0 at the nodes of a hydrogen atom 3*s* orbital are directly read off the text Figure 13.9c plot. At the two nodes we find the approximate values: $\boxed{r = 2.0\ a_0}$ and $\boxed{r = 7.2\ a_0}$. More accurate values can be determined with the analytical expression for the 3*s* hydrogen orbital, which is found in Atkins and de Paula, *Physical Chemistry*,

8th edition, 2006. Because $\psi_{3,0} \propto \rho^2 - 6\rho + 6$, we know that the radial nodes occur at $\rho^2 - 6\rho + 6 = 0$. Solving for ρ from the quadratic equation gives

$$\rho = \frac{6 \pm (6^2 - 4 \times 6)^{1/2}}{2} = 3 \pm 3^{1/2} = 1.27 \quad \text{and} \quad 4.73.$$

$$r = \left(\frac{na_0}{2Z}\right)\rho = \frac{3}{2}a_0\rho = \boxed{1.91a_0 \text{ and } 7.10a_0}.$$

(b) Because $\psi_{4,0} \propto 24 - 36\rho + 12\rho^2 - \rho^3$, the radial nodes occur at

$$24 - 36\rho + 12\rho^2 - \rho^3 = 0.$$

This is a cubic equation. The analytic solutions of cubic equations are cumbersome. Here we solve it numerically with a computer program using Newton's method. The roots found are $\rho = 0.936, 3.305$, and 7.759.

For a a $4s$ orbital of a hydrogen atom, radial nodes are at

$$r = na_0\rho/2Z = 2a_0\rho = \boxed{1.87a_0, 6.61a_0, \text{ and } 15.5a_0}.$$

13.25 Look for values of θ for which $\sin\theta$ or $\cos\theta$ go to zero. $\sin\theta$ goes to zero at $\theta = \boxed{0° \text{ and } 180°}$; $\cos\theta$ at $\boxed{90° \text{ and } 270°}$.

13.26 Identify l and use angular momentum $= \{l(l+l)\}^{1/2}\hbar$

(a) $l = 0$, so $\boxed{\text{ang. mom.} = 0}$.

(b) $l = 0$, so $\boxed{\text{ang. mom.} = 0}$.

(c) $l = 2$, so $\boxed{\text{ang. mom.} = \sqrt{6}\,\hbar}$.

(d) $l = 1$, so $\boxed{\text{ang. mom.} = \sqrt{2}\,\hbar}$.

(e) $l = 1$, so $\boxed{\text{ang. mom.} = \sqrt{2}\,\hbar}$.

The total number of nodes is equal to $n - 1$, and the number of angular nodes is equal to l; hence the number of radial nodes is equal to $n - 1 - l$. We can draw up the following table:

	1s	3s	3d	2p	3p
n, l	1,0	3,0	3,2	2,1	3,1
Ang. nodes	0	0	2	1	1
Rad. nodes	0	2	0	0	1

13.27 The energies are $E = -\dfrac{hcR_H}{n^2}$, and the orbital degeneracy g of an energy level of principal quantum number n is $g = n^2$

(a) $E = -hcR_H$ implies that $n = 1$, so $\boxed{g = 1}$ (the 1s orbital).

(b) $E = -\dfrac{hcR_H}{9}$ implies that $n = 3$, so $\boxed{g = 9}$ (the 3s orbital, the three p orbitals, and the five 3d orbitals).

(c) $E = -\dfrac{hcR_H}{49}$ implies that $n = 7$, so $\boxed{g = 49}$ (the 7s orbital, the three 7p orbitals, the five 7d orbitals, the seven 7f orbitals, the nine 7g orbitals).

13.28 For a given l there are $2l + 1$ values of m_l and hence $2l + 1$ orbitals. Each orbital may be occupied by two electrons. Therefore the maximum occupancy is $2(2l + 1)$.

	l	$2(2l+1)$
(a)	0	2
(b)	3	14
(c)	5	22

13.29 By analogy with eq 13.12 the ionization energy of an anion, I_1^-, is, for the reaction

$$E^-(g) \rightarrow E(g) + e^-(g), \qquad I_1^- = E(E) - E(E^-).$$

while the electron affinity of eq 13.13a is, for the reaction

$$E(g) + e^-(g) \rightarrow E^-(g), \qquad E_{ea} = E(E^-) - E(E).$$

Inspection of the two reactions indicates that each is the reverse of the other. Inspection of the energy equations indicates that one is the negative of the other. Both observations lead to the conclusion that $\boxed{I_1^- = -E_{ea}}$.

13.30 $$h\nu = \frac{1}{2}m_e v^2 + I$$

$$I = h\nu - \frac{1}{2}m_e v^2 = 6.626 \times 10^{-34}\,\text{J s} \times \frac{2.998 \times 10^8\,\text{m s}^{-1}}{58.4 \times 10^{-9}\,\text{m}}$$

$$- \frac{1}{2} \times (9.109 \times 10^{-31}\,\text{kg}) \times (1.59 \times 10^6\,\text{m s}^{-1})^2$$

$$= 2.25 \times 10^{-18}\,\text{J, corresponding to } \boxed{14.0\,\text{eV}}.$$

13.31 (a) Periodic table.

1s	H	He										
2s	Li	Be	B	C	N	O	F	Ne	Na	Mg	2p
3s	Al	Si	P	S	Cl	Ar	K	Ca	Sc	Ti	3p
4s	V	Cr								↑ 'Noble gases'	

13.32 The inner closed orbitals represented by [Ne]2s^2 have zero total spin and total orbital angular momentum so they need not be considered when evaluating the terms of the [Ne]$2s^2 2p^1 3d^1$ excited state. The two electrons of the $2p^1 3d^1$ orbitals have either parallel

($S=1$) or paired spins ($S=0$). Thus, triplet and singlet spin multiplicities are allowed because the spin multiplicity has the formula $2S+1$.

Since $l=1$ for the p orbital and $l=2$ for the d orbital, the allowed total orbital angular momentum terms are given by $L = l_1 + l_2, \ldots, |l_1 - l_2| = 3, 2, 1$. These are F, D, and P terms.

Triplet terms : ^{3}F, ^{3}D, ^{3}P; singlet terms : ^{1}F, ^{1}D, ^{1}P.

Possible values of J (using Russell–Saunders coupling) are 3, 2, and 1 ($S=0$) and 4, 3, 2, 1, and 0 ($S=1$). The term symbols are

$$\boxed{{}^1F_3;\ {}^3F_4,\ {}^3F_3,\ {}^3F_2,;\ {}^1D_2;\ {}^3D_3,\ {}^3D_2,\ {}^3D_1;\ {}^1P_1;\ {}^3P_2,\ {}^3P_1,\ {}^3P_0}.$$

Hund's rules state that the lowest energy level has maximum multiplicity. Consideration of spin–orbit coupling says the lowest energy level has the lowest value of $J(J+1) - L(L+1) - S(S+1)$. So the lowest energy level is $\boxed{{}^3F_2}$.

13.33 Possible values for four electrons in different orbitals are $\boxed{2, 1, \text{and } 0}$.

13.34 (a) $L=0, S=0, J=0$ $\boxed{1}$ level.

(b) $L=3, S=1, J=4,3,2$ $\boxed{3}$ levels.

(c) $L=0, S=2, J=2$ $\boxed{1}$ level.

(d) $L=1, S=2, J=3,2,1$ $\boxed{3}$ levels.

13.35 (a) The term of lowest energy will be a term that obeys Hund's rule, which states that the terms with largest multiplicity ($2S+1$) arising from terms with the maximum value of total spin, S, lie lowest in energy. Here

$$S = s_1 + s_2 = \frac{1}{2} + \frac{1}{2} = 1.$$

is the maximum allowed spin. The corresponding multiplicity is $2S+1=3$.

So our term of lowest energy will be a triplet term.

For terms of the same multiplicity, the term with the largest total orbital angular momentum, L, lies lowest in energy. Here $L = l_1 + l_2 = 2 + 2 = 4$ is the maximum total orbital angular momentum. For $L=4$ and $S=1$, the term symbol would be ^{3}G.

But for the configuration $3d^2$, this term is not allowed by the Pauli principle, for in order to achieve $L=4$, two electrons would have to occupy either the $3d_{+2}$ orbital or the $3d_{-2}$ orbital. According to the Pauli principle, this can only happen if the spins of the two electrons are antiparallel, namely $S = s_1 + s_2 = \left(+\frac{1}{2}\right) + \left(-\frac{1}{2}\right) = 0$

So a singlet G term ($2S+1=1$) can arise, but not a triplet G term.

This problem does not occur for F terms because here $L=3$, which can arise from $L = l_1 + l_2 = 2 + 1 = 3$

where the two electrons are in different orbitals, $3d_{+2}$ and $3d_{+1}$, for example.

Therefore, the term of lowest energy is $\boxed{{}^3\mathrm{F}}$.

The possible J values are obtained from the Clebsch–Gordon series and are

$J = 3 + 1 = 4$,

$J = 3 + 1 - 1 = 3$,

$J = 3 + 1 - 2 = 2$.

These values of J give rise to the terms

${}^3\mathrm{F}_4$, ${}^3\mathrm{F}_3$, ${}^3\mathrm{F}_2$.

Of these the lowest lying is ${}^3\mathrm{F}_2$.

(b) $\boxed{3}$.

13.36 The selection rules are $\Delta n =$ any integer, $\Delta l = \pm 1$.

(a) $\Delta l = 0$, $\boxed{\text{forbidden}}$.

(b) $\Delta l = -1$, $\boxed{\text{allowed}}$.

(c) $\Delta l = -1$, $\boxed{\text{allowed}}$.

(d) $\Delta l = -2$, $\boxed{\text{forbidden}}$.

(e) $\Delta l = -1$, $\boxed{\text{allowed}}$.

13.37 $\boxed{\text{All } d \text{ and } g \text{ orbitals}}$.

Chapter 14

The chemical bond

Answers to discussion questions

14.1 Our comparison of the two theories will focus on the manner of construction of the trial wavefunctions for the hydrogen molecule in the simplest versions of both theories. In the valence bond method, the trial function is a linear combination of two simple product wavefunctions, in which one electron resides totally in an atomic orbital(AO) on atom A, and the other totally in an orbital on atom B. See Derivation 14.1 and Figure 14.3 of the text. There is no contribution to the wavefunction from products in which both electrons reside on either atom A or B.

$$\psi_{A-B}(1,2) = \psi_A(1)\psi_B(2) + \psi_A(2)\psi_B(1) \quad [14.1].$$

So the valence bond approach undervalues, by totally neglecting, any ionic contribution to the trial function. It is a totally covalent function.

The one-electron molecular orbital (MO) extends throughout the molecule and is written as a linear combination of atomic orbitals (LCAO).

$$\psi_{MO}(1) = c_A\psi_A(1) + c_B\psi_B(1) \quad [14.6a].$$

The squares of the coefficients give the relative proportions of the AO contributing to the MO.

The two-electron molecular orbital function for the hydrogen molecule is a product of two one-electron MOs. That is,

$$\begin{aligned}\psi &= [c_A\psi_A(1) + c_B\psi_B(1)] \times [c_A\psi_A(2) + c_B\psi_B(2)] \\ &= c_A^2\psi_A(1)\psi_A(2) + c_B^2\psi_B(1)\psi_B(2) + c_Ac_B\psi_A(1)\psi_B(2) + c_Ac_B\psi_A(2)\psi_B(1).\end{aligned}$$

The first two terms are ionic forms for which both electrons are either on atom A or on atom B. The molecular orbital approach greatly overvalues the ionic contributions. At these crude levels of approximation, the valence bond method gives dissociation energies closer to the experimental values. However, more sophisticated versions of the molecular orbital approach are the method of choice for obtaining quantitative results on both diatomic and polyatomic molecules.

14.2 Consider the case of the carbon atom. Mentally we break the process of hybridization into two major steps. The first is promotion, in which we imagine that one of the electrons in the 2*s* orbital of carbon ($2s^2 2p^2$) is promoted to the empty 2*p* orbital giving the configuration $2s2p^3$. In the second step we mathematically mix the four orbitals by way of the specific linear combinations corresponding to the sp^3 hybrid orbitals described in Section 14.5. There is a principle of conservation of orbitals that enters here. If we mix four unhybridized atomic orbitals we must end up with four hybrid orbitals. In the construction of the sp^2 hybrids we start with the 2*s* orbital and two of the 2*p* orbitals, and after mixing we end up with three sp^2 hybrid orbitals. In the *sp* case we start with the 2*s* orbital and one of the 2*p* orbitals. The justification for all of this is in a sense the First Law of thermodynamics. Energy is a state function and therefore its value is determined only by the final state of the system, not by the path taken to achieve that state, and the path can even be imaginary.

14.3 Both the Pauling and Mulliken methods for measuring the attracting power of atoms for electrons seem to make good chemical sense. If we look at eqn 14.10a (the Pauling scale), we see that if $E(\text{A—B})$ were equal to $1/2[E(\text{A—A}) + E(\text{B—B})]$ the calculated electronegativity difference would be zero, as expected for completely non-polar bonds. Hence, any increased strength of the A—B bond over the average of the A—B and B—B bonds can reasonably be thought of as being due to the polarity of the A—B bond, which in turn is due to the difference in electronegativity of the atoms involved. Therefore, this difference in bond strengths can be used as a measure of electronegativity difference. To obtain numerical values for individual atoms, a reference state (atom) for electronegativity must be established. The value for fluorine is arbitrarily set at 4.0.

The Mulliken scale may be more intuitive than the Pauling scale because we are used to thinking of ionization energies and electron affinities as measures of the electron attracting powers of atoms. The choice of factor 1/2, however, is arbitrary, though reasonable, and no more arbitrary than the specific form that defines the Pauling scale.

14.4 In *ab initio* methods an attempt is made to evaluate the Schrödinger equation numerically. Approximations are employed, but these are mainly associated with the construction of the wavefunctions involved in the integrals. In semi-empirical methods, many of the integrals are expressed in terms of spectroscopic data or physical properties. Semi-empirical methods exist at several levels. At some levels, in order to simplify the calculations, many of the integrals are set equal to zero.

Density functional theory (DFT) is an *ab initio* method with focus upon the electron density, not upon the wavefunction as in earlier quantum methods. Whether focus is upon the electron density or the wavefunction, *ab initio* methods are iterative self-consistent methods in that the calculations are repeated until the energy and wavefunctions, or energy and electron density (DFT), are unchanged to within some acceptable tolerance.

Solutions to exercises

14.5 The valence bond description of P_2 is similar to that of N_2. One $\sigma(2p_{zA}, 2p_{zB})$; one $\pi(2p_{xA}, 2p_{xB})$; and one $\pi(2p_{yA}, 2p_{yB})$ bond; along with their antibonding counterparts.

In the tetrahedral P_4 molecule there are six single P—P bonds of roughly 200 kJ mol^{-1} bond enthalpy each. So the total bonding enthalpy is roughly 1200 kJ mol^{-1}. In the transformation

$$P_4 \rightarrow 2P_2$$

there is a loss of about 800 kJ mol^{-1} in σ-bond enthalpy. This loss is not likely to be made up by the formation of 4 P—P π-bonds. Period 3 atoms, such as P, are too large to get close enough to each other to form strong π-bonds.

14.6 The three valence bond wavefunctions for N_2 will all be of the form described by eqn 14.2.

$$\psi_1(\sigma\text{-bond}) = \psi_{2p_{zA}}(1)\psi_{2p_{zB}}(2) + \psi_{2p_{zA}}(2)\psi_{2p_{zB}}(1).$$

$$\psi_2(\pi\text{-bond}) = \psi_{2p_{xA}}(1)\psi_{2p_{xB}}(2) + \psi_{2p_{xA}}(2)\psi_{2p_{xB}}(1).$$

$$\psi_3(\pi\text{-bond}) = \psi_{2p_{yA}}(1)\psi_{2p_{yB}}(2) + \psi_{2p_{yA}}(2)\psi_{2p_{yB}}(1).$$

14.7 We use eqn 14.3 with R = 74.1 pm.

$$V_{\text{nuc,nuc}} = \frac{e^2}{4\pi\varepsilon_0 R} = \frac{(1.602 \times 10^{-19}\text{C})^2}{1.113 \times 10^{-10}\text{J}^{-1}\text{C}^2\text{m}^{-1} \times 74.1 \times 10^{-12}\text{m}}$$
$$= 3.11 \times 10^{-18}\text{J}.$$

The molar value is

$$6.022 \times 10^{23}\,\text{mol}^{-1} \times 3.11 \times 10^{-18}\,\text{J} = \boxed{1.87 \times 10^6\,\text{J mol}^{-1}}.$$

14.8 SO_2.

There are two localized S—O σ-bonds formed from S(sp^2) and O(p_z) orbitals. There is a π-bond that exhibits resonance and that can be described as the following superposition of wavefunctions:

$$\psi(\pi\text{-bond}) = (\psi_{2p_{xS}} + \psi_{2p_{xOA}}) + (\psi_{2p_{xS}} + \psi_{2p_{xOB}}).$$

The sulfur has a lone pair.

SO_3.

There are three localized S—O σ-bonds from S(sp^2) and O(p_z) orbitals. There is a π-bond that exhibits resonance and that can be described as the following superposition of wavefunctions:

$$\psi(\pi\text{-bond}) = (\psi_{2p_{xS}} + \psi_{2p_{xOA}}) + (\psi_{2p_{xS}} + \psi_{2p_{xOB}}) + (\psi_{2p_{xS}} + \psi_{2p_{xOC}}).$$

In the above OA, OB, and OC refer to oxygen atoms A, B, and C.

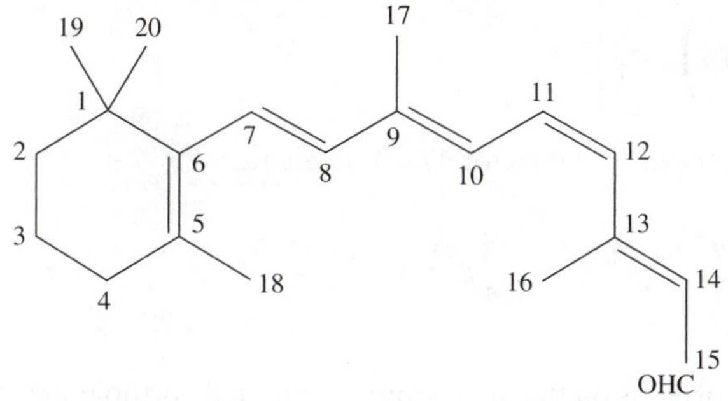

Fig 14.1

14.9 Refer to the structure shown in Figure 14.1 for the numbering of the carbon atoms in *cis*-retinal. Carbon atoms 5–15 each have three sp^2 hybrid atomic orbitals which form σ-bonds with their neighboring atoms. There are six conjugated π-bonds between these 11 C atoms and the one O atom. These six π-bonds are formed from 12 p_x atomic orbitals, one on each of the 12 atoms. They are resonance hybrids all of the form

$$\psi(\pi\text{-bond}) = \sum_{i=5}^{15} \psi_{2p_x\mathrm{C}_i} + \psi_{2p_x\mathrm{O}}.$$

All the remaining C atoms each have four sp^3 hybrid atomic orbitals, which form σ-bonds with their neighboring atoms.

14.10 $h_1 = s + p_x + p_y + p_z, \qquad h_2 = s - p_x - p_y + p_z.$

We need to evaluate

$$\int h_1 h_2 \mathrm{d}\tau = \int (s + p_x + p_y + p_z)(s - p_x - p_y + p_z)\mathrm{d}\tau.$$

We assume that the atomic orbitals are normalized hydrogenic orbitals. As stated, hydrogenic $2p_x$, $2p_y$, and $2p_z$ orbitals are mutually orthogonal as are all hydrogenic atomic orbitals on the same atom. Then

$$\int h_1 h_2 \mathrm{d}\tau = \int s^2 \mathrm{d}\tau - \int p_x^2 \mathrm{d}\tau - \int p_y^2 \mathrm{d}\tau + \int p_z^2 \mathrm{d}\tau = 1 - 1 - 1 + 1 = 0$$

[all functions normalized].

14.11 We need to demonstrate that $\int \psi^2 \mathrm{d}\tau = 1$, where $\psi = \dfrac{s + \sqrt{2}p}{\sqrt{3}}$.

$$\int \psi^2 \mathrm{d}\tau = \frac{1}{3}\int (s + \sqrt{2}p)^2 \mathrm{d}\tau = \frac{1}{3}\int (s^2 + 2p^2 + 2\sqrt{2}sp)^2 \mathrm{d}\tau = \frac{1}{3}(1 + 2 + 0) = 1$$

as $\int s^2 \,\mathrm{d}\tau = 1$, $\int p^2 \,\mathrm{d}\tau = 1$, and $\int sp \,\mathrm{d}\tau = 0$ [orthogonality].

14.12 Rewrite the sp^2 orbital of the preceding problem as

$$h_1 = \frac{s + \sqrt{2}p_x}{\sqrt{3}}.$$

Then $\boxed{h_2 = \left(s - \sqrt{\frac{1}{2}}p_x + \sqrt{\frac{3}{2}}p_y\right)/\sqrt{3}}$

is orthogonal to h_1. For simplicity drop the factor $1/\sqrt{3}$. Evaluate

$$\int h_1 h_2 \mathrm{d}\tau = \int \left(s + \sqrt{2}p_x\right)\left(s - \sqrt{\frac{1}{2}}p_x + \sqrt{\frac{3}{2}}p_y\right)\mathrm{d}\tau$$

and use the fact that all atomic orbitals on the same atom are mutually orthogonal. Then

$$\int h_1 h_2 \mathrm{d}\tau = \int s^2 \mathrm{d}\tau + \int (\sqrt{2}p_x)\left(-\sqrt{\frac{1}{2}}p_x\right)\mathrm{d}\tau$$
$$= 1 - 1 = 0 \text{ [orthogonality]}.$$

14.13 $$\int \psi^2 \mathrm{d}\tau = N^2 \int (\psi_{\text{cov}} + \lambda_{\text{ion}})^2 \mathrm{d}\tau = 1$$
$$= N^2 \int (\psi_{\text{cov}}^2 + \lambda^2\psi_{\text{ion}}^2 + 2\lambda\psi_{\text{cov}}\psi_{\text{ion}})\mathrm{d}\tau = 1$$
$$= N^2(1 + \lambda^2 + 2\lambda S)$$

where we have assumed that ψ_{cov} and ψ_{ion} are individually normalized and we have written

$$S = \int \psi_{\text{cov}}\psi_{\text{ion}}\mathrm{d}\tau.$$

Hence $\boxed{N = \left(\frac{1}{1 + 2\lambda S + \lambda^2}\right)^{1/2}}$.

14.14 The s orbital begins to spread into the region of negative amplitude of the p orbital. When their centers coincide, the region of positive overlap cancels the negative region. Draw up the following table.

R/a_0	0	1	2	3	4	5	6	7	8	9	10
S	0	0.858	1.173	1.046	0.757	0.483	0.283	0.155	0.081	0.041	0.02

As shown in Figure 14.2, the overlap is a maximum when $\boxed{R = 2.11a_0}$.

14.15 We seek an orbital of the form $aA + bB$, where a and b are constants, that is orthogonal to the orbital $N(0.145A + 0.844B)$. Orthogonality implies

$$\int (aA + bB)N(0.145A + 0.844B)\,\mathrm{d}\tau = 0,$$

$$N\int [0.145aA^2 + (0.145b + 0.844a)AB + 0.844bB^2]\,\mathrm{d}\tau = 0.$$

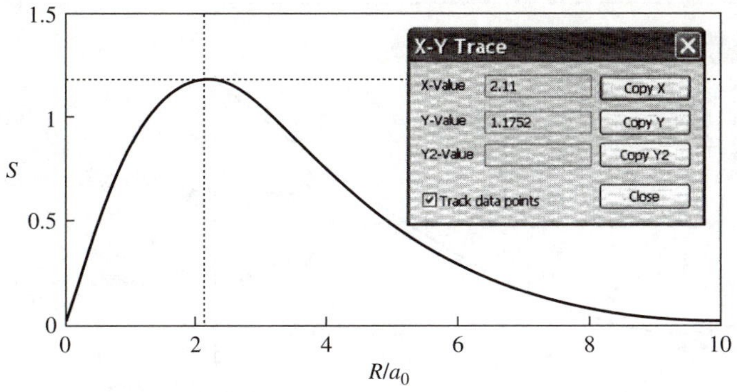

Fig 14.2

The integrals of squares of orbitals are 1 and the integral $\int AB\,d\tau$ is the overlap integral S, so

$$0 = (0.145 + 0.844S)a + (0.145S + 0.844)b \quad \text{so} \quad a = \boxed{-\frac{0.145S + 0.844}{0.145 + 0.844S}b}.$$

This would make the orbitals orthogonal, but not necessarily normalized. If $S = 0$, the expression simplifies to

$$a = -\frac{0.844}{0.145}b$$

and the new orbital would be normalized if $a = 0.844N$ and $b = -0.145N$. That is,

$$\boxed{N(0.844A - 0.145B)}.$$

14.16 In a normalized wavefunction of the form $\psi = c_1\psi_1 + c_2\psi_2$, the probabilities of the system being described by ψ_1 is c_1^2 and by ψ_2 in c_2^2. Therefore in this case

probability of $\psi_{\text{cov}} = 0.989^2 = 0.978$,

probability of $\psi_{\text{ion}} = 0.150^2 = 0.022$.

In 1000 inspections, both electrons will most likely be found in $\boxed{\psi_{\text{cov}} \text{ 978 times}}$ and in $\boxed{\psi_{\text{ion}} \text{ 22 times}}$.

14.17 $$E = \frac{3a\hbar^2}{2\mu} - \frac{e^2}{\varepsilon_0}\left(\frac{a}{2\pi^3}\right)^{1/2}, \qquad E_{1s} = \frac{\mu e^4}{32\pi^2\varepsilon_0^2\hbar^2}.$$

We have chosen to plot $E/(-E_{1s})$ against parameter a in Figure 14.3 so that it is visually evident that adjustments of the parameter can lead to a lower minimized energy but that the minimized energy of the trial wavefunction is always above the true energy. This is obvious because the curve is always higher than -1 and is an illustration of the variation theorem. As shown in Figure 14.3,

$$E_{\text{trial min}} = 0.84883\,E_{1s} = \boxed{-1.850\times10^{-18}\,\text{J}} \text{ at } a = 0.2829\,a_0^{-2} = \boxed{1.010\times10^{20}\,\text{m}^{-2}}.$$

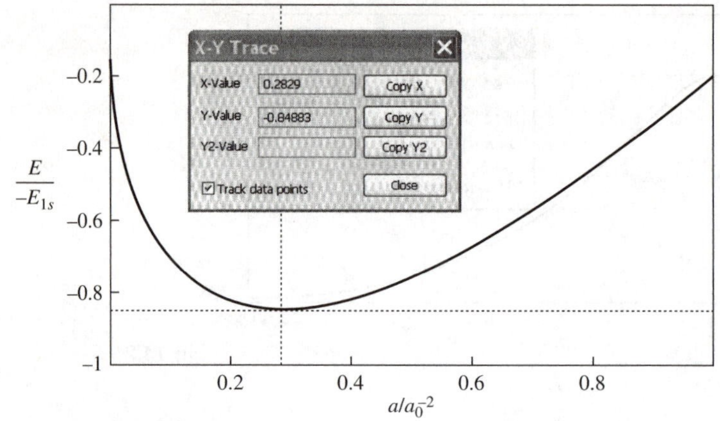

Fig 14.3

14.18 The variation theorem says that the minimum energy provided by the trial wavefunction is obtained by taking the derivative of the trial energy with respect to adjustable parameters, setting it equal to zero, and solving for the parameters. The result will always be greater than the true energy.

$$E_{\text{trial}} = \frac{3a\hbar^2}{2\mu} - \frac{e^2}{\varepsilon_0}\left(\frac{a}{2\pi^3}\right)^{1/2} \quad \text{so} \quad \frac{dE_{\text{trial}}}{da} = \frac{3\hbar^2}{2\mu} - \frac{e^2}{2\varepsilon_0}\left(\frac{1}{2\pi^3 a}\right)^{1/2} = 0.$$

Solving for a yields:

$$\frac{3\hbar^2}{2\mu} = \frac{e^2}{2\varepsilon_0}\left(\frac{1}{2\pi^3 a}\right)^{1/2} \quad \text{so} \quad a = \left(\frac{\mu e^2}{3\hbar^2\varepsilon_0}\right)^2\left(\frac{1}{2\pi^3}\right) = \frac{\mu^2 e^4}{18\pi^3\hbar^4\varepsilon_0^2}.$$

Substituting this back into the trial energy yields the minimum energy

$$E_{\text{trial}} = \frac{3\hbar^2}{2\mu}\left(\frac{\mu^2 e^4}{18\pi^3\hbar^4\varepsilon_0^2}\right) - \frac{e^2}{\varepsilon_0}\left(\frac{\mu^2 e^4}{18\pi^3\hbar^4\varepsilon_0^2 \cdot 2\pi^3}\right)^{1/2} = \boxed{\frac{-\mu e^4}{12\pi^3\varepsilon_0^2\hbar^2}}.$$

14.19 Covalent structures are shown in Figure 14.4(a), while ionic structures are shown in Figure 14.4(b).

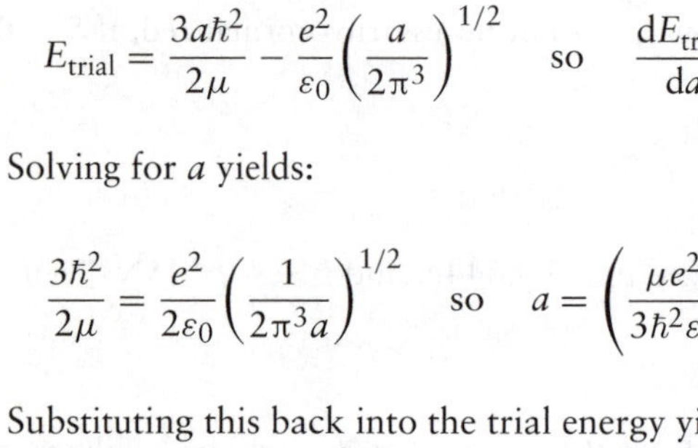

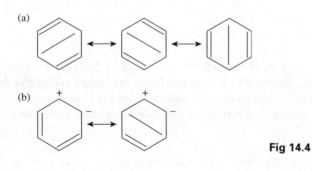

Fig 14.4

In addition there are many other possible ionic structures. These structures can be safely ignored in simple descriptions of the molecule because the coefficients of the wavefunction representing these structures in the linear combination of wavefunctions for the entire resonance hybrid are very small. Benzene is a very symmetrical molecule, and we expect that all the C atoms will be equivalent. Hence, those structures in which the C atoms are not equivalent should contribute little to the resonance hybrid.

14.20 In general we have $\psi = c_A\psi_A + c_B\psi_B$.

We need to determine the coefficients c_A and c_B.

A systematic way of finding the coefficients in the linear combinations used to build molecular orbitals is provided by the **variation principle**:

If an arbitrary wavefunction is used to calculate the energy, then the value calculated is never less than the true energy.

The arbitrary wavefunction is called the **trial wavefunction**. The principle implies that, if we vary the coefficients in the trial wavefunction until we achieve the lowest energy, then those coefficients will be the best. We might get a lower energy if we use a more complicated wavefunction (for example, by taking a linear combination of several atomic orbitals on each atom), but we shall have the optimum molecular orbital that can be built from the given set of atomic orbitals.

The method can be illustrated by the trial wavefunction

$$\psi = c_A\psi(A) + c_B\psi(B).$$

This function is real but not normalized (because the coefficients can take arbitrary values), so in the following we cannot assume that $\int \psi^2 d\tau = 1$. The energy of the orbital is the expectation value of the energy operator

$$E = \frac{\int \psi H \psi d\tau}{\int \psi^2 d\tau}.$$

We must search for values of the coefficients in the trial function that minimize the value of E. This is a standard problem in calculus, and is solved by finding the coefficients for which

$$\frac{\partial E}{\partial c_A} = 0 \quad \text{and} \quad \frac{\partial E}{\partial c_B} = 0.$$

The first step is to express the two integrals in terms of the coefficients. The denominator is

$$\begin{aligned}\int \psi^2 d\tau &= \int \{c_A\psi(A) + c_B\psi(B)\}^2 d\tau \\ &= c_A^2 \int \psi(A)^2 d\tau + c_B^2 \int \psi(B)^2 d\tau + 2c_A c_B \int \psi(A)\psi(B) d\tau \\ &= c_A^2 + c_B^2 + 2c_A c_B S \qquad (1)\end{aligned}$$

because the individual atomic orbitals are normalized and the third integral is the overlap integral S. The numerator is

$$\int \psi H \psi \mathrm{d}\tau = \int \{c_A\psi(A) + c_B\psi(B)\} H \{c_A\psi(A) + c_B\psi(B)\}\mathrm{d}\tau$$
$$= c_A^2 \int \psi(A)H\psi(A)\mathrm{d}\tau + c_B^2 \int \psi(B)H\psi(B)\mathrm{d}\tau + 2c_Ac_B \int \psi(A)H\psi(B)\mathrm{d}\tau.$$

There are some complicated integrals in this expression, but we can denote them by the constants

$$\alpha_A = \int \psi(A)H\psi(A)\mathrm{d}\tau, \qquad \alpha_B = \int \psi(B)H\psi(B)\mathrm{d}\tau, \qquad \beta = \int \psi(A)H\psi(B)\mathrm{d}\tau.$$

Then

$$\int \psi H \psi \mathrm{d}\tau = c_A^2\alpha_A + c_B^2\alpha_B + 2c_Ac_B\beta.$$

α is called a **Coulomb integral.** It is negative, and can be interpreted as the energy of the electron when it occupies $\psi(A)$ (for α_A) or $\psi(B)$ (for α_B). In a homonuclear diatomic molecule, $\alpha_A = \alpha_B$. β is called a **resonance integral** (for classical reasons). It vanishes when the orbitals do not overlap, and at equilibrium bond lengths it is normally negative.

The complete expression for E is

$$E = \frac{c_A^2\alpha_A + c_B^2\alpha_B + 2c_Ac_B\beta}{c_A^2 + c_B^2 + 2c_Ac_BS}.$$

Its minimum is found by differentiation with respect to the two coefficients. This involves elementary but slightly tedious work, the end result being the two **secular equations**

$$(\alpha_A - E)c_A + (\beta - ES)c_B = 0,$$
$$(\beta - ES)c_A + (\alpha_B - E)c_B = 0.$$

They have a solution if the determinant of the coefficients, the *secular determinant* vanishes; that is, if

$$\begin{vmatrix} \alpha_A - E & \beta - ES \\ \beta - ES & \alpha_B - E \end{vmatrix} = 0$$

This determinant expands to a quadratic equation in E, which may be solved. Its two roots give the energies of the bonding and antibonding MOs formed from the basis set and, according to the variation principle, these are the best energies for the given basis set. The corresponding values of the coefficients are then obtained by solving the secular equations using the two energies: the lower energy gives the coefficients for the bonding MO, the upper energy the coefficients for the antibonding MO. The secular equations give expressions for the *ratio* of the coefficients in each case, and so we need a further equation in order to find their individual values. This is obtained by demanding that the best wavefunction should be normalized, which means that we must also ensure (from eqn (1) above) that

$$\int \psi^2 \mathrm{d}\tau = c_A^2 + c_B^2 + 2c_A c_B S = 1.$$

The complete solutions are very cumbersome‡.

There are two cases where the roots can be written down very simply. First, when the two atoms are the same, and we can write $\alpha_A = \alpha_B = \alpha$, the solutions are

$$E_+ = \frac{\alpha+\beta}{1+S}, \quad c_A = \left\{\frac{1}{2(1+S)}\right\}^{1/2}, \quad c_B = c_A,$$

$$E_- = \frac{\alpha-\beta}{1-S}, \quad c_A = \left\{\frac{1}{2(1-S)}\right\}^{1/2}, \quad c_B = -c_A.$$

In this case, the best bonding function has the form

$$\psi_+ = \left\{\frac{1}{2(1+S)}\right\}^{1/2} \{\psi(A) + \psi(B)\}$$

and the corresponding antibonding function is

$$\psi_- = \left\{\frac{1}{2(1-S)}\right\}^{1/2} \{\psi(A) - \psi(B)\}.$$

(a) When it is justifiable to neglect overlap, the secular determinant is

$$\begin{vmatrix} \alpha_A - E & \beta \\ \beta & \alpha_B - E \end{vmatrix} = 0$$

and its solutions can be expressed in terms of the parameter θ, with

$$\tan 2\theta = \frac{2\beta}{\alpha_A - \alpha_B}.$$

The solutions are

$$E_- = \alpha_A - \beta\cot\theta, \qquad \psi_- = -\sin\theta\,\psi(A) + \cos\theta\,\psi(B),$$

$$E_+ = \alpha_B + \beta\cot\theta, \qquad \psi_+ = \cos\theta\,\psi(A) + \sin\theta\,\psi(B).$$

‡

$$E = (A \pm B)/(1-S^2),$$
$$A = \frac{1}{2}(\alpha_A + \alpha_B) - \beta S,$$
$$B = \left\{\frac{1}{4}(\alpha_A - \alpha_B)^2 - (\alpha_A + \alpha_B)\beta S + \alpha_A\alpha_B S^2 + \beta^2\right\}^{\frac{1}{2}},$$
$$c_A = (\beta - SE)/C,$$
$$C = \left\{(\beta - SE)^2 + (\alpha_A - E)^2 - 2S(\alpha_A - E)(\beta - SE)\right\}^{\frac{1}{2}}.$$

If $\theta = 0$, the wavefunction $\psi_+ = \psi(\mathrm{A})$; if $\theta = \frac{\pi}{2}$, the wavefunction $\psi_+ = \psi(\mathrm{B})$. So we see that this wavefunction can describe a polar covalent bond, the degree of polarity is dependent on θ. If $\theta = \frac{\pi}{4}$, the bond is completely covalent.

(b) We need to evaluate $\int \psi^2 \mathrm{d}\tau$ to see if it equals 1.

$$\begin{aligned}\int (\psi_\mathrm{A}\cos\theta + \psi_\mathrm{B}\sin\theta)^2\mathrm{d}\tau &= \int (\psi_\mathrm{A}^2\cos^2\theta + \psi_\mathrm{B}^2\sin^2\theta + 2\psi_\mathrm{A}\psi_\mathrm{B}\sin\theta\cos\theta)\mathrm{d}\tau \\ &= \cos^2\theta\int\psi_\mathrm{A}^2\mathrm{d}\tau + \sin^2\theta\int\psi_\mathrm{B}^2\mathrm{d}\tau \\ &\quad + 2\sin\theta\cos\theta\int\psi_\mathrm{A}\psi_\mathrm{B}\mathrm{d}\tau \\ &= \cos^2\theta + \sin^2\theta + 2\sin\theta\cos\theta S \\ &= 1.\end{aligned}$$

We have used the facts that ψ_A and ψ_B are each normalized and that S is assumed to be zero. Remember θ is a constant.

(c) In a homonuclear diatomic molecule we set $\alpha_\mathrm{A} = \alpha_\mathrm{B}$. Therefore $\tan 2\theta = \infty$, which solves to

$$\theta = \frac{1}{2}\arctan(\infty) = \frac{1}{2} \times 1.5708 = 0.7854 \text{ radian} = \boxed{45°}.$$

14.21 σ-bonding with $d_{x^2-y^2}$ orbital is shown in Figure 14.5.

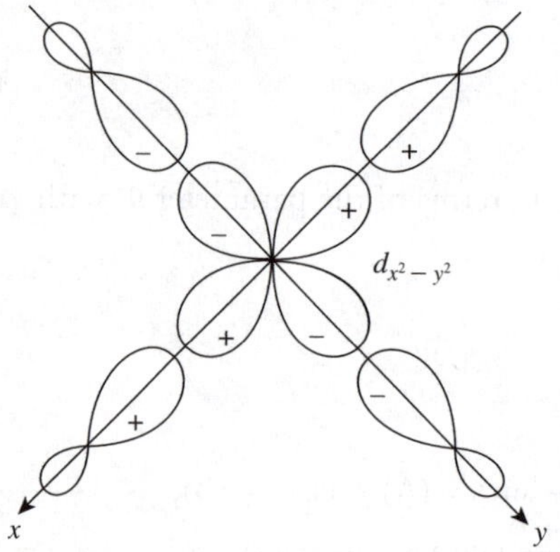

Fig 14.5

The σ-antibonding orbital looks the same but with the p-orbital lobes pointed in the opposite direction.

The σ-bonding and antibonding diagrams with the d_{xy}, d_{yz}, and d_{xz} orbitals have the same appearance as the diagram above except that the d-orbital lobes are pointed between the axes rather than along them.

σ-bonding with d_{z^2} orbital is shown in Figure 14.6.

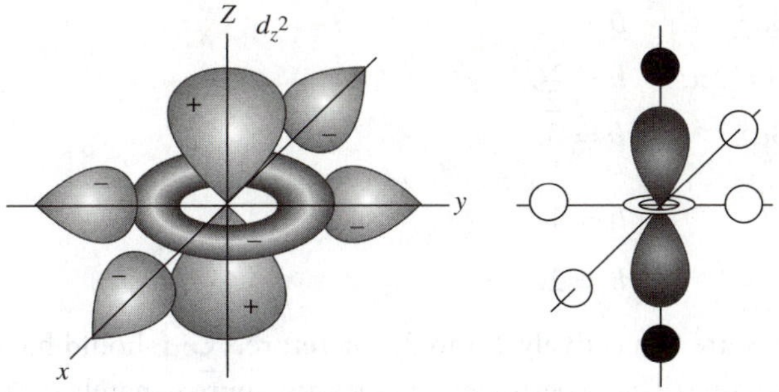

Fig 14.6

In this figure, only one of the p-orbital lobes is shown in each p-orbital. p-orbitals with positive lobes may also approach this orbital along the + and − z-direction as indicated in the smaller diagram on the right. The antibonding diagrams are similar, but with the signs of the p-orbital lobes reversed.

π-bonding (see Figure 14.7).

Only the d_{xy}, d_{yz}, and d_{xz} orbitals undergo π-bonding with p-orbitals on neighboring atoms. The bonding arrangement is pictured in Figure 14.7 with the d_{xy} orbital. The diagrams for the d_{yz} and d_{xz} orbitals are similar. The antibonding diagrams have the signs of the p-orbital lobes reversed.

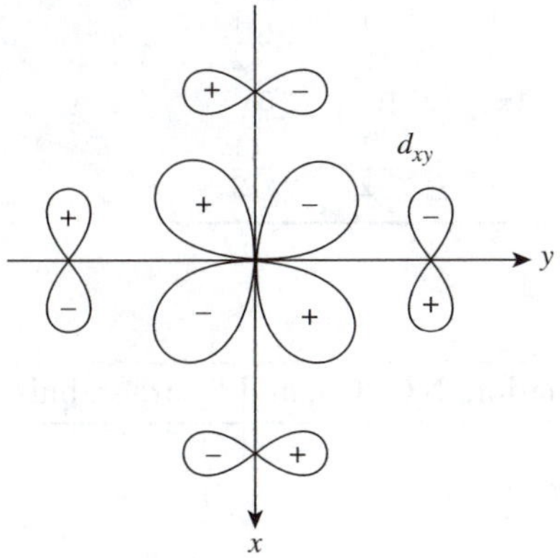

Fig 14.7

14.22 Configurations of valence electrons are shown in Fig. 14.26 of the text.

(a) Li_2	$1\sigma^2$	$b = 1.$
(b) Be_2	$1\sigma^2 1\sigma^{*2}$	$b = 0.$
(c) C_2	$1\sigma^2 1\sigma^{*2} 1\pi^4$	$b = 2.$

14.23

(a) H_2^-	$1\sigma^2 1\sigma^{*1}$	$b = \frac{1}{2}.$
(b) N_2	$1\sigma^2 1\sigma^{*2} 1\pi^4 2\sigma^2$	$b = 3.$
(c) O_2	$1\sigma^2 1\sigma^{*2} 2\sigma^2 1\pi^4 1\pi^{*2}$	$b = 2.$

14.24 (a) CO $1\sigma^2 2\sigma^{*2} 1\pi^4 3\sigma^2$ $b = 3.$

(b) NO $1\sigma^2 2\sigma^{*2} 1\pi^4 3\sigma^2 2\pi^{*1}$ $b = \frac{5}{2}.$

(c) CN^- $1\sigma^2 2\sigma^{*2} 1\pi^4 3\sigma^2$ $b = 3.$

14.25 B_2 $1\sigma^2 1\sigma^{*2} 1\pi^2$ $b = 1.$

C_2 $1\sigma^2 1\sigma^{*2} 1\pi^4$ $b = 2.$

The bond orders of B_2 and C_2 are respectively 1 and 2. Therefore, $\boxed{C_2}$ should have the greater bond dissociation enthalpy. The experimental values are approximately 4 eV and 6 eV, respectively.

14.26 Decide whether the electron added or removed increases or decreases the bond order. The simplest procedure is to decide whether the electron occupies or is removed from a bonding or antibonding orbital. The levels for the homonuclear diatomics are shown in Figures 14.22 and 14.23 of the text. The levels for the heteronuclear diatomics are shown in Figure 14.34.

The following table gives the orbital involved.

		N_2	NO	O_2	C_2	F_2	CN
(a)	AB^-	$1\pi^*$	$2\pi^*$	$1\pi^*$	2σ	$2\sigma^*$	3σ
	Δb	$-\frac{1}{2}$	$-\frac{1}{2}$	$-\frac{1}{2}$	$+\frac{1}{2}$	$-\frac{1}{2}$	$+\frac{1}{2}$
(b)	AB^+	2σ	$2\pi^*$	$1\pi^*$	1π	$1\pi^*$	3σ
	Δb	$-\frac{1}{2}$	$+\frac{1}{2}$	$+\frac{1}{2}$	$-\frac{1}{2}$	$+\frac{1}{2}$	$-\frac{1}{2}$

Therefore,

C_2 and CN are stabilized by anion formation. NO, O_2, and F_2 are stabilized by cation formation.

14.27 B_2 (6 valence electrons): $1\sigma^2 1\sigma^{*2} 1\pi^2$ $b = 1$ (See Figure 14.26)

C_2 (8 valence electrons): $1\sigma^2 1\sigma^{*2} 1\pi^4$ $b = 2$

The bond orders of B_2 and C_2 are respectively 1 and 2; so $\boxed{C_2}$ should have the greater bond dissociation enthalpy. The experimental values are approximately 4eV and 6eV, respectively.

14.28 We use Figure 14.34 of the text to construct the configuration of XeF Figure 14.8.

Because the bond order is increased when XeF^+ is formed from XeF (an electron is removed from an antibonding orbital), XeF^+ will have a shorter bond length than XeF.

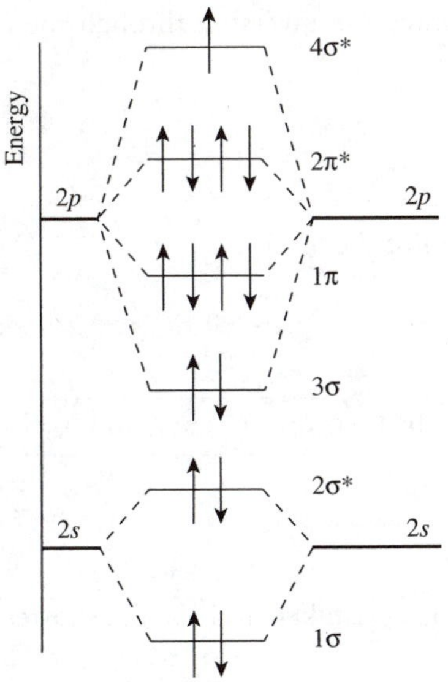

Fig 14.8

14.29 (a) 1 π* is $\boxed{\text{g}}$

(b) g, u is $\boxed{\text{inapplicable}}$ to a heteronuclear molecule because it has no center of inversion.

(c) $\boxed{\text{g}}$ (See Figure 14.9 (a)).

(d) $\boxed{\text{u}}$ (See Figure 14.9 (b)).

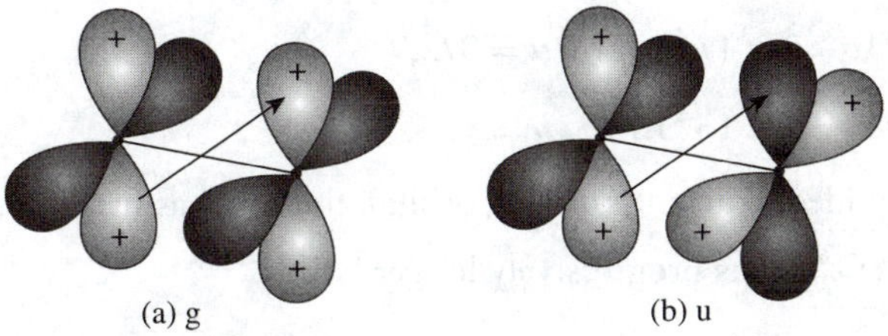

Fig 14.9

14.30 The wavefunctions are

$$\psi_n = \left(\frac{2}{L}\right)^{1/2} \sin\left(\frac{n\pi x}{L}\right).$$

The center is at $\frac{L}{2}$. We invert through $x = \frac{L}{2}$. For $n = 1$, when $x < \frac{L}{2}$, $\psi_1 = +$; when $x > \frac{L}{2}$, $\psi_1 = +$, therefore ψ_1 is $\boxed{\text{g}}$. In a similar fashion we determine

$n = 2$, $\boxed{\text{u}}$,

$n = 3$, $\boxed{\text{g}}$,

$n = 4$, $\boxed{\text{u}}$.

14.31 Refer to Figure 12.34 of the text. Examine the inversion through the center of these functions, i.e. replace x with $-x$.

(a) $v = 0$, g.
$v = 1$, u.
$v = 2$, g.
$v = 3$, u (not shown in Figure 12.34 of the text).

(b) If v is even, ψ_v is g.
If v is odd, ψ_v is u.

14.32 The parities are given in Figure 14.36 of the text $\boxed{a_{2u}, e_{1g}, e_{2u}, \text{and } b_{2g}}$.

14.33 NO $1\sigma^2 2\sigma^{*2} 1\pi^4 3\sigma^2 2\pi^{*1}$ $b = 5/2$.
N_2 $1\sigma^2 1\sigma^{*2} 1\pi^4 2\sigma^2$ $b = 3$.
Because the bond order of N_2 is greater, $\boxed{N_2}$ is likely to have the shorter bond length.

14.34 F_2^+ $1\sigma^2 1\sigma^{*2} 2\sigma^2 1\pi^4 1\pi^{*3}$ $b = \frac{3}{2}$,
F_2 $1\sigma^2 1\sigma^{*2} 2\sigma^2 1\pi^4 1\pi^{*4}$ $b = 1$,
F_2^- $1\sigma^2 1\sigma^{*2} 2\sigma^2 1\pi^4 1\pi^{*4} 2\sigma^{*1}$ $b = \frac{1}{2}$.
Therefore, order of bond lengths is $\boxed{F_2^+ < F_2 < F_2^-}$.

14.35 O_2^+ (11 electrons) : $1\sigma^2 1\sigma^{*2} 2\sigma^2 1\pi^4 1\pi^{*1}$ $b = 5/2$,
O_2 (12 electrons) : $1\sigma^2 1\sigma^{*2} 2\sigma^2 1\pi^4 1\pi^{*2}$ $b = 2$,
O_2^- (13 electrons) : $1\sigma^2 1\sigma^{*2} 2\sigma^2 1\pi^4 1\pi^{*3}$ $b = 3/2$,
O_2^{2-} (14 electrons) : $1\sigma^2 1\sigma^{*2} 2\sigma^2 1\pi^4 1\pi^{*4}$ $b = 1$.
Each electron added to O_2^+ is added to an antibonding orbital, thus increasing the length. So the sequence $\boxed{O_2^+, O_2, O_2^-, O_2^{2-}}$ has progressively longer bonds.

14.36 The molecular orbitals of the fragments and the molecular orbitals that they form are shown in Figure 14.10 (a) and 14.10 (b).

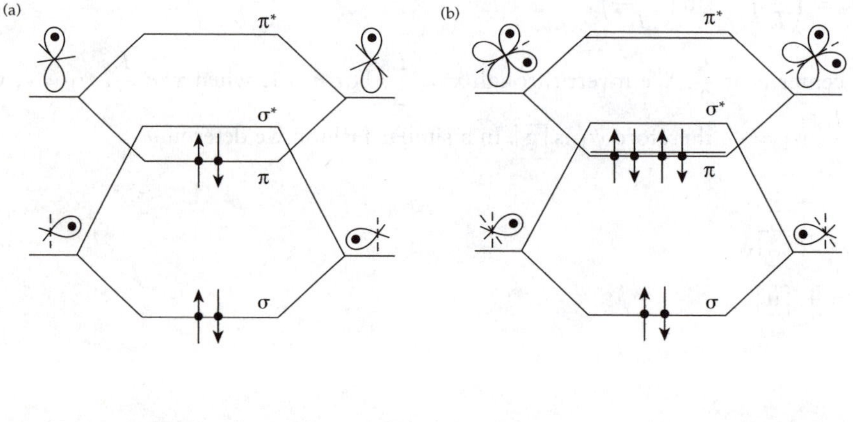

Fig 14.10

COMMENT Note that the π-bonding orbital must be lower in energy than the σ-antibonding orbital for π-bonding to exist in ethene.

QUESTION Would the ethene molecule exist if the order of the energies of the π and σ* orbitals were reversed?

14.37 (a) Benzene anion (7 π electrons): $1\pi^2 2\pi^4 3\pi$ or $a_{2u}{}^2 e_{1g}{}^4 e_{2u}$.

(b) Benzene anion (5 π electrons): $1\pi^2 2\pi^3$ or $a_{2u}{}^2 e_{1g}{}^3$.

14.38 $E_n = \dfrac{n^2 h^2}{8mL^2}$, $n = 1, 2, \ldots$ and $\psi_n = \left(\dfrac{2}{L}\right)^{1/2} \sin\left(\dfrac{n\pi x}{L}\right)$.

Two electrons occupy each level (by the Pauli principle), and so butadiene (in which there are four π electrons) has two electrons in ψ_1 and two electrons in ψ_2.

$$\psi_1 = \left(\frac{2}{L}\right)^{1/2} \sin\left(\frac{\pi x}{L}\right), \qquad \psi_2 = \left(\frac{2}{L}\right)^{1/2} \sin\left(\frac{2\pi x}{L}\right).$$

These orbitals are sketched in Figure 14.11.

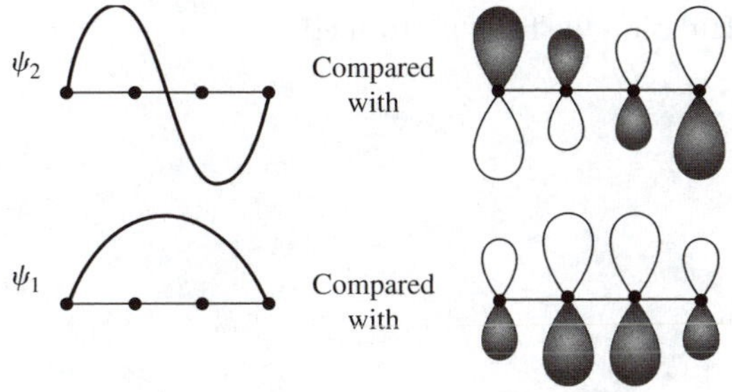

Fig 14.11

The minimum excitation energy is

$$\Delta E = E_3 - E_2 = 5\left(\frac{h^2}{8m_e L^2}\right).$$

In $CH_2{=}CH{-}CH{=}CH{-}CH{=}CH{-}CH{=}CH_2$ there are eight π electrons to accommodate, so the HOMO will be ψ_4 and the LUMO ψ_5. From the particle-in-a-box solutions (Chapter 12)

$$\Delta E = E_5 - E_4 = (25 - 16)\left(\frac{h^2}{8m_e L^2}\right) = \left(\frac{9h^2}{8m_e L^2}\right)$$

$$= \frac{(9) \times (6.626 \times 10^{-34}\,\text{J s})^2}{(8) \times (9.109 \times 10^{-31}\,\text{kg}) \times (1.12 \times 10^{-9}\,\text{m})^2} = 4.3 \times 10^{-19}\,\text{J}$$

which corresponds to $\boxed{2.7\ \text{eV}}$. The HOMO and LUMO are

$$\psi_n = \left(\frac{2}{L}\right)^{1/2} \sin\left(\frac{n\pi x}{L}\right)$$

with $n = 4, 5$ respectively; the two wavefunctions are sketched in Figure 14.12.

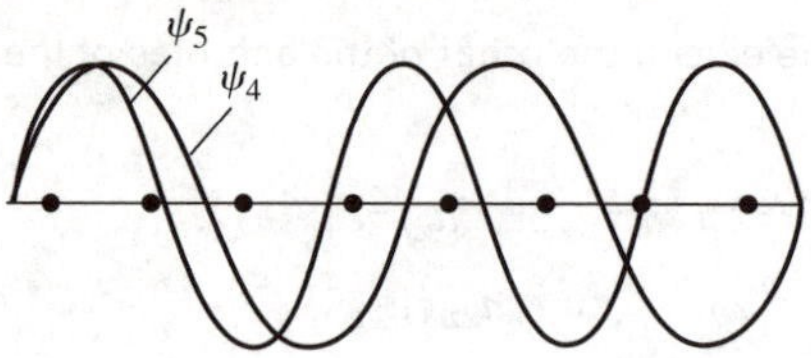

Fig 14.12

COMMENT It follows that

$$\lambda = \frac{hc}{\Delta E} = \frac{(6.626 \times 10^{-34}\,\mathrm{J\,s}) \times (2.998 \times 10^{8}\,\mathrm{m\,s^{-1}})}{4.3 \times 10^{-19}\,\mathrm{J}} = 4.6 \times 10^{-7}\,\mathrm{m}, \text{ or } \boxed{460\,\mathrm{nm}}.$$

The wavelength 460 nm corresponds to blue light, so on the basis of this calculation alone the molecule would appear orange in white light (because blue is subtracted). The experimental value of λ_{max} is 304 nm, and the compound is colorless. This illustrates the very approximate nature of the calculation we have performed.

Chapter 15

Metallic, ionic, and covalent solids

Answers to discussion questions

15.1 A metallic conductor is a substance with a conductivity that decreases as the temperature is raised. A semiconductor is a substance with a conductivity that increases as the temperature is raised. A semiconductor generally has a lower conductivity than that typical of metals, but the magnitude of the conductivity is not the criterion of the distinction. It is conventional to classify semiconductors with very low electrical conductivities, such as most synthetic polymers, as insulators. We shall use this term. But it should be appreciated that it is one of convenience rather than one of fundamental significance.

The conductivity of these three kinds of materials is explained by band theory. When each of N atoms of a metallic element contributes one atomic orbital to the formation of molecular orbitals, the resulting N molecular orbitals form an almost continuous band of levels. The orbital at the bottom of the band is fully bonding between all neighbors, and the orbital at the top of the band is fully antibonding between all immediate neighbors. If the atomic orbitals are *s*-orbitals, then the resulting band is called an *s*-band; if the original orbitals are *p*-orbitals, then they form a *p*-band. In a typical case, there is so large an energy difference between the *s* and *p* atomic orbitals that the resulting *s*- and *p*-bands are separated by a region of energy in which there are no orbitals. This region is called the band gap, and its width is denoted E_g.

When electrons occupy the orbitals in the bands, they do so in accord with the Pauli principle. If insufficient electrons are present to fill the band, the electrons close to the top of the band are mobile and the solid is a metallic conductor. An unfilled band is called a conduction band and the energy of the highest occupied orbital at $T = 0$ K is called the Fermi level. Only the electrons close to the Fermi level can contribute to conduction and to the heat capacity of a metal. If the band is full, then the electrons cannot transport a current readily, and the solid is an insulator; more formally, it is a species of semiconductor with a large band gap. A full band is called a valence band. The detailed population of the levels in a band, taking into account the role of temperature, is expressed by the Fermi–Dirac distribution.

The distinction between metallic conductors and semiconductors can be traced to their band structure: a metallic conductor has an incomplete band, its conductance band, and a semiconductor has full bands, and hence lacks a conductance band. The decreasing conductance of a metallic conductor with temperature stems from the scattering of electrons by the vibrating atoms of the metal lattice. The increasing conductance of a semiconductor arises from the increasing population of an upper empty band as the temperature is increased. Many substances, however, have such large band gaps that their ability to conduct an electric current remains very low at all temperatures: it is conventional to refer to such solids as insulators. The ability of a semiconductor to transport charge is enhanced by doping it, or adding substances in controlled quantities. If the dopant provides additional electrons, then the semiconductor is classified as n-type. If it removes electrons from the valence band and thereby increases the number of positive holes, it is classified as p-type.

15.2 The phase problem arises with the analysis of data in X-ray diffraction when seeking to perform a Fourier synthesis of the electron density. In order to carry out the sum it is necessary to know the signs of the structure factors; however, because diffraction intensities are proportional to the square of the structure factors, the intensities do not provide information on the sign. For non-centrosymmetric crystals, the structure factors may be complex, and the phase α in the expression $F_{hkl} = |F_{hkl}|\,e^{i\alpha}$ is indeterminate. The phase problem may be evaded by the use of a Patterson synthesis or tackled directly by using the so-called direct methods of phase allocation.

The Patterson synthesis is a technique of data analysis in X-ray diffraction which helps to circumvent the phase problem. In it, a function P is formed by calculating the Fourier transform of the squares of the structure factors (which are proportional to the intensities):

$$P(r) = \frac{1}{V}\sum_{hkl} |F_{hkl}|^2\, e^{2\pi i(hx+ky+lz)}.$$

The outcome is a map of the *separations* of the atoms in the unit cell of the crystal. If some atoms are heavy (perhaps because they have been introduced by isomorphous replacement), they dominate the Patterson function, and their locations can be deduced quite simply. Their locations can then be used in the determination of the locations of lighter atoms.

15.3 The majority of metals crystallize in structures, which can be interpreted as the closest packing arrangements of hard spheres. These are the cubic close-packed (ccp) and hexagonal close-packed (hcp) structures. In these models, 74% of the volume of the unit cell is occupied by the atoms (packing fraction = 0.74). Most of the remaining metallic elements crystallize in the body-centered cubic (bcc) arrangement, which is not too much different from the close-packed structures in terms of the efficiency of the use of space (packing fraction 0.68 in the hard sphere model). Polonium is an exception; it crystallizes in the simple cubic structure, which has a packing fraction of 0.52. See the solution to Exercise 15.27 for a derivation of the packing fraction in ccp systems. If atoms were truly hard spheres, we would expect that all metals would crystallize in either the ccp or hcp close-packed structures. The fact that a significant number crystallize in other structures is proof that a simple hard sphere model is an inaccurate representation of the interactions between the atoms. Covalent bonding between the atoms may influence the structure.

15.4 In a face-centered cubic close-packed lattice, there is an octahedral hole in the centre. The rock-salt structure can be thought of as being derived from an fcc structure of Cl^- ions in which Na^+ ions have filled the octahedral holes.

The cesium-chloride structure can be considered to be derived from the ccp structure by having Cl^- ions occupy all the primitive lattice points and octahedral sites, with all tetrahedral sites occupied by Cs^+ ions. This is exceedingly difficult to visualize and describe without carefully constructed figures or models. Refer to S.-M. Ho and B. E. Douglas, *J. Chem. Educ.* **46**, 208, 1969, for the appropriate diagrams.

Solutions to exercises

15.5 (a) P (group V) has one more valence electron than Ge (group IV); therefore Ge doped with P forms an **n-type** semiconductor.

(b) In (group III) has one less valence electron than Ge (group IV); therefore Ge doped with P forms an **p-type** semiconductor.

15.6 The resistance of metals increases with increasing temperature; the opposite is true of semiconductors. Therefore, this substance is a **metallic conductor**.

15.7 The valence electrons of Mg are 3*s* electrons, those of O are 2*p* electrons. We expect that the O2*p* electron will be much lower in energy than the Mg3*s* electrons. The energy level diagram is expected to look like Figure 15.1.

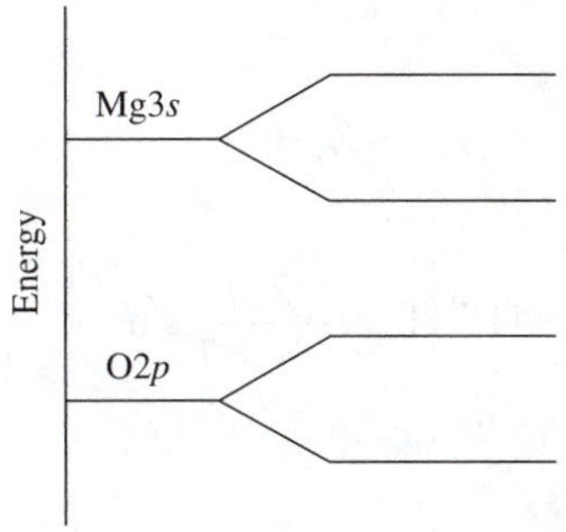

Fig 15.1

There are two degenerate O2*p* orbitals available for band formation, which results in a doubly degenerate O2*p* band in oxygen. These bands can hold 4*N* electrons, 2*N* of which are contributed by Mg. Only the lower band is occupied and because of the big band gap MgO is an insulator. Effectively electrons have been transferred from Mg to O as in the ionic model.

15.8 $E_k = \alpha + 2\beta\cos\left(\frac{k\pi}{N+1}\right), \quad k = 1, 2, \ldots, N \quad [15.1].$

$$E_{k+1} - E_k = \left[\alpha + 2\beta\cos\left(\frac{(k+1)\pi}{N+1}\right)\right] - \left[\alpha + 2\beta\cos\left(\frac{k\pi}{N+1}\right)\right]$$

$$= 2\beta\left\{\cos\left(\frac{(k+1)\pi}{N+1}\right) - \cos\left(\frac{k\pi}{N+1}\right)\right\}.$$

This is transformed into a simpler function of k with the trigonometry relationships

$$\cos A - \cos B = 2\sin\left(\frac{A+B}{2}\right)\sin\left(\frac{B-A}{2}\right) \text{ and } \sin(-A) = -\sin A.$$

$$E_{k+1} - E_k = 2\beta\sin\left(\frac{(2k+1)\pi}{2(N+1)}\right) \times \sin\left(\frac{-\pi}{2(N+1)}\right)$$

$$= -2\beta\sin\left(\frac{(2k+1)\pi}{2(N+1)}\right) \times \sin\left(\frac{\pi}{2(N+1)}\right).$$

Since $\beta < 0$ and $-2\beta = |2\beta|$,

$$\boxed{E_{k+1} - E_k = |2\beta|\sin\left(\frac{(2k+1)\pi}{2(N+1)}\right) \times \sin\left(\frac{\pi}{2(N+1)}\right)}.$$

Let us take the limit of the second factor as $N \to \infty$.

$$\lim_{N\to\infty}\left(\sin\left(\frac{\pi}{2(N+1)}\right)\right) = \lim_{N\to\infty}\left(\sin\left(\frac{\pi}{2N}\right)\right) = \sin(0) = 0.$$

The energy difference decreases to zero as N becomes very large.

15.9 The density of energy levels is

$$\rho = \frac{dE}{dk}$$

$$\text{where } \frac{dE}{dk} = \frac{d}{dk}\left(\alpha + 2\beta\cos\frac{k\pi}{N+1}\right) \quad [15.1] = -\frac{2\pi\beta}{N+1}\sin\frac{k\pi}{N+1}$$

$$\text{so } \rho(k) = -\frac{2\pi\beta}{N+1}\sin\frac{k\pi}{N+1}.$$

The sine factor is eliminated by application of the trigonometric identity

$$\sin^2\theta + \cos^2\theta = 1 \quad \text{or} \quad \sin\theta = \left(1 - \cos^2\theta\right)^{1/2}.$$

$$\rho(k) = -\frac{2\pi\beta}{N+1}\sin\frac{k\pi}{N+1} = -\frac{2\pi\beta}{N+1}\left(1 - \cos^2\frac{k\pi}{N+1}\right)^{1/2}.$$

But, from eqn 15.1, $\cos\dfrac{k\pi}{N+1}=\dfrac{E_k-\alpha}{2\beta}$, so

$$\boxed{\rho(E)=-\frac{2\pi\beta}{N+1}\left\{1-\left(\frac{E-\alpha}{2\beta}\right)^2\right\}^{1/2}}.$$

The density of states is greatest when $E=\alpha$ by this definition of density of states.

15.10 We need to find the enthalpy change for

$$MgO(s) \rightarrow Mg^{2+}(g)+O^{2-}(g), \quad \Delta H_L^{\ominus}.$$

We set up a Born–Haber cycle.

(1) Sublimation of Mg(s)	$+148\ \text{kJ mol}^{-1}$,
(2) Ionization of Mg(g) to $Mg^{2+}(g)$	$+2187\ \text{kJ mol}^{-1}$,
(3) Dissociation of $\frac{1}{2}O_2(g)$	$+248\ \text{kJ mol}^{-1}$,
(4) Electron attachment to O(g)	$+703\ \text{kJ mol}^{-1}$,
(5) Formation of MgO(s)	$-\Delta H_L^{\ominus}$.

The sum of the above 5 steps corresponds to

$$Mg(s)+\frac{1}{2}O_2(g) \rightarrow MgO(s), \qquad \Delta_f H^{\ominus}=-601.7\ \text{kJ mol}^{-1}.$$

Therefore

$$\Delta H_L^{\ominus}=(148+2187+248+703+601.7)\ \text{kJ mol}^{-1}$$
$$=\boxed{3888\ \text{kJ mol}^{-1}}$$

This does not agree exactly with the value in Table 15.1, which is $3850\ \text{kJ mol}^{-1}$.

15.11 We need to find the enthalpy change for

$$CaCl_2(s) \rightarrow Ca^{2+}(g)+2Cl^{-}(g), \qquad \Delta H_L^{\ominus}.$$

We set up a Born–Haber cycle.

(1) Sublimation of Ca(s)	$+178.2\ \text{kJ mol}^{-1}$,
(2) Ionization of Ca(g) to $Ca^{2+}(g)$	$+1740\ \text{kJ mol}^{-1}$,
(3) Dissociation of $Cl_2(g)$	$+242\ \text{kJ mol}^{-1}$,
(4) Electron attachment to 2Cl(g)	$-698\ \text{kJ mol}^{-1}$,
(5) Formation of $CaCl_2(s)$	$-\Delta H_L^{\ominus}$.

The sum of the above 5 steps corresponds to

$$Ca(s)+Cl_2(g) \rightarrow CaCl_2(s), \quad \Delta_f H^{\ominus}=-795.8\ \text{kJ mol}^{-1}.$$

Therefore

$$\Delta H_L^{\ominus}=(178+1740+242-698+796)\ \text{kJ mol}^{-1}$$
$$=\boxed{2258\ \text{kJ mol}^{-1}}.$$

15.12 Let there be N charges ze on the first sphere, of radius d; then there will be $N/2$ of charge $-ze$ on the second sphere of radius $2d$, $N/3$ of charge $+ze$ on the third sphere of radius $3d$, and so on. The total potential energy is therefore the sum

$$V = \frac{q}{4\pi\varepsilon_0}\left(\frac{Nze}{d} - \frac{Nze}{2\times 2d} + \frac{Nze}{3\times 3d} - \frac{Nze}{4\times 4d} + \ldots\right)$$

where q is the charge on the ion at the center. The above can be rewritten

$$V = \frac{qNze}{4\pi\varepsilon_0 d}\left(1 - \frac{1}{2^2} + \frac{1}{3^2} - \frac{1}{4^2} + \ldots\right) = \frac{qNze}{4\pi\varepsilon_0 d}\left(\frac{\pi^2}{12}\right) = \boxed{\frac{qNze\pi}{48\varepsilon_0 d}}.$$

15.13 We need to evaluate the ratio of the factors that depend on d for the two compounds.

$$\frac{\Delta H_L^{\ominus}(\text{CaO})}{\Delta H_L^{\ominus}(\text{SrO})} = \frac{\frac{1}{d}\times\left(1-\frac{d^*}{d}\right)(\text{CaO})}{\frac{1}{d}\times\left(1-\frac{d^*}{d}\right)(\text{SrO})} = \frac{\frac{1}{240}\times\left(1-\frac{34.5}{240}\right)}{\frac{1}{256}\times\left(1-\frac{34.5}{256}\right)} = \boxed{1.06}.$$

15.14 The relationship between critical temperature and critical magnetic field is given by

$$H_c(T) = H_c(0)\left(1 - \frac{T^2}{T_c^2}\right).$$

Solving for T gives the critical temperature for a given magnetic field.

$$T = T_c\left(1 - \frac{H_c(T)}{H_c(0)}\right)^{1/2} = (7.19\text{ K})\times\left(1 - \frac{20\times 10^3\text{ A m}^{-1}}{63901\text{ A m}^{-1}}\right)^{1/2} = \boxed{6.0\text{ K}}.$$

15.15 The points and planes are shown in Figure 15.2.

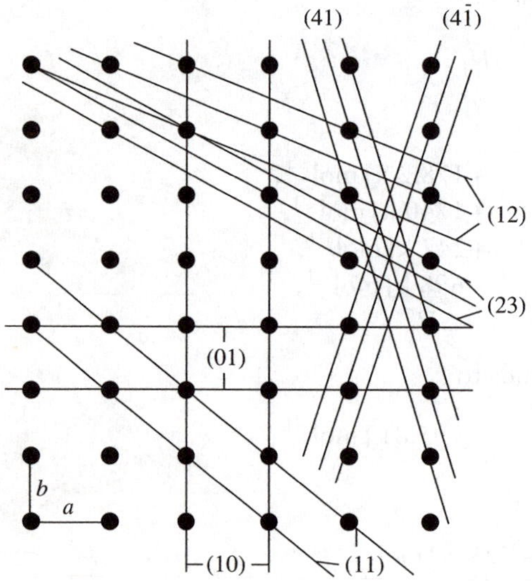

Fig 15.2

15.16 The points and planes are shown in Figure 15.3.

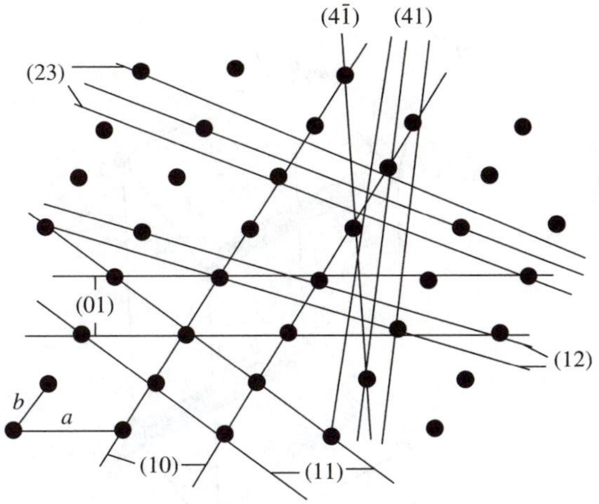

Fig 15.3

15.17 Draw up the following table, using the procedure set out in Section 15.11.

Original	Reciprocal	Clear fractions	Miller indices
$(2a, 3b, c)$ or (2, 3, 1)	$\left(\frac{1}{2}, \frac{1}{3}, 1\right)$	(3, 2, 6)	(326)
(a, b, c) or (1, 1, 1)	(1, 1, 1)	(1, 1, 1)	(111)
$(6a, 3b, 3c)$ or (6, 3, 3)	$\left(\frac{1}{6}, \frac{1}{3}, \frac{1}{3}\right)$	(1, 2, 2)	(122)
$(2a, -3b, -3c)$ or (2, −3, −3)	$\left(\frac{1}{2}, -\frac{1}{3}, -\frac{1}{3}\right)$	(3, −2, −2)	$(3\bar{2}\bar{2})$

15.18 See Figure 15.4(a) for the (100), (010), (001), (011), (101), and (111) planes. See Figure 15.4(b) for the $(10\bar{1})$ plane.

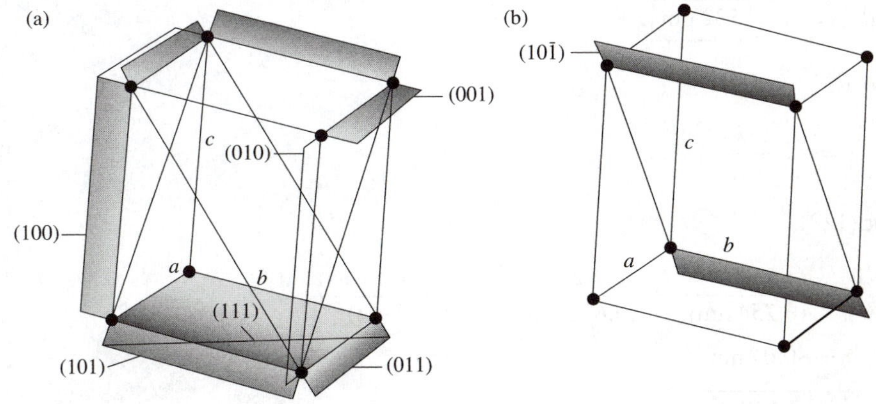

Fig 15.4

15.19 See Figure 15.5(a) for the (100), (010), (001), (011), (101), and (111) planes. See Figure 15.5(b) for the $(10\bar{1})$ plane.

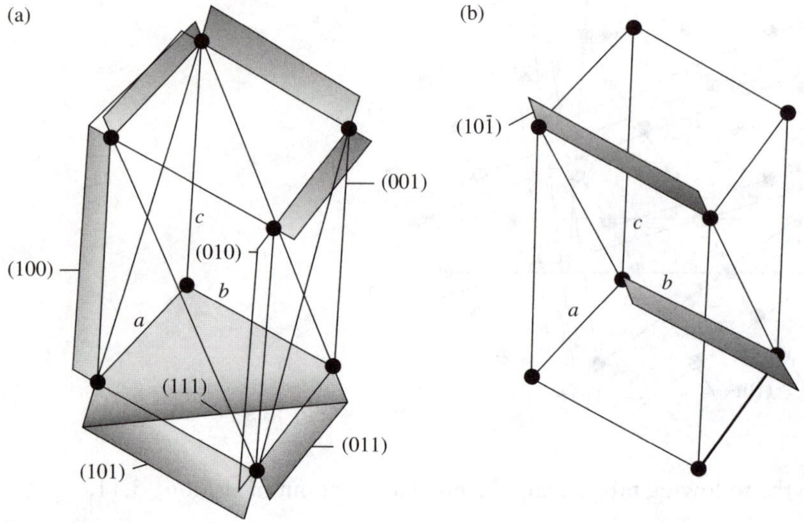

Fig 15.5

15.20 $$\frac{1}{d^2} = \frac{h^2}{a^2} + \frac{k^2}{b^2} + \frac{l^2}{c^2} \qquad [15.7].$$

For a cubic unit cell in which $a = b = c$,

$$d_{hkl} = \frac{a}{(h^2 + k^2 + l^2)^{1/2}}.$$

Therefore,

$$d_{111} = \frac{a}{3^{1/2}} = \frac{532\text{pm}}{3^{1/2}} = \boxed{307\,\text{pm}},$$

$$d_{211} = \frac{a}{6^{1/2}} = \frac{532\text{ pm}}{6^{1/2}} = \boxed{217\,\text{pm}},$$

$$d_{100} = a = \boxed{532\text{ pm}}.$$

15.21 We use

$$\frac{1}{d^2} = \frac{h^2}{a^2} + \frac{k^2}{b^2} + \frac{l^2}{c^2}.$$

For (123),

$$\frac{1}{d^2} = \frac{1^2}{(0.754\,\text{nm})^2} + \frac{2^2}{(0.623\,\text{nm})^2} + \frac{3^2}{(0.433\,\text{nm})^2}$$

$$= 60.07\,\text{nm}^{-2}$$

$$d = \boxed{0.129\,\text{nm}}.$$

For (236),

$$\frac{1}{d^2} = \frac{2^2}{(0.754\,\text{nm})^2} + \frac{3^2}{(0.623\,\text{nm})^2} + \frac{6^2}{(0.433\,\text{nm})^2}$$

$$= 222.2\,\text{nm}^{-2}$$

$$d = \boxed{0.0671\,\text{nm}}.$$

15.22 $\lambda = 2d \sin\theta \quad [15.8]$

$$= 2 \times (97.3\,\text{pm}) \times (\sin 19.85°) = \boxed{66.1\,\text{pm}}.$$

15.23 $d_{100} = a = 350$ pm.

$$\rho = \frac{NM}{VN_A}, \text{implying that}$$

$$N = \frac{\rho VN_A}{M} = \frac{(0.53 \times 10^6\,\text{g m}^{-3}) \times (350 \times 10^{-12}\text{m})^3 \times (6.022 \times 10^{23}\text{mol}^{-1})}{6.94\,\text{g mol}^{-1}}$$

$$= 1.97.$$

An fcc cubic cell has $N = 4$ and a bcc unit cell has $N = 2$. Therefore, lithium has a $\boxed{\text{bcc unit cell}}$.

15.24 (a) $\theta_{khl} = \arcsin\left\{\frac{\lambda}{2a}(h^2 + k^2 + l^2)^{1/2}\right\}.$

The systematic absences in an fcc structure are that (hkl) all over odd are the only permitted lines. Because $\frac{\lambda}{2a} = 0.213$, we expect the following lines.

(hkl)	111	200	220	311	...
θ	22°	25°	37°	45°	...

(b) The density is calculated from

$$\rho = \frac{NM}{VN_A} = \frac{4 \times 63.55\,\text{g mol}^{-1}}{(3.61 \times 10^{-8}\,\text{cm})^3 \times 6.022 \times 10^{23}\,\text{mol}^{-1}} = \boxed{8.97\,\text{g cm}^{-3}}.$$

15.25 We use

$$\rho(x) = \frac{1}{V}\left\{F_0 + 2\sum_{h=1}^{\infty} F_h \cos(2h\pi x)\right\}.$$

Because V is unknown, we work with

$$V\rho(x) = 30 + 16.4\cos(2\pi x) + 13.0\cos(4\pi x) + 8.2\cos(6\pi x)$$
$$+ 11\cos(8\pi x) - 4.8\cos(10\pi x) + 10.8\cos(12\pi x)$$
$$+ 6.4\cos(14\pi x) + 2.0\cos(16\pi x) + 2.2\cos(18\pi x)$$

$$+ 13.0\cos(20\pi x) + 10.4\cos(22\pi x) - 8.6\cos(24\pi x)$$
$$- 2.4\cos(26\pi x) + 0.2\cos(28\pi x) + 4.2\cos(30\pi x).$$

A plot of $V\rho(x)$ is shown in Figure 15.6.

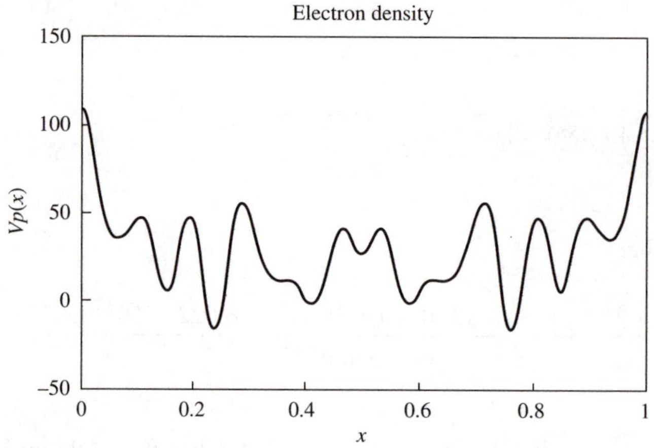

Fig 15.6

15.26 The hatched area shown in Figure 15.7 is $h \times 2R = 3^{1/2}R \times 2R = 2\sqrt{3}R^2$ where $h = 2R\cos 30°$. The net number of cylinders in a hatched area is 1, and the area of the cylinder's base is πR^2. The volume of the prism (of which the hatched area is the base) is $2\sqrt{3}R^2L$, and the volume occupied by the cylinders is $\pi R^2 L$. Hence, the packing fraction is

$$f = \frac{\pi R^2 L}{2\sqrt{3}R^2 L} = \frac{\pi}{2\sqrt{3}} = \boxed{0.9069}.$$

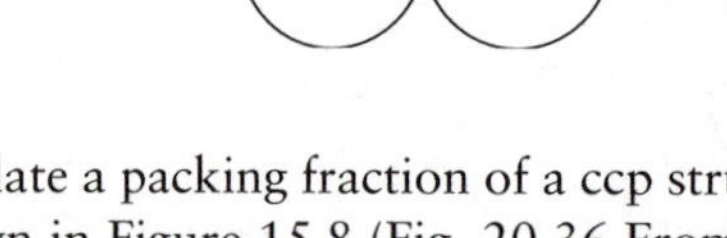

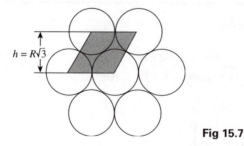

Fig 15.7

15.27 To calculate a packing fraction of a ccp structure, we first calculate the volume of a unit cell shown in Figure 15.8 (Fig. 20.36 From p.716 of P. Atkins and J.de Paula, *Physical Chemistry*, 8th edn, W.H. Freeman & Co., New York (2006)), and then calculate the total volume of the spheres that fully or partially occupy it. The first part of the calculation is a straightforward exercise in geometry. The second part involves counting the fraction of spheres that occupy the cell. Refer to the figure. Because a diagonal of any face passes completely through sphere and halfway through two other spheres, its length is $4R$.

The length of a side is therefore $8^{1/2}R$ and the volume of the unit cell is $8^{3/2}R^3$. Because each cell contains the equivalent of $6 \times \frac{1}{2} + 8 \times \frac{1}{8} = 4$ spheres, and the volume of each sphere is $\frac{16}{3}\pi R^3$ the total occupied volume is $8^{3/2}R^3$. The fraction of space occupied is therefore $\left(\frac{16}{3}\pi R^3\right)/(8^{3/2}R^3)$, or $\boxed{0.740}$. Because an hcp structure has the same coordination number, its packing fraction is the same.

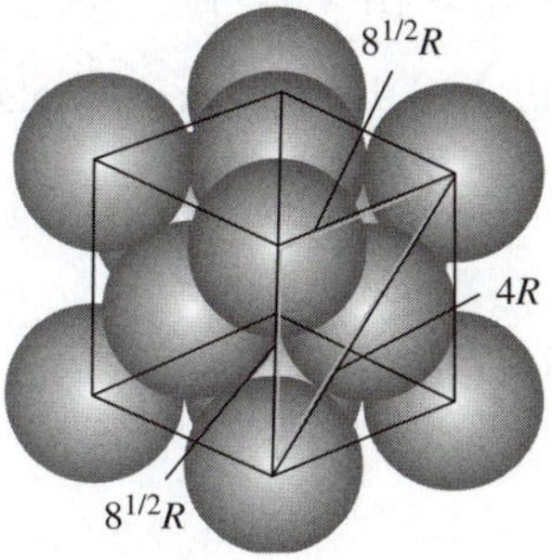

Fig 15.8

15.28 We need the packing fraction for hexagonal close-packing, which is 0.740. Therefore the density of the solid virus sample is

$$0.740 \times 1.00\,\text{g cm}^{-3} = \boxed{0.740\,\text{g cm}^{-3}}.$$

15.29 Refer to Figure 15.36 of the text.

(a) $\boxed{\text{Eight nearest neighbors}}$.

(b) $\boxed{\text{Six next-nearest neighbors}}$.

Nearest neighbors touch each along the body diagonal of the cube. If the side of the cube is a, then $\sqrt{3}\,a$ is the length of the body diagonal. As there are 2 atoms along the body diagonal, the distance between nearest neighbors is given by

$$2d = \sqrt{3}\,a \quad \text{or} \quad d = \frac{\sqrt{3}}{2}a = \frac{\sqrt{3}}{2} \times 500\text{nm}$$
$$= \boxed{433\,\text{nm}}.$$

The next nearest neighbors are the length of a side away, that is, $\boxed{500\,\text{nm}}$.

15.30 Refer to Figures 15.33 and 15.35 of the text.

(a) $\boxed{12}$ nearest neighbors.

(b) $\boxed{6}$ next nearest neighbors.

Nearest neighbors touch each other along the diagonal of a face; therefore with a = length of side of unit cell

$$(2d)^2 = 2a^2, \quad d = \text{distance between neighbors.}$$

$$2d = \sqrt{2}\,a.$$

$$d = a/\sqrt{2} = 500\,\text{nm}/\sqrt{2} = \boxed{354\,\text{nm}}.$$

For next nearest neighbors, $d = a = \boxed{500\,\text{nm}}$.

15.31 The packing fraction for cubic close-packing is 0.74; that for body-centered structures is 0.68. Therefore, (a) $\boxed{\text{less dense}}$, and (b) its density would decrease to 0.68/0.74 = 0.92, or $\boxed{92\%}$ of its former value.

15.32 $V = 651\,\text{pm} \times 651\,\text{pm} \times 934\,\text{pm} = \boxed{3.96 \times 10^{-28}\text{m}^3}$.

If we assume a simple tetragonal unit cell, then there is one formula unit per unit cell.

$$571.81\,\text{g}\,\text{mol}^{-1} \times (1\,\text{mol}/6.022 \times 10^{23}\ \text{atoms}) = 94.3 \times 10^{-23}\ \text{g atom}^{-1}.$$

$$d = m/V = 94.93 \times 10^{-23}\,\text{g}/3.96 \times 10^{-28}\ \text{m}^3$$

$$= \boxed{2.40 \times 10^6\,\text{g}\,\text{m}^{-3}}.$$

15.33 $\rho = \dfrac{\text{mass of unit cell}}{\text{volume of unit cell}} = \dfrac{m}{V}.$

$m = nM = \dfrac{N}{N_A}M$ [N is the number of formula units per unit cell].

Then, $\rho = \dfrac{NM}{VN_A}$

and $N = \dfrac{\rho V N_A}{M}$

$$= \frac{(3.9\times10^6\,\text{g}\,\text{m}^{-3}) \times (634) \times (784) \times (516\times10^{-36}\,\text{m}^3) \times (6.022\times10^{23}\,\text{mol}^{-1})}{154.77\,\text{g}\,\text{mol}^{-1}}$$

$$= 3.9.$$

Therefore, $\boxed{N = 4}$ and the true calculated density (in the absence of defects) is

$$\rho = \frac{(4) \times (154.77\,\text{g}\,\text{mol}^{-1})}{(634) \times (784) \times (516 \times 10^{-30}\,\text{cm}^3) \times (6.022 \times 10^{23})\,\text{mol}^{-1}} = \boxed{4.01\,\text{g}\,\text{cm}^{-3}}.$$

15.34 (a) $d_{hkl} = \left[\left(\dfrac{h}{a}\right)^2 + \left(\dfrac{k}{b}\right)^2 + \left(\dfrac{l}{c}\right)^2\right]^{-1/2}.$

$$d_{321} = \left\{\left(\frac{3}{812}\right)^2 + \left(\frac{2}{947}\right)^2 + \left(\frac{1}{637}\right)^2\right\}^{-1/2}\ \text{pm}$$

$$= \boxed{220\,\text{pm}}.$$

(b) $d_{642} = \frac{1}{2} d_{321} = \boxed{110\,\text{pm}}$.

15.35 Radius ratio $= \dfrac{72\,\text{pm}}{140\,\text{pm}} = 0.51.$

$$0.41 < 0.51 < 0.73.$$

Therefore, the $\boxed{\text{rock salt structure}}$ is predicted.

Chapter 16

Solid surfaces

Answers to discussion questions

16.1 *Langmuir isotherm.* This isotherm applies under the following conditions:

(1) Adsorption cannot proceed beyond monolayer coverage.
(2) All sites are equivalent and the surface is uniform.
(3) The ability of a molecule to adsorb at a given site is independent of the occupation of neighboring sites.

BET isotherm. Condition number 1 above is removed. This isotherm applies to multilayer coverage.

16.2 In the Langmuir–Hinshelwood mechanism of surface-catalyzed reactions, the reaction takes place by encounters between molecular fragments and atoms already adsorbed on the surface. We therefore expect the rate law to be second-order in the extent of surface coverage:

$$\mathrm{A} + \mathrm{B} \rightarrow \mathrm{P}, \quad \text{rate} = k\theta_{\mathrm{A}}\theta_{\mathrm{B}}.$$

Insertion of the appropriate isotherms for A and B then gives the reaction rate in terms of the partial pressures of the reactants. For example, if A and B follow Langmuir isotherms (eqn 16.3) and adsorb without dissociation, then it follows that the rate law is

$$\text{rate} = \frac{kK_{\mathrm{A}}K_{\mathrm{B}}p_{\mathrm{A}}p_{\mathrm{B}}}{(1 + K_{\mathrm{A}}p_{\mathrm{A}} + K_{\mathrm{B}}p_{\mathrm{B}})^2}.$$

The parameters K in the isotherms and the rate constant k are all temperature-dependent, so the overall temperature dependence of the rate may be strongly non-Arrhenius (in the sense that the reaction rate is unlikely to be proportional to $\exp(-E_{\mathrm{a}}/RT)$.

In the Eley–Rideal mechanism (ER mechanism) of a surface-catalyzed reaction, a gas-phase molecule collides with another molecule already adsorbed on the surface. The rate of formation of product is expected to be proportional to the partial pressure, p_{B} of the non-adsorbed gas B and the extent of surface coverage, θ_A, of the adsorbed gas A. It follows that the rate law should be

$$A + B \rightarrow P, \quad \text{rate} = kp_B\theta_A.$$

The rate constant, k, might be much larger than for the uncatalyzed gas-phase reaction because the reaction on the surface has a low activation energy and the adsorption itself is often not activated.

If we know the adsorption isotherm for A, we can express the rate law in terms of its partial pressure, p_A. For example, if the adsorption of A follows a Langmuir isotherm in the pressure range of interest, then the rate law would be

$$\text{rate} = \frac{kKp_Ap_B}{1 + Kp_A}.$$

If A were a diatomic molecule that adsorbed as atoms, we would substitute the isotherm given in eqn 16.5 instead.

According to eqn 16.14, when the partial pressure of A is high (in the sense $Kp_A \gg l$, there is almost complete surface coverage, and the rate is equal to kp_B. Now the rate-determining step is the collision of B with the adsorbed fragments. When the pressure of A is low ($Kp_A \ll 1$), perhaps because of its reaction, the rate is equal to kKp_Ap_B; and now the extent of surface coverage is important in the determination of the rate.

In the Mars van Krevelen mechanism of catalytic oxidation, for example, in the partial oxidation of propene to propenal, the first stage is the adsorption of the propene molecule with loss of a hydrogen to form the allyl radical, $CH_2{=}CHCH_2$. An O atom in the surface can now transfer to this radical, leading to the formation of acrolein (propenal, $CH_2{=}CHCHO$) and its desorption from the surface. The H atom also escapes with a surface O atom, and goes on to form H_2O, which leaves the surface. The surface is left with vacancies and metal ions in lower oxidation states. These vacancies are attacked by O_2 molecules in the overlying gas, which then chemisorb as O_2^- ions, so reforming the catalyst. This sequence of events involves great upheavals of the surface, and some materials break up under the stress.

16.3 The net current density at an electrode is j; j_0 is the exchange current density; α is the transfer coefficient; f is the ratio F/RT; and η is the overpotential.

(a) $j = j_0 f\eta$ is the current density in the low overpotential limit.

(b) $j = j_0 e^{(1-\alpha)f\eta}$ applies when the overpotential is large and positive.

(c) $j = -j_0 e^{-\alpha f\eta}$ applies when the overpotential is large and negative.

16.4 In cyclic voltammetry, the current at a working electrode is monitored as the applied potential difference is changed back and forth at a constant rate between pre-set limits (Figures 16.31 and 16.32). As the potential difference approaches $E^\ominus(\text{Ox, Red})$ for a solution that contains the reduced component (Red), current begins to flow as Red is oxidized. When the potential difference is swept beyond $E^\ominus(\text{Ox, Red})$, the current passes through a maximum and then falls as all the Red near the electrode is consumed and converted to Ox, the oxidized form. When the direction of the sweep is reversed and the potential difference passes through $E^\ominus(\text{Ox, Red})$, current flows in the reverse direction.

This current is caused by the reduction of the Ox formed near the electrode on the forward sweep. It passes through the maximum as Ox near the electrode is consumed. The forward and reverse current maxima bracket $E^\ominus$(Ox, Red), so the species present can be identified. Furthermore, the forward and reverse peak currents are proportional to the concentration of the couple in the solution, and vary with the sweep rate. If the electron transfer at the electrode is rapid, so that the ratio of the concentrations of Ox and Red at the electrode surface have their equilibrium values for the applied potential (that is, their relative concentrations are given by the Nernst equation), the voltammetry is said to be *reversible*. In this case, the peak separation is independent of the sweep rate and equal to $(59\,\text{mV})/n$ at room temperature, where n is the number of electrons transferred. If the rate of electron transfer is low, the voltammetry is said to be *irreversible*. Now, the peak separation is greater than $(59\,\text{mV})/n$ and increases with increasing sweep rate. If homogeneous chemical reactions accompany the oxidation or reduction of the couple at the electrode, the shape of the voltammogram changes, and the observed changes give valuable information about the kinetics of the reactions as well as the identities of the species present.

Solutions to exercises

16.5 $$Z_W = \frac{p}{(2\pi mkT)^{1/2}} \quad [16.1] = \frac{Z_0(p/\text{Pa})}{\{(T/\text{K})(M/(\text{g mol}^{-1}))\}^{1/2}}$$

where $Z_0 = 2.63 \times 10^{24}\,\text{m}^{-2}\,\text{s}^{-1}$.

Two other practical forms of this equation at 298 K are

$$Z_W = \frac{(2.03 \times 10^{21}\,\text{cm}^{-2}\text{s}^{-1})}{(M/\text{g mol}^{-1})^{1/2}} \qquad [100\text{ Pa} = 0.750\text{ Torr}]$$

and

$$Z_W = \frac{(2.03 \times 10^{25}\text{m}^{-2}\text{s}^{-1}) \times (p/\text{Torr})}{(M/\text{g mol}^{-1})^{1/2}}.$$

Hence, we can draw up the following table.

	H_2	C_3H_8
M/(g mol^{-1})	2.02	44.09
$Z_W/(\text{m}^{-2}\text{s}^{-1})$		
(i) 100 Pa	1.07×10^{25}	2.35×10^{24}
(ii) 10^{-7} Torr	1.4×10^{18}	3.1×10^{17}

16.6 $$p/\text{Pa} = \frac{\{Z_W/(\text{m}^{-2}\text{s}^{-1})\} \times \{(T/\text{K}) \times (M/\text{g mol}^{-1})\}^{1/2}}{2.63 \times 10^{24}} \quad [16.1]$$

$$= \frac{\{Z_W/\text{m}^{-2}\text{s}^{-1}\} \times (425 \times 39.95)^{1/2}}{2.63 \times 10^{24}}$$

$$= 4.95 \times 10^{-24} \times Z_W/(\text{m}^{-2}\text{s}^{-1}).$$

The collision rate required is

$$Z_W = \frac{4.5 \times 10^{20}\,\mathrm{s^{-1}}}{\pi \times (0.075\,\mathrm{cm})^2} = 2.5\bar{5} \times 10^{22}\,\mathrm{cm^{-2}\,s^{-1}} = 2.5\bar{5} \times 10^{26}\,\mathrm{cm^{-2}\,s^{-1}}.$$

Hence $p = (4.95 \times 10^{-23}\,\mathrm{Pa}) \times (2.5\bar{5} \times 10^{26}) = \boxed{1.3 \times 10^4\,\mathrm{Pa}}$.

16.7 $$Z_W = (2.63 \times 10^{24}\,\mathrm{m^{-2}\,s^{-1}}) \times \left(\frac{p/\mathrm{Pa}}{\{(T/\mathrm{K}) \times (M/\mathrm{g\,mol^{-1}})\}^{1/2}}\right) \quad [16.1]$$

$$= (2.63 \times 10^{24}\,\mathrm{m^{-2}\,s^{-1}}) \times \left(\frac{35}{(80 \times 4.00)^{1/2}}\right) = 5.1 \times 10^{24}\,\mathrm{m^{-2}\,s^{-1}}.$$

The area occupied by a Cu atom is $\left(\frac{1}{2}\right) \times (3.61 \times 10^{-10}\,\mathrm{m})^2 = 6.52 \times 10^{-20}\,\mathrm{m}^2$ (in an fcc unit cell, there is the equivalent of two Cu atoms per face). Therefore,

$$\text{rate per Cu atom} = (5.2 \times 10^{24}\,\mathrm{m^{-2}\,s^{-1}}) \times (6.52 \times 10^{-2}\,\mathrm{m}^2) = \boxed{3.4 \times 10^5\,\mathrm{s^{-1}}}.$$

16.8 (a) $v = Ae^{-d/l} = (5 \times 10^{14}\,\mathrm{s^{-1}})e^{-(750\,\mathrm{pm})/(70\,\mathrm{pm})} = \boxed{1.\bar{1} \times 10^{10}\,\mathrm{s^{-1}}}$.

(b) $$\frac{v(d_2)}{v(d_1)} = \frac{Ae^{-d_2/l}}{Ae^{-d_2/l}} = e^{-(d_2-d_1)/l} = e^{-(850\,\mathrm{pm}-750\,\mathrm{pm})/(70\,\mathrm{pm})} = \boxed{0.240}.$$

16.9 $$F = -\frac{dV}{dr} = -\frac{d}{dr}\left(\frac{q_1q_2}{4\pi\varepsilon_0 r}\right) = \frac{q_1q_2}{4\pi\varepsilon_0 r^2} = \frac{e^2}{4\pi\varepsilon_0 r^2},$$ for the repulsion of two electrons

$$= \frac{(1.602 \times 10^{-19}\,\mathrm{C})^2}{4\pi(8.854 \times 10^{-12}\,\mathrm{J^{-1}\,C^2\,m^{-1}})(0.50 \times 10^{-9}\,\mathrm{m})^2} = \boxed{9.2 \times 10^{-10}\,\mathrm{N}}.$$

$$\Delta F = F_{r=0.60\,\mathrm{nm}} - F_{r=0.50\,\mathrm{nm}}$$

$$= \frac{(1.602 \times 10^{-19}\,\mathrm{C})^2}{4\pi(8.854 \times 10^{-12}\,\mathrm{J^{-1}\,C^2\,m^{-1}})(0.60 \times 10^{-9}\,\mathrm{m})^2}$$

$$- \frac{(1.602 \times 10^{-19}\,\mathrm{C})^2}{4\pi(8.854 \times 10^{-12}\,\mathrm{J^{-1}\,C^2\,m^{-1}})(0.50 \times 10^{-9}\,\mathrm{m})^2}$$

$$= \boxed{-2.8 \times 10^{-10}\,\mathrm{N}}.$$

16.10 The number of CO molecules adsorbed on the catalyst is

$$N = nN_A = \frac{pVN_A}{RT} = \frac{(1.00\,\mathrm{atm}) \times (4.25 \times 10^{-3}\,\mathrm{dm^3}) \times (6.022 \times 10^{23}\,\mathrm{mol^{-1}})}{(0.08206\,\mathrm{dm^3\,atm\,K^{-1}\,mol^{-1}}) \times (273\,\mathrm{K})}.$$

$$= 1.14 \times 10^{20}.$$

The area of the surface must be the same as that of the molecules spread into a monolayer, namely, the number of molecules times each one's effective area.

$$A = Na = (1.14 \times 10^{20}) \times (0.165 \times 10^{-18}\,\mathrm{m}^2) = \boxed{18.8\,\mathrm{m}^2}.$$

16.11 $\theta = \dfrac{Kp}{1+Kp}$ [16.3], which implies that $p = \left(\dfrac{\theta}{1-\theta}\right)\dfrac{1}{K}$.

(a) $p = \left(\dfrac{0.15}{0.85}\right) \times \left(\dfrac{1}{0.85\,\mathrm{kPa}^{-1}}\right) = \boxed{0.12\,\mathrm{kPa}}$.

(b) $p = \left(\dfrac{0.95}{0.05}\right) \times \left(\dfrac{1}{0.85\,\mathrm{kPa}^{-1}}\right) = \boxed{22\,\mathrm{kPa}}$.

16.12 $\theta = \dfrac{Kp}{1+Kp}$ [Langmuir isotherm [16.3]].

Taking the inverse of the expression gives

$$\frac{1}{\theta} = \frac{1+Kp}{Kp} = 1 + \left(\frac{1}{K}\right)\left(\frac{1}{p}\right).$$

Thus, a plot of $1/\theta$ against $1/p$ is expected to be linear with a slope equal to $1/K$ when the Langmuir isotherm conditions are appropriate. The equilibrium constant equals the inverse of the linear regression slope.

16.13 We follow Example 16.2 of the text and draw up the following table.

p/Pa	25	129	253	540	1000	1593
$V/10^{-2}\,\mathrm{cm}^3$	4.2	16.3	22.1	32.1	41.1	47.1
$pV^{-1}/10^2\,\mathrm{Pa\,cm}^{-3}$	5.95	7.91	11.4	16.8	24.3	33.8

pV^{-1} is plotted against p in Figure 16.1.

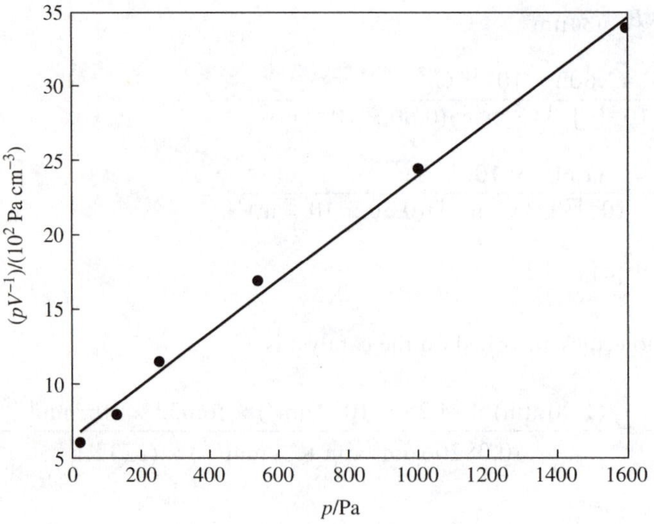

Fig 16.1

The low-pressure points fall on a straight line with intercept 629 Pa cm^{-3} and slope 1.77 cm^{-3}. It follows (Example 16.2) that

$$\frac{1}{V_\infty} = \text{slope} = 1.77\,\text{cm}^{-3} \quad \text{or} \quad \boxed{V_\infty = 0.56\,\text{cm}^3}.$$

$$\frac{1}{KV_\infty} = \text{intercept} \quad \text{or} \quad K = \frac{1}{V_\infty \times (\text{intercept})} = \frac{1}{(0.56\,\text{cm}^3) \times (629\,\text{Pa cm}^{-3})}$$

$$= \boxed{2.8 \times 10^{-3}\,\text{Pa}^{-1}}.$$

COMMENT It is unlikely that low-pressure data can be used to obtain an accurate value of the volume corresponding to complete coverage.

16.14 $\ln K' - \ln K = \dfrac{\Delta_{\text{ads}}H^\ominus}{R}\left(\dfrac{1}{T} - \dfrac{1}{T'}\right)$ [van't Hoff equation [7.15]].

$$\Delta_{\text{ads}}H^\ominus = \frac{R(\ln K' - \ln K)}{\left(\frac{1}{T} - \frac{1}{T'}\right)} = \frac{R\ln\left(\frac{K'}{K}\right)}{\left(\frac{1}{T} - \frac{1}{T'}\right)}$$

$$= \frac{(8.3145\,\text{J mol}^{-1}\,\text{K}^{-1})\ln\left(\dfrac{1.0 \times 10^{-3}\,\text{Torr}^{-1}}{2.7 \times 10^{-3}\,\text{Torr}^{-1}}\right)}{\left(\dfrac{1}{250\,\text{K}} - \dfrac{1}{273\,\text{K}}\right)} = \boxed{-2\bar{5}\,\text{kJ mol}^{-1}}.$$

16.15 Example 16.3 of the text demonstrates that a plot of $\ln p$ against $1/T$ should be a straight line of slope $\Delta_{\text{ads}}H^\ominus/R$ so we draw up the following table, prepare a plot to check linearity (Figure 16.2), and perform a linear least squares fit in order to acquire the slope.

T/K	200	210	220	230	240	250
p/kPa	4.32	5.59	7.07	8.80	10.67	12.80
$10^3/(T/\text{K})$	5.00	4.76	4.55	4.35	4.17	4.00
$\ln(p/\text{kPa})$	1.46	1.72	1.96	2.17	2.37	2.55

We find that

$$\Delta_{\text{ads}}H^\ominus/R = \text{slope} = -1.08\bar{7} \times 10^3\,\text{K}.$$

$$\Delta_{\text{ads}}H^\ominus = (\text{slope}) \times R = (-1.08\bar{7} \times 10^3\,\text{K}) \times (8.3145\,\text{J mol}^{-1}\,\text{K}^{-1})$$

$$= \boxed{-9.04\,\text{kJ mol}^{-1}}.$$

16.16 The van't Hoff equation for adsorption equilibrium is

$\dfrac{d\ln K}{dT} = \dfrac{\Delta_{\text{ads}}H^\ominus}{RT^2}$, at fixed surface coverage, θ. $\Delta_{\text{ads}}H^\ominus$ is called the isosteric enthalpy of adsorption.

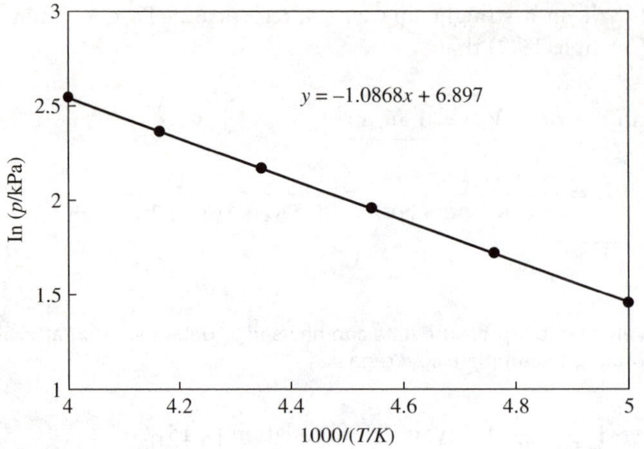

Fig 16.2

Solving the Langmuir isotherm for $\ln K$ and taking the derivative with respect to T at fixed surface coverage, we find

$$\theta = \frac{Kp}{1+Kp} \quad \text{or} \quad \theta(1+Kp) = Kp,$$

$$\theta = (1-\theta)Kp.$$

$$Kp = \frac{\theta}{1-\theta}.$$

$$\ln(Kp) = \ln\left(\frac{\theta}{1-\theta}\right).$$

$$\ln K + \ln p = \ln\left(\frac{\theta}{1-\theta}\right).$$

$$\frac{\mathrm{d}\ln K}{\mathrm{d}T} + \frac{\mathrm{d}\ln p}{\mathrm{d}T} = \frac{\mathrm{d}\ln\left(\frac{\theta}{1-\theta}\right)}{\mathrm{d}T} = 0 \quad \text{because } \theta \text{ is constant.}$$

$$\frac{\mathrm{d}\ln K}{\mathrm{d}T} = -\frac{\mathrm{d}\ln p}{\mathrm{d}T}.$$

Substitution into the van't Hoff equation yields

$$\boxed{\frac{\mathrm{d}\ln p}{\mathrm{d}T} = -\frac{\Delta_{\mathrm{ads}}H^{\ominus}}{RT^2}} \text{ at fixed surface coverage, } \theta.$$

With $\mathrm{d}(1/T)/\mathrm{d}T = -1/T^2$, this expression rearranges to $\boxed{\frac{\mathrm{d}\ln p}{\mathrm{d}T} = \frac{\Delta_{\mathrm{ads}}H^{\ominus}}{R}}$.

Therefore, a plot of $\ln p$ against $1/T$ should be a straight line of slope $\frac{\Delta_{\mathrm{ads}}H^{\ominus}}{R}$.

16.17 $\theta = \dfrac{(Kp)^{1/2}}{1+(Kp)^{1/2}}$ [16.5].

$p(\theta)$ is found by first solving the above expression for $(Kp)^{1/2}$.

$\theta\{1+(Kp)^{1/2}\} = (Kp)^{1/2}$ or $(Kp)^{1/2} = \dfrac{\theta}{1-\theta}$. Therefore, $\boxed{p = \dfrac{1}{K}\left(\dfrac{\theta}{1-\theta}\right)^2}$.

16.18 Derivation 16.1 provides a model for deriving equations [16.6]. However, with A and B competing for N sites, the number of sites not occupied equals $(1-\theta_A-\theta_B)N$ where θ_A and θ_B are the fraction of sites occupied by A and B, respectively. The rate at which A adsorbs to the surface is proportional to the partial pressure of A, p_A, and to $(1-\theta_A-\theta_B)N$.

Rate of adsorption of A $= k_{a,A}N(1-\theta_A-\theta_B)p_A$.

The rate at which adsorbed A molecules leave the surface is proportional to the number currently on the surface, $N\theta_A$.

Rate of desorption of A $= k_{d,A}\,N\theta_A$.

At equilibrium the two rates are equal:

$$k_{a,A}N(1-\theta_A-\theta_B)p_A = k_{d,A}N\theta_A.$$

Using $K_A = k_{a,A}/k_{d,A}$, this rearranges to

$$K_A(1-\theta_A-\theta_B)p_A = \theta_A.$$

Similarly,

$$K_B(1-\theta_A-\theta_B)p_B = \theta_B.$$

Division of these two equations yields

$$\theta_B = (K_Bp_B/K_Ap_A)\theta_A$$

and, upon substitution into the first equation and solving for θ_A, it is found that

$$\boxed{\theta_A = \frac{K_Ap_A}{1+K_Ap_A+K_Bp_B}}.$$

Substitution of $\theta_A = (K_Ap_A/K_Bp_B)\theta_B$ into the second equation and solving for θ_B, yields

$$\boxed{\theta_B = \frac{K_Bp_B}{1+K_Ap_A+K_Bp_B}}.$$

16.19 $\dfrac{V}{V_{mon}} = \dfrac{cz}{(1-z)\{1-(1-c)z\}}$ $\left[\text{[16.7], BET isotherm, } z = \dfrac{p}{p^*}\right]$

This rearranges to

$$\frac{z}{(1-z)V} = \frac{1}{cV_{mon}} + \frac{(c-1)z}{cV_{mon}}.$$

Therefore, a plot of the left-hand side against z should result in a straight line if the data obeys the BET isotherm. We draw up the following tables where, at 0°C, $p^* = 429.6\,\text{kPa}$.

p/kPa	14.0	37.6	65.6	79.2	82.7	100.7	106.4
V/cm^3	11.1	13.5	14.9	16.0	15.5	17.3	16.5
$10^3 z$	32.6	87.5	152.7	184.4	192.4	234.3	247.7
$\dfrac{10^3 z}{(1-z)(V/\text{cm}^3)}$	3.04	7.10	12.1	14.1	15.4	17.7	20.0

The points are plotted in Figure 16.3, but we analyze the data by a least-squares procedure.

The intercept is 4.66×10^{-4}. Hence,

$$\frac{1}{cV_{\text{mon}}} = 4.66 \times 10^{-4}\,\text{cm}^{-3}.$$

The slope is 0.07610. Hence

$$\frac{c-1}{cV_{\text{mon}}} = .07610\,\text{cm}^{-3}.$$

Solving the equations gives

$$c - 1 = 163.\bar{3}$$

and hence

$$c = \boxed{164.\bar{3}} \quad \text{and} \quad V_{\text{mon}} = \boxed{13.1\text{cm}^3}.$$

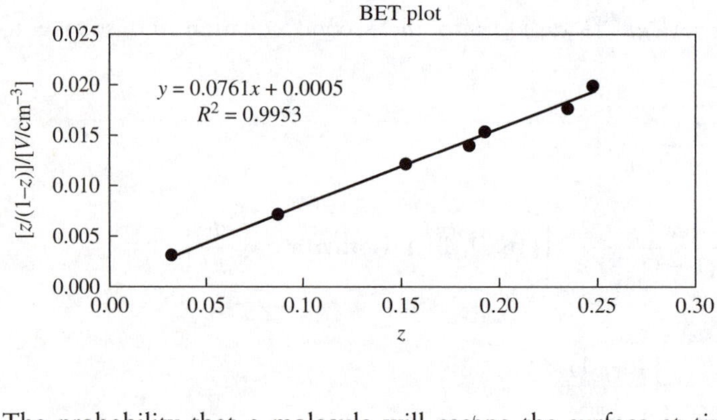

Fig 16.3

16.20 The probability that a molecule will escape the surface at time t is proportional to $\theta(t)/\theta(0)$, which equals $e^{-k_d t}$ for first-order rate of desorption. Consequently, the mean lifetime for the presence of a molecule on a surface is given by

$$\langle t\rangle = (\text{normalization constant}) \times \int_0^\infty t \times \left(\frac{\theta(t)}{\theta(0)}\right) dt = \frac{\int_0^\infty t \times \left(\frac{\theta(t)}{\theta(0)}\right) dt}{\int_0^\infty \left(\frac{\theta(t)}{\theta(0)}\right) dt} = \frac{\int_0^\infty t e^{-k_d t}\, dt}{\int_0^\infty e^{-k_d t} dt}.$$

Standard methods of integration yield the relationships

$$\int_0^\infty e^{-k_d t} dt = \frac{1}{k_d} \quad \text{and} \quad \int_0^\infty t e^{-k_d t} dt = \frac{1}{k_d^2}.$$

Thus, $\langle t\rangle = \dfrac{1}{k_d} = \dfrac{t_{1/2}}{\ln 2}\,[16.10] = \dfrac{\tau_0\, e^{E_d/RT}}{\ln 2}\,[16.10] \approx \dfrac{\tau_0\, e^{\Delta_{ads}H^\ominus/RT}}{\ln 2}$ (see Section 16.5)

With the guess $\tau_0 \approx 10^{-14}$s, the estimated mean lifetime is

$$\langle t\rangle \approx \frac{(10^{-14}\,\text{s}) \times \exp\left(\dfrac{155 \times 10^3\,\text{J mol}^{-1}}{(8.3145\,\text{J K}^{-1}\,\text{mol}^{-1}) \times (500\,\text{K})}\right)}{\ln 2} = \boxed{22\bar{5}\,\text{s}}$$

16.21 The average residence time, $\langle t\rangle$, for particle adsorbed on the surface is

$$\langle t\rangle = \frac{1}{k_d} = \frac{t_{1/2}}{\ln 2}[16.10] = \frac{\tau_0 e^{E_d/RT}}{\ln 2}[16.10] \text{(see exercise 16.20)}$$

$$\ln\langle t\rangle - \ln\langle t\rangle' = \frac{E_d}{R} \times \left(\frac{1}{T} - \frac{1}{T'}\right)$$

$$E_d = \frac{R \times (\ln\langle t\rangle - \ln\langle t\rangle')}{\left(\dfrac{1}{T} - \dfrac{1}{T'}\right)} = \frac{R \times \ln\left(\dfrac{\langle t\rangle}{\langle t\rangle'}\right)}{\left(\dfrac{1}{T} - \dfrac{1}{T'}\right)}$$

$$\text{(a) } E_d = \frac{(8.3145\,\text{J K}^{-1}\,\text{mol}^{-1}) \times \ln\left(\dfrac{0.36}{3.49}\right)}{\left(\dfrac{1}{2548\,\text{K}} - \dfrac{1}{2548\,\text{K}}\right)} = \boxed{61\bar{1}\,\text{kJ mol}^{-1}}$$

$$\text{(b) } \langle t\rangle = \frac{\tau_0 e^{E_d/RT}}{\ln 2} \text{ (see exercise 16.20)} = A^{-1} e^{E_d/RT}\ [16.10]$$

$$A = \frac{e^{E_d/RT}}{\langle t\rangle} = \frac{e^{61\bar{1} \times 10^3\,\text{J mol}^{-1}/(8.3145\,\text{J K}^{-1}\,\text{mol}^{-1} \times 2362\,\text{K})}}{3.49\,\text{s}} = \boxed{9.3 \times 10^{12}\,\text{s}^{-1}}$$

16.22 The desorption time for a given volume is proportional to the half-life of the absorbed species and, since

$$t_{1/2} = \tau_0 e^{E_d/RT} \quad [16.10],$$

we can write

$$E_d = \frac{R\ln\left(\frac{t_{1/2}}{t_{1/2}}\right)}{\left(\frac{1}{T}-\frac{1}{T'}\right)} = \frac{R\ln\left(\frac{t}{t'}\right)}{\frac{1}{T}-\frac{1}{T'}}$$

where t and t' are the two desorption times. We evaluate E_d from the data for the two temperatures.

$$E_d = \frac{8.3145\,\mathrm{J\,K^{-1}\,mol^{-1}}}{\left(\frac{1}{1856\,\mathrm{K}}-\frac{1}{1978\,\mathrm{K}}\right)} \times \ln\frac{27}{2.0} = \boxed{65\bar{0}\ \mathrm{kJ\,mol^{-1}}}.$$

We write

$$t = t_0\,e^{65\bar{0}\times 10^3/(8.314\times 1856)} = t_0 \times (1.9\bar{6} \times 10^{18}).$$

Therefore, since $t = 27$ min, $t_0 = 1.3\bar{8} \times 10^{-17}$ min. Consequently,

(a) At 298 K,

$$t = (1.3\bar{8} \times 10^{-17}\ \mathrm{min}) \times e^{65\bar{0}\times 10^3/(8.314\times 298)} = \boxed{1.1 \times 10^{97}\ \mathrm{min}},$$

which is just about forever.

(b) At 3000 K,

$$t = (1.3\bar{8} \times 10^{-17}\ \mathrm{min}) \times e^{65\bar{0}\times 10^3/(8.314\times 3000)} = \boxed{2.9 \times 10^{-6}\ \mathrm{min}}.$$

16.23 $\text{rate} = k\theta = \dfrac{kKp}{1+Kp}$ [16.11].

On gold, $\theta \approx 1$, and rate $= k\theta \approx$ constant, a $\boxed{\text{zero-order}}$ reaction.

On platinum, $\theta \approx Kp$(as $Kp \ll 1$), so rate $= kKp$ and the reaction is $\boxed{\text{first-order}}$.

16.24 (a) $\text{rate} = \dfrac{kK_AK_Bp_Ap_B}{(1+K_Ap_A+K_Bp_B)^2}$ [16.13].

(b) $$\lim_{p_A,p_B\to 0}\left(\frac{kK_AK_Bp_Ap_B}{(1+K_Ap_A+K_Bp_B)^2}\right) = \boxed{kK_AK_Bp_Ap_B}.$$

(c) $\boxed{\text{Yes}}$, this mechanism can exhibit zero-order kinetics should the factors containing partial pressures in the denominator of the rate law cancel with the partial pressure factor in the numerator. This requires that

$$p_Ap_B = \text{constant} \times (1+K_Ap_A+K_Bp_B)^2 \qquad \text{for a set of experiments.}$$

16.25 In the high overpotential limit,

$$j = j_0e^{(1-\alpha)f\eta} \quad \text{so} \quad \frac{j_1}{j_2} = e^{(1-\alpha)f(\eta_1-\eta_2)} \quad \text{where} \quad f = \frac{F}{RT} = \frac{1}{25.69\,\mathrm{mV}}.$$

The overpotential η_2 is

$$\eta_2 = \eta_1 + \frac{1}{f(1-\alpha)} \ln \frac{j_2}{j_1} = 105\,\text{mV} + \left(\frac{25.69\,\text{mV}}{1-0.42}\right) \times \ln\left(\frac{72.55\,\text{mA cm}^{-2}}{17.0\,\text{mA cm}^{-2}}\right)$$

$$= \boxed{17\bar{1}\,\text{mV}}.$$

16.26 In the high overpotential limit,

$$j = j_0\,\mathrm{e}^{(1-\alpha)f\eta} \quad \text{so} \quad j_0 = j\mathrm{e}^{(\alpha-1)f\eta}$$

$$j_0 = (17.0\,\text{mA cm}^{-2}) \times \mathrm{e}^{\{(0.42-1)\times(105\,\text{mV})/(25.69\,\text{mV})\}} = \boxed{1.6\text{ mA cm}^{-2}}.$$

16.27 $$\frac{j}{j_0} = \mathrm{e}^{(1-\alpha)f\eta} - \mathrm{e}^{-\alpha f\eta}\ [16.15] = \mathrm{e}^{(1/2)f\eta} - \mathrm{e}^{-(1/2)f\eta} \quad [\alpha = 0.5]$$

$$= 2\,\sinh\left(\frac{1}{2}f\eta\right) \quad \left[\sinh x = \frac{e^x - e^{-x}}{2}\right]$$

and we use $\frac{1}{2}f\eta = \frac{1}{2} \times \frac{\eta}{25.69\,\text{mV}} = 0.01946\,(\eta/\text{mV})$

or

$$j = 2j_0 \sinh\left(\frac{1}{2}f\eta\right) = (1.58\,\text{mA cm}^{-2}) \times \sinh\left(\frac{0.01946\,\eta}{\text{mV}}\right).$$

(a) $\eta = 10\,\text{mV}$,

$$j = (1.58\,\text{mA cm}^{-2}) \times (\sinh 0.1946) = \boxed{0.31\text{mA cm}^{-2}}.$$

(b) $\eta = 100\,\text{mV}$,

$$j = (1.58\,\text{mA cm}^{-2}) \times (\sinh 1.946) = \boxed{5.41\text{ mA cm}^{-2}}.$$

(c) $\eta = -0.5\,\text{V}$,

$$j = (1.58\,\text{mA cm}^{-2}) \times (\sinh -0.973) \approx \boxed{-2.19\text{ A cm}^{-2}}.$$

16.28 The current density of electrons is $\frac{j_0}{e}$ because each one carries a charge of magnitude e. Therefore,

(1) $Pt|H_2|H^+; j_0 = 0.79\,\text{mA cm}^{-2}$[Table 16.5].

$$\frac{j_0}{e} = \frac{0.79\text{ mA cm}^{-2}}{1.602 \times 10^{-19}\,\text{C}} = \boxed{4.9 \times 10^{15}\,\text{cm}^{-2}\text{s}^{-1}}.$$

(2) $Pt|Fe^{3+}, Fe^{2+}; j_0 = 2.5\,\text{mA cm}^{-2}$

$$\frac{j_0}{e} = \frac{2.5\text{ mA cm}^{-2}}{1.602 \times 10^{-19}\,\text{C}} = \boxed{1.6 \times 10^{16}\,\text{cm}^{-2}\text{s}^{-1}}.$$

(3) $Pb|H_2|H^+; j_0 = 5.0 \times 10^{-12}\,\text{A cm}^{-2}$

$$\frac{j_0}{e} = \frac{5.0 \times 10^{-12}\,\text{A cm}^{-2}}{1.602 \times 10^{-19}\,\text{C}} = \boxed{3.1 \times 10^{7}\,\text{cm}^{-2}\text{s}^{-1}}.$$

There are approximately

$$\frac{1.0\,\text{cm}^2}{(280\,\text{pm})^2} = 1.3 \times 10^{15} \text{ atoms in each square centimeter of surface.}$$

The numbers of electrons per atom are therefore $\boxed{3.8\,\text{s}^{-1}}$, $\boxed{12\,\text{s}^{-1}}$, and $\boxed{2.4\times10^{-8}\,\text{s}^{-1}}$, respectively.

The last corresponds to less than one event per year.

16.29 $\ln j = \ln j_0 + (1-\alpha)f\eta \quad [16.17a]$

Draw up the following table.

η/mV	50	100	150	200	250
$\ln(j/\text{mA cm}^{-2})$	0.98	2.19	3.40	4.61	5.81

The points are plotted in Figure 16.4.

The intercept is at -0.25, and so $j_0/(\text{mAcm}^{-2}) = \text{e}^{-0.25} = \boxed{0.78}$. The slope is 0.0243, and so $\dfrac{(1-\alpha)F}{RT} = 0.0243\ \text{mV}^{-1}$. It follows that $1-\alpha = 0.62$, and so $\alpha = \boxed{0.38}$. If η were large but negative,

$$|j| \approx j_0\text{e}^{-\alpha f\eta}\ [16.17b] = (0.78\ \text{mAcm}^{-2}) \times (\text{e}^{-0.38\eta/25.7\,\text{mV}})$$

$$= \left(0.78\,\text{mA cm}^{-2}\right) \times (\text{e}^{-0.015(\eta/\text{mV})})$$

and we draw up the following table.

η/mV	−50	−100	−150	−200	−250
$j/(\text{mA cm}^{-2})$	1.65	3.50	7.40	15.7	33.2

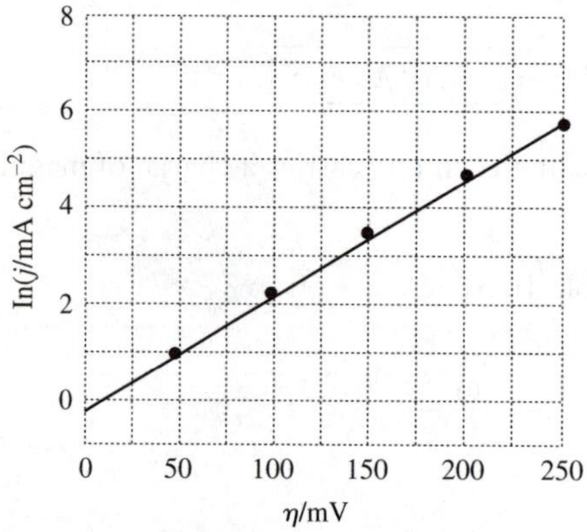

Fig 16.4

16.30 This problem differs somewhat from the simpler one-electron transfers considered in the text. In place of $\text{Ox} + \text{e}^- \rightarrow \text{Red}$ we have here

$$In^{3+} + 3e^- \rightarrow In,$$

namely, a three-electron transfer. Therefore, the Butler–Volmer equation [16.15] and the Tafel equations [16.16 and 16.17] need to be modified by including the factor z (in this case 3). Thus,

$$\ln j = \ln j_0 + z(1-\alpha)f\eta \quad \text{anode},$$
$$\ln(-j) = \ln j_0 - z\alpha f\eta \quad \text{cathode}.$$

We draw up the following table.

$j/(\mathrm{A\,m^{-2}})$	$-E/\mathrm{V}$	η/V	$\ln(j/(\mathrm{A\,m^{-2}}))$
0	0.388	0	
0.590	0.365	0.023	−0.5276
1.438	0.350	0.038	0.3633
3.507	0.335	0.053	1.255

We now do a linear regression of $\ln j$ against η with the following results (see Figure 16.5).

$$z(1-\alpha)f = 59.42\mathrm{V}^{-1}, \quad \text{standard deviation} = 0.0154,$$
$$\ln j_0 = -1.894, \quad \text{standard deviation} = 0.0006,$$
$$R = 1 \text{ (almost exact)}.$$

Thus, although there are only three data points, the fit to the Tafel equation is almost exact. Solving for α from $z(1-\alpha)f = 59.42\,\mathrm{V}^{-1}$, we obtain

$$\alpha = 1 - \frac{59.42\,\mathrm{V}^{-1}}{3f} = 1 - \left(\frac{59.42\,\mathrm{V}^{-1}}{3}\right) \times (0.025262\,\mathrm{V})$$
$$= 0.49\overline{96} = \boxed{0.50},$$

which matches the usual value of α exactly.

$$j_0 = \mathrm{e}^{-1.894} = \boxed{0.150\,\mathrm{A\,m^{-2}}}.$$

The cathodic current density is obtained from

$$\ln(-j_c) = \ln j_0 - z\alpha f\eta \quad \eta = 0.023\,\mathrm{V} \text{ at } -E/\mathrm{V} = 0.365$$
$$= -1.894 - (3 \times 0.49\overline{96} \times 0.023)/(0.025262)$$
$$= -3.2\overline{59},$$
$$-j_c = \mathrm{e}^{-3.2\overline{59}} = 0.038\overline{4}\,\mathrm{A\,m^{-2}},$$
$$-j_c = \boxed{0.038\overline{4}\,\mathrm{A\,m^{-2}}}.$$

16.31 At large positive values of the overpotential the current density is anodic.

$$j = j_0\left[\mathrm{e}^{(1-\alpha)f\eta} - \mathrm{e}^{-\alpha f\eta}\right] \quad [16.15]$$
$$\approx j_0\mathrm{e}^{(1-\alpha)f\eta} = j_a.$$
$$\ln j = \ln j_0 + (1-\alpha)f\eta \quad [16.17].$$

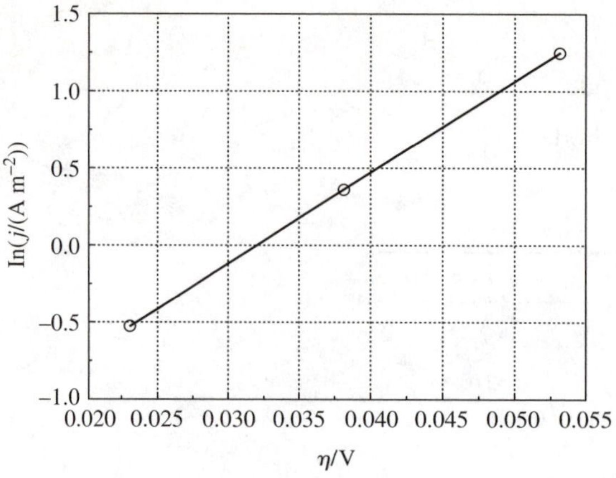

Fig 16.5

Performing a linear regression analysis of ln j against η, we find

$$\ln(j_0/(\text{mA m}^{-2})) = -10.826, \quad \text{standard deviation} = 0.287.$$

$$(1-\alpha)f = 19.550\,\text{V}^{-1}, \quad \text{standard deviation} = 0.355.$$

$$\boxed{R = 0.99901}.$$

$$j_0 = e^{-10.826}\text{mA m}^{-2} = \boxed{2.00 \times 10^{-5}\text{mA m}^{-2}}.$$

$$\alpha = 1 - \frac{19.550\,\text{V}^{-1}}{f} = 1 - \frac{19.550\,\text{V}^{-1}}{(0.025693\,\text{V})^{-1}}.$$

$$\boxed{\alpha = 0.498}.$$

The linear regression explains 99.90 per cent of the variation in a ln j against η plot and standard deviations are low. There are $\boxed{\text{no}}$ observable deviations from the Tafel equation/plot.

16.32 (a) The electroactive species is reduced during the forward sweep of potential. The process is reversible because the reduced species shows oxidization upon reversing the direction of the potential sweep. The peak reduction current and peak oxidation current lie symmetrically about the standard reduction potential.

(b) The electroactive species experiences a second reduction at high potential during the forward sweep of potential. Both reductions are reversible and, consequently, show oxidation currents upon reversing the potential scan toward low potentials.

(c) Reduction of the electroactive species is observed during the forward potential sweep. However, the process is not reversible. No oxidation current is observed upon reversing the sweep.

(d) The electroactive species experiences a second reduction at high potential during the forward sweep of potential. Upon reversing the potential sweep, the highly reduced species is reversibly oxidized with the loss of one electron but loss of a second electron is not observed at low potential. The complete reoxidation of the species is irreversible.

Chapter 17

Molecular interactions

Answers to discussion questions

17.1 Molecules with a permanent separation of electric charge have a permanent dipole moment. In molecules containing atoms of differing electronegativity, the bonding electrons may be displaced in such a way as to produce a net separation of charge in the molecule. Separation of charge may also arise from a difference in atomic radii of the bonded atoms. The separation of charges in the bonds is usually, though not always, in the direction of the more electronegative atom but depends on the precise bonding situation in the molecule as described in Section 17.2. A heteronuclear diatomic molecule necessarily has a dipole moment if there is a difference in electronegativity between the atoms, but the situation in polyatomic molecules is more complex. A polyatomic molecule has a permanent dipole moment only if at least one of its x, y, z components is non-zero (see eqn 17.4a).

An external electric field can distort the electron density in both polar and non-polar molecules and this results in an induced dipole moment that is proportional to the field. The constant of proportionality is called the polarizability.

17.2 (a) The electrostatic attraction arises when an end of the dipole moment of a water molecule is attracted to the opposite end of the dipole moment on a neighboring water molecule. Alternatively, think of the attraction between the partial positive charge of the hydrogen toward the partial negative charge of the oxygen in an adjacent water molecule.

(b) If the A—H bond in the A—H $\cdots$ B arrangement is regarded as formed from the overlap of an orbital on A, ψ_A, and a hydrogen 1s orbital ψ_H, and if the lone pair on B occupies an orbital on B, ψ_B, then, when the two molecules are close together, we can build three molecular orbitals from the three basis orbitals:

$$\psi = c_A\psi_A + c_H\psi_H + c_B\psi_B.$$

One of the molecular orbitals is bonding, one almost nonbonding, and the third antibonding. These three orbitals need to accommodate four electrons, two from the A—H bond and two from the lone pair on B. Two enter the bonding orbital and two the nonbonding orbital, so the net effect is a lowering of the energy, that is, a bond has formed.

17.3 The increase in entropy of a solution when hydrophobic molecules or groups in molecules cluster together and reduce their structural demands on the solvent (water) is the origin of the hydrophobic interaction that tends to stabilize clustering of hydrophobic groups in solution. A manifestation of the hydrophobic interaction is the clustering together of hydrophobic groups in biological macromolecules. For example, the side chains of amino acids that are used to form the polypeptide chains of proteins are hydrophobic, and the hydrophobic interaction is a major contributor to the tertiary structure of polypeptides. At first thought, this clustering would seem to be a nonspontaneous process as the clustering of the solute results in a decrease in entropy of the solute. However, the clustering of the solute results in greater freedom of movement of the solvent molecules and an accompanying increase in disorder and entropy of the solvent. The total entropy of the system has increased and the process is spontaneous.

17.4 In the Monte Carlo method, the particles in the box are moved through small but otherwise random distances, and the change in total potential energy of the N particles in the box, ΔV_N, is calculated using one of the intermolecular potentials discussed in this chapter. Whether this new configuration is accepted is then judged from the following rules.

1 If the potential energy is not greater than before the change, then the configuration is accepted.
2 If the potential energy is greater than before the change, the Boltzmann factor $e^{-\Delta V_N/kT}$ is compared with a random number between 0 and 1; if the factor is larger than the random number, the configuration is accepted; if the factor is not larger, the configuration is rejected. This procedure ensures that at equilibrium the probability of occurrence of any configuration is proportional to the Boltzmann factor.

In the molecular dynamics approach, the history of an initial arrangement is followed by calculating the trajectories of all the particles under the influence of the intermolecular potentials. Newton's laws are used to predict where each particle will be after a short time interval (about 1 fs which is shorter than the average time between collisions), and then the calculation is repeated for tens of thousands of such steps. The time-consuming part of the calculation is the evaluation of the net force on the molecule arising from all the other molecules present in the system. The calculation gives a series of snapshots of the liquid.

Solutions to exercises

17.5 $\chi(\mathrm{H}) = 2.1, \quad \chi(\mathrm{Cl}) = 3.0, \quad \Delta\chi = 9.0.$

We use

$$\mu = \Delta\chi\ \mathrm{D}[17.2] = \boxed{0.9\ \mathrm{D}}.$$

$$\mu = 0.9 \times 3.3 \times 10^{-30}\,\text{C m} = \boxed{3.0 \times 10^{-30}\ \text{C m}}.$$

The experimental value is 1.08 D.

17.6 The Lewis structure of sulfur tetrafluoride is shown in Figure 17.1.

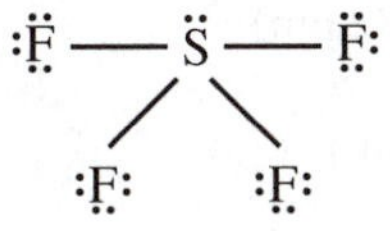

Fig 17.1

This leads to a distorted see-saw geometry using the VSEPR model. In this geometry, the dipole moments of the bonds do not cancel; hence the molecule is $\boxed{\text{polar}}$.

17.7 Refer to diagram **4** of the text and eqn 17.3. Here $\mu_1 = \mu_2$; therefore

$$\mu_{\text{res}} = (\mu_1^2 + \mu_2^2 + 2\mu_1\mu_2\cos\theta)^{1/2} = (2\mu_1^2 + 2\mu_1^2\cos\theta)^{1/2} = \sqrt{2}(1+\cos\theta)^{1/2}\mu_1.$$

1 Ortho-xylene, $\mu_{\text{res}} = \sqrt{2}(1+\cos 60°)^{1/2}\mu_1 = \sqrt{3}\,\mu_1 = \sqrt{3} \times 0.4\,\text{D} = \boxed{0.7\ \text{D}}$.

2 Meta-xylene, $\mu_{\text{res}} = \sqrt{2}\,(1+\cos 120°)^{1/2}\mu_1 = \mu_1 = \boxed{0.4\ \text{D}}$.

3 Para-xylene, $\mu_{\text{res}} = \boxed{0}$. The para-xylene value is exact by symmetry.

17.8 Add the dipole moments vectorially using the dipole moments calculated in Exercise 17.7.

(a) 1,2,3-trimethylbenzene, $\mu_{\text{res}} = m\text{-xylene} + \text{toluene} = 0.4\,\text{D} + 0.4\,\text{D} = \boxed{0.8\ \text{D}}$.

(b) 1,2,4-trimethylbenzene, $\mu_{\text{res}} = p\text{-xylene} + \text{toluene} = 0 + 0.4\,\text{D} = \boxed{0.4\ \text{D}}$.

(c) 1,3,5-trimethylbenzene, $\mu_{\text{res}} = m\text{-xylene} - \text{toluene} = 0.4\,\text{D} - 0.4\,\text{D} = \boxed{0}$. (These cancel exactly by symmetry.)

17.9 $\mu = (\mu_1^2 + \mu_2^2 + 2\mu_1\mu_2\cos\theta)^{1/2}$ [17.3]

$$= \left\{(1.5)^2 + (0.08)^2 + 2 \times 1.5 \times 0.08 \times (\cos 109.5°)\right\}^{1/2}\text{D} = \boxed{1.4\ \text{D}}.$$

17.10 The dipole moments vanish in the x- and z-directions, leaving only the y-component (vertical direction on the page). For the y-component, the dipole moments of the three structures are:

for Diagram **13**, $\mu = 0\,\text{D}$,

for Diagrams **14** and **15**, $\mu = (1.5^2 + 1.5^2 + 2 \times 1.5 \times 1.5 \times \cos 60°)\,\text{D} = 2.6\,\text{D}$.

(a) $\mu_{\text{res}} = \dfrac{1}{3}(0 + 2.6\,\text{D} + 2.6\,\text{D}) = \boxed{1.7\ \text{D}}$.

(b) $\mu_{\text{res}} = \boxed{2.6\ \text{D}}$.

(c) $\mu_{\text{res}} = \frac{1}{4}(2 \times 0 + 2.6\,\text{D} + 2.6\,\text{D}) = \boxed{1.3\,\text{D}}$.

(d) $\mu_{\text{res}} = \frac{1}{5}(0 + 2 \times 2.6\,\text{D} + 2 \times 2.6\,\text{D}) = \boxed{2.1\,\text{D}}$.

17.11

$$\begin{aligned}\mu_x = {} & 0.02e \times (-86\,\text{pm}) + 0.02e \times (34\,\text{pm}) + 0.06e \times (-195\,\text{pm}) \\ & + 0.18e \times (-199\,\text{pm}) - 0.36e \times (-101\,\text{pm}) + 0.45e \times (82\,\text{pm}) \\ & - 0.38e \times (199\,\text{pm}) + 0.18e \times (-80\,\text{pm}) - 0.38e \times (49\,\text{pm}) \\ & + 0.42e \times (129\,\text{pm}) \\ = {} & -29.76e\,\text{pm}.\end{aligned}$$

$$\begin{aligned}\mu_y = {} & 0.02e \times (118\,\text{pm}) + 0.02e \times (146\,\text{pm}) + 0.06e \times (70\,\text{pm}) \\ & + 0.18e \times (-1\,\text{pm}) - 0.36e \times (-11\,\text{pm}) + 0.45e \times (-15\,\text{pm}) \\ & - 0.38e \times (16\,\text{pm}) + 0.18e \times (-110\,\text{pm}) + 0.38e \times (-107\,\text{pm}) \\ & + 0.42e \times (-146\,\text{pm}) \\ = {} & +21.29\,e\,\text{pm}.\end{aligned}$$

$$\begin{aligned}\mu_z = {} & 0.02e \times (37\,\text{pm}) + 0.02e \times (-98\,\text{pm}) + 0.06e \times (-38\,\text{pm}) \\ & + 0.18e \times (-100\,\text{pm}) - 0.36e \times (-126\,\text{pm}) + 0.45e \times (34\,\text{pm}) \\ & - 0.38e \times (-38\,\text{pm}) + 0.18e \times (-111\,\text{pm}) - 0.38e \times (88\,\text{pm}) \\ & + 0.42e \times (126\,\text{pm}) \\ = {} & +53.1\,e\,\text{pm}.\end{aligned}$$

The magnitute of the dipole moment is given by

$$\begin{aligned}\mu &= (\mu_x^2 + \mu_y^2 + \mu_z^2)^{1/2} \qquad [17.4\text{a}] \\ &= [(-29.76)^2 + (21.29)^2 + (53.1)^2]^{1/2}\ e\,\text{pm} \\ &= 64.5e\,\text{pm} \\ &= 64.5 \times 1.602 \times 10^{-19}\,\text{C} \times 10^{-12}\,\text{m} \\ &= 1.03 \times 10^{-29}\,\text{C m} \\ &= \boxed{3.10\,\text{D}}.\end{aligned}$$

17.12 First, determine the dipole moment of the OH fragment, $\mu_{\text{O—H}}$, of the H_2O molecule by considering the total dipole moment of the molecule, $\mu_{\text{H—O—H}} = 1.85\,\text{D}$, to be the resultant of the dipoles of two identical OH fragments at an angle θ equal to 104.5° with respect to each other. This is the bond angle in water. Use eqn 17.3 with $\mu_1 = \mu_2 = \mu_{\text{O—H}}$. When this is the case, eqn 17.3 simplifies to

$$\mu_{\text{H—O—H}} = 2\mu_1 \cos\left(\frac{1}{2}\theta\right) = 2\mu_{\text{O—H}}\cos(52.25°),$$

$$\mu_{\text{O—H}} = \frac{\mu_{\text{H—O—H}}}{2\cos(52.25°)} = \frac{1.85\,\text{D}}{2\cos(52.25°)} = 1.51\,\text{D}.$$

Then, for H_2O_2

$$\mu_{\text{H—O—O—H}} = \boxed{2\mu_{\text{O—H}}\cos\left(\frac{1}{2}\theta\right)}.$$

(a) $\mu_{\text{H—O—O—H}}$ is plotted as function of ϕ in Figure 17.2. At 90°, $\mu_{\text{H—O—O—H}}$ is $\boxed{2.13\,\text{D}}$, which is the experimental value.

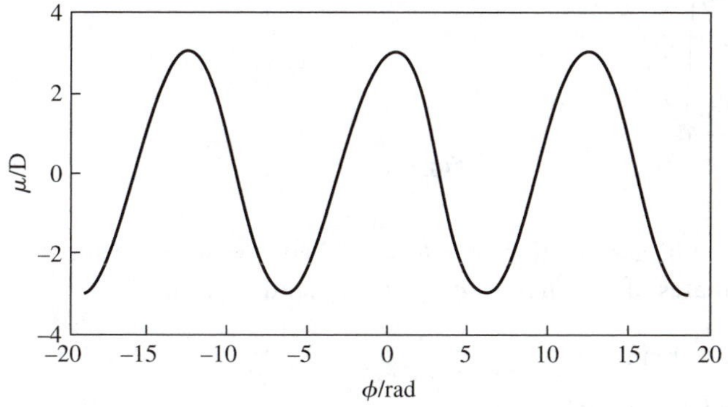

Fig 17.2

(b) The angle can be related to the dipole moment as follows:

$$\phi = 2\arccos(\mu_{\text{H—O—O—H}}/2\mu_{\text{O—H}}) = \boxed{2\arccos(\mu_{\text{H—O—O—H}}/3.02\,\text{D})}.$$

17.13 We assume that the dipole of the water molecule and the Li^+ ion are collinear and that the separation of charges in the dipole is smaller than the distance to the ion. This is certainly justified at 300 pm, but may not be at 100 pm. With these assumptions we can use eqn 17.5 (a) of the text. To flip the water molecule over requires twice the energy of interaction given by eqn 17.5(a):

$$E = 2V = \frac{2q_2\mu_1}{4\pi\varepsilon_0 r^2} \qquad |q_2| = e = 1.602\times10^{-19}\text{C}$$

$$= \frac{2\times1.602\times10^{-19}\,\text{C}\times1.85\times3.336\times10^{-30}\,\text{C m}}{4\pi\times8.854\times10^{-12}\,\text{J}^{-1}\,\text{C}^2\,\text{m}^{-1}\times r^2}$$

$$= \frac{1.777\times10^{-38}\text{J m}^2}{r^2}.$$

(a) $E = \dfrac{1.777 \times 10^{-38}\,\mathrm{J\,m^2}}{(1.00 \times 10^{-10}\,\mathrm{m})^2} = 1.777 \times 10^{-18}\,\mathrm{J}.$

$$\text{Molar energy} = N_A \times E = \boxed{1070\,\mathrm{kJ\,mol^{-1}}}.$$

(b) $E = \dfrac{1.777 \times 10^{-38}\,\mathrm{J\,m^2}}{(3.00 \times 10^{-10}\,\mathrm{m})^2} = 1.975 \times 10^{-19}\,\mathrm{J}.$

$$\text{Molar energy} = N_A \times E = \boxed{119\,\mathrm{kJ\,mol^{-1}}}.$$

17.14 Figure 17.3 defines symbols for distances.

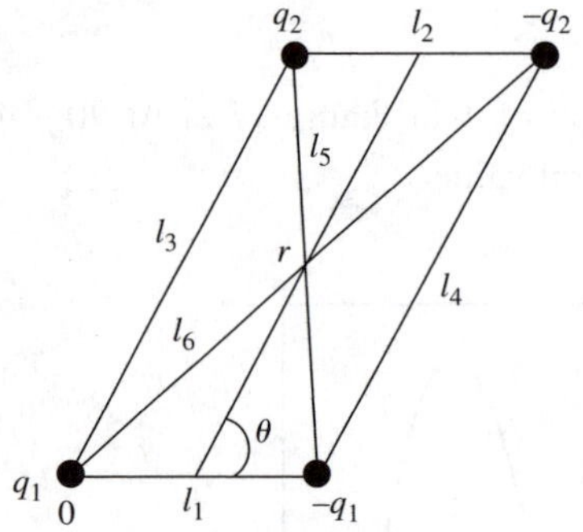

Fig 17.3

To simplify the analysis let us assume that $l_1 = l_2 = l$. Then $l_3 = l_4 = r$. Take the origin to be 0. Then the coordinates of the charges q_2, $-q_2$, q_1, and $-q_1$ are

q_2 : $x = r\cos\theta, y = r\sin\theta,$

$-q_2$: $x = r\cos\theta + l, y = r\sin\theta,$

q_1 : $x = 0,\ y = 0,$

$-q_1$: $x = l, y = 0.$

Then $l_6 = [(r\cos\theta + l)^2 + (r\sin\theta)^2]^{1/2},$

$l_5 = [(r\cos\theta - l)^2 + (r\sin\theta)^2]^{1/2}.$

Let $\lambda = \dfrac{l}{r}$, then $l = \lambda r$ and we can write

$$(r\cos\theta + l)^2 = (r\cos\theta + \lambda r)^2 = r^2(\cos\theta + \lambda)^2,$$

$$(r\cos\theta - l)^2 = (r\cos\theta - \lambda r)^2 = r^2(\cos\theta - \lambda)^2.$$

$$l_6 = [r^2(\cos\theta + \theta)^2 + r^2\sin^2\theta]^{1/2} = r[\cos\theta + \lambda)^2 + \sin^2\theta]^{1/2}$$
$$= r[1 + 2\lambda\cos\theta + \lambda^2]^{1/2}.$$

$$l_5 = r(1 - 2\lambda\cos\theta + \lambda^2)^{1/2}.$$

Let $4\pi\varepsilon_0 = K$; then the Coulomb interactions are

$$V = \frac{q_1q_2}{Kr} + \frac{q_1q_2}{Kr} - \frac{q_1q_2}{Kr(1-2\lambda\cos\theta+\lambda^2)^{1/2}} - \frac{q_1q_2}{Kr(1+2\lambda\cos\theta+\lambda^2)^{1/2}}$$

$$= -\frac{q_1q_2}{Kr}\left(-2+\frac{1}{(1-2\lambda\cos\theta+\lambda^2)^{1/2}}+\frac{1}{(1+2\lambda\cos\theta+\lambda^2)^{1/2}}\right).$$

We will assume that $2\lambda\ \cos\ \theta+\lambda^2 \ll 1$; then we can expand the denominator in a Taylor series expansion.

$$\frac{1}{(1+2(\cos\theta)\lambda+\lambda^2)^{1/2}} = 1-(\cos\theta)\lambda+\left(\frac{3\cos^2\theta}{2}-\frac{1}{2}\right)\lambda^2+\ldots,$$

$$\frac{1}{(1-2(\cos\theta)\lambda+\lambda^2)^{1/2}} = 1+(\cos\theta)\lambda+\left(\frac{3\cos^2\theta}{2}-\frac{1}{2}\right)\lambda^2+\cdots.$$

$$V = \frac{-q_1q_2}{Kr}\left[-2+1+(\cos\theta)\lambda+\left(\frac{3\cos^2\theta-1}{2}\right)\lambda^2+1-(\cos\theta)\lambda\right.$$

$$\left.+\left(\frac{3\cos^2\theta-1}{2}\right)\lambda^2\right].$$

$$V = \frac{-q_1q_2}{Kr}[(3\cos^2\theta-1)\lambda^2] = -\frac{q_1q_2}{Kr}\left[(3\cos^2\theta-1)\left(\frac{1}{r}\right)^2\right].$$

Now use $\mu_1 = q_1l, \mu_2 = q_2l, K = 4\pi\varepsilon_0$, and rearrange to get

$$V = \frac{\mu_1\mu_2(1-3\cos^2\theta)}{4\pi\varepsilon_0r^3}.\text{Q.E.D.}$$

17.15 (a) The kinetic energy per mole is given by

$$E = 3/2RT = \boxed{3.7\,\text{kJ mol}^{-1}} \quad \text{at } 298\,\text{K}.$$

(b) For a mole of molecules in a $10\,\text{dm}^3$ volume, the volume occupied per molecule is on average

$$v = \frac{10\times10^{-3}\,\text{m}^3\,\text{mol}^{-1}}{6.02\times10^{23}\,\text{mol}^{-1}} = 1.66\times10^{-26}\,\text{m}^3.$$

This places the molecules at an average distance of $r = v^{1/3}$ with respect to each other.

$$r = (1.66\times10^{-26}\,\text{m}^3)^{1/3} = 2.55\times10^{-9}\,\text{m} = 2.55\,\text{nm}.$$

Then per pair of molecules

$$V = -\frac{2\times(1.08\,\text{D})^4\times(3.336\times10^{-30}\text{Cm/D})^4}{3\times(4\pi\times8.854\times10^{-12}\,\text{J}^{-1}\text{C}^2\,\text{m}^{-1})^2\times1.38\times10^{-23}\,\text{J}\,\text{K}^{-1}\times298\,\text{K}\times(2.55\times10^{-9}\,\text{m})^6}$$

$$= -8.03\times10^{-27}\,\text{J}.$$

Each molecule has an average of 6 nearest neighbors, but we count a pair interaction only once. Then the total potential energy per mole in this sample is

$$V = -3 \times N_A \times 8.03 \times 10^{-27}\,\text{J} = \boxed{-0.014\,\text{J}}.$$

This potential energy is exceedingly small compared to 3.7 kJ mol^{-1}, so the kinetic theory of gases $\boxed{\text{is justifiable}}$ for this sample.

17.16 (a) Polarizability, $\alpha = \mu^*/\mathcal{E}$ [17.8] where μ^*, the induced dipole moment has the SI unit 'C m'. The electric field strength, $\mathcal{E}$, is the force per unit charge experienced by a charge. It has the unit 'N C^{-1}' or 'J C^{-1} m^{-1}'. Consequently, polarizability has the SI unit

$$(\text{C m})/(\text{J C}^{-1}\,\text{m}^{-1}) = \text{C}^2\,\text{m}^2\,\text{J}^{-1}.$$

(b) Polarizability volume, $\alpha' = \alpha/4\pi\varepsilon_0$ [17.9] where α has the SI unit 'C^2 m^2 J^{-1}' and ε_0 has the 'C^2 J^{-1} m^{-1}' unit. Consequently, polarizability volume has the SI unit

$$(\text{C}^2\,\text{m}^2\,\text{J}^{-1})/(\text{C}^2\,\text{J}^{-1}\,\text{m}^{-1}) = \boxed{\text{m}^3}.$$

17.17 $$\mu^* = \alpha\mathcal{E}\quad [17.8] = 4\pi\varepsilon_0\alpha'\mathcal{E}\quad [17.9] = 4\pi\varepsilon_0\alpha'\left(\frac{e}{4\pi\varepsilon_0 r^2}\right) = 1.85\text{D}.$$

$$\frac{\alpha' e}{r^2} = 1.85\,\text{D}.$$

Solve for r,

$$r = \left(\frac{\alpha' e}{1.85\,\text{D}}\right)^{1/2}$$
$$= \left(\frac{1.48 \times 10^{-30}\,\text{m}^3 \times 1.602 \times 10^{-19}\,\text{C}}{1.85 \times 3.336 \times 10^{-30}\,\text{C m}}\right)^{1/2}$$
$$= 1.96 \times 10^{-10}\,\text{m} = \boxed{196\,\text{pm}}.$$

17.18 $$V = -\frac{\mu_1^2\alpha_2'}{4\pi\varepsilon_0 r^6}\quad [17.10]$$
$$= -\frac{(2.7 \times 3.336 \times 10^{-30}\,\text{C m})^2 \times 10.4 \times 10^{-30}\,\text{m}^3}{(4\pi \times 8.854 \times 10^{-12}\,\text{J}^{-1}\,\text{C}^2\,\text{m}^{-1}) \times (4.0 \times 10^{-9}\,\text{m})^6}\quad [\text{Table17.2}]$$
$$= -1.46 \times 10^{-27}\,\text{J}$$
$$= \boxed{-8.81 \times 10^{-4}\,\text{J mol}^{-1}}.$$

This value seems exceedingly small. The distance suggested in the exercise may be too large compared to typical values.

17.19 $$V = \frac{2}{3} \times \frac{\alpha_1'\alpha_2'}{r^6} \times \frac{I_1 I_2}{I_1 + I_2} \quad [17.11] = \frac{2}{3}\frac{(\alpha')^2}{r^6} \times \frac{I}{2}$$

$$= \frac{2}{3} \times \frac{(10.4 \times 10^{-30}\,\text{m}^3)^2}{(4.0 \times 10^{-9}\,\text{m})^6} \times \frac{5.0\,\text{eV}}{2}$$

$$= 4.4 \times 10^{-8}\,\text{eV/molecule}$$

$$= \boxed{4.2 \times 10^{-3}\,\text{J mol}^{-1}}.$$

17.20 The geometry at the hydrogen bond is linear (see Figure 17.4), that is, θ in structure (12) is 0°. The partial charges are as given in Table 17.1 and Exercise 17.28. Distances in structure **19** are not given, but we may take $r_{\text{O—H}}$ to be 95.7 pm and $R_{\text{O....N}}$ to be 200 pm as a typical value.

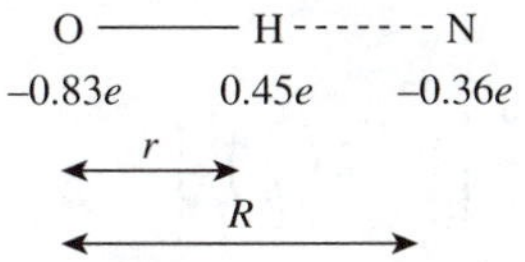

Fig 17.4

Then

$$V_{\text{O...N}} = \frac{N_A \delta_O \delta_N e^2}{4\pi\varepsilon_0 R} \quad [17.1a] \quad \text{and} \quad V_{\text{N...H}} = \frac{N_A \delta_N \delta_H e^2}{4\pi\varepsilon_0 (R - r)} \quad [17.1a].$$

$$\frac{N_A e^2}{4\pi\varepsilon_0} = 1.389 \times 10^{-4}\,\text{J m}.$$

$$V = V_{\text{O....N}} + V_{\text{N...H}} = 1.389 \times 10^{-4}\,\text{J m}\left[\frac{\delta_O \delta_N}{R} + \frac{\delta_N \delta_H}{(R - r)}\right]$$

$$= 1.389 \times 10^{-4}\,\text{J m}\left[\frac{0.83 \times 0.36}{2.00 \times 10^{-10}\,\text{m}} - \frac{0.36 \times 0.45}{(2.00 - 0.957) \times 10^{-10}\,\text{m}}\right]$$

$$= 1.389 \times 10^{-4}\,\text{J m} \times (0.1494 - 0.1553) \times 10^{10}\,\text{m}$$

$$= \boxed{-8.2\,\text{kJ mol}^{-1}}.$$

17.21 We assume that the Lennard-Jones potential (eqn 17.15), adequately represents the potential energy in this case.

$$V(R) = 4\varepsilon\left\{\left(\frac{\sigma}{R}\right)^{12} - \left(\frac{\sigma}{R}\right)^{6}\right\} = \frac{4\varepsilon\sigma^6}{R^6}\left\{\left(\frac{\sigma}{R}\right)^{6} - 1\right\},$$

$$V(R + \delta R) = 4\varepsilon\left\{\left(\frac{\sigma}{R + \delta R}\right)^{12} - \left(\frac{\sigma}{R + \delta R}\right)^{6}\right\}$$

$$= \frac{4\varepsilon\sigma^6}{R^6}\left\{\frac{\sigma^6}{R^6}\left(\frac{1}{1 + \frac{\delta R}{R}}\right)^{12} - \left(\frac{1}{1 + \frac{\delta R}{R}}\right)^{6}\right\}.$$

Since $(1+x)^{-12} \simeq 1-12x$ and $(1+x)^{-6} \simeq 1-6x$ when $x \ll 1$, and since $\delta R/R \ll 1$,

$$V(R+\delta R) = \frac{4\varepsilon\sigma^6}{R^6}\left\{\frac{\sigma^6}{R^6}\left(1-12\frac{\delta R}{R}\right)-\left(1-6\frac{\delta R}{R}\right)\right\}.$$

$$\Delta V = V(R+\delta R) - V(R) = \frac{4\varepsilon\sigma^6}{R^6}\left\{\frac{\sigma^6}{R^6}\left(1-12\frac{\delta R}{R}\right)-\left(1-6\frac{\delta R}{R}\right)\right\}$$

$$-\frac{4\varepsilon\sigma^6}{R^6}\left\{\left(\frac{\sigma}{R}\right)^6-1\right\}$$

$$=\frac{4\varepsilon\sigma^6}{R^6}\left(\frac{\delta R}{R}\right)\left\{1-2\left(\frac{\sigma}{R}\right)^6\right\}$$

$$F = -\frac{\Delta V}{\Delta r} = -\left[\frac{\frac{4\varepsilon\sigma^6}{R^6}\left(\frac{\delta R}{R}\right)\left\{1-2\left(\frac{\sigma}{R}\right)^6\right\}}{(R+\delta R)-R}\right] = -\frac{4\varepsilon\sigma^6}{R^7}\left\{1-2\left(\frac{\sigma}{R}\right)^6\right\}.$$

Thus, F equals zero when the factor $1-2\left(\frac{\sigma}{R}\right)^6$ equals zero or $\boxed{R = 2^{1/6}\sigma}$.

17.22 We assume that the Lennard-Jones potential (eqn 17.15), adequately represents the potential energy in this case.

$$V = 4\varepsilon\left\{\left(\frac{\sigma}{r}\right)^{12}-\left(\frac{\sigma}{r}\right)^6\right\}.$$

$$F = -\frac{\mathrm{d}V}{\mathrm{d}r} = -4\varepsilon\left\{\left(\frac{12\sigma}{r^{13}}\right)^{12}-\left(\frac{6\sigma}{r^7}\right)^6\right\}.$$

The minimum occurs when

$$\frac{12\sigma^{1/2}}{r^{13}} = \frac{6\sigma^6}{r^7}$$

$$\boxed{r = 2^{1/6}\sigma}.$$

17.23 In the dimer, the dipole moments will tend to cancel, at least partially, depending on the exact orientation. As the temperature increases, collisions and internal thermal agitation will break up the dimers and the dipole moments will not cancel.

17.24 The distances that appear in the denominator of the Coulombic interaction are shown in Figure 17.5. They are deduced with the standard formula $d = \{(x_1-x_2)^2+(y_1-y_2)^2\}^{1/2}$ where the nth atom has the coordinate (x_n, y_n). The value of the vertical height above the C—C axis (y = 107.4 pm) is based upon the assumption that the O—C—O angle equals 120°. The magnitude of the partial charges is assumed to be identical for all relevant atoms ($\pm\,\delta$).

$$d_{CC} = R \qquad\qquad d_{OO} = \{(R-124)^2 + 4y^2\}^{1/2}\text{ pm}$$

$$d_{OH} = \{(R-240)^2 + 4y^2\}^{1/2}\text{ pm} \qquad d_{HH} = \{(R-336)^2 + 4y^2\}^{1/2}\text{ pm}$$

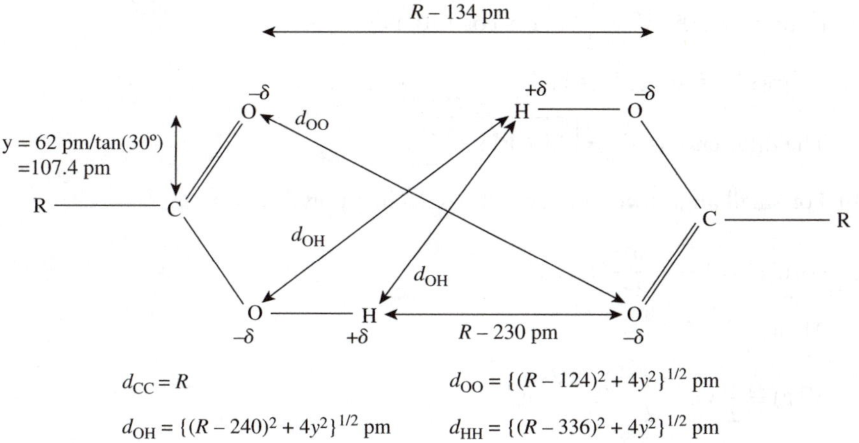

Fig 17.5

Each Coulombic interaction has the form $V_{1,2} = \pm \dfrac{\delta^2}{4\pi\varepsilon_0 d_{1,2}}$ where the positive sign designates repulsion between atoms 1 and 2 and the negative designates attraction. The total potential is the sum of all such terms. It is most conveniently analyzed as the function

$$f(R) = \frac{4\pi\varepsilon_0 V_{total}}{\delta^2} = -\frac{2}{R-230\text{ pm}} + \frac{2}{R-134\text{ pm}} + \frac{2}{\{(R-124\text{ pm})^2 + 4y^2\}^{1/2}}$$
$$-\frac{2}{\{(R-240\text{ pm})^2 + 4y^2\}^{1/2}} + \frac{1}{\{(R-336\text{ pm})^2 + 4y^2\}^{1/2}}.$$

V_{total} becomes attractive at the value of R for which $f(R) = 0$. This value may be determined either with the numeric solve function on a scientific calculator or it may be determined graphically. The Figure 17.6 plot yields $\boxed{R = 461.2\text{ pm}}$.

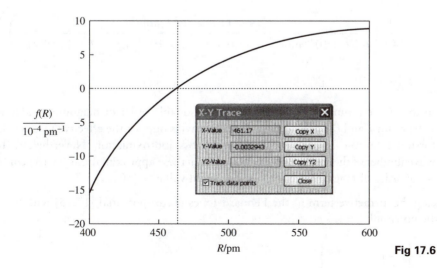

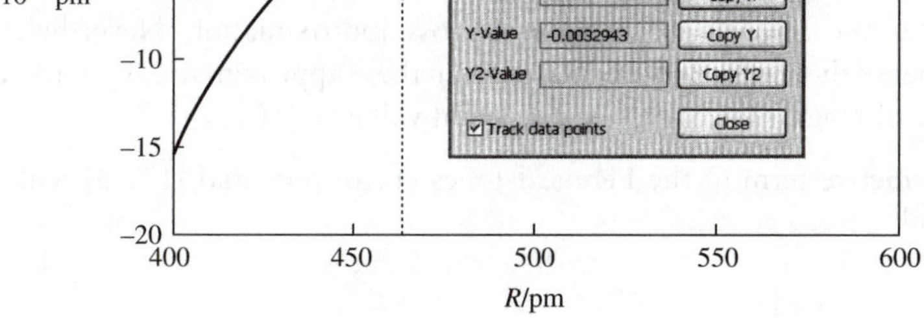

Fig 17.6

17.25 (a) trans: $\phi = 60°$, $V = \frac{1}{2}V_0(1 + \cos 180°) = 0$,

eclipsed: $\phi = 0°$, $V = V_0$,

The difference is $V_0 = \boxed{11.6\,\text{kJ mol}^{-1}}$.

(b) For small angles we can expand $\cos 3\phi$ in a power series

$$\cos(3\phi) = 1 - \frac{(3\phi)^2}{2} + \ldots.$$

Then

$$V(\phi) \cong \frac{1}{2}V_0\frac{(9\phi)^2}{2} \cong \frac{9V_0}{4}\phi^2.$$

This potential can be transformed into the harmonic oscillator form

$$V = \frac{1}{2}ky^2$$

for small ϕ by using

$$\sin\phi = \frac{y}{r} = \phi - \left(\frac{\phi^3}{6}\right) + \ldots \cong \phi.$$

Then $V(\phi) = \left(\frac{9V_0}{4r^2}\right)y^2$ where y is the linear displacement and $9V_0/2r^2$ is the force constant, k.

(c) We use

$$\nu = \frac{1}{2\pi}\left(\frac{k}{m}\right)^{1/2} \quad [12.18] = \frac{1}{2\pi}\left(\frac{9V_0}{2mr^2}\right)^{1/2}$$

$$= \frac{1}{2\pi}\left(\frac{9 \times 11.6 \times 10^3\,\text{J mol}^{-1}}{6.02 \times 10^{23}\,\text{mol}^{-1} \times 15.0 \times 1.661 \times 10^{-27}\,\text{kg} \times (1 \times 10^{-10}\,\text{m})^2}\right)^{1/2}$$

$$= \boxed{4 \times 10^{12}\,\text{s}^{-1}}$$

In the above estimate we have used the reasonable order of magnitude value of 100 pm for r and the actual mass of the methyl group for the effective mass of the torsional motion. The latter estimate is a gross approximation. Nevertheless, the wavenumber of the transitions calculated from these approximations, $\approx 133\,\text{cm}^{-1}$, is the order of magnitude of the experimental value $\approx 250\,\text{cm}^{-1}$.

17.26 Replacing the attractive term of the Lennard-Jones (12,6)-potential [17.15] with $e^{-r/\sigma}$ gives the potential

$$V = 4\varepsilon\left\{\left(\frac{\sigma}{r}\right)^{12} - e^{-r/\sigma}\right\}.$$

It is plotted in Figure 17.7. The minimum is at $\boxed{r = 1.34\sigma}$.

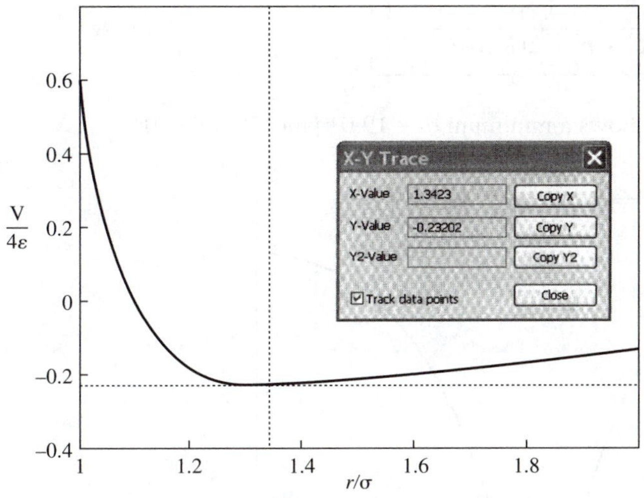

Fig 17.7

17.27 Replacing the attractive term of the Lennard-Jones (12,6)-potential [17.15] with $e^{-r/\sigma}$ gives the potential

$$V = 4\varepsilon\left\{\left(\frac{\sigma}{r}\right)^{12} - e^{-r/\sigma}\right\} = 4\varepsilon(x^{-12} - e^{-x}) \quad \text{where} \quad x = r/\sigma.$$

The minimum occurs where

$$\frac{dV}{dx} = 4\varepsilon\frac{d}{dx}(x^{-12} - e^{-x}) = 4\varepsilon(-12x^{-13} + e^{-x}) = 0,$$

which occurs when $-12x^{-13} + e^{-x} = 0$. Solve this equation with the numeric solve function of a scientific calculator to find a minimum at $r = \boxed{1.34\sigma}$.

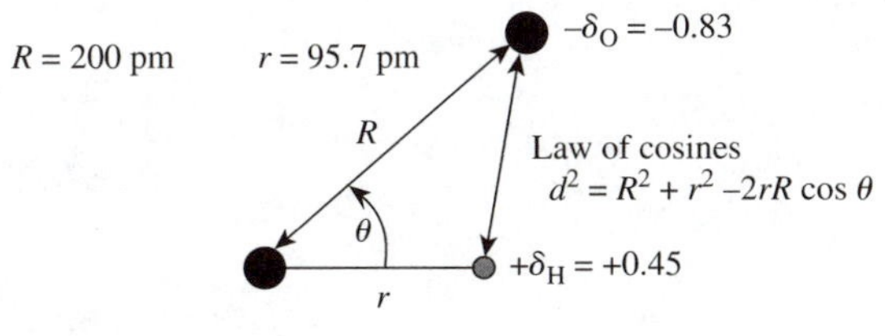

Fig 17.8

Using the definitions shown in Figure 17.8, the molar potential energy of the hydrogen bonding interaction as a function of the OOH angle, θ, is

$$\boxed{V = \frac{e^2 N_A}{4\pi\varepsilon_0}\left\{\frac{\delta_O^2}{R} - \frac{\delta_O\delta_H}{(R^2 + r^2 - 2rR\cos\theta)^{1/2}}\right\}}.$$

The plot in Figure 17.9 shows a minimum of $-19.0\,\text{kJ}\,\text{mol}^{-1}$ at $\theta = 0°$.

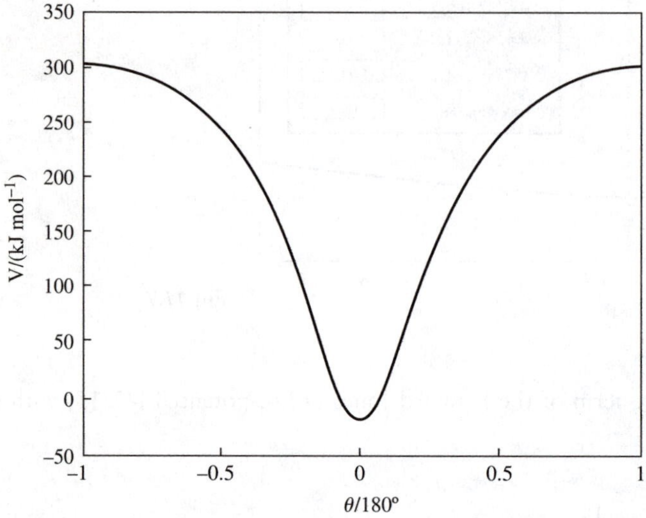

Fig 17.9

Chapter 18

Macromolecules and aggregates

Answers to discussion questions

18.1 Number average molar mass, $\overline{M}_n$, is the value obtained by weighting each molar mass by the number of molecules with that mass.

$$\overline{M}_n = \frac{1}{N}\sum_i N_i M_i \qquad [18.1].$$

In this expression, N_i is the number of molecules of molar mass M_i and N is the total number of molecules. Measurements of the osmotic pressures of macromolecular solutions yield the number average molar mass.

Weight average molar mass, $\overline{M}_w$, is the value obtained by weighting each molar mass by the mass of each one present.

$$\overline{M}_w = \frac{1}{m}\sum_i m_i M_i = \frac{\sum_i N_i M_i^2}{\sum_i N_i M_i} \qquad [18.2].$$

18.2 Contour length: the length of the macromolecule measured along its backbone, the length of all its monomer units placed end to end. This is the stretched-out length of the macromolecule, but with bond angles maintained within the monomer units. It is proportional to the number of monomer units, N, and to the length of each unit (eqn 18.5b).

Root mean square separation: one measure of the average separation of the ends of a random coil. It is the square root of the mean value of R^2, where R is the separation of the two ends of the coil. This mean value is calculated by weighting each possible value of R^2 with the probability, f (eqn 18.4), of that value of R occurring. It is proportional to $N^{1/2}$ and the length of each unit (eqn 18.5a).

Radius of gyration: the radius of a thin hollow spherical shell of the same mass and moment of inertia as the macromolecule. In general, it is not easy to visualize this distance geometrically. However, for the simple case of a molecule consisting of a chain of identical atoms, this quantity can be visualized as the root mean square distance of the atoms from the center of mass. It also depends on $N^{1/2}$, but is smaller than the root mean square separation by a factor of $(1/6)^{1/2}$ (eqn 18.5c).

18.3 (a) ΔS is the change in conformational entropy of a random coil of a polymer chain. It is the statistical entropy arising from the arrangement of bonds, when a coil containing N bonds of length l is stretched or compressed by nl, where n is a numerical factor giving the amount of stretching in units of l. The amount of stretching relative to the number of monomer units in the chain is $v = n/N$.

(b) F is the restoring force of a one-dimensional perfect elastomer at temperature T. The elastomer has N bonds of length l and the polymer is stretched or compressed by nl. For small strains, F is proportional to the extension, corresponding to Hooke's law.

(c) R_{rms} is one of several measures of the size of a random coil. For a polymer of N monomer units each of length l, the root mean square separation, R_{rms}, is a measure of the average separation of the ends of a random coil. It is the square root of the average value of R^2, calculated by weighting each possible value of R^2 with the probability that R occurs.

(d) R_G, the radius of gyration, is another measure of the size of a random coil. It is the radius of a thin hollow spherical shell of the same mass and moment of inertia as the polymer molecule.

All of these expressions are derived for the freely jointed random coil model of polymer chains, which is the simplest possibility for the conformation of identical units not capable of forming hydrogen bonds or any other type of specific bond. In this model, any bond is free to make any angle with respect to the preceding one (Figure 18.3 of the text). We assume that the residues occupy zero volume, so different parts of the chain can occupy the same region of space. We also assume in the derivation of the expression for the probability of the ends of the chain being a distance nl apart, that the chain is compact in the sense that $n << N$. This model is obviously an oversimplification because a bond is actually constrained to a cone of angles around a direction defined by its neighbor. In a hypothetical one-dimensional freely jointed chain all the residues lie in a straight line, and the angle between neighbors is either 0° or 180°. The residues in a three-dimensional freely jointed chain are not restricted to lie in a line or a plane.

The random coil model ignores the role of the solvent: a poor solvent will tend to cause the coil to tighten; a good solvent does the opposite. Therefore, calculations based on this model are best regarded as lower bounds to the dimensions of a polymer in a good solvent and as an upper bound for a polymer in a poor solvent. The model is most reliable for a polymer in a bulk solid sample, where the coil is likely to have its natural dimensions.

18.4 Washing a cotton shirt disturbs the secondary structure of the cellulose. The water tends to break the hydrogen bonds between the cellulose chains by forming its own hydrogen bonds to the chains. Upon drying, the hydrogen bonds between chains reform, but in a random manner, causing wrinkles. The wrinkles are removed by moistening the shirt, which again breaks the hydrogen bonds, making the fibre more flexible and plastically deformable. A hot iron shapes the cloth and causes the water to evaporate so new hydrogen bonds are formed between the chains while they are held in place by the pressure of the iron.

18.5 Kevlar is a polyaromatic amide. Phenyl groups provide aromaticity and a planar, rigid structure. The amide group is expected to be like the peptide bond that connects amino acid residues within protein molecules. This group is also planar because resonance produces a partial double bond character between the carbon and nitrogen atoms. There is a substantial energy barrier preventing free rotation about the C—N bond. The two bulky phenyl groups on the ends of an amide group are trans because steric hinderance makes the cis conformation unfavorable.

The flatness of the Kevlar polymeric molecule makes it possible to process the material so that many molecules with parallel alignment form highly ordered, untangled crystal bundles. The alignment makes possible both considerable van der Waals attractions between adjacent molecules and strong hydrogen bonding between the polar amide groups on adjacent molecules. These bonding forces create the high thermal stability and mechanical strength observed in Kevlar.

Kevlar is able to absorb great quantities of energy, such as the kinetic energy of a spreading bullet, through hydrogen bond breakage and the transition to the cis conformation.

18.6 It is hard to imagine that the stacking of disk-like molecules would result in 'wires' capable of conducting electricity. Electrical conductivity mechanisms of metals are well known as is the conductivity of delocalized π-electrons found in the lamellar atomic planes of a graphite crystal. However, there is little electric conductivity normal to the covalent atomic planes of crystalline graphite. The nearest neighbor atomic distances are about 0.2–0.3 nm in metals and about 0.1–0.2 nm in molecules that have delocalized π-electrons. The lamellar planes of graphite are separated by 0.34 nm and held together by the weak van der Waals force. Stacks of disk-like organic molecules seem to mimic the stacking of graphite planes but with greater separation (0.5 nm). No electric conductivity is expected along this type of discotic liquid crystal wire.

However, it may be possible to design a disk-like molecule that has:

(1) filled, or partially filled, delocalized bonding molecular orbitals;

(2) relatively low energy, unfilled, delocalized antibonding molecular orbitals; and

(3) a relatively small stacking distance.

A potential difference, when applied to the ends of a molecular stack, might cause a π-electron within one molecule to jump into an antibonding π-molecular orbital of an adjacent molecule. In such a case, the 'wire' would conduct electricity.

18.7 A surfactant is a species that is active at the interface of two phases or substances, such as the interface between hydrophilic and hydrophobic phases. A surfactant accumulates at the interface and modifies the properties of the surface, in particular, decreasing its surface tension. A typical surfactant consists of a long hydrocarbon tail and other non-polar materials, and a hydrophilic head group, such as the carboxylate group, $-CO_2^-$, that dissolves in a polar solvent, typically water. In other words, a surfactant is an amphipathic substance, meaning that it has both hydrophobic and hydrophilic regions.

How does the surfactant decrease the surface tension? Surface tension is a result of cohesive forces and the solute molecules must weaken the attractive forces between solvent molecules. Thus, molecules with bulky hydrophobic regions such as fatty acids can decrease the surface tension because they attract solvent molecules less strongly than solvent molecules attract each other.

18.8 The formation of micelles is favored by the interaction between hydrocarbon tails and is opposed by charge repulsion of the polar groups, which are placed close together at the micelle surface. As salt concentration is increased, the repulsion of head groups is reduced because their charges are partly shielded by the ions of the salt. This favors micelle formation causing the micelles to be larger and the critical micelle concentration to be smaller.

18.9 The increase in temperature with the hydrophobic chain length is a result of the increased strength of the van der Waals interaction between long unsaturated portions of the chains that can interlock well with each other. The introduction of double bonds in the chains can affect the interlocking of the parallel chains by putting kinks in the chains, thereby decreasing the strength of the van der Waals interactions between chains. Double bonds can be either cis or trans. Only cis-double bonds produce a kink, but most fatty acids are the cis-isomer. So we expect that the transition temperatures will decrease in rough proportion to the number of C=C bonds.

Solutions to exercises

18.10 Equal amounts imply equal numbers of molecules. Hence the number-average is (eqn 18.1)

$$\overline{M}_{\mathrm{n}} = \frac{N_1M_1 + N_2M_2}{N} = \frac{n_1M_1 + n_2M_2}{n} = \frac{1}{2}(M_1 + M_2)\left[n_1 = n_2 = \frac{1}{2}n\right]$$

$$= \frac{62 + 78}{2}\ \mathrm{kg\,mol^{-1}}$$

$$= \boxed{70\ \mathrm{kg\,mol^{-1}}},$$

and the weight-average is (eqn 18.2)

$$\overline{M}_{\mathrm{w}} = \frac{m_1M_1 + m_2M_2}{m} = \frac{n_1M_1^2 + n_2M_2^2}{n_1M_1 + n_2M_2} = \frac{M_1^2 + M_2^2}{M_1 + M_2}\quad [n_1 = n_2]$$

$$= \frac{62^2 + 78^2}{62 + 78}\ \mathrm{kg\,mol^{-1}}$$

$$= \boxed{71\ \mathrm{kg\,mol^{-1}}}.$$

18.11 (a) Osmometry gives the number-average molar mass, so

$$\overline{M}_n = \frac{N_1M_1 + N_2M_2}{N_1 + N_2} = \frac{\left(\frac{m_1}{M_1}\right)M_1 + \left(\frac{m_2}{M_2}\right)M_2}{\left(\frac{m_1}{M_1}\right) + \left(\frac{m_2}{M_2}\right)} = \frac{m_1 + m_2}{\left(\frac{m_1}{M_1}\right) + \left(\frac{m_2}{M_2}\right)}$$

$$= \frac{100\,g}{\left(\frac{30\,g}{30\,\text{kg mol}^{-1}}\right) + \left(\frac{70g}{15\,\text{kg mol}^{-1}}\right)} \text{[assume 100 g of solution]}$$

$$= \boxed{18\,\text{kg mol}^{-1}}.$$

(b) Light-scattering gives the weight-average molar mass, so

$$\overline{M}_w = \frac{m_1M_1 + m_2M_2}{m_1 + m_2} = \frac{(30) \times (30) + (70) \times (15)}{100}\,\text{kg mol}^{-1}$$

$$= \boxed{20\,\text{kg mol}^{-1}}.$$

18.12 Example 18.1 illustrates the methodology for calculating the heterogeneity index ($\overline{M}_w/\overline{M}_n$).

Interval/(kg mol^{-1})	5–10	10–15	15–20	20–25	25–30	30–35	Total
Molar mass/(kg mol^{-1})	7.5	12.5	17.5	22.5	27.5	32.5	
Mass in interval/g	18.0	22.0	17.5	14.2	9.7	4.5	85.9
Amount/mmol	2.40	1.76	1.00	0.631	0.353	0.138	6.28

$$\overline{M}_n = \frac{1}{N}\sum_i N_iM_i \quad [18.1] = \frac{1}{n}\sum_i n_iM_i$$

$$= \frac{1}{6.28}(2.40 \times 7.5 + 1.76 \times 12.5 + 1.00 \times 17.5 + 0.631 \times 22.5$$

$$+0.353 \times 27.5 + 0.138 \times 32.5)\,\text{kg mol}^{-1}$$

$$= 13.\bar{7}\,\text{kg mol}^{-1}.$$

$$\overline{M}_w = \frac{1}{m}\sum_i m_iM_i \quad [18.2]$$

$$= \frac{1}{85.9}(18.0 \times 7.5 + 22.0 \times 12.5 + 17.5 \times 17.5 + 14.2 \times 22.5$$

$$+9.7 \times 27.5 + 4.5 \times 32.5)\,\text{kg mol}^{-1}$$

$$= 16.\bar{9}\,\text{kg mol}^{-1}.$$

$$\text{Heterogeneity index} = \overline{M}_w/\overline{M}_n = \frac{16.\bar{9}}{13.\bar{7}} = \boxed{1.2}.$$

18.13 The peaks are separated by $104\,\mathrm{g\,mol^{-1}}$, so this is the molar mass of the repeating unit of the polymer. This peak separation is consistent with the identification of the polymer as polystyrene, for the repeating group of $CH_2CH(C_6H_5)$ (8 C atoms and 8 H atoms) has a molar mass of $8 \times (12+1)\,\mathrm{g\,mol^{-1}} = 104\,\mathrm{g\,mol^{-1}}$. A consistent difference between peaks suggests a pure system and points away from different numbers of subunits of different molecular weight (such as the *t*-butyl initiators) being incorporated into the polymer molecules. The most intense peak has a molar mass equal to that of *n* repeating groups plus that of a silver cation plus that of terminal groups:

$$M(\text{peak}) = nM(\text{repeat}) + M(\mathrm{Ag^+}) + M(\text{terminal}).$$

If both ends of the polymer have terminal *t*-butyl groups, then

$$M(\text{terminal}) = 2M(t\text{–butyl}) = 2(4 \times 12 + 9)\,\mathrm{g\,mol^{-1}} = 114\,\mathrm{g\,mol^{-1}}$$

and $n = \dfrac{M(\text{peak}) - M(\mathrm{Ag^+}) - M(\text{terminal})}{M(\text{repeat})} = \dfrac{25598 - 108 - 114}{104} = \boxed{244}$.

18.14 The number of solute molecules with potential energy E is proportional to $\mathrm{e}^{-E/kT}$. Hence,

$$c \propto N \propto \mathrm{e}^{-E/kT}, \quad E = \frac{1}{2}m_{\text{eff}}r^2\omega^2.$$

Therefore, $c \propto e^{Mb\omega^2r^2/2RT}$ $[m_{\text{eff}} = bm, M = mN_{\mathrm{A}}]$

$$\ln c = \text{constant} + \frac{Mbw^2r^2}{2RT}\,[b = 1 - \rho v_{\mathrm{s}}],$$

and slope of ln c against r^2 is equal to $\dfrac{Mb\omega^2}{2RT}$. Therefore,

$$M = \frac{2RT \times \text{slope}}{b\omega^2} = \frac{(2) \times (8.314\,\mathrm{J\,K^{-1}\,mol^{-1}}) \times (300\,\mathrm{K}) \times (729 \times 10^4\,\mathrm{m^{-2}})}{(1 - 0.997 \times 0.61) \times \left(\dfrac{(2\pi) \times (50000)}{60\,\mathrm{s}}\right)^2}$$

$$= \boxed{3.4 \times 10^3\,\mathrm{kg\,mol^{-1}}}.$$

18.15 From eqn 18.3, we can relate concentration ratios to the molar mass

$$\ln\frac{c_1}{c_2} = \frac{\overline{M}_{\mathrm{w}}b\omega^2(r_1^2 - r_2^2)}{2RT} = \frac{2\pi^2\overline{M}_{\mathrm{w}}bv^2(r_1^2 - r_2^2)}{RT}$$

and hence

$$v = \left(\frac{RT \ln\left(\frac{c_1}{c_2}\right)}{2\pi^2 \overline{M}_\mathrm{w} b \left(r_1^2 - r_2^2\right)} \right)^{1/2}$$

$$= \left(\frac{\left(8.3145\,\mathrm{J\,K^{-1}\,mol^{-1}}\right) \times (298\,\mathrm{K}) \times (\ln 5)}{2\pi^2 \times \left(1 \times 10^2\,\mathrm{kg\,mol^{-1}}\right) \times (1 - 0.75) \times \left(7.0^2 - 5.0^2\right) \times 10^{-4}\,\mathrm{m^2}} \right)^{1/2}$$

$$= 58\,\mathrm{Hz}, \text{or } \boxed{3500\,\mathrm{rpm}}.$$

QUESTION What would the concentration gradient be in this system with a speed of operation of 7.0×10^4 rpm in an ultracentrifuge?

18.16 The equation for scattered radiation indicates that a plot of R_θ^{-1} against $R_\theta^{-1} \sin^2\left(\frac{1}{2}\theta\right)$ is linear with an intercept equal to $(Kc_\mathrm{p}\overline{M}_\mathrm{w})^{-1}$ and a slope equal to $16\pi^2 R_\mathrm{G}^2/3\lambda^2$. Thus, we prepare a table of tranformed data for the plot, check plot linearity (Figure 18.1), perform a linear regression fit of the plot, and use the intercept and slope to calculate $\overline{M}_\mathrm{w}$ and R_G, respectively,

$$\overline{M}_\mathrm{w} = \frac{1}{Kc_\mathrm{p} \times \text{intercept}} \quad \text{and} \quad R_\mathrm{G} = (3 \times \text{slope})^{1/2} \left(\frac{\lambda}{4\pi}\right).$$

$\theta/°$	15	45	70	85	90
R_θ/m^2	23.8	22.9	21.6	20.7	20.4
$100\,\mathrm{m}^2/R_\theta$	4.20	4.37	4.63	4.83	4.90
$100\,\mathrm{m}^2/R_\theta \times \sin^2(\theta/2)$	0.0716	0.640	1.52	2.20	2.45

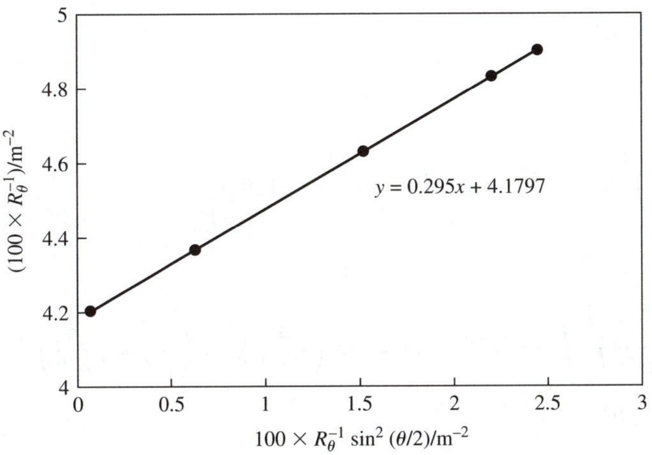

Fig 18.1

$$\overline{M}_w = \frac{1}{(2.40\times10^{-2}\,\mathrm{mol\,m^3\,kg^{-2}})\times(2.0\,\mathrm{kg\,m^{-3}})\times(0.04180)} = \boxed{5.0\times10^2\,\mathrm{kg\,mol^{-1}}}.$$

$$R_G = (3\times0.295)^{1/2}\times\left(\frac{532\,\mathrm{nm}}{4\pi}\right) = \boxed{39.8\,\mathrm{nm}}.$$

18.17 $R_{rms} = N^{1/2}l$ [18.5*a*] $= (700)^{1/2}\times(0.90\,\mathrm{nm}) = \boxed{24\,\mathrm{nm}}$.

18.18 The repeating unit (monomer) of polyethylene is ($-CH_2-CH_2-$) which has a molar mass of 28 g mol^{-1}. The number of repeating units, N, is therefore

$$N = \frac{280000\,\mathrm{g\,mol^{-1}}}{28\,\mathrm{g\,mol^{-1}}} = 1.00\times10^4$$

and $l = 2R(\mathrm{C{-}C})$. [Add half a bond-length on either side of monomer.]

Apply the formulas for the contour length [18.5b] and rms separation [18.5a]:

$$R_c = Nl = 2\times(1.00\times10^4)(154\,\mathrm{pm}) = 3.08\times10^6\,\mathrm{pm} = \boxed{3.08\times10^{-6}\,\mathrm{m}}$$

and $$R_{rms} = N^{1/2}l = 2\times(1.00\times10^4)^{1/2}\times(154\,\mathrm{pm}) = 3.08\times10^4\,\mathrm{pm}$$

$$= \boxed{3.08\times10^{-8}\,\mathrm{m}}.$$

18.19 For a random coil, the radius of gyration is [18.5c]

$$R_G = \left(\frac{N}{6}\right)^{1/2} l \quad \text{so} \quad N = 6\left(\frac{R_G}{l}\right)^2 = (6)\times\left(\frac{7.3\,\mathrm{nm}}{0.154\,\mathrm{nm}}\right)^2 = \boxed{1.4\times10^4}.$$

18.20 (a) Following Derivation 18.1, we have

$$R_{rms}^2 = \int_0^\infty R^2 f\,\mathrm{d}R$$

with

$$f = 4\pi\left(\frac{a}{\pi^{1/2}}\right)^3 R^2 e^{-a^2R^2}, \quad a = \left(\frac{3}{2Nl^2}\right)^{1/2} \quad [18.4].$$

Therefore,

$$R_{rms}^2 = 4\pi\left(\frac{a}{\pi^{1/2}}\right)^3\int_0^\infty R^4 e^{-a^2R^2}\,\mathrm{d}R = 4\pi\left(\frac{a}{\pi^{1/2}}\right)^3\times\left(\frac{3}{8}\right)\times\left(\frac{\pi}{a^{10}}\right)^{1/2}$$

$$= \frac{3}{2a^2} = Nl^2.$$

Hence, $R_{rms} = \boxed{lN^{1/2}}$.

(b) The mean separation is

$$R_{\text{mean}} = \int_0^\infty Rf\,dr = 4\pi\left(\frac{a}{\pi^{1/2}}\right)^3 \int_0^\infty R^3 e^{-a^2R^2}\,dR$$

$$= 4\pi\left(\frac{a}{\pi^{1/2}}\right)^3 \times \left(\frac{1}{2a^4}\right) = \frac{2}{a\pi^{1/2}} = \boxed{\left(\frac{8}{3\pi}\right)^{1/2} l}.$$

(c) The most probable separation is the value of R for which f is a maximum, so set $\frac{df}{dR} = 0$ and solve for R.

$$\frac{df}{dR} = 4\pi\left(\frac{a}{\pi^{1/2}}\right)^3 \{2R - 2a^2R^3\}e^{-a^2R^2} = 0 \quad \text{when} \quad a^2R^2 = 1.$$

Therefore, the most probable separation is

$$R^* = \frac{1}{a} = \boxed{l\left(\frac{2}{3}N\right)^{1/2}}.$$

When $N = 4000$ and $l = 154$ pm,

(a) $R_{\text{rms}} = \boxed{9.74\,\text{nm}}$, (b) $R_{\text{mean}} = \boxed{8.97\,\text{nm}}$, (c) $R^* = \boxed{7.95\,\text{nm}}$.

18.21 A simple procedure is to generate numbers in the range 1 to 8, and to step north for a 1 or 2, east for 3 or 4, south for 5 or 6, and west for 7 or 8 on a uniform grid. One such walk is shown in Figure 18.2. Roughly, the mean and most probable separations of the ends appear to vary as $N^{1/2}$.

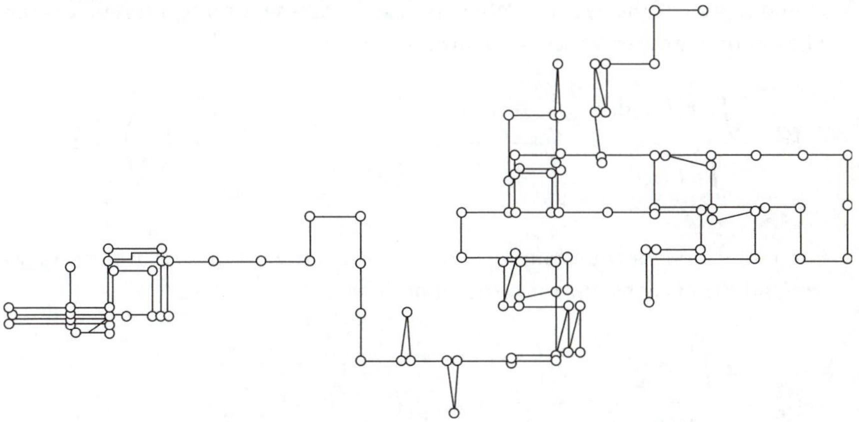

Fig 18.2

18.22 Assume the solute particles are solid spheres and see how well R_G calculated based on that assumption agrees with experimental values. For a sphere of uniform density the radius of gyration is given by

$$R_G/\text{nm} = 0.05690 \times \{[v_s/(\text{cm}^3\text{g}^{-1})] \times (M/\text{g mol}^{-1})^{1/3}\} \qquad (1)[\text{derived below}].$$

Draw up the following table.

	$M/(\mathrm{g\,mol^{-1}})$	$v_s/(\mathrm{cm^3\,g^{-1}})$	$(R_G/\mathrm{nm})_{\mathrm{calc}}$	$(R_g/\mathrm{nm})_{\mathrm{expt}}$
Serum albumin	66×10^3	0.752	2.09	2.98
Bushy stunt virus	10.6×10^6	0.741	11.3	12.0
DNA	4×10^6	0.556	7.43	117.0

Therefore, serum albumin and bushy stunt virus resemble solid spheres, but DNA does not.

Derivation of eqn (1). We use the following definition of the radius of gyration.

$$R_G^2 = \frac{1}{N}\sum_j R_j^2.$$

(a) For a sphere of uniform density, the center of mass is at the center of the sphere. We may visualize the sphere as a collection of a very large number, N, of small particles distributed with equal number density throughout the sphere. Then the summation above may be replaced with an integration.

$$R_G^2 = \frac{1}{N}\,\frac{N\int_0^a r^2 P(r)\mathrm{d}r}{\int_0^a P(r)\mathrm{d}r}$$

$P(r)$ is the probability per unit distance that a small particle will be found at distance r from the center, that is, within a spherical shell of volume $4\pi r^2\mathrm{d}r$. Hence, $P(r) = 4\pi r^2\mathrm{d}r$. If $P(r)$ were normalized, the integral in the numerator would represent the average value of r^2, so N times that integral replaces the sum. The denominator enforces normalization. Hence

$$R_g^2 = \frac{\int_0^a r^2 P(r)\mathrm{d}r}{\int_0^a P(r)\mathrm{d}r} = \frac{\int_0^a 4\pi r^4\mathrm{d}r}{\int_0^a 4\pi r^2\mathrm{d}r} = \frac{\frac{1}{5}a^5}{\frac{1}{3}a^3} = \frac{3}{5}a^2, \quad \boxed{R_g = \left(\frac{3}{5}\right)^{1/2} a}.$$

(b) For a long straight rod of uniform density the center of mass is at the center of the rod and $P(z)$ is constant for a rod of uniform radius; hence,

$$R_g^2 = \frac{2\int_0^{l/2} z^2\mathrm{d}z}{2\int_0^{l/2}\mathrm{d}z} = \frac{\frac{1}{3}\left(\frac{1}{2}l\right)^3}{\frac{1}{2}l} = \frac{1}{12}l^2, \quad \boxed{R_g = \frac{l}{2\sqrt{3}}}.$$

COMMENT The radius of the rod does not enter into the result. In fact, the distribution function is $P(r, z)$, the probability that a small particle will be found at a distance r from the central axis of the rod and z along that axis from the center, that is, within a squat cylindrical shell of volume $2\pi r\mathrm{d}r\mathrm{d}z$. Integration radially outward from the axis is the same in numerator and denominator.

For a spherical macromolecule, the specific volume is:

$$v_s = \frac{V}{m} = \frac{4\pi a^3}{3} \times \frac{N_A}{M} \quad \text{so} \quad a = \left(\frac{3v_s M}{4\pi N_A}\right)^{1/3}$$

and

$$R_G = \left(\frac{3}{5}\right)^{1/2} \times \left(\frac{3\,v_s M}{4\pi N_A}\right)^{1/3}$$

$$= \left(\frac{3}{5}\right)^{1/2} \times \left(\frac{(3\,v_s/\text{cm}^3\,\text{g}^{-1}) \times \text{cm}^3\text{g}^{-1} \times (M/\text{g}\,\text{mol}^{-1}) \times \text{g}\,\text{mol}^{-1}}{(4\pi) \times (6.022 \times 10^{23}\,\text{mol}^{-1})}\right)^{1/3}$$

$$= (5.690 \times 10^{-9}) \times (v_s/\text{cm}^3\text{g}^{-1})^{1/3} \times (M/\text{g}\,\text{mol}^{-1})^{1/3}\,\text{cm}$$

$$= (5.690 \times 10^{-11}\text{m}) \times \{(v_s/\text{cm}^3\text{g}^{-1}) \times (M/\,\text{g}\,\text{mol}^{-1})\}^{1/3}.$$

That is, $R_G/\text{nm} = \boxed{0.05690 \times \{(v_s/\text{cm}^3\,\text{g}^{-1}) \times (M/\text{g}\,\text{mol}^{-1})^{1/3}\}}$.

Sample calculation for a spherical macromolecule:

When $M = 100\,\text{kg}\,\text{mol}^{-1}$ and $v_s = 0.750\,\text{cm}^3\,\text{g}^{-1}$,

$$R_G/\text{nm} = (0.05690) \times \{0.750 \times 1.00 \times 10^5\}^{13} = \boxed{2.40}.$$

Example calculation for a rod: $v_{\text{mol}} = \pi a^2 l$, so

$$R_G = \frac{v_{\text{mol}}}{2\pi a^2\sqrt{3}} = \frac{v_s M}{N_A} \times \frac{1}{2\pi a^2\sqrt{3}}$$

$$= \frac{(0.750\,\text{cm}^3\,\text{g}^{-1}) \times (1.00 \times 10^5\,\text{g}\,\text{mol}^{-1})}{(6.022 \times 10^{23}\,\text{mol}^{-1}) \times (2\pi) \times (0.5 \times 10^{-7}\,\text{cm})^2 \times \sqrt{3}}$$

$$= 4.6 \times 10^{-6}\,\text{cm} = \boxed{46\,\text{nm}}.$$

COMMENT R_G may also be defined through the relation

$$R_G^2 = \frac{\sum_i m_i r_i^2}{\sum_i m_i}.$$

QUESTION Does this definition lead to the same formulas for the radii of gyration of the sphere and the rod as those derived above?

18.23 A 10% expansion from the coiled state corresponds to a 10% increase in the radius of gyration. Thus, the expansion is

$$nl = 0.1\,R_G = 0.1\left(\frac{N}{6}\right)^{1/2} l \quad [18.5c] \qquad \text{or} \qquad n = 0.1\left(\frac{N}{6}\right)^{1/2}.$$

$$\nu = \frac{n}{N} = \frac{1}{N}\left\{0.1\left(\frac{N}{6}\right)^{1/2}\right\} = 0.1\left(\frac{1}{6N}\right)^{1/2} = 0.1\left(\frac{1}{6 \times 1000}\right)^{1/2} = 0.0129.$$

Since $\nu \ll 1$, (eqn 18.8) estimates the requisite extension force. Assume l = 154, which is the C—C bond length in a typical polymer.

$$F = \frac{\nu kT}{l} \quad [18.8] = \frac{(0.0129)\left(1.381 \times 10^{-23}\,\mathrm{J\,K^{-1}}\right)(300\,\mathrm{K})}{154 \times 10^{-12}\,\mathrm{m}} = \boxed{3.47 \times 10^{-13}\,\mathrm{N}}.$$

18.24 (a) The data set, plotted in Figure 18.3, gives rise to a tolerably linear curve, so we estimate the melting point by interpolation using the best-fit straight line.

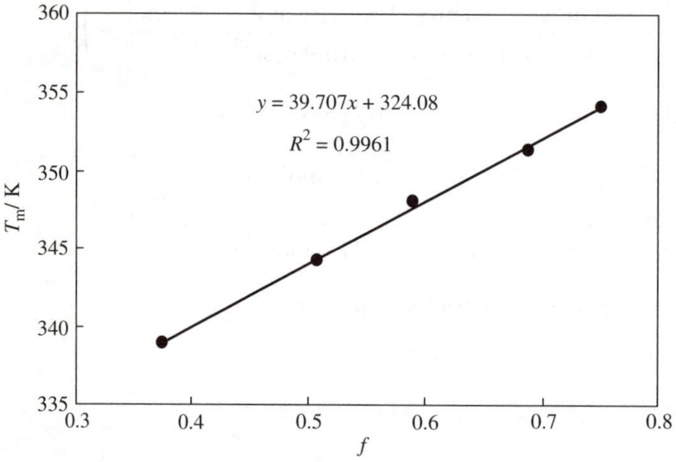

Fig 18.3

The best-fit equation has the form $T_\mathrm{m}/\mathrm{K} = mf + b$, and we want T_m when f = 0.40:
$c_\mathrm{salt} = 1.0 \times 10^{-2}\,\mathrm{mol\,dm^{-3}}$: $T_\mathrm{m} = (39.7 \times 0.40 + 324)\,\mathrm{K} = \boxed{340\,\mathrm{K}}$.

(b) T_m increases with the number of G—C base pairs because this pair is held together with three hydrogen bonds in the double helical structure whereas the A—T pair is held with two hydrogen bonds (see Section 18.3). The ΔH_m contribution is greater for the G—C pair.

18.25 The glass transition temperature T_g is the temperature at which internal bond rotations freeze. In effect, the easier such rotations are, the lower T_g. Internal rotations are more difficult for polymers that have bulky side chains than for polymers without such chains because the side chains of neighboring molecules can impede each other's motion. Of the four polymers in this problem, polystyrene has the largest side chain (phenyl) and the largest T_g. The chlorine atoms in poly(vinyl chloride) interfere with each other's motion more than the smaller hydrogen atoms that hang from the carbon backbone of polyethylene. Poly(oxymethylene), like polyethylene, has only hydrogen atoms protruding from its backbone; however, poly(oxymethylene) has fewer hydrogen protrusions and a still lower T_g than polyethylene.

Chapter 19

Molecular rotations and vibrations

Answers to discussion questions

19.1 *Doppler broadening*. This contribution to the linewidth is due to the Doppler effect which shifts the frequency of the radiation emitted or absorbed when the atoms or molecules involved are moving towards or away from the detecting device. Molecules have a wide range of speeds in all directions in a gas and the detected spectral line is the absorption or emission profile arising from all the resulting Doppler shifts.

Lifetime broadening. The Doppler broadening is significant in gas-phase samples, but lifetime broadening occurs in all states of matter. This kind of broadening is a quantum mechanical effect related to the uncertainty principle in the form of eqn 19.7a and is due to the finite lifetimes of the states involved in the transition. When τ is finite, the energy of the states is smeared out and hence the transition frequency is broadened as shown in eqn 19.7b.

Pressure broadening or collisional broadening. The actual mechanism affecting the lifetime of energy states depends on various processes, one of which is collisional deactivation and another is spontaneous emission. Lowering the pressure can reduce the first of these contributions; the second cannot be changed and results in a natural linewidth.

19.2 (a) For *microwave rotational spectroscopy*, the allowed transitions depend on the existence of an oscillating dipole moment, which can stir the electromagnetic field into oscillation (and vice versa for absorption). This implies that the molecule must have a permanent dipole moment, which is equivalent to an oscillating dipole when the molecule is rotating. See Figure 19.19 of the text.

(b) The gross selection rule for *rotational Raman spectroscopy* is that the molecule must be anisotropically polarizable, which is to say that its polarizability, α, depends upon the direction of the electric field relative to the molecule. Non-spherical rotors satisfy this condition. Therefore, linear and symmetric rotors are rotationally Raman active.

(c) In the case of *infrared vibrational spectroscopy*, the physical basis of the gross selection rule is that the molecule must have a structure that allows for the existence of an oscillating dipole moment when the molecule vibrates. Polar molecules necessarily satisfy this requirement, but non-polar molecules may also have a fluctuating dipole moment upon vibration. See Figure 19.25 of the text.

(d) The gross selection rule for *vibrational Raman spectroscopy* is that the polarizability of the molecule must change as the molecule vibrates. All diatomic molecules satisfy this condition as the molecules swell and contract during a vibration, the control of the nuclei over the electrons varies, and the molecular polarizability changes. Hence both homonuclear and heteronuclear diatomics are vibrationally Raman active. In polyatomic molecules it is usually quite difficult to judge by inspection whether or not the molecule is anisotropically polarizable; hence group theoretical methods are relied on for judging the Raman activity of the various normal modes of vibration. The procedure is an advanced topic but the **exclusion rule** can be useful. It states that, if the molecule has a center of inversion, then no modes can be both infrared and Raman active.

19.3 The answer to this question depends precisely on what is meant by equilibrium bond length. The angular velocity increases with the quantum number J. Thus, centrifugal distortion and bond length r_c is greater for higher rotational energy levels. But the equilibrium bond length r_e remains constant if by that term one means the value of r corresponding to a vibrating non-rotating molecule with $J = 0$.

19.4 The exclusion rule applies to the benzene molecule because it has a center of inversion. Consequently, none of the normal modes of vibration of benzene can be both infrared and Raman active. If we wish to characterize all the normal modes, we must obtain both kinds of spectra.

Solutions to exercises

19.5 (a) $\nu = \dfrac{c}{\lambda} = \dfrac{2.998 \times 10^8\ \text{m s}^{-1}}{6.70 \times 10^{-7}\ \text{m}} = \boxed{4.47 \times 10^{14}\ \text{s}^{-1}}$.

(b) $\tilde{\nu} = \dfrac{1}{\lambda} = \dfrac{1}{6.70 \times 10^{-7}} = 1.49 \times 10^6\ \text{m}^{-1} = \boxed{1.49 \times 10^4\ \text{cm}^{-1}}$.

19.6 (a) $\tilde{\nu} = \dfrac{\nu}{c} = \dfrac{92.0 \times 10^6\ \text{s}^{-1}}{2.998 \times 10^8\ \text{m s}^{-1}} = 0.307\ \text{m}^{-1} = \boxed{3.07 \times 10^{-3}\ \text{cm}^{-1}}$.

(b) $\lambda = \dfrac{1}{\tilde{\nu}} = \dfrac{1}{0.307\ \text{m}^{-1}} = \boxed{3.26\ \text{m}}$.

19.7 $\log \dfrac{I_0}{I} = \varepsilon\,[\text{J}]l$ [19.3]

$= 743\ \text{mol}^{-1}\ \text{dm}^3\ \text{cm}^{-1} \times (3.25 \times 10^{-3}\ \text{mol dm}^{-3}) \times 0.25\ \text{cm}$

$= 0.604.$

$\dfrac{I}{I_0} = 0.249.$

The reduction is $\boxed{75.1\%}$.

19.8 $\log T = -\varepsilon[\text{J}]l$ [19.3 and 19.4].

Solve for ε,

$$\begin{aligned}\varepsilon &= \frac{1}{[\mathrm{J}]} \log T \\ &= -\frac{\log 0.715}{4.33 \times 10^{-4}\,\mathrm{mol\,dm^{-3}} \times 0.25\,\mathrm{cm}} \\ &= \boxed{1.35 \times 10^{3}\,\mathrm{dm^{3}\,mol^{-1}\,cm^{-1}}}.\end{aligned}$$

19.9 $\mathrm{d}I = -k[\mathrm{J}]_0 \mathrm{e}^{-x/\lambda} I\,\mathrm{d}x.$

$$\frac{\mathrm{d}I}{I} = -k[\mathrm{J}]_0 \mathrm{e}^{-x/\lambda}.$$

$$\int_{I_0}^{I} \frac{\mathrm{d}I}{I} = -k[\mathrm{J}]_0 \int_0^l \mathrm{e}^{-x/\lambda}.$$

$$\begin{aligned}\ln\frac{I}{I_0} &= -k[\mathrm{J}]_0(-\lambda \mathrm{e}^{-x/\lambda})\Big|_0^l \\ &= k\lambda[\mathrm{J}]_0(\mathrm{e}^{-l/\lambda} - 1).\end{aligned}$$

The exponential can be expanded

$$\mathrm{e}^{-l/\lambda} = 1 - \frac{l}{\lambda} + \frac{l^2}{\lambda^2} + \ldots.$$

If $l \ll \lambda$, we need only the first two terms; then

$$\ln\frac{I}{I_0} = k\lambda[\mathrm{J}]_0\left(-\frac{l}{\lambda}\right) = -k[\mathrm{J}]_0 l.$$

$$\log\frac{I}{I_0} = -\varepsilon[\mathrm{J}]_0 l.$$

which is the same as eqn 19.3.

If $l \gg \lambda$, then $\mathrm{e}^{-l/\lambda}$ can be safely ignored relative to -1 and we obtain

$$\ln\frac{I}{I_0} = -k\lambda[\mathrm{J}]_0 \quad \text{or} \quad \boxed{\log\frac{I}{I_0} = -\varepsilon\lambda[\mathrm{J}]_0},$$

which is independent of the length of the tube.

19.10 $\varepsilon = -\dfrac{1}{[\mathrm{J}]l}\log\dfrac{I}{I_0}$ with $l = 0.20\,\mathrm{cm}$.

We use this formula to draw up the following table.

$[Br_2]/(mol\,dm^{-3})$	0.0010	0.0050	0.0100	0.0500	
I/I_0	0.814	0.356	0.127	3.0×10^{-5}	
$\varepsilon/(dm^3\,mol^{-1}\,cm^{-1})$	447	449	448	452	mean :$44\overline{9}$

Hence, the molar absorption coefficient is $\varepsilon = \boxed{450\ \mathrm{dm^3\ mol^{-1}\ cm^{-1}}}$.

19.11 $\log \dfrac{I}{I_0} = -\varepsilon(\mathrm{J})l.$

Solve for ε; then

$$\varepsilon = \frac{1}{[\mathrm{J}]l}\log\frac{I}{I_0} = \frac{-1}{(0.010\,\mathrm{mol\,dm^{-3}}) \times (0.20\,\mathrm{cm})}\log 0.48 = \boxed{159\,\mathrm{dm^3\,mol^{-1}\,cm^{-1}}}.$$

$$T = \frac{I}{I_0} = 10^{-[\mathrm{J}]\varepsilon l}$$

$$= 10^{\left(-0.010\,\mathrm{mol\,dm^{-3}}\right)\times(159\,\mathrm{dm^3\,mol^{-1}\,cm^{-1}})\times(0.40)} = 10^{-0.63\overline{6}} = 0.23, \text{or} \boxed{23 \text{ per cent}}.$$

19.12 $l = \dfrac{-1}{\varepsilon[\mathrm{J}]}\log\dfrac{I}{I_0}.$

For water, $[H_2O] \approx \dfrac{1.00\,\mathrm{kg/dm^3}}{18.02\,\mathrm{g\,mol^{-1}}} = 55.5\,\mathrm{mol\,dm^{-3}}$

and $\varepsilon[\mathrm{J}] = (55.5\ \mathrm{M}) \times (6.2 \times 10^{-5}\ \mathrm{M^{-1}\ cm^{-1}}) = 3.4 \times 10^{-3}\ \mathrm{cm^{-1}} = 0.34\ \mathrm{m^{-1}}$,

so $\dfrac{1}{\varepsilon[\mathrm{J}]} = 2.9\,\mathrm{m}.$

Hence, $l/\mathrm{m} = -2.9 \times \log\dfrac{I}{I_0}.$

(a) $\dfrac{I}{I_0} = 0.5, \quad l = -2.9\,\mathrm{m} \times \log 0.5 = \boxed{0.9\,\mathrm{m}}.$

(b) $\dfrac{I}{I_0} = 0.1, \quad l = -2.9\,\mathrm{m} \times \log 0.1 = \boxed{3\,\mathrm{m}}.$

19.13 55 mph = $24.6\ \mathrm{m\ s^{-1}}$.

We use $\nu' = \left(\dfrac{1 - s/c}{1 + s/c}\right)^{1/2}\nu \quad [19.5a]$

or $\lambda' = \left(\dfrac{1 - s/c}{1 + s/c}\right)^{1/2}\lambda$

$$= \left(\frac{1 - 24.6/2.998 \times 10^8}{1 + 24/2.998 \times 10^8}\right)^{1/2}\lambda$$

$$= \boxed{0.999\,999\,918 \times 660\,\mathrm{nm}}.$$

The formula for λ' can be rewritten after expanding the numerator and denominator

$$(1 \pm x)^{1/2} = 1 \pm \frac{x}{2} + \ldots \text{Then}$$

$$\lambda' \approx (1 - s/c)\lambda.$$

Now solve for s such that $\lambda' = 520$ nm.

$$s = \left(1 - \frac{\lambda'}{\lambda}\right) c$$

$$= \left(1 - \frac{520}{660}\right) \times 2.998 \times 10^{8}\ \text{m s}^{-1}$$

$$= \boxed{6.36 \times 10^{7}\ \text{m s}^{-1}} = 2.3 \times 10^{8}\ \text{kph} = 1.4 \times 10^{8}\ \text{mph}.$$

19.14 The star is receding so we use

$$\nu' = \left(\frac{1 - s/c}{1 + s/c}\right)^{1/2} \nu$$

or

$$\lambda' = \left(\frac{1 + s/c}{1 - s/c}\right)^{1/2} \lambda$$

or $\lambda' \approx (1 + s/c)\lambda$.

Solve for s.

$$s = \left(\frac{\lambda'}{\lambda} - 1\right) c$$

$$= \left(\frac{706.5}{654.2} - 1\right) \times 2.998 \times 10^{8}\ \text{m s}^{-1} = \boxed{2.397 \times 10^{7}\ \text{m s}^{-1}}.$$

$$\delta\lambda = \frac{2\lambda}{c}\left(\frac{2kT}{m}\ln 2\right)^{1/2} \quad [19.6] \text{ which implies that}$$

$$T = \frac{m}{2k \ln 2}\left(\frac{c\delta\lambda}{2\lambda}\right)^{2}$$

$$= \frac{48 \times (1.6605 \times 10^{-27}\ \text{kg})}{2 \times (1.381 \times 10^{-23}\ \text{J K}^{-1}) \times \ln 2}\left(\frac{2.998 \times 10^{8}\ \text{m s}^{-1} \times (61.8 \times 10^{-12}\ \text{m})}{2 \times (654.2 \times 10^{-9}\ \text{m}^{2})}\right)^{2}$$

$$= \boxed{8.4 \times 10^{5}\ \text{K}}.$$

19.15 $\delta\tilde{\nu} = \dfrac{5.3\ \text{cm}^{-1}}{\tau/\text{ps}}$ implying that $\tau = \dfrac{5.3\text{ps}}{\delta\tilde{\nu}/\text{cm}^{-1}}$.

(a) $\tau = \dfrac{5.3\text{ps}}{0.1} = \boxed{53\ \text{ps}}$.

(b) $\tau = \dfrac{5.3\text{ps}}{1} = \boxed{5\text{ ps}}$.

(c) $\lambda = \dfrac{2.998 \times 10^{8}\text{ m s}^{-1}}{1.0 \times 10^{9}\text{ s}^{-1}} = 0.300\text{ m} = 30.0\text{ cm}.$

$\tilde{\nu} = \dfrac{1}{\lambda} = 0.0333\text{ cm}^{-1}.$

$\tau = \dfrac{5.3}{0.0333} = \boxed{1.6 \times 10^{2}\text{ ps}}$.

19.16 $\delta\tilde{\nu} = \dfrac{5.3\text{ cm}^{-1}}{\tau/\text{ps}}.$

(a) $\tau \approx 1 \times 10^{-13}$ s = 0.1 ps, implying that $\boxed{\delta\tilde{\nu} = 53\text{ cm}^{-1}}$.

(b) $\tau \approx 200 \times (1 \times 10^{-13}$ s) = 20 ps, implying that $\boxed{\delta\tilde{\nu} = 0.27\text{ cm}^{-1}}$.

19.17 $E_J = hBJ(J+1) \approx hBJ^2$ for large J.

$$J = \left(\frac{E_J}{hB}\right)^{1/2}, \quad B = \frac{\hbar}{4\pi I}, \quad I = mR^2.$$

$$J = \frac{\pi R}{h}(8mE_J)^{1/2}$$

$$= \frac{\pi \times 0.07\text{ m}}{6.63 \times 10^{-34}\text{ J s}} \times (8 \times 0.75\text{ kg} \times 0.2\text{ J})^{1/2}$$

$$\approx \boxed{4 \times 10^{33}}.$$

19.18 (a) $I = \mu R^2, \quad \mu = \dfrac{m_A m_B}{m_A + m_B} = \dfrac{m}{2}$ for H_2.

$$I = \frac{m}{2}R^2 = \frac{1.0078 \times 1.661 \times 10^{-27}\text{ kg} \times (74.14 \times 10^{-12}\text{ m})^2}{2}$$

$$= \boxed{4.601 \times 10^{-48}\text{ kg m}^2}.$$

(b) $I = \dfrac{2.0140 \times 1.661 \times 10^{-27}\text{ kg} \times (74.15 \times 10^{-12}\text{ m})^2}{2} = \boxed{9.196 \times 10^{-48}\text{ kg m}^2}$.

(c) For a linear AB_2 molecule, $I = 2m_B R^2$.

$$I = 2 \times 15.9949 \times 1.661 \times 10^{-27}\text{ kg} \times (112 \times 10^{-12}\text{ m})^2$$

$$= \boxed{6.67 \times 10^{-46}\text{ kg m}^2}.$$

(d) $I = \boxed{6.67 \times 10^{-46}\text{ kg m}^2}$.

19.19 Unit of B = unit of $\left(\dfrac{\hbar}{4\pi I}\right) = \dfrac{\text{J s}}{\text{kg m}^2} = \dfrac{\text{kg m}^2\text{ s}^{-2}\text{ s}}{\text{kg m}^2} = \text{s}^{-1}.$

So we see that the unit of B is $s^{-1} = Hz$

(a) $B = \dfrac{1.0546 \times 10^{-34}\,\text{J s}}{4\pi \times 4.599 \times 10^{-48}\,\text{kg m}^2} = \boxed{1.825 \times 10^{12}\,\text{Hz}}$.

(b) $B = \dfrac{1.0546 \times 10^{-34}\,\text{J s}}{4\pi \times 9.194 \times 10^{-48}\,\text{kg m}^2} = \boxed{9.128 \times 10^{11}\,\text{Hz}}$.

(c) $B = \dfrac{1.0546 \times 10^{-34}\,\text{J s}}{4\pi \times 6.67 \times 10^{-46}\,\text{kg m}^2} = \boxed{1.26 \times 10^{10}\,\text{Hz}}$.

(d) $B = \boxed{1.26 \times 10^{10}\,\text{Hz}}$.

19.20 (a) An octahedral AB_6 molecule is a spherical rotor and all moments of inertia are equal.

$$I = 4m_B R^2.$$

(b) $B = \dfrac{\hbar}{4\pi I} = \dfrac{1.0546 \times 10^{-34}\,\text{J s}}{4\pi \times 4 \times 31.97 \times 1.661 \times 10^{-27}\,\text{kg} \times (158 \times 10^{-12}\,\text{m})^2}$

$= \boxed{1.583 \times 10^9\,\text{Hz}}$.

19.21 $\boxed{I_{\parallel} = 4mR^2 \text{ and } I_{\perp} = 2mR^2}$ See the definitions in Figure 19.1.

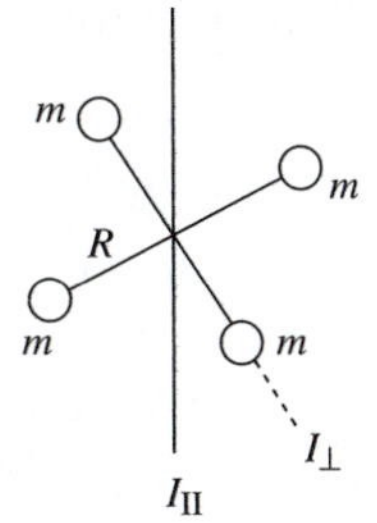

Fig 19.1

19.22 (a) SO_3 is a trigonal planar molecule. $I_{\parallel}$ is perpendicular to the plane of the molecule and along the vertical axis that passes through the S atom. $I_{\perp}$ is shown in the Figure 19.2. The bond angles are 120°.

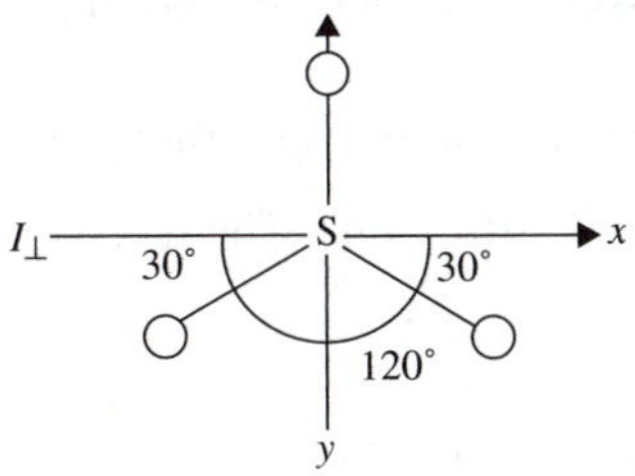

Fig 19.2

$$I_{\parallel} = 2m_{\mathrm{O}}R^2(1 - \cos\theta), \quad \theta = 120°.$$

$$I_{\perp} = m_{\mathrm{O}}R^2(1 - \cos\theta).$$

$$I_{\parallel} = 2 \times 15.995 \times 1.661 \times 10^{-27}\,\mathrm{kg} \times (143 \times 10^{-12}\,\mathrm{m})^2 \times (1 - \cos 120°)$$

$$= 1.629 \times 10^{-45}\,\mathrm{kg\,m^2}.$$

$$A = \frac{\hbar}{4\pi I_{\parallel}} = \frac{1.0546 \times 10^{-34}\,\mathrm{J\,s}}{4\pi \times 1.629 \times 10^{-45}\,\mathrm{kg\,m^2}} = \boxed{5.152 \times 10^9\,\mathrm{Hz}}.$$

$$I_{\perp} = \frac{1}{2}I_{\parallel} = 8.147 \times 10^{-46}\,\mathrm{kg\,m^2}.$$

$$B = \frac{\hbar}{4\pi I_{\parallel}} = 1.030 \times 10^{10}\,\mathrm{Hz}.$$

(b) The mass of the sulfur atom does not affect the rotational constants, so microwave spectroscopy could not be used to distinguish between $^{32}S^{16}O_3$ and $^{33}S^{16}O_3$.

19.23 Polar molecules show a pure rotational spectrum. Therefore, select the polar molecules based on their well-known structures.

(a), (b), (c), and (d) have dipole moments and will have a pure rotational spectrum. (e) will not.

19.24 None of the molecules listed are spherical rotors; therefore all will show a rotational Raman spectrum.

19.25 Methane is a spherical rotor; hence $I_{\parallel} = I_{\perp}$ and $A = B$ and

$$E_{\mathrm{J,K}} = hcBJ(J+1)$$

and there is one energy level with $J = 10$.

COMMENT There is a degeneracy of $2J + 1 = 21$ associated with this level, so there is a total of 21 quantum states.

19.26 $E_{J,K} = hBJ(J+1) + h(A - B)K^2$ [19.11].

$J = 0, 1, 2 \ldots. \quad K = J, J-1, \ldots, -J \quad$ ($2J + 1$ possible K values for a given J)

There are 21 states for $J = 10$ but, as K^2 determines the energy, there are only 11 energy levels, and only one, $K = 0$, will be at energy $hBJ(J+1)$. The rest will be at higher ($K > 0$) or lower ($K < 0$) energies.

19.27 For a spherical rotor, $P_J \propto (2J+1)^2 e^{-hcBJ(J+1)/kT}$ $[g_J = (2J+1)^2]$

and the greatest population occurs when

$$\frac{\mathrm{d}P_J}{\mathrm{d}J} \propto \left(8J + 4 - \frac{hcB(2J+1)^3}{kT}\right) \mathrm{e}^{-hcBJ(J+1)/kT} = 0$$

which occurs when

$$4(2J+1) = \frac{hcB(2J+1)^3}{kT} \quad \text{or at} \quad J_{\max} = \boxed{\left(\frac{kT}{hcB}\right)^{1/2} - \frac{1}{2}}.$$

19.28 $E_J = hBJ(J+1)$.

$$\Delta E = E_{J+1} - E_J = hB(J+1)(J+2) - hBJ(J+1)$$
$$= 2hB(J+1).$$

The separations of the lines are therefore

$2hB, 4hB, 6hB.$

(a) The frequencies of the lines are

$2B, 4B, 6B, \ldots$

$= \boxed{636\,\text{GHz}, 1272\,\text{GHz}, 1908\,\text{GHz}\ldots}.$

(b) The wavenumbers of the lines are given by

$$\tilde{\nu} = \frac{\nu}{c} = \frac{636 \times 10^9\,\text{s}^{-1}}{2.998 \times 10^8\,\text{m}\,\text{s}^{-1}} = 2121\,\text{m}^{-1} = \boxed{21.21\,\text{cm}^{-1}}.$$

So the set of lines corresponds to

$\boxed{21.21\,\text{cm}^{-1}, 42.42\,\text{cm}^{-1}, 63.63\,\text{cm}^{-1}, \ldots}.$

19.29 (a) $\Delta(\Delta E) = \Delta E_{J+1\to J+3} - \Delta E_{J\to J+2}$ is the separation between lines in the rotational Raman spectrum because the selection rule is $\Delta J = \pm 2$. Since $\Delta E_{J\to J+2} = 2hB(2J+3)$ [19.18].

$$\Delta(\Delta E) = 2hB\{[2(J+1)+3] - [2J+3]\} = 4hB.$$

$$\Delta(\Delta E)/h = 4B = 4(318.0\,\text{GHz}) = \boxed{1.272\,\text{THz}}.$$

(b) $$\Delta(\Delta E)/hc = \frac{4B}{c} = \frac{1.272 \times 10^{12}\,\text{s}^{-1}}{2.998 \times 10^8\,\text{m}\,\text{s}^{-1}} \left(\frac{10^{-2}\,\text{m}}{1\,\text{cm}}\right) = \boxed{42.4\,\text{cm}^{-1}}.$$

19.30 $$B = \frac{\hbar}{4\pi I}, \quad \nu = 2B, \quad \tilde{\nu} = \frac{2B}{c}.$$

$$I = \mu R^2, \quad \mu = \frac{m_{\mathrm{H}} m_{\mathrm{Cl}}}{m_{\mathrm{H}} + m_{\mathrm{Cl}}}.$$

If m_{H} increases as it does when ^{2}H replaces ^{1}H, then μ increases. When μ increases. I increases. When I increases, B decreases, and hence $\tilde{\nu}$ $\boxed{\text{decreases (shifts to lower wavenumber)}}$.

19.31 $B = \dfrac{\hbar}{4\pi I} = 11.70\,\text{GHz} = 11.70 \times 10^9\,\text{s}^{-1}.$

$I = 2m_0R^2.$

Solve for R^2 in terms of B.

$$R^2 = \frac{\hbar}{8\pi m_0 B}.$$

$$R^2 = \frac{1.0546 \times 10^{-34}\,\text{Js}}{8\pi \times 15.995 \times 1.66054 \times 10^{-27}\,\text{kg} \times 11.70 \times 10^9\,\text{s}^{-1}}$$

$$= 1.350 \times 10^{-20}\,\text{m}^2.$$

$$R = 1.162 \times 10^{-10}\,\text{m} = \boxed{116.2\,\text{pm}}.$$

19.32 $2B = 384\,\text{GHz} =$ separation of lines.

$$B = 192\,\text{GHz}$$

$$B = \frac{\hbar}{4\pi I} = \frac{\hbar}{4\pi\mu R^2}.$$

$$R^2 = \frac{\hbar}{4\pi\mu B}.$$

$$\mu(^1\text{H}^{127}\text{I}) = \frac{1.0078 \times 126.9045}{1.0078 + 126.9045} \times 1.66054 \times 10^{-27}\,\text{kg}$$

$$= 1.6603 \times 10^{-27}\,\text{kg}.$$

$$R^2 = \frac{1.05457 \times 10^{-34}\,\text{J s}}{4\pi \times 1.6603 \times 10^{-27}\,\text{kg} \times 192 \times 10^9\,\text{s}^{-1}}$$

$$= 2.632 \times 10^{-20}\,\text{m}^2.$$

$$R = 1.622 \times 10^{-10}\,\text{m} = \boxed{162.2\,\text{pm}}.$$

$$\mu(^2\text{H}^{127}\text{I}) = 3.2921 \times 10^{-21}\,\text{kg}.$$

$$B = \frac{\hbar}{4\pi\mu R^2} = \frac{1.05457 \times 10^{-34}\,\text{J s}}{4\pi \times 3.2921 \times 10^{-27}\,\text{kg} \times 2.632 \times 10^{-20}\,\text{m}^2}$$

$$= 96.9 \times 10^9\,\text{s}^{-1} = 969.9\,\text{GHz}.$$

$$2B = \boxed{194\,\text{GHz}} = \text{separation of lines}.$$

19.33 Since the spectrum is reported in wavenumber units (cm^{-1}), it is convenient to use this unit throughout the analysis. Let $\tilde{\nu}_J = \nu_J/c$ [19.2b], $\tilde{B} = B/c$, and $\tilde{D} = D/c$. With these definitions eqn 19.17 becomes

$$\frac{\tilde{\nu}_J}{J+1} = 2\tilde{B} - 4\tilde{D}(J+1)^2.$$

This equation indicates that a plot of $\frac{\tilde{\nu}_J}{J+1}$ against $(J+1)^2$ should be linear with intercept equal to $2\tilde{B}$ and slope equal to $-4\tilde{D}$. The transition assignments are determined by guessing assignments and checking that the plot is linear. For example, guessing the lowest energy observed line has the assignment $J = 0 \rightarrow 1$ and subsequent lines have the assignments $J = 1 \rightarrow 2, J = 2 \rightarrow 3$, and $J = 3 \rightarrow 4$ yields a very non-linear plot. The assignments $J = 2 \rightarrow 3, J = 3 \rightarrow 4, J = 4 \rightarrow 5$, and $J = 5 \rightarrow 6$ also yield a non-linear plot. Only the assignments $J = 1 \rightarrow 2, J = 2 \rightarrow 3, J = 3 \rightarrow 4$, and $J = 4 \rightarrow 5$ yield a highly linear plot (see Figure 19.3). The intercept of the linear plot equals 0.405704 $\mathrm{cm^{-1}}$ and the slope equals $-2.493 \times 10^{-7}\,\mathrm{cm^{-1}}$.

$$\tilde{B} = \frac{\text{intercept}}{2} = \frac{0.405704\,\mathrm{cm^{-1}}}{2} = 0.202852\,\mathrm{cm^{-1}}.$$

$$\tilde{D} = -\frac{\text{slope}}{4} = -\frac{(-2.493 \times 10^{-7}\,\mathrm{cm^{-1}})}{4} = 6.2 \times 10^{-8}\,\mathrm{cm^{-1}}.$$

$$B = c\tilde{B} = (2.997925 \times 10^{10}\,\mathrm{cm\,s^{-1}}) \times (0.405704\,\mathrm{cm^{-1}}) = \boxed{1.21627 \times 10^{10}\,\mathrm{Hz}}.$$

$$D = c\tilde{D} = (2.997925 \times 10^{10}\,\mathrm{cm\,s^{-1}}) \times (6.2 \times 10^{-8}\,\mathrm{cm^{-1}}) = \boxed{1.96 \times 10^{3}\,\mathrm{Hz}}.$$

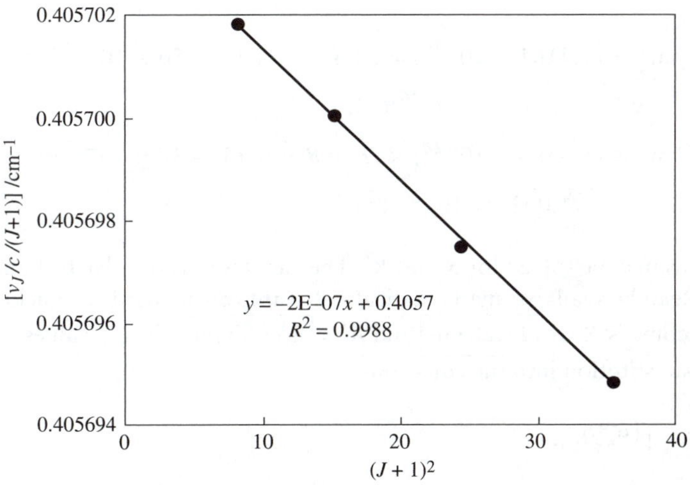

Fig 19.3

19.34 From the equation for a linear rotor in Table 19.1 it is possible to show that

$$I_m = m_A m_C (R + R')^2 + m_A m_B R^2 + m_B m_C R'^2$$

where $R = R_{AB} = R_{OC}$ and $R' = R_{BC} = R_{CS}$

Thus,

$$I(^{16}\text{O}^{12}\text{C}^{32}\text{S}) = \left(\frac{m(^{16}\text{O})m(^{32}\text{S})}{m(^{16}\text{O}^{12}\text{C}^{32}\text{S})}\right) \times (R+R')^2 + \left(\frac{m(^{12}\text{C})\{m(^{16}\text{O})R^2 + m(^{32}\text{S})R'^2\}}{m(^{16}\text{O}^{12}\text{C}^{32}\text{S})}\right),$$

$$I(^{16}\text{O}^{12}\text{C}^{34}\text{S}) = \left(\frac{m(^{16}\text{O})m(^{34}\text{S})}{m(^{16}\text{O}^{12}\text{C}^{34}\text{S})}\right) \times (R+R')^2 + \left(\frac{m(^{12}\text{C})\{m(^{16}\text{O})R^2 + m(^{34}\text{S})R'^2\}}{m(^{16}\text{O}^{12}\text{C}^{34}\text{S})}\right),$$

$m(^{16}\text{O}) = 15.9949\,\text{u}$, $m(^{12}\text{C}) = 12.0000\,\text{u}$, $m(^{32}\text{S}) = 31.9721\,\text{u}$
and $m(^{34}\text{S}) = 33.9679\,\text{u}$. Hence,

$$I(^{16}\text{O}^{12}\text{C}^{32}\text{S})/\text{u} = (8.5279) \times (R+R')^2 + (0.20011) \times (15.9949R^2 + 31.9721R'^2),$$

$$I(^{16}\text{O}^{12}\text{C}^{34}\text{S})/\text{u} = (8.7684) \times (R+R')^2 + (0.19366) \times (15.9949R^2 + 33.9679R'^2).$$

The spectral data provide the experimental values of the moments of inertia based on the relation $\nu = 2B(J+1)$ [19.8] with $B = \dfrac{\hbar}{4\pi I}$ [19.9]. These values are set equal to the above equations which are then solved for R and R'. The mean values of I obtained from the data are

$$I(^{16}\text{O}^{12}\text{C}^{32}\text{S}) = 1.37998 \times 10^{-45}\,\text{kg m}^2,$$

$$I(^{16}\text{O}^{12}\text{C}^{34}\text{S}) = 1.41460 \times 10^{-45}\,\text{kg m}^2.$$

Therefore, after conversion of the atomic mass units to kg, the equations we must solve are

$$1.37998 \times 10^{-45}\,\text{m}^2 = (1.4161 \times 10^{-26}) \times (R+R')^2 + (5.3150 \times 10^{-27}R^2) + (1.0624 \times 10^{-26}R'^2),$$

$$1.41460 \times 10^{-45}\,\text{m}^2 = (1.4560 \times 10^{-26}) \times (R+R')^2 + (5.1437 \times 10^{-27}R^2) + (1.0923 \times 10^{-26}R'^2).$$

These two equations may be solved for R and R'. They are tedious to solve by hand, but straightforward. Readily available mathematical software can be used to quickly give the result. The outcome is $R = \boxed{116.28\,\text{pm}}$ and $R' = \boxed{155.97\,\text{pm}}$. These values may be checked by direct substitution into the equations.

19.35 $\nu = \dfrac{1}{2\pi}\left(\dfrac{k}{\mu}\right)^{1/2}$ [19.20b].

(a) $\mu = \dfrac{m_\text{C} m_\text{O}}{m_\text{C} + m_\text{O}} = \dfrac{12.0000 \times 15.9949}{12.0000 + 15.9949} \times 1.66054 \times 10^{-27}\,\text{kg}$

$= 1.1385 \times 10^{-26}\,\text{kg}.$

$$\nu = \frac{1}{2\pi}\left(\frac{908\,\mathrm{N\,m^{-1}}}{1.1385 \times 10^{-26}\,\mathrm{kg}}\right)^{1/2}$$

$$= 4.49 \times 10^{13}\,\mathrm{s^{-1}} = \boxed{4.49 \times 10^{13}\,\mathrm{Hz}}.$$

(b) $\mu = 1.1910 \times 10^{-26}$ kg.

$$\nu = 4.39 \times 10^{13}\,\mathrm{s^{-1}} = \boxed{4.39 \times 10^{13}\,\mathrm{Hz}}.$$

19.36 $$\tilde{\nu} = \frac{\nu}{c} = \frac{1}{2\pi c}\left(\frac{k}{\mu}\right)^{1/2}, \quad \mu = \frac{1}{2}m\,(^{35}\mathrm{Cl}).$$

Solve the above for k

$$k = (2\pi c\tilde{\nu})^2\mu = (2\pi c\tilde{\nu})^2 \times \frac{1}{2}m\,(^{35}\mathrm{Cl})$$

$$= (2\pi \times 2.998 \times 10^8\,\mathrm{m\,s^{-1}} \times 5.65 \times 10^4\,\mathrm{m^{-1}})^2 \times \frac{1}{2} \times 34.9688 \times 1.66054 \times 10^{-27}\,\mathrm{kg}$$

$$= \boxed{329\,\mathrm{N\,m^{-1}}}.$$

19.37 As shown in the solution to Exercise 19.36,

$$k = (2\pi c\tilde{\nu})^2\mu.$$

$$\mu(\mathrm{HF}) = \frac{1.0078 \times 18.9908}{1.0078 + 18.9908}\,\mathrm{u} = 0.9570\,\mathrm{u}.$$

$$\mu(\mathrm{H^{35}C1}) = \frac{1.0078 \times 34.9688}{1.0078 + 34.9688}\,\mathrm{u} = 0.9796\,\mathrm{u}.$$

$$\mu(\mathrm{H^{81}Br}) = \frac{1.0078 \times 80.9163}{1.0078 + 80.9163}\,\mathrm{u} = 0.9954\,\mathrm{u}.$$

$$\mu(\mathrm{H^{127}I}) = \frac{1.0078 \times 126.9045}{1.0078 + 126.9045}\,\mathrm{u} = 0.9999\,\mathrm{u}.$$

Using the above equation draw up the following table.

	HF	HCl	HBr	HI
$\tilde{\nu}$	4141.3	2988.9	2649.7	2309.5
μ/u	0.9570	0.9796	0.9954	0.9999
$k/(\mathrm{N\,m^{-1}})$	967.1	515.6	411.8	314.2

19.38 Form $\tilde{\nu} = \frac{1}{2\pi c}\left(\frac{k}{\mu}\right)^{1/2}$ with the values of k from Exercise 19.37 and the reduced masses, for example:

$$\mu(^2\text{HF}) = \frac{2.0141 \times 18.9908}{2.0141 + 18.9908}\,\text{u} = 1.8210\,\text{u} \text{ and similarly for the other halides.}$$

Draw up the following table.

	^{2}HF	^{2}HCl	^{2}HBr	^{2}HI
$\tilde{\nu}/\text{cm}^{-1}$	3002	2144	1886	1640
μ/u	1.8210	1.9044	1.9652	1.9826
$k/(\text{N m}^{-1})$	967.1	515.6	411.8	314.2

19.39 The R branch obeys the relation

$\tilde{\nu}_R(J) = \tilde{\nu} + 2B(J+1)$ (See Section 19.14).

$\tilde{\nu} = 2648.98\,\text{cm}^{-1}$ (See Exercise 19.37).

$I = \mu R^2$ [Table 19.1]

$$= \left(\frac{1.0078 \times (126.9045)\,\text{g mol}^{-1}}{1.0078 + 126.9045}\right) \times \left(\frac{1\,\text{kg}}{10^3\,\text{g}}\right) \times \left(\frac{1\,\text{mol}}{6.022 \times 10^{23}}\right)$$

$$\times \left(141 \times 10^{-12}\,\text{m}\right)^2 (\text{Table 19.2})$$

$$= 3.30 \times 10^{-47}\,\text{kg m}^2.$$

$$B = \frac{\hbar}{4\pi c I}\,[19.9] = \frac{1.054 \times 10^{-34}\,\text{J s}}{4\pi(2.998 \times 10^{10}\,\text{cm s}^{-1}) \times (3.30 \times 10^{-47}\,\text{kg m}^2)} = 8.48\,\text{cm}^{-1}.$$

Hence, $\tilde{\nu}_R(2) = \tilde{\nu} + 6B = (2648.98) + (6) \times (8.48\,\text{cm}^{-1}) = \boxed{2699.86\,\text{cm}^{-1}}$.

19.40 Select those molecules in which a vibration gives rise to a change in dipole moment. It is helpful to write down the structural formulas of the compounds.

The molecules that show infrared absorption are

(b) HCl, (c) CO_2, (d) H_2O, (e) CH_3CH_3, (f) CH_4, and (g) CH_3Cl.

19.41 For non-linear molecules, $3N - 6$.

For linear molecules, $3N - 5$.

We need to establish the linearity of the molecules listed. (c) and (d) are clearly non-linear. From the Lewis structures of (a) and (b) and VSEPR we decide that (a) is non-linear, and (b) is linear

(a) $3N - 6 = 9 - 6 = \boxed{3}$.

(b) $3N - 5 = 9 - 5 = \boxed{4}$.

(c) $C_6H_{12}, 3N - 6 = 3 \times 18 - 6 = \boxed{48}$.

(d) $C_6H_{14}, 3N - 6 = 3 \times 20 - 6 = \boxed{54}$.

19.42 The uniform expansion mode is depicted in Figure 19.4. Benzene is centrosymmetric and it has a center of inversion. Consequently, the exclusion rule applies (Section 19.15). The mode is $\boxed{\text{infrared inactive}}$ (symmetric breathing leaves the molelcular dipole moment unchanged at zero), and therefore the mode may be $\boxed{\text{Raman active}}$ (and is).

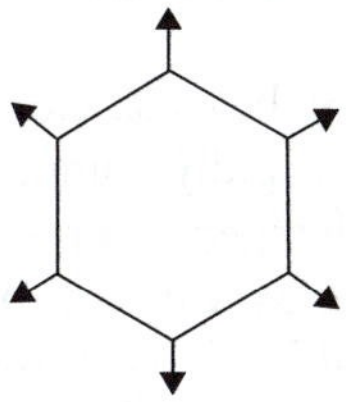

Fig 19.4

Chapter 20

Electronic transitions and photochemistry

Answers to discussion questions

20.1 The Franck–Condon principle states that because electrons are so much lighter than nuclei an electronic transition occurs so rapidly compared to vibrational motions that the internuclear distance is relatively unchanged as a result of the transition. This implies that the most probable transitions $\nu_f \leftarrow \nu_i$ are vertical. This vertical line will, however, intersect any number of vibrational levels ν_f in the upper electronic state. Hence transitions to many vibrational states of the excited state will occur with transition probabilities proportional to the Frank–Condon factors, which are in turn proportional to the overlap integral of the wavefunctions of the initial and final vibrational states. A vibrational progression is observed, the shape of which is determined by the relative horizontal positions of the two electronic potential energy curves. The most probable transitions are those to excited vibrational states with wavefunctions having a large amplitude at the internuclear position R_e.

Q **QUESTION** You might check the validity of the assumption that electronic transitions are so much faster than vibrational transitions by calculating the time scale of the two kinds of transitions. How much faster is the electronic transition, and is the assumption behind the Franck–Condon principle justified?

20.2 Color can arise by emission, absorption, or scattering of electromagnetic radiation by an object. Many molecules have electronic transitions that have wavelengths in the visible portion of the electromagnetic spectrum. When a substance emits radiation the perceived color of the object will be that of the emitted radiation and it may be an additive color resulting from the emission of more than one wavelength of radiation. When a substance absorbs radiation its color is determined by the subtraction of those wavelengths from white light. For example, absorption of red light results in the object being perceived as green. Scattering, including the diffraction that occurs when light falls on a material with a grid of variation in texture or refractive index having dimensions comparable to the wavelength of light, for example, a bird's plumage, may also form color.

20.3 The transition dipole moment, $\boldsymbol{\mu}_{fi}$, between states i and f determines the intensity of a transition. If it has a high absolute value, the transition probability and intensity are high (see Section 19.3). We need to determine how it depends on the length of the chain. We assume that wavefunctions of the conjugated electrons in the linear polyene can be approximated by the wavefunctions of a particle in a one-dimensional box. Then, for a transition between the states n' and n,

$$\mu_x = -e\int_0^L \psi_{n'}(x)\, x\, \psi_n(x)\mathrm{d}x, \qquad \psi_n = \left(\frac{2}{L}\right)^{1/2} \sin\left(\frac{n\pi x}{L}\right)$$

$$= -\frac{2e}{L}\int_0^L x\sin\left(\frac{n'\pi x}{L}\right)\sin\left(\frac{n\pi x}{L}\right)\mathrm{d}x$$

$$= \begin{cases} 0 & \text{if } n' = n+2 \\ \left(\dfrac{8eL}{\pi^2}\right)\dfrac{n(1+1)}{(2n+1)^2} & \text{if } n' = n+1. \end{cases}$$

Thus, the selection rule for radiation absorption is $\Delta n = +1$ and the transition integral is proportional to L.

Longer lengths of the dye's conjugated electronic π system are expected to yield greater absorption intensity.

To examine the effect that changing the length has on the apparent color, consider the energy of the absorption.

$$h\nu = E_{n+1} - E_n = \frac{(n+1)^2 h^2}{8m_e L^2} - \frac{n^2 h^2}{8m_e L^2} = (2n+1)\frac{h^2}{8m_e L^2}.$$

Therefore, since L appears in the denominator, increasing the length L of the polyene shifts the absorption to lower energy. This is a blue shift.

20.4 A typical n-to-π^* transition at about 4 eV (32000 cm^{-1}) characterizes an absorption in which a carbonyl lone pair electron on oxygen is excited to a π^* orbital while an unconjugated alkene exhibits a π-to-π^* electron transition at about 7 eV (56000 cm^{-1}). Thus, the 30000 cm^{-1} absorption of $CH_3CH{=}CHCHO$ is a n-to-π^* transition and the 46950 cm^{-1} absorption is a π-to-π^* transition, which lies lower than the norm because of conjugation between the alkene and carbonyl π bonds.

20.5 Tryptophan (Trp) and tyrosine (Tyr) show the characteristic absorption of a phenyl group at about 280 nm. Cysteine (Cys) and glycine (Gly) lack the phenyl group as is evident from their spectra.

20.6 There are three isosbestic wavelengths (or wavenumbers). The presence of two or more isosbestic points is good evidence that only two solutes in equilibrium with each other are present. The solutes here being $\mathrm{Her(CNS)_8}$ and $\mathrm{Her(OH)_8}$.

20.7 The fluorescence spectrum gives the vibrational splitting of the lower state. The wavelengths stated correspond to the wavenumbers 22 730, 24 390, 25 640, 27 030 cm^{-1}, indicating spacings of 1660, 1250, and 1390 cm^{-1}. The absorption spectrum spacing gives the separation of the vibrational levels of the upper state. The wavenumbers of the absorption peaks are 27 800, 29 000, 30 300, and 32 800 cm^{-1}. The vibrational spacings are therefore 1200, 1300, and 2500 cm^{-1}.

20.8 See Section 20.6 for a detailed description of both the theory and experiment involved in laser action. Here we restrict our discussion to only the most fundamental concepts. The basic requirement for a laser is that it has at least three energy levels. Of these levels, the highest lying state must be capable of being efficiently populated above its thermal equilibrium value by a pulse of radiation. A second state, lower in energy, must be a metastable state with a long enough lifetime for it to accumulate a population greater than its thermal equilibrium value by spontaneous transitions from the higher overpopulated state.

The metastable state must than be capable of undergoing stimulated transitions to a third lower lying state. This last requirement implies not only that the metastable state must have more than its thermal equilibrium population, but also that it must have a higher population than the third lower lying state, namely, that it achieve population inversion. The amplification process occurs when low intensity radiation of frequency equal to the transition frequency between the metastable state and the lower lying state stimulates the transition to the lower lying state and many more photons (higher intensity of the radiation) of that frequency are created. Examples of practical lasers are featured in Figures 20.20–20.24 of the text.

Two important applications of lasers in chemistry have been to Raman spectroscopy and to the development of time-resolved spectroscopy. Prior to the invention of lasers the source of intense monochromatic radiation required for Raman spectroscopy was a large spiral discharge tube with liquid mercury electrodes. The intense heat generated by the large current required to produce the radiation had to be dissipated by clumsy water-cooled jackets and exposures of several weeks were sometimes necessary to observe the weaker Raman lines. These problems have been eliminated with the introduction of lasers as the source of the required monochromatic radiation. As a consequence, Raman spectroscopy has been revitalized and is now almost as routine as infrared spectroscopy. See Section 19.1(b). Time-resolved laser spectroscopy can be used to study the dynamics of chemical reactions. Laser pulses are used to obtain the absorption, emission, and Raman spectrum of reactants, intermediates, products, and even transition states of reactions. When we want to study the rates at which energy is transferred from one mode to another in a molecule, we need femotosecond and picosecond pulses. These time-scales are available from mode-locked lasers and their development has opened up the possibility of examining the details of chemical reactions at a level that would have been unimaginable before.

20.9 The overall process associated with fluorescence involves the following steps. The molecule is first promoted from the vibrational ground state of a lower electronic level to a higher vibrational–electronic energy level by absorption of energy from a radiation field. Because of the requirements of the Franck–Condon principle, the transition is to excited vibrational levels of the upper electronic state. See Fig. 20.15 of the text. Therefore, the absorption spectrum shows a vibrational structure characteristic of the upper state. The excited state molecule can now lose energy to the surroundings through radiationless transitions and decay to the lowest vibrational level of the upper state. A spontaneous radiative transition now occurs to the lower electronic level and this fluorescence spectrum has a vibrational structure characteristic of the lower state. The fluorescence spectrum is not the mirror image of the absorption spectrum because the vibrational frequencies of the upper and lower states are different due to the difference in their potential energy curves.

The first steps of phosphorescence are the same as in fluorescence: a singlet state absorbs energy in a transition to an excited singlet state and non-radiative vibrational relaxation occurs within the excited singlet state. The presence of a heavy atom, which can provide angular momentum via strong spin–orbit coupling, and a triple state of energy similar to the excited singlet, provides the critical step needed for phosphorescence. It is the intersystem crossing step in which an electron undergoes a spin-flip from the excited singlet to the triplet state. The electron may remain in the triple state for an unusually long period because emission transition from a triple state to a single state is spin-forbidden.

20.10 (a) Vibrational energy spacings of the $\boxed{\text{lower}}$ state are determined by the spacing of the emission peaks marked A. From the spectrum, $\tilde{\nu} \approx 1800\,\text{cm}^{-1}$.

(b) Nothing can be said about the spacing of the upper state levels (without a detailed analysis of the intensities of the lines). For the second part of the question, we note that after some vibrational decay the benzophenone (which does absorb near 360 nm) can transfer its energy to naphthalene. The latter then emits the energy radiatively.

Solutions to exercises

20.11 $$[\text{J}] = \frac{n}{V} = \frac{0.0302\,\text{g}/(602\,\text{g}\,\text{mol}^{-1})}{0.500\,\text{dm}^3} = 1.00 \times 10^{-4}\,\text{mol}\,\text{dm}^{-3}.$$

(a) $A = \varepsilon[\text{J}]l \qquad [19.3].$

Solve for ε,

$$\varepsilon = \frac{A}{[\text{J}]l} = \frac{1.011}{1.00 \times 10^{-4}\,\text{mol}\,\text{dm}^{-3} \times 1.00\,\text{cm}}$$

$$= \boxed{1.01 \times 10^4\,\text{dm}^3\,\text{mol}^{-1}\,\text{cm}^{-1}}.$$

(b) $\log T = -\varepsilon[\text{J}]l$

$$= -1.01 \times 10^4\,\text{dm}^3\,\text{mol}^{-1}\,\text{cm}^{-1} \times 2.00 \times 10^{-4}\,\text{mol}\,\text{dm}^{-3} \times 1.00\,\text{cm}$$

$$= -2.015.$$

$$T = 0.00965 = \boxed{0.965\%}.$$

20.12 $T_\text{k} = T_\text{u}$,

$\log T_\text{k} = \log T_\text{u}$

$-\varepsilon[\text{k}]l_\text{k} = -\varepsilon[\text{u}]l_\text{u}.$

k = solution of known concentration, u = solution of unknown concentration.

Solve for [u].

$$[\text{u}] = [\text{k}]\left(\frac{l_\text{k}}{l_\text{u}}\right) = 25\,\mu\text{g}\,\text{dm}^{-3} \times \left(\frac{1.55\,\text{cm}}{1.18\,\text{cm}}\right) = \boxed{33\,\mu\text{g}\,\text{dm}^{-3}}.$$

20.13 $E = 5\text{eV} = 1.602 \times 10^{-19}\,\text{C} \times 5.0\,\text{V} = \boxed{8.0 \times 10^{-19}\,\text{J}}$ (or 5.0 eV).

20.14 Ionization energy $= 11.0\,\text{eV} = 1.76 \times 10^{-18}\,\text{J}$.

$$\text{Energy of photon} = h\nu = 6.63 \times 10^{-34}\,\text{J s} \times \frac{3.00 \times 10^{8}\,\text{ms}^{-1}}{1.00 \times 10^{-7}\,\text{m}}$$

$$= 1.99 \times 10^{-18}\,\text{J}.$$

The difference $E - I$ is the kinetic energy of the ejected electron.

$$E - I = \boxed{2.3 \times 10^{-19}\,J}.$$

20.15 $I = h\nu - \text{KE}$ [20.3] where KE is the kinetic energy.

$= 21.21\text{eV} - \text{KE}.$

$I = \boxed{10.20\,\text{eV},\ 12.98\,\text{eV, and } 15.99\,\text{eV}}.$

See Figure 20.1 for a sketch of the ionizations.

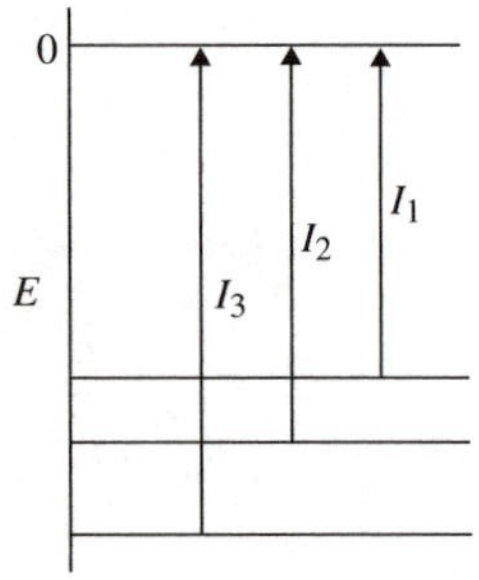

Fig 20.1

20.16 The lifetime of the unimolecular photochemical reaction is $\tau = \frac{1}{k} = \frac{1}{1.7 \times 10^{4}\,\text{s}^{-1}} = 5.9 \times 10^{-5}$ s. This is a much longer lifetime than that of fluorescence (1.0×10^{-9} s) so we conclude that the excited singlet state decays too rapidly to be the precursor of this photochemical reaction. The lifetime of phosphorescence (1.0×10^{-3} s) is longer than the reaction lifetime making the $\boxed{\text{triplet state}}$ the likely reaction precursor.

20.17 $A \rightarrow 2R,\ I_{abs}$,

$A + R \rightarrow R + B,\ k_2$,

$R + R \rightarrow R_2,\ k_3$,

$$\frac{d[A]}{dt} = \boxed{-I_{abs} - k_2\,[A][R]}, \qquad \frac{d[R]}{dt} = 2I_{abs} - 2k_3[R]^2 = 0.$$

The latter implies that $[R] = \left(\frac{I_{abs}}{k_3}\right)^{1/2}$, and so

$$\frac{d[A]}{dt} = \boxed{-I_{abs} - k_2\left(\frac{I_{abs}}{k_3}\right)^{1/2}[A]},$$

$$\frac{d[B]}{dt} = k_2[A][R] = k_2\left(\frac{I_{abs}}{k_3}\right)^{1/2}[A].$$

Therefore, only the combination $\frac{k_2}{k_3^{1/2}}$ may be determined if the reaction attains a steady state.

COMMENT If the reaction can be monitored at short enough times so that termination is negligible compared to initiation, then $[R] \approx 2I_{abs}t$ and $\frac{d[B]}{dt} \approx k_2 I_{abs} t[A]$. So monitoring B sheds light on just k_2.

20.18 Number of photons absorbed $= \phi^{-1} \times$ number of molecules that react (Section 20.8). Therefore,

$$\text{Number absorbed} = \frac{(1.14 \times 10^{-3}\,\text{mol}) \times (6.022 \times 10^{23}\,\text{einstein}^{-1})}{2.1 \times 10^{2}\,\text{mol einstein}^{-1}}$$

$$= \boxed{3.3 \times 10^{18}}.$$

20.19 For a source of power P and wavelength λ, the amount of photons (n_λ) generated in a time t is

$$n_\lambda = \frac{Pt}{h\nu N_A} = \frac{P\lambda t}{hcN_A}$$

$$= \frac{(100\,\text{W}) \times (45) \times (60\,\text{s}) \times (490 \times 10^{-9}\,\text{m})}{(6.626 \times 10^{-34}\,\text{J s}) \times (2.998 \times 10^{8}\,\text{m s}^{-1}) \times (6.022 \times 10^{23}\,\text{mol}^{-1})}$$

$$= 1.11\,\text{mol}.$$

The amount of photons absorbed is 60 per cent of this incident flux, or 0.664 mol. Therefore,

$$\phi = \frac{0.344\,\text{mol}}{0.664\,\text{mol}} = \boxed{0.518}.$$

Alternatively, expressing the amount of photons in einsteins [1 mol photons = 1 einstein],

$$\phi = 0.518\,\text{mol einstein}^{-1}.$$

20.20 $M + h\nu_i \rightarrow M^*, \qquad I_a[M = \text{benzophenone}],$

$M^* + Q \rightarrow M + Q, \qquad k_q,$

$M^* \rightarrow M + h\nu_f, \qquad k_f,$

$$\frac{d[M^*]}{dt} = I_a - k_f[M^*] - k_q[Q][M^*] \approx 0 \text{ [steady state]}$$

and hence $[M^*] = \frac{I_a}{k_f + k_q[Q]}$.

Then $I_f = k_f[M^*] = \dfrac{k_f I_{abs}}{k_f + k_q[Q]}$

and so $\boxed{\dfrac{1}{I_f} = \dfrac{1}{I_{abs}} + \dfrac{k_q[Q]}{k_f I_a}}$.

If the exciting light is extinguished, $[M^*]$, and hence I_f, decays as $e^{-k_f t}$ in the absence of a quencher. Therefore we can measure $k_q/k_f I_{abs}$ from the slope of $1/I_f$ plotted against [Q], and then use k_f to determine k_q.

We draw up the following table.

10^3[Q]/M	1	5	10
$\frac{1}{I_f}$	2.4	4.0	6.3

The points are plotted in Figure 20.2.

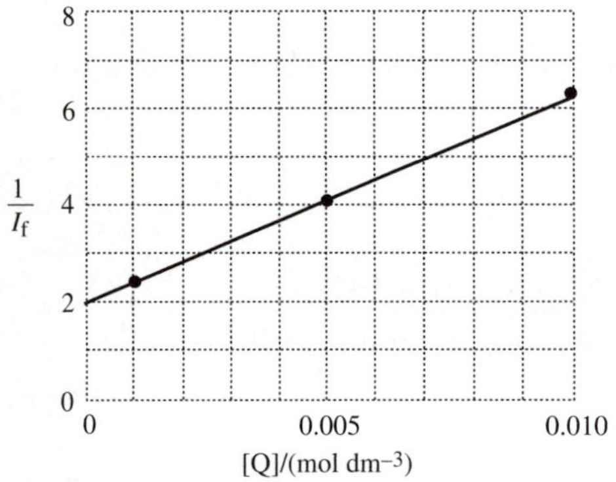

Fig 20.2

The intercept lies at 2.0, and so $I_{abs} = \dfrac{1}{2.0} = 0.50$. The slope is 430, and so

$$\frac{k_q}{k_f I_{abs}} = 430\,\text{dm}^3\text{mol}^{-1}.$$

Then, since $I_{abs} = 0.50$ and $k_f = \dfrac{\ln 2}{t_{1/2}}$,

$$k_q = (0.50) \times (430\,\text{dm}^3\,\text{mol}^{-1}) \times \left(\frac{\ln 2}{29 \times 10^{-6}\,\text{s}}\right) = \boxed{5.1 \times 10^6\,\text{dm}^3\,\text{mol}^{-1}\,\text{s}^{-1}}.$$

20.21 $\dfrac{\phi_f}{\phi} = \dfrac{\tau_0}{\tau} = 1 + \tau_0 k_Q[Q]$ Stern–Volmer equation [20.12].

τ_0 is the lifetime in the absence of quenching; τ is the lifetime in the presence of quenching.

A plot of τ_0/τ against $[O_2]$ with $\tau_0 = 2.6$ ns should be linear with slope equal to $\tau_0 k_Q$. We draw up a table for the plot, perform a linear regression fit with an intercept fixed at 1 (see Figure 20.3), and calculate k_Q.

$[O_2]/10^{-2}$ mol dm^{-3}	2.3	5.5	8.0	10.8
τ/ns	1.5	0.92	0.71	0.57
τ_0/τ	1.73	2.83	3.66	4.56

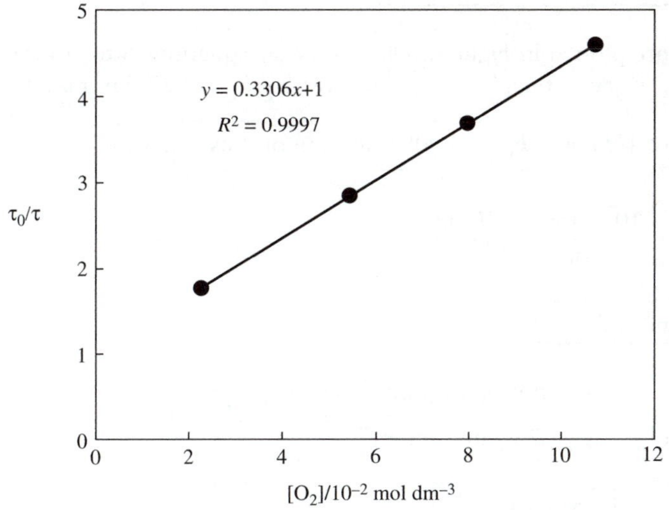

Fig 20.3

$$k_Q = \frac{\text{slope}}{\tau_0} = \left(\frac{0.3306}{10^{-2}\,\text{mol dm}^{-3}}\right)\left(\frac{1}{2.6\times10^{-9}\,\text{s}}\right) = \boxed{1.2\bar{7}\times10^{10}\,\text{mol}^{-1}\,\text{dm}^3\,\text{s}^{-1}}.$$

NOTE There is an alternative to doing a linear regression fit in this particular case. The Stern–Volmer equation rearranges to $\frac{(\tau_0/\tau)-1}{[Q]} = \tau_0 k_Q$. Thus, the ratio $\frac{(\tau_0/\tau)-1}{[Q]}$ may be computed for the data set and, since it equals a constant, averaged. The quenching rate constant equals this average divided by the fluorescence lifetime in the absence of quenching agent.

20.22 The Stern–Volmer equation may be written in the form

$$\frac{I_{abs}}{I_f} = 1 + \left(\frac{k_Q}{k_f}\right)[Q] \quad [20.12].$$

This expression shows that, if we plot I_{abs}/I_f against the quencher concentration, which is called a Stern–Volmer plot, then we should get a straight line with the intercept 1 and slope k_Q/k_f. Furthermore, since

$$\frac{1}{\tau} = \frac{1}{\tau_0} + k_Q[Q] \quad [20.13],$$

a plot of $1/\tau$ against the quencher concentration should give a straight line with a slope equal to k_Q.

First, we plot I_{abs}/I_f against [Q] to determine the ratio k_Q/k_f, then we plot $1/\tau$ against [Q] to determine k_Q from the slope. The half-life is determined from $t_{1/2} = \ln 2/k_f$. Draw up the following table of data.

[Q]/(mmol dm^{-3})	1.0	2.0	3.0	4.0	5.0
I_{abs}/I_f	3.2	5.5	7.7	10	12.3
$\left(\frac{1}{\tau}\right)/\text{ns}^{-1}$	0.013	0.022	0.031	0.040	0.050

The two sets of data are plotted in Figure 20.4. The linear equations fitted to the data are also displayed. The slope of the plot of I_{abs}/I_f gives $k_Q/k_f = 2.27\ \text{dm}^3\ \text{mmol}^{-1}$.

The slope of the plot of $1/\tau$ gives $\boxed{k_Q = 0.0092\ \text{dm}^3\ \text{mmol}^{-1}\ \text{ns}^{-1}}$. Thus,

$$k_Q = \frac{0.0092\ \text{dm}^3}{\text{mmol ns}} \times \frac{10^3\ \text{mmol}}{\text{mol}} \times \frac{10^9\ \text{ns}}{\text{s}}$$

$$= \boxed{9.2 \times 10^9\ \text{dm}^3\ \text{mol}^{-1}\ \text{s}^{-1}}$$

$$k_f = \frac{k_Q}{2.27\ \text{dm}^3\ \text{mmol}^{-1}} = \frac{0.0092\ \text{dm}^3\ \text{mmol}^{-1}\ \text{ns}^{-1}}{2.27\ \text{dm}^3\ \text{mmol}^{-1}} = \boxed{4.1 \times 10^6\ \text{s}^{-1}}$$

$$t_{1/2} = \frac{\ln 2}{4.1 \times 10^6\ \text{s}^{-1}} = \boxed{1.7 \times 10^{-7}\ \text{s}}.$$

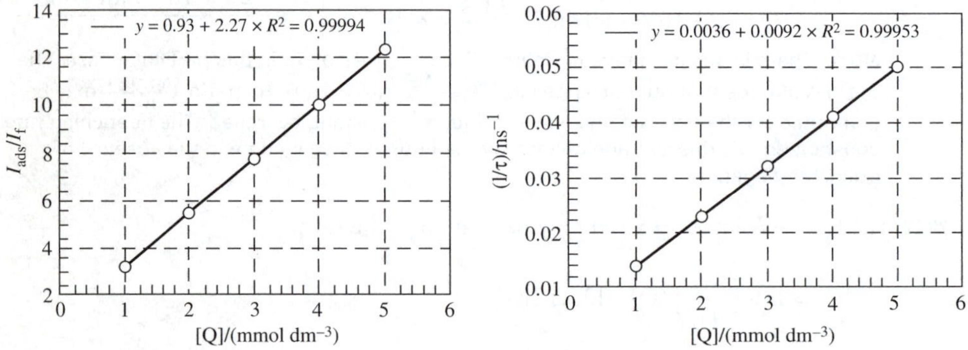

Fig 20.4

20.23 $E_T = \frac{R_0^6}{R_0^6 + R^6}$ or $\frac{1}{E_T} = 1 + (R/R_0)^6$ [20.15].

Since a plot of E_T^{-1} values against R^6 (Figure 20.5) appears to be linear with an intercept equal to 1, we conclude that eqn 20.15 adequately describes the data. Solving eqn 20.15 for R_0 gives $R_0 = R(E_T^{-1} - 1)^{1/6}$. R_0 may be evaluated by taking the mean of experimental data in this expression. The two data points at lowest R must be excluded from the mean as they are highly uncertain. $\boxed{R_0 = 3.5\bar{2}\ \text{nm}}$ with a standard deviation of $0.17\bar{3}$ nm.

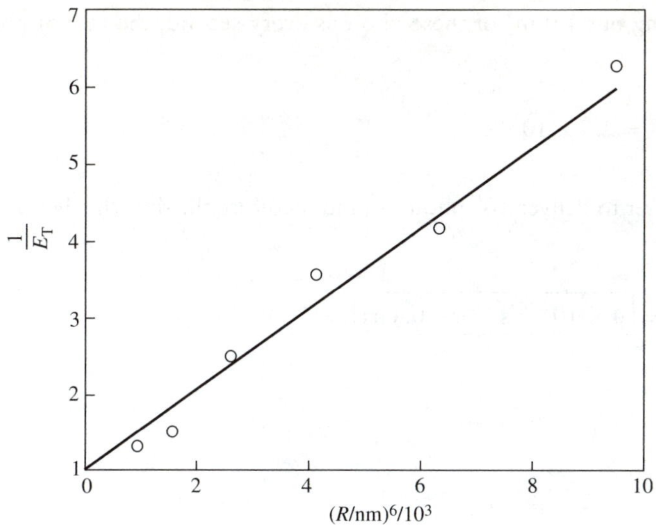

Fig 20.5

20.24 (a) The Beer–Lambert law is

$$A = \log \frac{I_0}{I} = \varepsilon[\mathrm{J}]l.$$

The absorbed intensity is

$$I_{abs} = I_0 - I \quad \text{so} \quad I = I_0 - I_{abs}.$$

Substitute this expression into the Beer–Lambert law and solve for I_{abs}.

$$\log \frac{I_0}{I_0 - I_{abs}} = \varepsilon[\mathrm{J}]l \quad \text{so} \quad I_0 - I_{abs} = I_0 \times 10^{-\varepsilon[\mathrm{J}]l},$$

$$\text{and} \qquad I_{abs} = \boxed{I_0 \times (1 - 10^{-\varepsilon[\mathrm{J}]l})}.$$

(b) The problem states that $I_f(\tilde{\nu}_f)$ is proportional to ϕ_f and to $I_{abs}(\tilde{\nu})$, so:

$$I_f(\tilde{\nu}_f) \propto \phi_f I_0(\tilde{\nu}) \times (1 - 10^{\varepsilon[\mathrm{J}]l}).$$

If the exponent is small, we can expand $1 - 10^{-\varepsilon[\mathrm{J}]l}$ in a power series

$$10^{-\varepsilon[\mathrm{J}]l} = (e^{\ln 10})^{-\varepsilon[\mathrm{J}]l} \approx 1 - \varepsilon\,[\mathrm{J}]\,l \ln 10 + \cdots,$$

$$\text{and} \qquad I_f(\tilde{\nu}_f) \propto \boxed{\phi_f I_0(\tilde{\nu})\varepsilon[\mathrm{J}]l \ln 10}.$$

20.25 The laser is delivering photons of energy

$$E = h\nu = \frac{hc}{\lambda} = \frac{(6.626 \times 10^{-34}\,\mathrm{J\,s}) \times (2.998 \times 10^{8}\,\mathrm{m\,s^{-1}})}{488 \times 10^{-9}\,\mathrm{m}} = 4.07 \times 10^{-19}\,\mathrm{J}.$$

Since the laser is putting out 1.0 mJ of these photons every second, the rate of photon emission is

$$r = \frac{1.0 \times 10^{-3}\,\mathrm{J\,s^{-1}}}{4.07 \times 10^{-19}\,\mathrm{J}} = 2.5 \times 10^{15}\,\mathrm{s^{-1}}.$$

The time it takes the laser to deliver 10^6 photons (and therefore the time the dye remains fluorescent) is

$$t = \frac{10^6}{2.5 \times 10^{15}\,\mathrm{s^{-1}}} = \boxed{4 \times 10^{-10}\,\mathrm{s} \quad \text{or} \quad 0.4\,\mathrm{ns}}.$$

Chapter 21

Magnetic resonance

Answers to discussion questions

21.1 Discussions of the origins of the local, neighboring group, and solvent contributions to the shielding constant can be found in Section 21.3. The local contribution is essentially the contribution of the electrons in the atom that contains the nucleus being observed. It can be expressed as a sum of a diamagnetic and paramagnetic parts, that is, $\sigma(\text{local}) = \sigma_{\text{d}} + \sigma_{\text{p}}$. The diamagnetic part arises because the applied field generates a circulation of charge in the ground state of the atom. In turn, the circulating charge generates a magnetic field. The direction of this field can be found through Lenz's law, which states that the induced magnetic field must be opposite in direction to the field producing it. Thus, it shields the nucleus. The diamagnetic contribution is roughly proportional to the electron density on the atom and it is the only contribution for closed shell free atoms and for distributions of charge that have spherical or cylindrical symmetry. The local paramagnetic contribution is somewhat harder to visualize since there is no simple and basic principle analogous to Lenz's law that can be used to explain the effect. The applied field adds a term to the hamiltonian of the atom which mixes in excited electronic states into the ground state and any theoretical calculation of the effect requires detailed knowledge of the excited state wave functions. It is to be noted that the paramagnetic contribution does not require that the atom or molecule be paramagnetic. It is paramagnetic only in the sense in that it results in an induced field in the same direction as the applied field.

The neighboring group contributions arise in a manner similar to the local contributions. Both diamagnetic and paramagnetic currents are induced in the neighboring atoms and these currents result in shielding contributions to the nucleus of the atom being observed. However, there are some differences. The magnitude of the effect is much smaller because the induced currents in neighboring atoms are much farther away. It also depends on the anisotropy of the magnetic susceptibility (Section 15.9) of the neighboring group. Only anisotropic susceptibilities result in a contribution.

Solvents can influence the local field in many different ways. Detailed theoretical calculations of the effect are difficult due to the complex nature of the solute–solvent interaction. Polar solvent–polar solute interactions are an electric field effect that usually causes deshielding of the solute protons. Solvent magnetic antisotropy can cause shielding or deshielding, for example, for solutes in benzene solution. In addition, there is a variety of specific chemical interactions between solvent and solute that can affect the chemical shift.

21.2 Both spin–lattice and spin–spin relaxation are caused by fluctuating magnetic and electric fields at the nucleus in question and these fields result from the random thermal motions present in the solution or other form of matter. These random motions can be a result of a number of processes and it is hard to summarize all that could be important. In theory, every known nuclear interaction coupled with every type of motion can contribute to relaxation and detailed treatments can be exceedingly complex. However, they all depend on the magnetogyric ratio of the atom in question and the magnetogyric ratio of the proton is much larger than that of ^{13}C. Hence, the interaction of the proton with fluctuating local magnetic fields caused by the presence of neighboring magnetic nuclei will be greater, and the relaxation will be quicker, corresponding to a shorter relaxation time for protons. Another consideration is the structure of compounds containing carbon and hydrogen. Typically, the C atoms are in the interior of the molecule bonded to other C atoms, 99% of which are non-magnetic, so the primary relaxation effects are due to bonded protons. Protons are on the outside of the molecule and are subject to many more interactions and hence faster relaxation.

21.3 Oxygen (O_2) is a paramagnetic molecule and, as a result, its presence in solution will cause strong local fluctuating magnetic fields at the positions of the nuclei. The interaction of these fluctuating local fields with the nuclei is one of the principal contributors to relaxation. Removing the oxygen eliminates this contribution. It is to be noted that the strength of the local field caused by an electronic magnetic moment is considerably greater than that caused by nuclear moments, so even a small amount of dissolved oxygen can have a major effect on the relaxation time.

21.4 At, say, room temperature, the tumbling rate of benzene, the small molecule, in a mobile solvent, may be close to the Larmor frequency, and hence its spin–lattice relaxation time will be short. As the temperature increases, the tumbling rate may increase well beyond the Larmor frequency, resulting in an increased spin–lattice relaxation time.

For the larger oligopeptide at room temperature, the tumbling rate may be well below the Larmor frequency, but with increasing temperature, it will approach the Larmor frequency due to the increased thermal motion of the molecule combined with the decreased viscosity of the solvent. Therefore, the spin–lattice relaxtion time may decrease.

21.5 Spin–spin couplings in NMR are due to a polarization mechanism, which is transmitted through bonds. The following description applies to the coupling between the protons in H_X—C—H_Y group as is typically found in organic compounds. See Figure 21.18 of the text. On H_X, the Fermi contact interaction causes the spins of its proton and electron to be aligned antiparallel. The spin of the electron from C in the H_X—C bond is then aligned antiparallel to the electron from H_X due to the Pauli exclusion principle. The spin of the C electron in the bond H_Y is then aligned parallel to the electron from H_X because of Hund's rule. Finally, the alignment is transmitted through the second bond in the same manner as the first. This progression of alignments (antiparallel × antiparallel × parallel × antiparallel × antiparallel) yields an overall energetically favorable parallel alignment of the two proton nuclear spins. Therefore, in this case the coupling constant, ${}^2J_{\mathrm{HH}}$ is negative in sign.

21.6 The molecular orbital occupied by the unpaired electron in an organic radical can be identified through the observation of hyperfine splitting in the EPR spectrum of the radical. The magnitude of this splitting is proportional to the spin density of the unpaired electron at those positions in the radical having atoms with nuclear moments. In addition, the spin density on carbon atoms adjacent to the magnetic nuclei can be determined indirectly through the McConnell relation. Thus, for example, in the benzene negative ion, unpaired spin densities on both the carbon atoms and hydrogen atoms can be determined from the EPR hyperfine splittings. The next step then is to construct a molecular orbital that will theoretically reproduce these experimentally determined spin densities. A good match indicates that we have found a good molecular orbital for the radical.

Solutions to exercises

21.7 $E_{m_s} = g_e\mu_B\mathcal{B}m_s$ [21.4] where $m_s = +\frac{1}{2}$ or $-\frac{1}{2}$ (α and β spin states, respectively).

$$\Delta E = E_\alpha - E_\beta = E_{m_s=1/2} - E_{m_s=-1/2} = g_e\mu_B\mathcal{B}\left(\frac{1}{2} - \left(-\frac{1}{2}\right)\right) = g_e\mu_B\mathcal{B} \quad [21.8]$$

$$= (2.0023) \times (9.274 \times 10^{-24}\,\mathrm{J\,T^{-1}}) \times (0.300\,\mathrm{T}) = \boxed{5.57 \times 10^{-24}\,\mathrm{J}}.$$

21.8 $E_{m_I} = -\gamma_N\hbar\mathcal{B}m_I = -g_N\mu_N\mathcal{B}m_I$ [21.5, 21.7]

$$m_I = \frac{3}{2}, \frac{1}{2}, -\frac{1}{2}, -\frac{3}{2}.$$

$$E_{m_I} = 0.4289 \times (5.051 \times 10^{-27}\,\mathrm{J\,T^{-1}}) \times (7.500\,\mathrm{T}) \times m_I$$

$$= \boxed{-1.625 \times 10^{-26}\,\mathrm{J} \times m_I}.$$

21.9 (a) Unit of γ_N = unit of $\left(\frac{g_I\mu_N}{\hbar}\right) = \frac{\mathrm{J\,T^{-1}}}{\mathrm{J\,s}} = \mathrm{T^{-1}\,s^{-1}} = \boxed{\mathrm{T^{-1}\,Hz}}$.

(b) $1\,\mathrm{T} = 1\,\mathrm{kg\,s^{-2}\,A^{-1}}$,

$$\gamma_N = \mathrm{kg^{-1}\,s^2\,A \times s^{-1}} = \boxed{\mathrm{A\,s\,kg^{-1}}}.$$

21.10 $\gamma_N\hbar = g_I\mu_N$ [21.5, 21.7].

Therefore,

$$g_I = \frac{\gamma_N\hbar}{\mu_N} = \frac{1.0840 \times 10^8\,\mathrm{T^{-1}s^{-1}} \times 1.05457 \times 10^{-34}\,\mathrm{J\,s}}{5.051 \times 10^{-27}\,\mathrm{J\,T^{-1}}} = \boxed{2.263}.$$

21.11 We assume a temperature of 300 K.

$$\frac{N_\beta - N_\alpha}{N} = \frac{N_\beta - N_\alpha}{N_\beta + N_\alpha} \simeq \frac{g_e\mu_B\mathcal{B}}{2kT} \quad \text{[see Derivation21.1].}$$

(a) $$\frac{N_\beta - N_\alpha}{N} = \frac{(2.0023) \times (9.274 \times 10^{-24}\,\mathrm{J\,T^{-1}}) \times (0.300\,\mathrm{T})}{2(1.381 \times 10^{-23}\,\mathrm{J\,K^{-1}}) \times (300\,\mathrm{K})} = \boxed{6.27 \times 10^{-4}}.$$

(b) $$\frac{N_\beta - N_\alpha}{N} = \frac{(2.0023) \times (9.274 \times 10^{-24}\,\mathrm{J\,T^{-1}}) \times (1.1\,\mathrm{T})}{2(1.381 \times 10^{-23}\,\mathrm{J\,K^{-1}}) \times (300\,\mathrm{K})} = \boxed{2.47 \times 10^{-3}}.$$

21.12 $\nu = \dfrac{g_e\mu_B\mathcal{B}}{h}$ [21.10] where $g_e = 2.0023$ and μ_B is the Bohr magneton.

$$\nu = \frac{2.0023 \times 9.274 \times 10^{-24}\,\mathrm{J\,T^{-1}} \times 0.330\mathrm{T}}{6.626 \times 10^{-34}\,\mathrm{J\,s}}$$

$$= 9.248 \times 10^9\,\mathrm{s^{-1}} = \boxed{9.248\,\mathrm{GHz}}.$$

$$\lambda = \frac{c}{\nu} = \frac{2.998 \times 10^8\,\mathrm{m\,s^{-1}}}{9.248 \times 10^9\,\mathrm{s^{-1}}} = \boxed{0.0324\,\mathrm{m}}.$$

EPR employs microwave radiation, rather than the radiofrequency radiation of NMR.

21.13 We assume a temperature of 300 K.

$$\frac{N_\alpha - N_\beta}{N} = \frac{N_\alpha - N_\beta}{N_\alpha + N_\beta} \simeq \frac{\gamma_N\hbar\mathcal{B}}{2kT} \quad \text{[21.12] for spin-}\tfrac{1}{2}\text{ nuclei like } {}^1\mathrm{H} \text{ and } {}^{13}\mathrm{C}.$$

Nuclear magnetogyric ratios are found in Table 21.2.

(a) ^{1}H: $$\frac{N_\alpha - N_\beta}{N} = \frac{(26.752\times10^7\,\mathrm{T^{-1}\,s^{-1}})\times(1.055\times10^{-34}\,\mathrm{J\,s})\times(10\,\mathrm{T})}{2(1.381\times10^{-23}\,\mathrm{J\,K^{-1}})\times(300\,\mathrm{K})}$$

$$= \boxed{3.5 \times 10^{-5}}.$$

(b) ^{13}C: $$\frac{N_\alpha - N_\beta}{N} = \frac{(6.7272\times10^7\,\mathrm{T^{-1}\,s^{-1}})\times(1.055\times10^{-34}\,\mathrm{J\,s})\times(10\,\mathrm{T})}{2(1.381\times10^{-23}\,\mathrm{J\,K^{-1}})\times(300\,\mathrm{K})}$$

$$= \boxed{8.6\times10^{-6}}.$$

21.14 $$\nu = \frac{\gamma_N\mathcal{B}}{2\pi}\ [21.13] = \frac{2.5177 \times 10^8\,\mathrm{T^{-1}\,s^{-1}} \times 8.200\,\mathrm{T}}{2\pi}$$

$$= 3.286 \times 10^8\,\mathrm{s^{-1}} = \boxed{328.6\,\mathrm{MHz}}.$$

21.15 $$\nu = \frac{\gamma_N\mathcal{B}}{2\pi}\ [21.13] = \frac{g_N\mu_N}{h}\mathcal{B}.$$

$$\nu = \frac{0.4036 \times 5.051 \times 10^{-27}\,\mathrm{J\,T^{-1}} \times 15.00\,\mathrm{T}}{6.626 \times 10^{-34}\,\mathrm{J\,s}} = 4.615 \times 10^7\,\mathrm{s^{-1}} = \boxed{46.15\,\mathrm{MHz}}.$$

21.16 $\mathcal{B}=\dfrac{h\nu}{\gamma_N\hbar}$ [21.11] $=\dfrac{2\pi\nu}{\gamma_N}$ where $\gamma_N=26.752\times10^7\,T^{-1}\,s^{-1}$ for ^1H (Table 21.2)

$$=\frac{2\pi(550.0\times10^6\,Hz)}{26.752\times10^7\,T^{-1}\,s^{-1}}=\boxed{12.92\,T}.$$

21.17 $\delta=\dfrac{\nu-\nu^o}{\nu^o}\times10^6$ [21.17].

$$\text{Shift}=\nu-\nu^o=\frac{\delta\times\nu^o}{10^6}=\frac{6.33\times420\times10^6\,Hz}{10^6}$$

$$=2.66\times10^3\,Hz=\boxed{2.66\,kHz}.$$

21.18 $\mathcal{B}_{loc}=(1-\sigma)\mathcal{B}$ [21.15].

$$|\Delta\mathcal{B}_{loc}|=|(\Delta\sigma)|\mathcal{B}\approx|[\delta(CH_3)-\delta(CHO)]|\mathcal{B}\left[|\Delta\sigma|\approx\left|\frac{\nu-\nu^o}{\nu^o}\right|\right]$$

$$=|(2.20-9.80)|\times10^{-6}\mathcal{B}=7.60\times10^{-6}\mathcal{B}.$$

(a) $\mathcal{B}=1.5\,T,\quad |\Delta\mathcal{B}_{loc}|=7.60\times10^{-6}\times1.5\,T=\boxed{11\,\mu T}$.

(b) $\mathcal{B}=6.0\,T,\quad |\Delta\mathcal{B}_{loc}|=7.60\times10^{-6}\times6.0\,T=\boxed{46\,\mu T}$.

21.19 $\nu-\nu^o=\dfrac{\nu^o\delta}{10^6}$ [21.17],

$$\Delta(\nu-\nu^o)=\frac{\nu^o\Delta\delta}{10^6}.$$

(a) $\Delta(\nu-\nu^o)=300\,Hz\times(9.5-1.5)\approx\boxed{2.4\,kHz}$.

(b) $\Delta(\nu-\nu^o)=550\,Hz\times(9.5-1.5)\approx\boxed{4.4\,kHz}$.

21.20 For identical nuclei with spin 1/2, there will be $N+1$ lines from the splitting. In this case 8 lines. The lines will have relative intensities of $\boxed{1{:}7{:}21{:}35{:}35{:}21{:}7{:}1}$.

These relative intensities can be determined by extending Pascals' triangle shown in (**1**) of the text three more rows to $N+1=8$. Alternatively, the intensities can also be determined from the coefficients in the expansion of $(1+x)^N$.

21.21 Because each resonance is split into three lines by a single N nucleus, the result will be:

(a) $\boxed{\text{quintet 1:2:3:2:1}}$ and (b) $\boxed{\text{septet 1:3:6:7:6:3:1}}$.

21.22 $E=-\gamma_N\hbar(1-\sigma_A)\,\mathcal{B}\,m_A-\gamma_N\hbar(1-\sigma_X)\,\mathcal{B}\,m_{X_1}-\gamma_N\hbar(1-\sigma_X)\,\mathcal{B}\,m_{X_2}$

As m_{X_1} and m_{X_2} can each be $\pm\dfrac{1}{2}$, there are a total of 6 energy levels, two of which are two-fold degenerate, for a total of eight levels. These are shown on the left of Figure 21.1. The allowed transitions are indicated by arrows. There are 7 transitions, but only 2 transition frequencies. This follows from the selection rule for magnetic resonance transitions, which is $\Delta(m_1+m_2)=\pm1$. The shorter arrows represent the X transitions, the larger arrows the A transitions. Spin–spin splitting perturbs these levels as follows:

$$E_{\text{spin–spin}} = hJm_A m_{X_1} + hJm_A m_{X_2},$$

	$\alpha\alpha_1\alpha_2$	$\alpha\alpha_1\beta_2$	$\alpha\beta_1\alpha_2$	$\alpha\beta_1\beta_2$
$E_{\text{spin–spin}}$	$\frac{1}{2}hJ$	0	0	$-\frac{1}{2}hJ$.
	$\beta\beta_1\beta_2$	$\beta\alpha_1\beta_2$	$\beta\beta_1\alpha_2$	$\beta\alpha_1\alpha_2$

There are again a total of 6 energy levels (two of which are two-fold degenerate), but they are perturbed by the amounts in the above chart. The perturbed levels are shown on the right in Figure 21.1. The frequencies of the X transitions are changed by $\pm\frac{1}{2}J$, the frequencies of the A transitions by $-J$, 0, $+J$. A stick diagram representing the spectrum is shown in Figure 21.2.

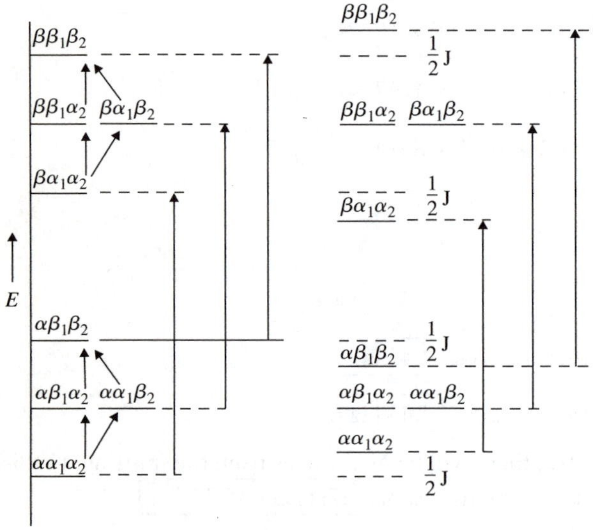

Fig 21.1

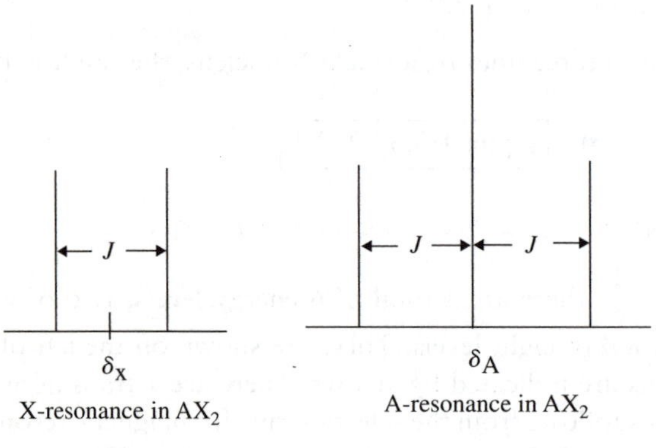

Fig 21.2

21.23 $\nu - \nu^{o} = \nu^{o}\delta \times 10^{-6}$.

$$|\Delta\nu| \equiv (\nu - \nu^{o})(\text{CHO}) - (\nu - \nu^{o})(\text{CH}_3)$$
$$= \nu(\text{CHO}) - \nu(\text{CH}_3)$$
$$= \nu^{o}[\delta(\text{CHO}) - \delta(\text{CH}_3)] \times 10^{-6}$$
$$= (9.80 - 2.20) \times 10^{-6}\nu^{o} = 7.60 \times 10^{-6}\nu^{o}$$

(a) $\nu^{o} = 300\,\text{MHz} \quad |\Delta\nu| = 7.60 \times 10^{-6} \times 300\,\text{MHz} = \boxed{2.28\,\text{kHz}}$.

(b) $\nu^{o} = 550\,\text{MHz} \quad |\Delta\nu| = 7.60 \times 10^{-6} \times 550\,\text{MHz} = \boxed{4.18\,\text{kHz}}$.

(a) The spectrum is shown in Figure 21.3.
(b) When the frequency is changed to 550 MHz, the separation of the CH_3 and CHO resonance increases (4.18 kHz), the fine structure remains unchanged, and the intensity increases.

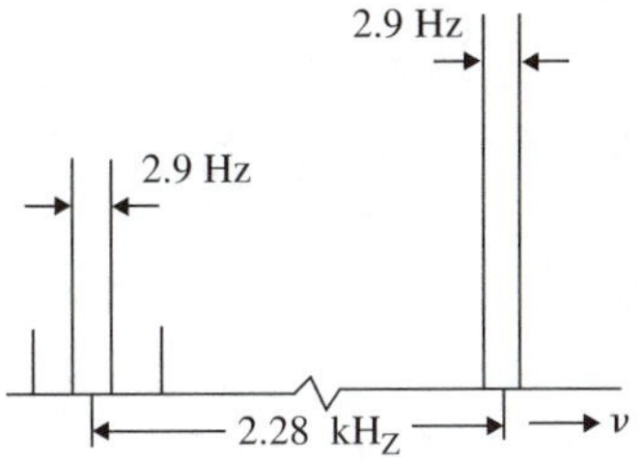

Fig 21.3

21.24 The four equivalent ^{19}F nuclei $\left(I = \frac{1}{2}\right)$ give a single line. However, the ^{10}B nucleus ($I = 3$, 19.6 per cent abundant) splits this line into $2 \times 3 + 1 = 7$ lines and the ^{11}B nucleus ($I = \frac{3}{2}$, 80.4 per cent abundant) into $2 \times \frac{3}{2} + 1 = 4$ lines. The splitting arising from the ^{11}B nucleus will be larger than that arising from the ^{10}B (because its magnetic moment is larger, by a factor of 1.5). Moreover, the total intensity of the four lines due to the ^{11}B nuclei will be greater (by a factor of $80.4/19.6 \approx 4$) than the total intensity of the seven lines due to the ^{10}B nuclei. The individual line intensities will be in the ratio $\frac{7}{4} \times 4 = 7$. The spectrum is sketched in Figure 21.4.

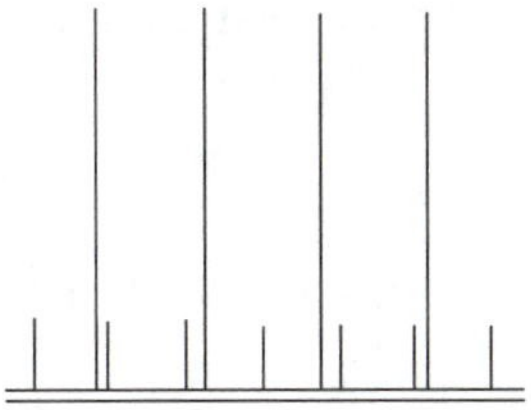

Fig 21.4

21.25 The A, M, and X resonances lie in distinctly different groups. The A resonance is split into a 1:2:1 triplet by the M nuclei, and each line of that triplet is split into a 1:4:6:4:1 quintet by the X nuclei (with $J_{AM} > J_{AX}$). The M resonance is split into a 1:3:3:1 quartet by the A nuclei and each line is split into a quintet by the X nuclei (with $J_{AM} > J_{MX}$). The X resonance is split into a quartet by the A nuclei and then each line is split into a triplet by the M nuclei (with $J_{AX} > J_{MX}$). The spectrum is sketched in Figure 21.5.

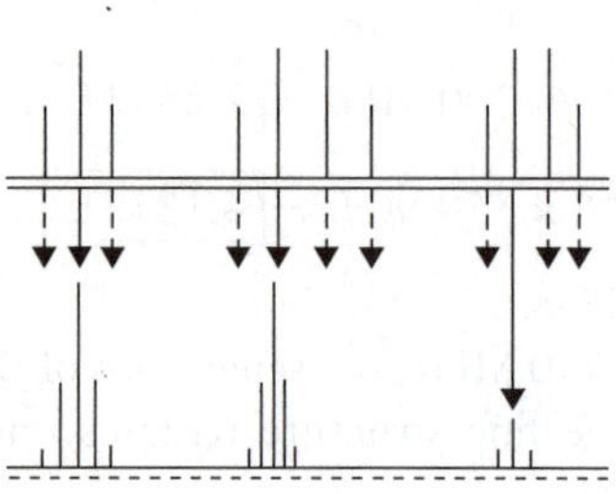

Fig 21.5

21.26 $^3J_{HH} = A + B\cos\phi + C\cos 2\phi$ [21.20].

$$\frac{d}{d\phi}(^3J_{HH}) = -B\sin\phi - 2C\sin 2\phi = 0.$$

This equation has a number of solutions:

$$\phi = 0, \quad \phi = n\pi, \quad \phi = \pi - \arccos\left(\frac{B}{4C}\right) = \arccos\left(\frac{B}{4C}\right).$$

The first two are trivial solutions.

If $\phi = \arccos\left(\frac{B}{4C}\right)$ then, $\sin\phi = \sqrt{1 - \frac{B}{16C^2}}$.

$$\sin 2\phi = 2\sin\phi\cos\phi = 2\sqrt{1 - \frac{B^2}{16C^2}}\left(\frac{B}{4C}\right).$$

$$B\sin\phi + 2C\sin 2\phi = B\sqrt{1 - \frac{B^2}{16C^2}} + 4C\sqrt{1 - \frac{B^2}{16C^2}}\left(\frac{B}{4C}\right) = 0.$$

So $\frac{B}{4C} = \cos\phi$ clearly satisfies the condition for an extremum.

The second derivative is

$$\frac{d^2}{d\phi^2}(^3J_{HH}) = -B\cos\phi - 4C\cos 2\phi = -B\cos\phi - 4C(2\cos^2\phi - 1)$$

$$= -B\left(\frac{B}{4C}\right) - 4C\left(2\frac{B^2}{16C^2} - 1\right) = -\frac{B^2}{4C} - \frac{2B}{4C} + 4C.$$

This quantity is positive if

$$16C^2 > 3B^2.$$

This is certainly true for typical values of B and C, namely $B = -1\,\text{Hz}$ and $C = 5\,\text{Hz}$. Therefore the condition for a minimum is as stated, namely, $\boxed{\cos\phi = B/4C}$.

21.27 The frequency difference between the two signals is

$$\Delta\nu = [\nu^{\circ} + \nu^{\circ} \times 10^{-6}\delta'] - [\nu^{\circ} + \nu^{\circ} \times 10^{-6}\delta] = \nu^{\circ} \times 10^{-6}(\delta' - \delta) \quad [21.18].$$

$$\tau = \frac{2^{1/2}}{\pi\Delta\nu}\ [21.21] = \frac{2^{1/2}}{\pi\nu^{\circ} \times 10^{-6}(\delta' - \delta)}$$

$$= \frac{2^{1/2}}{\pi(550 \times 10^{6}\,\text{s}^{-1}) \times 10^{-6}(4.8 - 2.7)} = 3.9 \times 10^{-4}\ \text{s}.$$

Therefore, the signals merge when the conversion rate is greater than about $1/\tau = \boxed{2.6 \times 10^{3}\,\text{s}^{-1}}$.

21.28 $g = \dfrac{h\nu}{\mu_B \mathcal{B}}$ [21.23].

We shall often need the value

$$\frac{h}{\mu_B} = \frac{6.62608 \times 10^{-34}\,\text{J Hz}^{-1}}{9.27402 \times 10^{-24}\,\text{J T}^{-1}} = 7.14478 \times 10^{-11}\,\text{T Hz}^{-1}.$$

Then, in this case,

$$g = \frac{(7.14478 \times 10^{-11}\,\text{T Hz}^{-1}) \times (9.2231 \times 10^{9}\,\text{Hz})}{329.12 \times 10^{-3}\,\text{T}} = \boxed{2.0022}.$$

21.29 $a = \mathcal{B}(\text{line } 3) - \mathcal{B}(\text{line } 2) = \mathcal{B}(\text{line } 2) - \mathcal{B}(\text{line } 1).$

$$\left.\begin{aligned} \mathcal{B}_3 - \mathcal{B}_2 &= (334.8 - 332.5)\,\text{mT} = 2.3\,\text{mT} \\ \mathcal{B}_2 - \mathcal{B}_1 &= (332.5 - 330.2)\,\text{mT} = 2.3\,\text{mT} \end{aligned}\right\} a = \boxed{2.3\,\text{mT}}.$$

Use the center line to calculate g.

$$g = \frac{h\nu}{\mu_B \mathcal{B}}\ [21.23] = (7.14478 \times 10^{-11}\,\text{T Hz}^{-1}) \times \frac{9.319 \times 10^{9}\,\text{Hz}}{332.5 \times 10^{-3}\,\text{T}} = \boxed{2.002\bar{5}}.$$

21.30 We construct Figure 21.6(a) for CH_3 and Figure 21.6(b) for CD_3. The predicted intensity distribution is determined by counting the number of overlapping lines of equal intensity from which the hyperfine line is constructed.

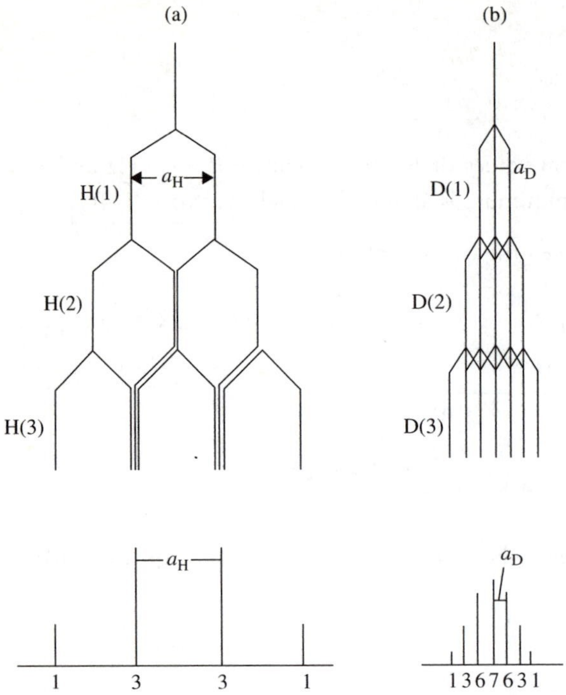

Fig 21.6

21.31 $\mathcal{B} = \dfrac{h\nu}{g\mu_B}$ [21.23] $= \dfrac{7.14478 \times 10^{-11}}{2.0025}\ \mathrm{T\,Hz^{-1}} \times \nu$ [Exercise 21.28] $= 35.68\ \mathrm{mT} \times (\nu/\mathrm{GHz})$.

(a) $\nu = 9.302\ \mathrm{GHz}$, $\mathcal{B}_0 = \boxed{331.9\ \mathrm{mT}}$.

(b) $\nu = 33.67\ \mathrm{GHz}$, $\mathcal{B}_0 = 1201\ \mathrm{mT} = \boxed{1.201\ \mathrm{T}}$.

21.32 If a radical contains N equivalent nuclei with spin quantum number I, then there are $2NI + 1$ hyperfine lines in the EPR spectrum. For $N = 2$ and five hyperfine lines, $\boxed{I = 1}$. The intensity ratio for the lines is 1:2:3:2:1 as demonstrated in Figure 21.7 (also see Figures 21.32 and 21.33 of text).

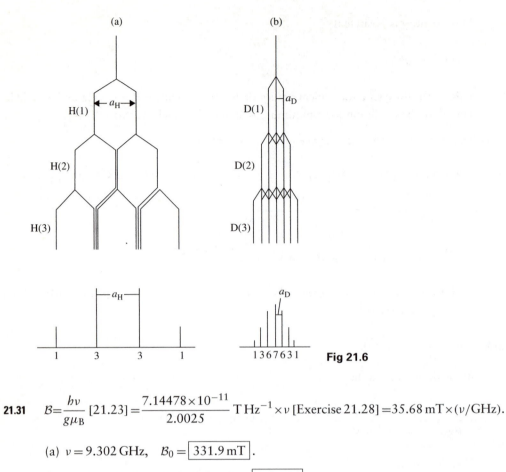

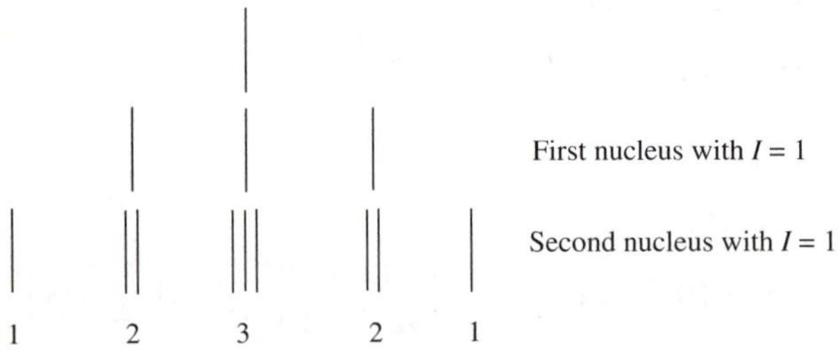

Fig 21.7

21.33 For $C_6H_6^-$, $a = Q\rho$ with $Q = 2.25\ \mathrm{mT}$ [21.26]. If we assume that the value of Q does not change from this value (a good assumption in view of the similarity of the anions), we may write

$$\rho = \frac{a}{Q} = \frac{a}{2.25\,\mathrm{mT}}.$$

Hence, we can construct the spin density maps shown in Figure 21.8.

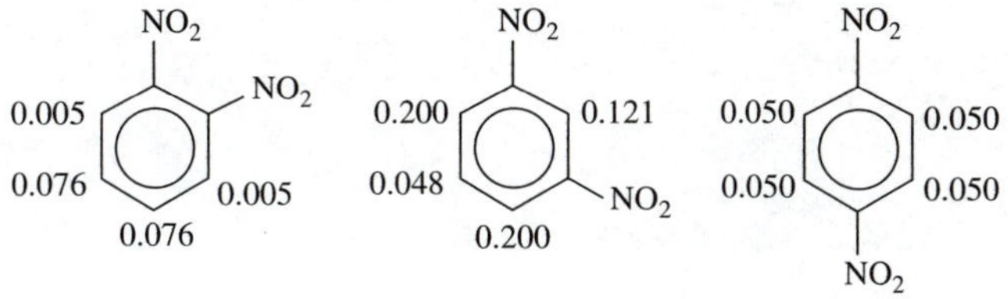

Fig 21.8

Chapter 22

Statistical thermodynamics

Answers to discussion questions

22.1 Consider the value of the partition function at the extremes of temperature. The limit of q as T approaches zero, is simply g_0, the degeneracy of the ground state. As T approaches infinity, each term in the sum is simply the degeneracy of the energy level. If the number of levels is infinite, the partition function is infinite as well. In some special cases where we can effectively limit the number of states, the upper limit of the partition function is just the number of states. In general, we see that the molecular partition function gives an indication of the average number of states thermally accessible to a molecule at the temperature of the system.

22.2 An approximation involved in the derivation of all of these expressions is the assumption that the contributions from the different modes of motion are separable. The expression $q^{\mathrm{R}} = kT/hcB$ is the high temperature approximation to the rotational partition function for non-symmetrical linear rotors. The expression $q^{\mathrm{V}} = kT/hc\tilde{\nu}$ is the high temperature form of the partition function for one vibrational mode of the molecule in the harmonic approximation. The expression $q^{\mathrm{E}} = g^{\mathrm{E}}$ for the electronic partition function applies at normal temperatures to atoms and molecules with no low lying excited electronic energy levels.

22.3 See Figure 22.3 of text, Example 22.1 of text, and Self-test 22.1 for the discussion of the partition function of a two state system. For convenience, the lower state is assigned an energy equal to 0 with the upper state having an energy that is $hc\tilde{\nu}$ higher. Each state is assumed to have a degeneracy equal to 1. Furthermore, to explore the variation of both the internal energy and entropy with temperature, it is convenient to use the definition for a unitless 'temperature', $x = kT/hcv$. Being proportional, x and T vary in the same way. For this simple system of non-interacting states the average energy E of the occupied states and the internal energy are identical. Thus,

$$U = E = \frac{NkT^2}{q}\frac{\mathrm{d}q}{\mathrm{d}T} \quad \text{[22.10 and Derivation22.1]}$$

$$= \frac{NKT^2}{1+\mathrm{e}^{-hc\tilde{\nu}/kT}}\frac{\mathrm{d}}{\mathrm{d}T}\left(1+\mathrm{e}^{-hc\tilde{\nu}/kT}\right) \quad \text{[Example 22.1]},$$

$$U = \frac{NkT^2}{1+e^{-1/x}} \frac{dx}{dT} \frac{d}{dx}\left(1+e^{-1/x}\right) = \left(\frac{NkT^2}{1+e^{-1/x}}\right) \times \left(\frac{k}{hc\tilde{\nu}}\right) \times \left(\frac{1}{x^2}e^{-1/x}\right)$$

$$= Nhc\tilde{\nu}\left(\frac{e^{-1/x}}{1+e^{-1/x}}\right),$$

$$\boxed{U = \frac{Nhc\tilde{\nu}}{1+e^{1/x}}}.$$

The equation for $U(x)$ indicates that, at the absolute zero of temperature (i.e. $x=0$), $U = 0$. At absolute zero the upper state is unpopulated. All particles are in the ground state. At very high temperature the term $e^{1/x}$ approaches $e^0 = 1$ and the internal energy approaches $Nhc\tilde{\nu}/2$. The two levels are equally populated as T approaches infinity and the internal energy equals the average of the two state energies.

The entropy variation of distinguishable particles is explored with eqn 22.13a.

$$S = \frac{U - U(0)}{T} + Nk \ln q \text{ [22.13a]} = \frac{U}{T} + Nk \ln q$$

$$= \frac{Nhc\tilde{\nu}/T}{1+e^{1/x}} + Nk \ln\left(1+e^{-1/x}\right),$$

$$\boxed{S = Nk\left\{\frac{1}{x\left(1+e^{1/x}\right)} + \ln\left(1+e^{-1/x}\right)\right\}}.$$

The equation for $S(x)$ indicates that, at the absolute zero of temperature (i.e. $x=0$) $S=0$ and the system is perfectly ordered with all particles in the lowest level. At very high temperature the term $e^{1/x}$ approaches $e^0 = 1$ and the internal energy approaches $Nk \ln 2$. Comparing this result with the Boltzmann formula for entropy ($S = k \ln W$ [22.12]) reveals that $W = 2^N$. This happens because each particle has an equal probability of being in either the lower or the upper level at very high temperature.

22.4 The statistical entropy may be defined in terms of the Boltzmann formula, $S = k \ln W$, where W is the statistical weight of the most probable configuration of the system. Entropy defined through this formula has the properties we expect of the entropy. W can be thought of as a measure of disorder; hence, the greater W, the greater the entropy. The logarithmic form is consistent with the additive properties of the entropy. We expect the total disorder of a combined system to be the product of the individual disorders and $S = k \ln W = k \ln W_1 W_2 = k \ln W_1 + k \ln W_2 = S_1 + S_2$.

The partition function, as seen in eqns 22.13 and 22.14, plays a role of special importance in statistical thermodynamics. Eqn 22.14 reveals that $\ln q$ is proportional to non-pV work in processes that change the thermodynamic state of matter. This provides the molecular interpretation of work as a change in the quantum mechanical energy states while keeping the molecular population over the states fixed. Details of the derivation of eqn 22.13a are found in P. Atkins and J. de Paula, *Physical Chemistry*, 8th edn., W. H. Freeman & Co., New York (2006), Ch. 16). It begins by demonstrating that W, the number of different ways in which the molecules of a system can be arranged with fixed total energy, is a maximum when

$$\ln W = -N\sum_i p_i \ln p_i \tag{1}$$

where p_i is the fraction of molecules in the ith energy state and it is described by the Boltzmann distribution

$$p_i = \frac{g_i e^{-E_i/kT}}{\sum_i g_i e^{-E_i/kT}} = \frac{g_i e^{-E_i/kT}}{q}. \tag{2}$$

Taking the natural logarithm of eqn (2) yields $\ln p_i = -E_i/kT - \ln q$ and substitution of this result into eqn (1) gives a working equation for statistical thermodynamics.

$$\ln W = -N\sum_i p_i \{-E_i/kT - \ln q\} = \frac{N\sum_i p_i E_i}{kT} + N \ln q \sum_i p_i.$$

But $\sum_i p_i = 1$ and $N\sum_i p_i E_i$ equals the thermodynamic energy of the system as measured from the zero point energy. Thus,

$$\ln W = \frac{U - U(0)}{kT} + N \ln q.$$

Substitution into the Boltzmann entropy formula yields eqn 22.13a of the text.

$$\boxed{S = \frac{U - U(0)}{T} + Nk \ln q}.$$

22.5 Residual entropy is due to the presence of some disorder in the system even at $T = 0$. It is observed in systems where there is very little energy difference, or none, between alternative arrangements of the molecules at very low temperatures. Consequently, the molecules cannot lock into a preferred orderly arrangement and some disorder persists.

22.6 See Further information 22.2 for a derivation of the general expression (eqn 22.18) for the equilibrium constant in terms of the partition functions and difference in molar energy, ΔE, of the products and reactants in a chemical reaction. The partition functions are functions of temperature and the ratio of partition functions in eqn 22.18 will therefore vary with temperature. However, the most direct effect of temperature on the equilibrium constant is through the exponential term $e^{-\Delta E/RT}$. The manner in which both factors affect the magnitudes of the equilibrium constant and its variation with temperature is described in detail for a simple $R \rightleftharpoons P$ gas phase equilibrium in Example 22.5.

The molecular partition function gives an indication of the number of states that are thermally accessible to a molecule at the temperature of the system. The fact that a ratio of partition functions for products and reactants appears in eqn 22.18 reveals the important method of comparing molecular distributions over available energy levels when evaluating an equilibrium constant. This ratio accounts for the role of entropy in the establishment of equilibrium. The exponential factor accounts for the role of energy. See Section 22.7 for a detailed discussion of these features.

Solutions to exercises

22.7 We apply the method of Illustration 22.1.

$$\frac{N_{\text{stretched}}}{N_{\text{coil}}} = e^{-\Delta E/RT} = e^{\frac{2.4\times10^{3}\,\text{J mol}^{-1}}{(8.3145\,\text{J K}^{-1}\text{mol}^{-1})\times(293\,\text{K})}}$$

$$= \boxed{0.37}.$$

22.8 The energy difference is $\Delta E = \gamma_{\text{N}}\hbar\mathcal{B}$ [21.13].

$\gamma_N = \gamma_{\text{hydrogen}-1} = 26.752 \times 10^{7}\,\text{T}^{-1}\text{s}^{-1}$ (Table 21.2).

$$\frac{\Delta E}{kT} = \frac{\gamma_{\text{N}}\hbar\mathcal{B}}{kT} = \frac{26.752 \times 10^{7}\,\text{T}^{-1}\,\text{s}^{-1} \times 1.05457 \times 10^{-34}\,\text{J s} \times \mathcal{B}}{1.38066 \times 10^{-23}\,\text{J K}^{-1} \times 293.15\,\text{K}}$$

$$= (6.97 \times 10^{-6}\,\text{T}^{-1}) \times \mathcal{B}$$

(a) $\dfrac{N_\beta}{N_\alpha} = e^{-\Delta E/kT} = e^{-97 \times 10^{-6}\,\text{T}^{-1} \times 1.5\,\text{T}} \quad e^{-6.97 \times 10^{-6}\,\text{T}^{-1} \times 1.5\,\text{T}}$

$$= \boxed{0.9999895}$$

(b) $\dfrac{N_\beta}{N_\alpha} = e^{-6.97 \times 0^{-6}\,\text{T}^{-1} \times 15\text{T}}$

$$= \boxed{0.9998955}$$

22.9 $\Delta E = g_e\mu_B\mathcal{B}$ [21.8] where $g_e = 2.0023$ and μ_B is the Bohr magneton.

$$\frac{N_\alpha}{N_\beta} = e^{-\Delta E/kT} = e^{-g_e\mu_B\mathcal{B}/kT} = e^{-\frac{2.0023\times9.274\times10^{-24}\text{JT}^{-1}\times 0.33\text{T}}{1.38066\times10^{-23}\text{JK}^{-1}\times 293\,\text{K}}}$$

$$= \boxed{0.99849}.$$

22.10 $E_J = hBJ(J+1)$ [19.8].

$$\Delta E = E_5 = -E_1 = hB[5(5+1) - 1(1+1)]$$

$$= 28\,hB.$$

$$\frac{N_5}{N_1} = \frac{g_5}{g_1}\,e^{-\Delta E/kT} = \frac{g_5}{g_1}\,e^{-28hB/kT}, \quad g_J = 2J+1$$

$$= \frac{11}{3}\,e^{-\frac{28 \times 6.626\times 10^{-34}\text{Js} \times 11.70 \times 10^{9}\text{s}^{-1}}{1.38066 \times 10^{-23}\text{JK}^{-1} \times 293\text{K}}}$$

$$= \boxed{3.475}.$$

22.11 $$\frac{N_5}{N_1}=\frac{g_5}{g_1}\,\mathrm{e}^{-28hB/kT}\qquad g_J=(2J+1)^2$$

$$=\frac{121}{9}\,\mathrm{e}^{-\frac{28\times 6.62\times 10^{-34}\,\mathrm{J\,s}\times 157\times 10^{9}\,\mathrm{s}^{-1}}{1.38066\times 10^{-23}\,\mathrm{J\,K}^{-1}\times 293\,\mathrm{K}}}$$

$$=\left(\frac{121}{9}\right)0.487=\boxed{6.55}.$$

22.12 (a) $q=\sum_i g_i\mathrm{e}^{-E_i/kT}$ [22.2] with degeneracy factor g_i inserted.

$$q=g_0\mathrm{e}^{-E_0/kT}+g_1\mathrm{e}^{-E_1/kT}+g_2\mathrm{e}^{-E_2/kT}$$

$$=\boxed{1+5\mathrm{e}^{-\varepsilon/kT}+3\mathrm{e}^{3\varepsilon/kT}}.$$

(b) Since $\lim_{T\to 0}(\mathrm{e}^{-\varepsilon/kT})=0$ $\boxed{q=1}$ in the limit of the absolute zero of temperature.

(c) Since $\lim_{T\to\infty}(\mathrm{e}^{-\varepsilon/kT})=1,\ q=1+5+3=\boxed{9}$.

22.13 $$q=\sum_i g_i\mathrm{e}^{-E_i/kT}=\sum_i g_i\mathrm{e}^{-hc\tilde{\nu}_i/kT}.$$

$$\frac{hc}{k}=\frac{6.626\times 10^{-34}\,\mathrm{J\,s}\times 2.998\times 10^{8}\,\mathrm{m\,s}^{-1}}{1.381\times 10^{-23}\,\mathrm{J\,K}^{-1}}$$

$$=0.014387\,\mathrm{m\,K}=1.4387\,\mathrm{cm\,K}.$$

$$q=1+3\,\mathrm{e}^{-1.4387\,\mathrm{cm\,K}\times 16.4\,\mathrm{cm}^{-1}/T}+5\mathrm{e}^{-1.4387\,\mathrm{cm\,K}\times 43.5\,\mathrm{cm}^{-1}/T}$$

$$=1+3\mathrm{e}^{-23.59\,\mathrm{K}/T}+5\mathrm{e}^{-62.58\,\mathrm{K}/T}.$$

(a) $q=1+3\mathrm{e}^{-23.59\,\mathrm{K}/10\,\mathrm{K}}+5\mathrm{e}^{-62.58\,\mathrm{K}/10\,\mathrm{K}}$

$$=\boxed{1.29}.$$

(b) $q=1+3\mathrm{e}^{-23.59\,\mathrm{K}/298\,\mathrm{K}}+5\mathrm{e}^{-62.58\,\mathrm{K}/298\,\mathrm{K}}$

$$=\boxed{7.28}.$$

22.14 $$q=\sum_i g_i\mathrm{e}^{-E_i/kT}=\sum_i g_i\mathrm{e}^{-hc\tilde{\nu}_i/kT}$$

$$=\sum_i g_i\mathrm{e}^{-1.4387\,\mathrm{cm\,K}(\tilde{\nu}_i/T)}$$ [solution to Exercise 22.13].

(a) $q=(2\times 2+1)+(2\times 1+1)+1=\boxed{9}$.

(b) $q=5+3\mathrm{e}^{-1.4387\,\mathrm{cm\,K}(158.5\,\mathrm{cm}^{-1}/298\mathrm{K})}+\mathrm{e}^{-1.4387\,\mathrm{cm\,K}(226.5\mathrm{cm}^{-1}/298\mathrm{K})}=\boxed{6.731}$.

22.15 $$q = \frac{1}{1 - e^{-hc\tilde{\nu}/kT}}$$

$$= \frac{1}{1 - e^{-1.4387\,\mathrm{cm\,K}(\tilde{\nu}/T)}} = \frac{1}{1 - e^{-1.4387\,\mathrm{cm\,K} \times 560\,\mathrm{cm^{-1}}/298\,\mathrm{K}}}$$

$$= \boxed{1.072}.$$

22.16 (a) $$V = \frac{4}{3}\pi r^3 = \frac{4}{3}\pi(0.5\,\mathrm{nm})^3 = 0.52\,\mathrm{nm}^3 = 5.2 \times 10^{-28}\,\mathrm{m}^3.$$

$$q = \frac{(2\pi mkT)^{3/2} V}{h^3} \quad [22.5].$$

$$q = \frac{(2\pi \times 16.04 \times 1.66054 \times 10^{-27}\,\mathrm{kg} \times 1.381 \times 10^{-23}\,\mathrm{J\,K^{-1}} \times 298\,\mathrm{K})^{3/2} \times 5.2 \times 10^{-28}\,\mathrm{m}^3}{(6.626 \times 10^{-34}\,\mathrm{J\,s})^3}$$

$$= \boxed{3.2 \times 10^4}.$$

(b) $V = 100\,\mathrm{cm}^3 = 1.00 \times 10^{-4}\,\mathrm{m}^3.$

$$q = \boxed{6.2 \times 10^{27}}.$$

22.17 (a) $$q = \frac{kT}{\sigma hB}[22.6] = \frac{1.381 \times 10^{-23}\,\mathrm{J\,K^{-1}} \times 298\,\mathrm{K}}{1 \times 6.626 \times 10^{-34}\,\mathrm{J\,s} \times 318 \times 10^9\,\mathrm{s^{-1}}} = \boxed{19.5}.$$

(b) $$q = \frac{1.381 \times 10^{-23}\,\mathrm{J\,K^{-1}} \times 298\mathrm{K}}{2 \times 6.626 \times 10^{-34}\,\mathrm{J\,s} \times 11.70 \times 10^9\,\mathrm{s^{-1}}} = \boxed{265}.$$

22.18 N_2O is linear, but not symmetrical. The structure is NNO. Thus, the symmetry number is one for NNO, but 2 for CO_2.

22.19 (a) $$q = \frac{1}{1 - e^{-h\nu/kT}} \; [22.4; E_0 \equiv 0], \quad E_{v=0,1,2,\ldots} = vh\nu \; [12.18; E_0 \equiv 0].$$

$$E = \frac{N}{q}\sum_{v=0} E_v e^{-E_v/kT} \; [22.9] = \frac{N}{q}\sum_{v=0} vh\nu\, e^{-vh\nu/kT}$$

$$= \frac{Nh\nu}{q}(0 \times e^{-0 \times h\nu/kT} + 1 \times e^{-1 \times h\nu/kT} + 2 \times e^{-2 \times h\nu/kT} + \cdots)$$

$$= \frac{Nh\nu\, e^{-h\nu/kT}}{q}(1 + 2e^{-h\nu/kT} + 3e^{-2h\nu/kT} + 4e^{-3h\nu/kT} + \cdots)$$

$$= \frac{Nh\nu\, e^{-h\nu/kT}}{q}\left\{\frac{1}{(1 - e^{-h\nu/kT})^2}\right\}$$

$$= \frac{Nh\nu\, e^{-h\nu/kT}}{q}\left(\frac{q}{1 - e^{-h\nu/kT}}\right) = \boxed{\frac{Nh\nu}{e^{h\nu/kT} - 1}}.$$

22.20 $q=\sum_i g_i e^{-E_i/kT} = 1e^{-0\times\varepsilon/kT} + 5e^{-1\times\varepsilon/kT} + 3e^{-3\times\varepsilon/kT} = \boxed{1+5e^{-\varepsilon/kT} + 3e^{-3\varepsilon/kT}}$.

22.21 (a) From the solution to Exercise 22.13 we have

$$q = 1 + 3e^{-23.59\,\mathrm{K}/T} + 5\,e^{-62.58\,\mathrm{K}/T}.$$

$$E = \frac{NkT^2}{q}\frac{dq}{dT}.$$

$$\frac{dq}{dT} = 3\left(\frac{23.59\,\mathrm{K}}{T^2}\right)e^{-23.59\mathrm{K}/T} + 5\left(\frac{62.58}{T^2}\right)e^{-62.58\,\mathrm{K}/T}.$$

$$E = \frac{R(70.77\,\mathrm{K})e^{-23.59\mathrm{K}/T} + R(312.9\,\mathrm{K})e^{-62.58\,\mathrm{K}/T}}{1 + 3e^{-23.59\,\mathrm{K}/T} + 5e^{-62.58\,\mathrm{K}/T}}.$$

The plot of E against T is shown in Figure 22.1.

(b) Substitute $T = 298\,\mathrm{K}$ into the expression for E and obtain $E = \boxed{339\,\mathrm{J\,mol^{-1}}}$.

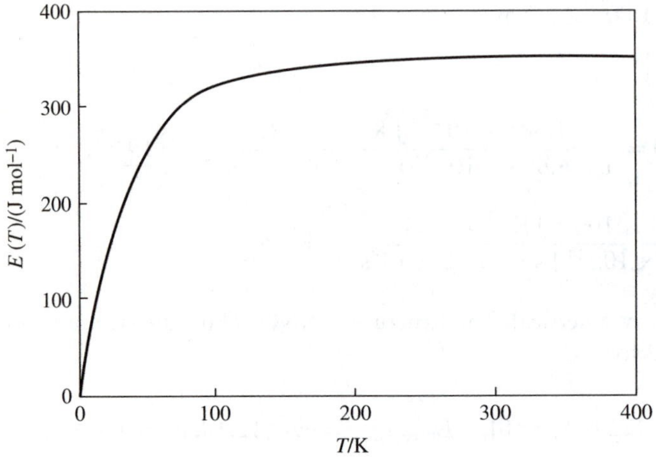

Fig 22.1

22.22 (a) From Exercise 22.14 we have

$$q = 5 + 3e^{-228.0\,\mathrm{K}/T} + e^{-325.9\,\mathrm{K}/T}.$$

$$E = \frac{NkT^2}{q}\frac{dq}{dT}.$$

$$\frac{dq}{dT} = 3\left(\frac{228.0\,\mathrm{K}}{T^2}\right)e^{-228.0\,\mathrm{K}/T} + \left(\frac{325.9\,\mathrm{K}}{T^2}\right)e^{-325.9\,\mathrm{K}/T}.$$

$$E = \frac{R(684.0\,\mathrm{K}\,e^{-228.0\,\mathrm{K}/T} + 325.9\,\mathrm{K}\,e^{-325.9\,\mathrm{K}/T})}{5 + 3e^{-228.0\,\mathrm{K}/T} + e^{-325.9\,\mathrm{K}/T}}.$$

$$C_{V,\mathrm{m}} = \frac{dU}{dT} = \frac{dE}{dT}.$$

The expression for $C_{V,\mathrm{m}}$ is derived below and plots are presented of $E_{\mathrm{m}}(T)$ in Figure 22.2 and $C_{V,\mathrm{m}}(T)$ in Figure 22.3.

$a = 228\,\mathrm{K}, \quad b = 325.9\,\mathrm{K}, \quad R = 8.3145\,\mathrm{J\,K^{-1}\,mol^{-1}}.$

$$E_{\mathrm{m}}(T) = R\,\frac{\left(3a\mathrm{e}^{\frac{-3a}{T}} + b\mathrm{e}^{\frac{-b}{T}}\right)}{\left(5 + 3\mathrm{e}^{\frac{-a}{T}} + \mathrm{e}^{\frac{-b}{T}}\right)}.$$

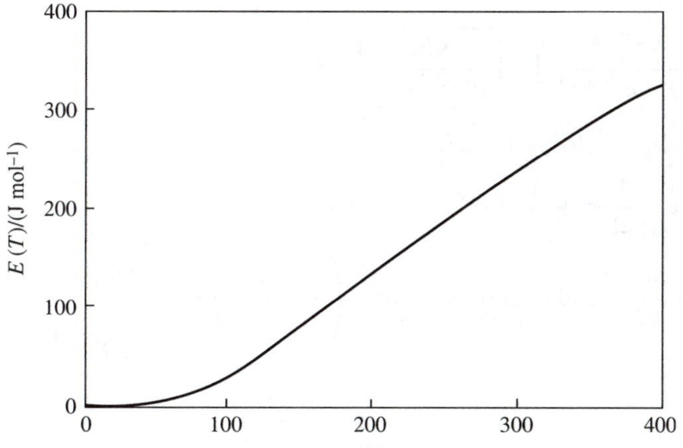

Fig 22.2

$$C_{V,\mathrm{m}}(T) = R\,\frac{\left(9\frac{a^2}{T^2}\exp\left(-3\frac{a}{T}\right) + \frac{b^2}{T^2}\exp\left(\frac{-b}{T}\right)\right)}{\left(5 + 3\exp\left(\frac{-a}{T}\right) + \exp\left(\frac{-b}{T}\right)\right)} - R\,\frac{\left(3a\exp\left(-3\frac{a}{T}\right) + b\exp\left(\frac{-b}{T}\right)\right)}{\left(5 + 3\exp\left(\frac{-a}{T}\right) + \exp\left(\frac{-b}{T}\right)\right)^2} \times \left(3\frac{a}{T^2}\exp\left(\frac{-a}{T}\right) + \frac{b}{T^2}\exp\left(\frac{-b}{T}\right)\right).$$

(b) $C_{V,\mathrm{m}}(298\,\mathrm{K}) = \boxed{0.993\,\mathrm{J\,K^{-1}\,mol^{-1}}}$

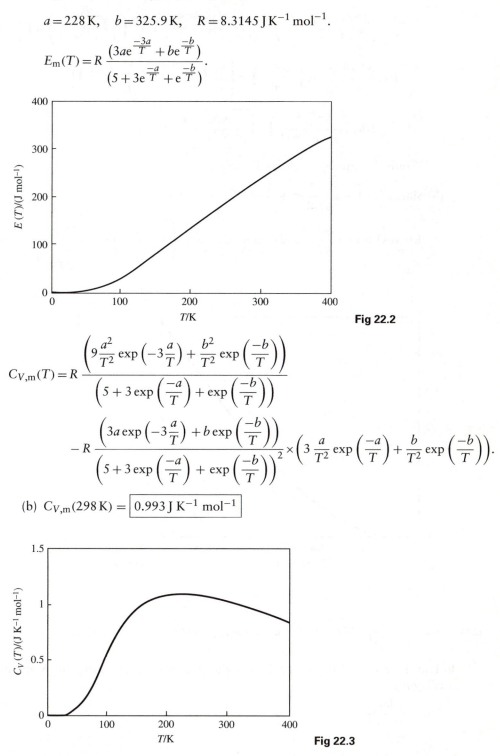

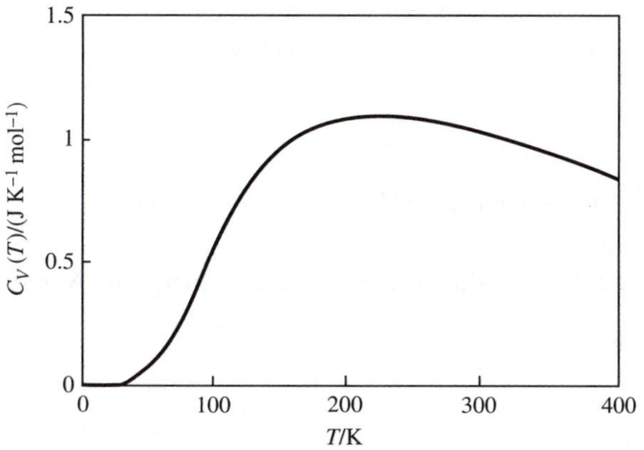

Fig 22.3

22.23 (a) $q = \dfrac{1}{1 - e^{-h\nu/kT}}$ [22.4; $E_0 \equiv 0$], $E = \dfrac{NkT^2}{q}\dfrac{dq}{dT}$ [[22.10] and Derivation 22.1].

$$\frac{dq}{dT} = \frac{d}{dT}\left(\frac{1}{1 - e^{-h\nu/kT}}\right) = \frac{(h\nu/k)e^{-h\nu/kT}}{(1 - e^{-h\nu/kT})^2 T^2} = \frac{h\nu}{k}\left(\frac{e^{-h\nu/kT}}{1 - e^{-h\nu/kT}}\right) \times \left(\frac{q}{T^2}\right).$$

$$E = \frac{NkT^2}{q}\frac{dq}{dT} = \left(\frac{NkT^2}{q}\right) \times \left\{\frac{h\nu}{k}\left(\frac{e^{-h\nu/kT}}{1 - e^{-h\nu/kT}}\right) \times \left(\frac{q}{T^2}\right)\right\}$$

$$= Nh\nu\left(\frac{e^{-h\nu/kT}}{1 - e^{-h\nu/kT}}\right) = \boxed{\frac{Nh\nu}{e^{h\nu/kT} - 1}}.$$

Figure 22.4 presents a plot of E against T.

(b) Since $e^x = 1 + x + \frac{1}{2!}x^2 + \frac{1}{3!}x^3 \cdots \simeq 1 + x$ when $|x| \ll 1$

$$\lim_{T\to\infty}\left(e^{\frac{h\nu}{kT}}\right) = 1 + \frac{h\nu}{kT} \quad \text{and} \quad \lim_{T\to\infty} E = \lim_{T\to\infty}\left(\frac{Nh\nu}{e^{\frac{h\nu}{kT}} - 1}\right) = \frac{Nh\nu}{1 + \frac{h\nu}{kT} - 1} = \boxed{NkT}.$$

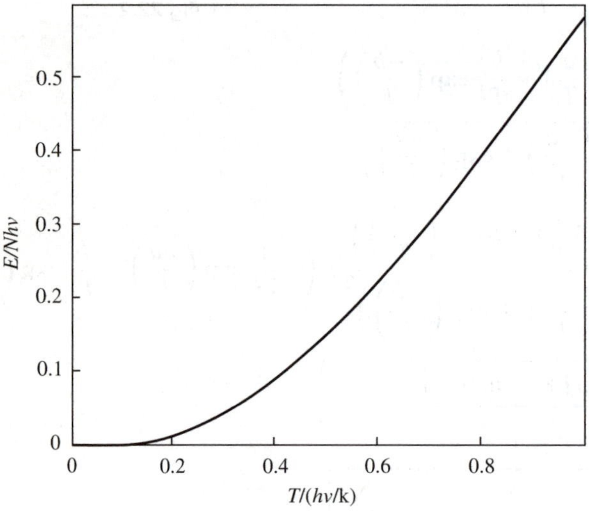

Fig 22.4

22.24 $q = \dfrac{1}{1 - e^{-h\nu/kT}}$ [22.4; $E_0 \equiv 0$], $E_{v=0,1,2,\ldots} = vh\nu$ [12.18; $E_0 \equiv 0$].

In both Exercises 22.19 and 22.23 it is shown that the mean energy of N harmonic oscillators is

$$E = \frac{Nh\nu}{e^{h\nu/kT} - 1}.$$

For a system of non-interacting harmonic oscillators the internal energy U equals the mean energy. Consequently,

$$C_V = \frac{dU}{dT} \text{ at constant} V = \frac{d}{dT}\left(\frac{Nh\nu}{e^{h\nu/kT}-1}\right) = Nh\nu\left\{\frac{-\left(-\frac{h\nu}{kT^2}\right)e^{h\nu/kT}}{(e^{h\nu/kT}-1)^2}\right\}$$

$$= Nk\left(\frac{h\nu}{kT}\right)^2\left(\frac{e^{h\nu/2kT}}{e^{h\nu/kT}-1}\right)^2$$

$$= Nk\boxed{\left(\frac{h\nu}{kT}\right)^2 \times \left(\frac{e^{-h\nu/2kT}}{1-e^{-h\nu/kT}}\right)^2}.$$

A plot of C_V against T is presented in Figure 22.5.

Since $e^x = 1 + x + \frac{1}{2!}x^2 + \frac{1}{3!}x^3 \cdots \simeq 1 + x$ when $|x| \ll 1$,

$$\lim_{T\to\infty}(e^{-\frac{h\nu}{kT}}) = 1 - \frac{h\nu}{kT} \quad \text{when} \boxed{T \gg h\nu/k}.$$

$$\lim_{T\to\infty} C_V = \lim_{T\to\infty}\left(Nk\left(\frac{h\nu}{kT}\right)^2\left(\frac{e^{-h\nu/2kT}}{1-e^{-h\nu/kT}}\right)^2\right)$$

$$= Nk\left(\frac{h\nu}{k}\right)^2 \lim_{T\to\infty}\left(\frac{1}{T^2}\left(\frac{1-\frac{h\nu}{kT}}{1-1+\frac{h\nu}{kT}}\right)^2\right)$$

$$= Nk\left(\frac{h\nu}{k}\right)^2 \lim_{T\to\infty}\left(\frac{1}{T^2}\left(\frac{1}{\frac{h\nu}{kT}}\right)^2\right) = \boxed{Nk}.$$

Thus, the high temperature limit of C_V is Nk and the condition for 'high' temperature is $T \gg h\nu/k$.

Furthermore, if $N = N_A$, $\lim_{T\to\infty} C_V = N_A k = \boxed{R}$

22.25 $S_m = k \ln 4^{N_A} = N_A k \ln 4 = R \ln 4$

$$= \boxed{11.5\,\text{J K}^{-1}\ \text{mol}^{-1}}.$$

22.26 $S = k \ln W$ [22.12] $= k \ln 4^N = Nk \ln 4$

$$= (5\times10^8)\times(1.38\times10^{-23}\,\text{J K}^{-1})\times\ln 4 = \boxed{9.57\times10^{-15}\,\text{J K}^{-1}}.$$

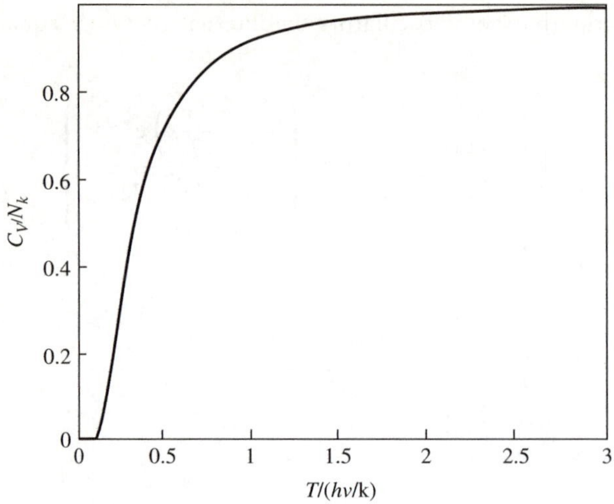

Fig 22.5

QUESTION Is this a large residual entropy? The answer depends on what comparison is made. Multiply the answer by Avogadro's number to obtain the molar residual entropy, 5.76×10^{9} J K^{-1} mol^{-1}, surely a large number—but then DNA is a macromolecule. The residual entropy per mole of base pairs may be a more reasonable quantity to compare to molar residual entropies of small molecules. To obtain that answer, divide the molecule's entropy by the number of base pairs before multiplying by N_A. The result is 11.5 J K^{-1} mol^{-1}, a quantity more in line with examples discussed in Section 22.6.

22.27 We use eqn 22.13b for indistinguishable molecules.

$$S_m = \frac{U - U(0)}{T} + Nk \ln q_m - Nk(\ln N - 1) \quad \text{with } N = N_A.$$

For N_2 (g) at 298 K we assume that the high-temperature limit (equipartion value) for the energy applies. Then $U_m - U_m(0) = 3/2RT + RT = 5/2RT$ (see Self-test 22.2).

$$S_m = 5/2R + R\ln q_m - R\ln N_A + R = 5/2R + R\left(\ln\frac{q_m}{N_A} + 1\right).$$

$$q_m = q_m^{\text{Trans}} q^{\text{Rot}}.$$

$$q_m^{\text{Trans}} = (2\pi mkT)^{3/2} V_m/h^3.$$

$$q^{\text{Rot}} = \frac{kT}{\sigma hB}.$$

We replace V_m by $RT/p^{\ominus}$ and substitute in the values of the various constants and obtain

$$\frac{q_m^{\text{Trans}}}{N_A} = 2.561 \times 10^{-2}\left(\frac{T}{\text{K}}\right)^{5/2} (M/g\,\text{mol}^{-1})^{3/2}.$$

$$q^{\text{Rot}} = \frac{0.6950}{\sigma} \times \frac{T/K}{(\tilde{B}/\text{cm}^{-1})}, \quad \tilde{B} = 1.987\,\text{cm}^{-1} \text{ for } N_2.$$

Then

$$\frac{q_m^{\text{Trans}}}{N_A} = (2.561 \times 10^{-2}) \times (298)^{5/2} \times (28.02)^{3/2} = 5.823 \times 10^6,$$

$$q^{\text{Rot}} = \frac{1}{2} \times 0.6950 \times \frac{298}{1.987} = 51.81,$$

and

$$\frac{q_m^{\ominus}}{N_A} = (5.823 \times 10^6) \times 51.81 = 3.02 \times 10^8.$$

$$S_m = \frac{5}{2}R + R(\ln 3.02 \times 10^8 + 1) = 23.03\,R$$

$$= \boxed{191.4\ \text{J K}^{-1}\ \text{mol}^{-1}}.$$

22.28 The change in entropy per micelle is

$$\Delta S = 100\,k \ln \frac{V_{\text{solution}}}{V_{\text{micelle}}}.$$

In the absence of specific data on the volumes involved we can arrive at a rough value of ΔS by making some reasonable estimates. Let us assume that the micelle is spherical in shape. Let us also assume that the radius of this sphere is roughly the same as the length of the hydrocarbon chain of the amphiphile. Assume that the chain consists of 10 zig-zag carbon atoms with an average C–C–C length of 250 pm, or about 125 pm per carbon atom. Then the radius of the micelle is about 1.25 nm and the volume is

$$V_{\text{micelle}} = \frac{4}{3}\pi r^3 \approx 4r^3 = 4 \times \left(1.25 \times 10^{-9}\,\text{m}\right)^3$$

$$\approx 1 \times 10^{-26}\,\text{m}^3.$$

Assume that the volume of the solution is that of a typical 100 cm^3 beaker. 100 $\text{cm}^3 = 10^{-4}\ \text{m}^3$.

$$\Delta S = 100\,\text{k}\ \ln\left(\frac{10^{-4}\,\text{m}^3}{10^{-26}\text{m}^3}\right)$$

$$= 100 \times 1.38 \times 10^{-23}\,\text{J K}^{-1} \times \ln 10^{22}$$

$$= 7 \times 10^{-20}\,\text{J K}^{-1}(\text{per micelle}).$$

For one mole of micelles

$$\Delta S \approx N_A \times 7 \times 10^{-20}\text{J K}^{-1}$$

$$\approx 4 \times 10^4\text{J K}^{-1}\ \text{mol}^{-1}$$

$$\approx \boxed{40\ \text{kJ K}^{-1}\ \text{mol}^{-1}}.$$

This value seems high and is probably a result of underestimating V_{micelle} for a micelle consisting of 100 amphiphiles. End effects contributing to the length of the amphiphile were neglected and the number of carbon atoms for amphiphiles that could form a micelle of 100 amphiphiles may be larger than the 10 assumed. Experimental evidence for the volume of micelles with 100 amphiphiles indicates typical values of about 1×10^{25} m^3. For that volume

$$\Delta S = 4\,\text{kJ}\,\text{K}^{-1}\,\text{mol}^{-1}$$

22.29 $$\Delta S = R \ln \frac{V_f}{V_i} = R \ln \frac{r_f^3}{r_i^3}$$

$$= R \ln \frac{l^3}{(N^{1/2} l)^3} = R \ln N^{-3/2}.$$

Plots of ΔS against N are shown in Figures 22.6 and 22.7.

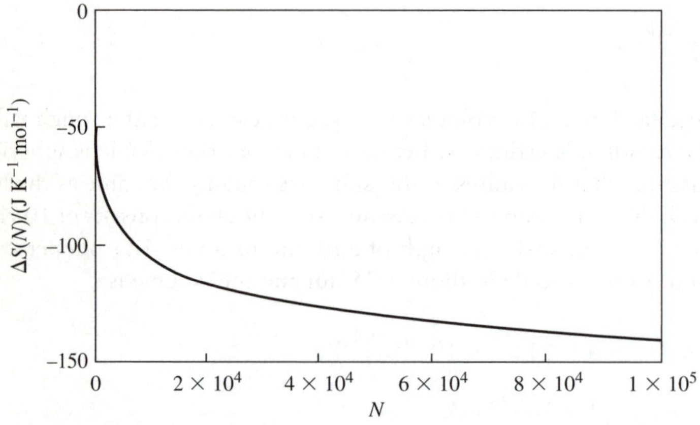

Fig 22.6

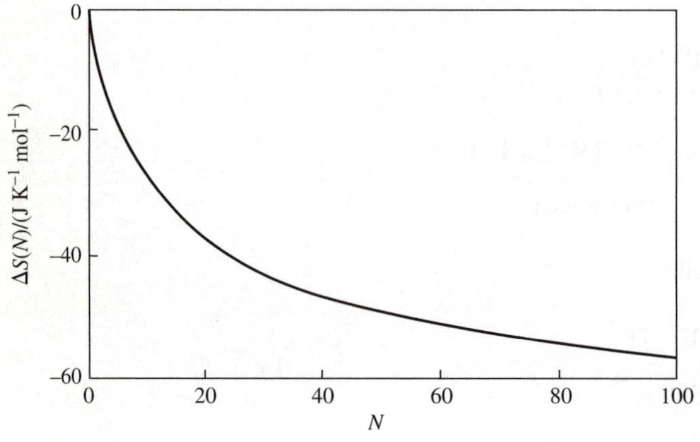

Fig 22.7

22.30 $G_m^\ominus - G_m^\ominus(0) = -RT \ln \dfrac{q_m^\ominus}{N_A}$.

In exercise 22.27 we calculated the value of $q_m^\ominus/N_A$ for N_2 as 3.02×10^8. Then,

$$G_m^\ominus - G_m^\ominus(0) = -8.3145\,\mathrm{J\,K^{-1}\,mol^{-1}} \times 298\,\mathrm{K} \times \ln(3.02 \times 10^8)$$

$$= \boxed{-48.4\,\mathrm{kJ\,mol^{-1}}}.$$

22.31 Follow the procedure of Example 22.5 in the text.

$$K = \frac{q_m^\ominus(\mathrm{Na^+})q_m^\ominus(\mathrm{e^-})}{q_m^\ominus(\mathrm{Na})N_A}\mathrm{e}^{-\Delta E/RT}.$$

$$q_m^\ominus(\mathrm{e^-}) = \frac{2(2\pi m_e kT)^{3/2}RT}{p^\ominus h^3}.$$

$$q_m^\ominus(\mathrm{Na^+}) = \frac{(2\pi m_{\mathrm{Na}} kT)^{3/2}RT}{p^\ominus h^3}.$$

$$q_m^\ominus(\mathrm{Na}) = 2q_m^\ominus(\mathrm{Na^+}).$$

$$K = \frac{(2\pi m_e kT)^{3/2}kT}{p^\ominus h^3}\mathrm{e}^{-I/RT} \qquad I = \Delta E = 495.8\,\mathrm{kJ\,mol^{-1}}$$

$$= \frac{(2\pi m_e)^{3/2}(kT)^{5/2}}{p^\ominus h^3}\mathrm{e}^{-I/RT}$$

$$= \frac{(2\pi \times 9.11 \times 10^{-31}\,\mathrm{kg})^{3/2}\,(1.381 \times 10^{-23}\,\mathrm{J\,K^{-1}} \times 1000\,\mathrm{K})^{5/2}}{10^5\,\mathrm{Pa} \times (6.626 \times 10^{-34}\,\mathrm{J\,s})^3}$$

$$\times\, \mathrm{e}^{-(4.958\times10^5/8.3145\times1000)}$$

$$= \boxed{1.37 \times 10^{-25}}.$$

22.32 Assume $\Delta E \approx \Delta H(\mathrm{I{-}I}) = 151\ \mathrm{kJ\ mol^{-1}}$.

$$K = \left[\frac{\left(q_{\mathrm{I,m}}^\ominus\right)^2}{\left(q_{\mathrm{I,m}}^\ominus\right)^2 N_A}\right]\mathrm{e}^{-\Delta E/RT}.$$

$$q_{\mathrm{I,m}}^\ominus = q_m^{\mathrm{Trans}}(\mathrm{I})q^{\mathrm{Elec}}(\mathrm{I}), \quad q^{\mathrm{Elec}}(\mathrm{I}) = 4.$$

$$(q_{\mathrm{I.m}}^\ominus)^2 = q_m^{\mathrm{Trans}}(\mathrm{I_2})q^{\mathrm{Rot}}(\mathrm{I_2})q^{\mathrm{Vib}}(\mathrm{I_2})q^{\mathrm{Elec}}(\mathrm{I_2})q^{\mathrm{Elec}}(\mathrm{I_2}) = 1.$$

See the solution to Exercise 22.27 for the partition function formulas.

$$\frac{q_m^{\mathrm{Trans}}}{N_A} = 2.561 \times 10^{-2}(T/\mathrm{K})^{5/2} \times (M/\mathrm{g\,mol^{-1}})^{3/2}.$$

$$\frac{q_m^{\text{Trans}}(I_2)}{N_A} = 2.561 \times 10^{-2} \times 500^{5/2} \times 253.8^{3/2} = 5.79 \times 10^8.$$

$$\frac{q_m^{\text{Trans}}(I)}{N_A} = 2.561 \times 10^{-2} \times 500^{5/2} \times 126.9^{3/2} = 2.05 \times 10^8.$$

$$q^{\text{Rot}}(I_2) = \frac{0.6950}{\sigma} \times \frac{T/K}{B/\text{cm}^{-1}} = \frac{1}{2} \times 0.6950 \times \frac{500}{0.0373}$$

$$= 4.66 \times 10^3.$$

$$q^{\text{Vib}}(I_2) = \frac{1}{1 - e^{-a}} \quad a = 1.4388\frac{\tilde{\nu}/\text{cm}^{-1}}{T/K}$$

$$= \frac{1}{1 - e^{-1.4388 \times 214.36/500}} = 2.17.$$

$$K = \frac{(2.05 \times 10^8)^2 \times 4^2 \times e^{-151 \times 10^3/8.3145 \times 500}}{5.78 \times 10^8 \times 4.66 \times 103 \times 2.17}$$

$$= \boxed{1.93 \times 10^{-11}}.$$

Solutions to box exercises

Box 1.1 *Exercise 1*

Air is roughly 80% $N_2(g)$ and 20% $O_2(g)$. There are also some minor components, but they do not much affect the average molar mass.

Molar mass air $= (0.80 \times 28 + 0.20 \times 32)\ \text{g mol}^{-1} \approx 29\ \text{g mol}^{-1}$.

Molar mass H_2 (g) $= 2.0\ \text{g mol}^{-1}$.

Hence,

$$\frac{\text{density } H_2(g)}{\text{density air}} = \frac{2.0\,\text{g mol}^{-1}}{29\,\text{g mol}^{-1}} = \boxed{0.069}.$$

In the same volume (the volume of the balloon), the mass of air would be $29\,\text{g mol}^{-1}/2.0\ \text{g mol}^{-1} = 14.5$ times the mass of hydrogen.

Mass of air displaced $= 14.5 \times 10\,\text{kg} = 145\,\text{kg}$.

The payload is the difference between the mass of displaced air and the mass of the balloon (here assumed to be the mass of hydrogen).

Payload $= 145\ \text{kg} - 10\ \text{kg} = \boxed{135\,\text{kg}}$.

Box 1.1 *Exercise 2*

We use the perfect gas law, $pV = nRT$, and solve for V. We calculate n from the mass of SO_2.

$$n = \frac{\text{mass}}{\text{molar mass}} = \frac{250 \times 10^3\,\text{kg}}{64\,\text{g mol}^{-1}} = 3.9 \times 10^6\text{mol}.$$

$$V = \frac{nRT}{p} = \frac{3.9 \times 10^6\text{mol} \times 0.082\,\text{dm}^3\,\text{atm}\,\text{K}^{-1}\text{mol}^{-1} \times 1100\,\text{K}}{1.0\,\text{atm}}$$

$$= \boxed{3.5 \times 10^8\ \text{dm}^3}.$$

Box 1.2 *Exercise 1*

We use

$$p = \frac{nRT}{V} = \frac{\rho RT}{M}$$

where M is the average molar mass of the particles. For each C^{6+} ion, there are 6 electrons. The average molar mass is therefore

$$M = \frac{12\,\text{g mol}^{-1} + 6 \times 0\,\text{g mol}^{-1}}{7} = 1.7\,\text{g mol}^{-1}.$$

$$p = \frac{1.20 \times 10^{3}\,\text{kg m}^{-3} \times 8.3145\,\text{J K}^{-1}\text{mol}^{-1} \times 3.5 \times 10^{3}\,\text{K}}{1.7\,\text{g mol}^{-1}}$$

$$= \boxed{2.1 \times 10^{7}\,\text{Pa}}.$$

Box 1.2 *Exercise 2*

The average molar mass is now 12 g mol^{-1}; then, if the mass density were the same, the pressure would be

$$p = \boxed{3.0 \times 10^{6}\,\text{Pa}}.$$

However, the mass density would not be the same. It is hard to say what it would be under these circumstances, but a best guess might be $\frac{12}{1.7} \times 1.20\,\text{g cm}^{-3}$. Then the pressure would be 2.1×10^{7} Pa, as in Box 1.2 Exercise 1.

Box 2.1 *Exercise 1*

The physical or chemical change associated with either exothermic or endothermic processes alters the heat capacity of the matter. This alters the thermogram baseline of the DSC.

Box 2.1 *Exercise 2*

The area of a DSC peak for a physical transformation is proportional to the number of moles undergoing the transformation. The ratio of the observed area to the area expected for a pure sample equals the mole fraction composition of the sample. See Figure B1 for an example of how impurities can change the shape of a thermogram curve.

Impurities change the slope of melting transformation by moving the maximum to a lower value and skewing the curve toward lower temperatures. Software of DSC instruments is designed to compute the mole percentage composition from the observed transformation curve.

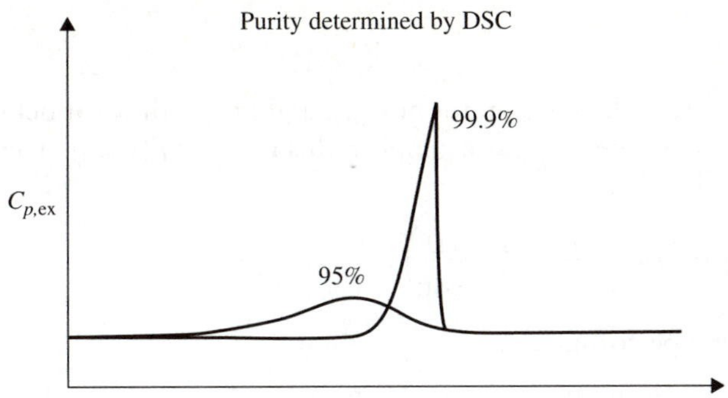

Fig B1

Box 3.1 *Exercise 1*

(a) $1.0\,\text{dm}^3 \simeq 1000\,\text{g}$ water.

$$n = \frac{1000\,\text{g}}{18\,\text{g mol}^{-1}} = 56\,\text{g}.$$

$$\Delta H_{\text{water}} = n\Delta_{\text{vap}}H = (56\,\text{mol}) \times \left(40.7\,\text{kJ mol}^{-1}\right) = 2.3 \times 10^3\,\text{kJ}.$$

$$\Delta H_{\text{runner}} = \boxed{-2.3 \times 10^3\,\text{kJ}}.$$

(b) $\Delta H = mC_p\Delta T$.

$$\Delta T = \frac{\Delta H}{mC_p} = \frac{2.3 \times 10^6\,\text{J}}{6.0 \times 10^4\,\text{g} \times 4.184\,\text{JK}^{-1}\text{g}^{-1}}$$

$$= \boxed{9.2\,\text{K}}.$$

Box 3.1 *Exercise 2*

The carbohydrates in pasta break down into glucose during the digestion process.

$$n = \frac{\text{mass}}{\text{molar mass}} = \frac{40\,\text{g}}{180\,\text{g mol}^{-1}} = 0.22\,\text{mol}.$$

$$\Delta H = n\Delta_{\text{c}}H(\text{glucose})$$

$$= 0.22\,\text{mol} \times \left(-2808\,\text{kJ mol}^{-1}\right) = -620\,\text{kJ}.$$

$$1\,\text{kcal} = 4.184\,\text{kJ}.$$

$$\Delta H = -\frac{-620\,\text{kJ}}{4.184\,\text{kJ/kcal}} = -148\,\text{kcal}.$$

$$\frac{148\,\text{kcal}}{2200\,\text{kcal}} \times 100\% = \boxed{6.7\%}.$$

Box 4.1 *Exercise 1*

Let $|q|$ be the heat extracted from the refrigerator at T_{cold} and $|q'|$ be the heat delivered to environment at T_{hot}. The work needed to accomplish this is $w = |q'| - |q|$. The heat transfer occurs most efficiently when

$$\Delta S_{\text{total}} = \frac{|q'|}{T_{\text{hot}}} - \frac{|q|}{T_{\text{cold}}} = 0 \quad \text{or} \quad \frac{|q|}{|q'|} = \left(\frac{T_{\text{cold}}}{T_{\text{cold}}}\right).$$

The best coefficient of cooling performance is

$$c_{\text{cool}} = \frac{|q|}{w} = \frac{|q|}{|q'| - |q|} = \frac{(|q|/|q'|)}{1 - (|q|/|q'|)} = \frac{(T_{\text{cold}}/T_{\text{hot}})}{1 - (T_{\text{cold}}/T_{\text{hot}})} = \boxed{\frac{T_{\text{cold}}}{T_{\text{hot}} - T_{\text{cold}}}}.$$

Sample calculation:

Since $|q| = c_{\text{cool}}\, w$, the rate of extracting heat equals c_{cool} multiplied by the refrigerator power rating.

$$\text{Rate of heat extraction} = \left(\frac{T_{\text{cold}}}{T_{\text{hot}} - T_{\text{cold}}}\right) \times \text{power rating}$$

$$= \left(\frac{278\,\text{K}}{295\,\text{K} - 278\,\text{K}}\right) \times 200\text{W} = \boxed{3.27\,\text{kW}}.$$

Box 4.1 *Exercise 2*

Let $|q|$ be the heat extracted by the heat pump at T_{cold} and $|q'|$ be the heat delivered at T_{hot}. The work needed to accomplish this is $w = |q'| - |q|$. The heat transfer occurs most efficiently when

$$\Delta S_{\text{total}} = \frac{|q'|}{T_{\text{hot}}} - \frac{|q|}{T_{\text{cold}}} = 0 \quad \text{or} \quad \frac{|q|}{|q'|} = \left(\frac{T_{\text{cold}}}{T_{\text{cold}}}\right).$$

The best coefficient of heating performance is

$$c_{\text{warm}} = \frac{|q'|}{w} = \frac{|q'|}{|q'| - |q|} = \frac{1}{1 - (|q|/|q'|)} = \frac{1}{1 - (T_{\text{cold}}/T_{\text{hot}})} = \boxed{\frac{T_{\text{hot}}}{T_{\text{hot}} - T_{\text{cold}}}}.$$

Sample calculation:

Since $|q'| = c_{\text{warm}}\, w$, the rate of heat delivery equals c_{warm} multiplied by the heat pump power rating.

$$\text{Rate of heat delivery} = \left(\frac{T_{\text{hot}}}{T_{\text{hot}} - T_{\text{cold}}}\right) \times \text{power rating} = \left(\frac{295\,\text{K}}{295\,\text{K} - 291\,\text{K}}\right) \times 2.5\,\text{kW}$$

$$= \boxed{184\,\text{kW}}.$$

Practical heat pumps are not reversible so the heat gain is less.

Box 5.1 *Exercise 1*

The supercritical fluid extractor consists of a pump to pressurize the solvent (e.g. CO_2), an oven with extraction vessel, and a trapping vessel. Extractions are performed dynamically or statically. Supercritical fluid flows continuously through the sample within the extraction vessel when operating in dynamic mode. Analytes extracted into the fluid are released through a pressure-maintaining restrictor into a trapping vessel. In static mode the supercritical fluid circulates repetitively through the extraction vessel until being released into the trapping vessel after a period of time. Supercritical carbon dioxide volatilizes when decompression occurs upon release into the trapping vessel.

Advantages	Disadvantages	Current uses
Dissolving power of SCF can be adjusted with selection of T and p	Elevated pressures are required and the necessary apparatus expensive	Extraction of caffeine, fatty acids, spices, aromas, flavors, and biological materials from natural sources
Select SCFs are inexpensive and non-toxic. They reduce pollution	Cost may prohibit large-scale applications	Extraction of toxic salts (with a suitable chelation agent) and organics from contaminated water
Thermally unstable analytes may be extracted at low temperature	Modifiers like methanol (1–10%) may be required to increase solvent polarity	Extraction of herbicides from soil
The volatility of $scCO_2$ makes it easy to isolate analyte	$scCO_2$ is toxic to whole cells in biological applications (CO_2 is not toxic to the environment.)	scH_2O oxidation of toxic, intractable organic waste during water treatment
SCFs have high diffusion rates, low viscosity, and low surface tension		Synthetic chemistry, polymer synthesis and crystallization, textile processing
O_2 and H_2 are completely miscible with $scCO_2$. This reduces multi-phase reaction problems		Heterogeneous catalysis for green chemistry processes

Box 5.1 *Exercise 2*

The critical point corresponds to a point of zero slope which is simultaneously a point of inflection in a plot of pressure versus molar volume. A critical point exists if there are values of p, V, and T that result in a point which satisfies these conditions. Letting $V_m = V/n$, the equation of state becomes

$$p = \frac{RT}{V_m} - \frac{a}{V_m^2} + \frac{b}{V_m^3}.$$

$$\left.\begin{aligned} \left(\frac{\partial p}{\partial V_m}\right)_T &= -\frac{RT}{V_m^2} + \frac{2a}{V_m^3} - \frac{3b}{V_m^4} = 0 \\ \left(\frac{\partial^2 p}{\partial V_m^2}\right)_T &= \frac{2RT}{V_m^3} - \frac{6a}{V_m^4} + \frac{12b}{V_m^5} = 0 \end{aligned}\right\} \text{at the critical point.}$$

$$\text{That is,} \quad \left.\begin{aligned} -RT_c V_c^2 + 2aV_c - 3b = 0 \\ RT_c V_c^2 - 3aV_c + 6b = 0 \end{aligned}\right\}.$$

These two independent equations describe a critical point if they have a real solution. Solving these simultaneous equations for V_c and T_c gives

$$\boxed{V_c = \frac{3b}{a}} \quad \text{and} \quad \boxed{T_c = \frac{a^2}{3Rb}}.$$

Now use the equation of state to find p_c.

$$p_c = \frac{RT_c}{V_c} - \frac{a}{V_c^2} + \frac{b}{V_c^3} = \left(\frac{Ra^2}{3Rb}\right) \times \left(\frac{a}{3b}\right) - a\left(\frac{a}{3b}\right)^2 + b\left(\frac{a}{3b}\right)^3 = \boxed{\frac{a^3}{27b^2}}.$$

It follows that $Z_c = \dfrac{p_c V_c}{RT_c} = \left(\dfrac{a^3}{27b^2}\right) \times \left(\dfrac{3b}{a}\right) \times \left(\dfrac{1}{R}\right) \times \left(\dfrac{3Rb}{a^2}\right) = \boxed{\dfrac{1}{3}}.$

Having found a real inflection point, we conclude that this equation of state does describe a critical point and that the critical constants are related to the parameters a and b by the above equations.

Box 6.1 *Exercise 1*

The 97 per cent saturated hemoglobin in the lungs releases oxygen in the capillary until the hemoglobin is 75 per cent saturated.

100 cm^3 of blood in the lung containing 15 g of Hb at 97 per cent saturated with O_2 binds

$$1.34\,cm^3 g^{-1} \times 15\,g = 20\,cm^3 O_2.$$

The same 100 cm^3 of blood in the arteries would contain

$$20\,cm^3 O_2 \times \frac{75\%}{97\%} = 15.5\,cm^3.$$

Therefore, about $(20 - 15.5)\,cm^3$ or $\boxed{4.5\,cm^3}$ of O_2 is given up in the capillaries to body tissue.

Box 6.1 *Exercise 2*

In this case, we write the Henry's law expression as

mass of $N_2 = p_{N_2} \times$ mass of $H_2O \times K_{N_2}$.

(1) At $p_{N_2} = 0.78 \times 4.0\,atm = 3.1\,atm$,

mass of $N_2 = 3.1\,atm \times 100\,gH_2O \times 1.8 \times 10^{-4}\,mg\,N_2/(g\,H_2O\,atm)$

$= \boxed{0.056\text{ mg }N_2}$.

(2) At $p_{N_2} = 0.78\,atm$

mass of $N_2 = \boxed{0.0014\text{ mg }N_2}$.

(3) In fatty tissue the increase in N_2 concentration from 1 atm to 4 atm is $4 \times (0.056 - 0.014)$ mg $N_2 = \boxed{0.17\text{ mg }N_2}$.

Box 6.2 *Exercise 1*

To examine the process of zone leveling with the phase diagram shown in Figure B2, consider a solid on the isopleth through a_1 and heat the sample without coming to overall equilibrium. If the temperature rises to a_2, a liquid of composition b_2 forms and the remaining solid is at a'_2. Heating that solid down an isopleth passing through a'_2 forms a liquid of composition b_3 and leaves the solid at a'_3. This sequence of heater passes shows that in a pass the impurities at the end of a sample are reduced while being transferred to the liquid phase, which moves with the heater down the length of the sample. With enough passes the dopant, which is initially at the end of the sample, is distributed evenly throughout.

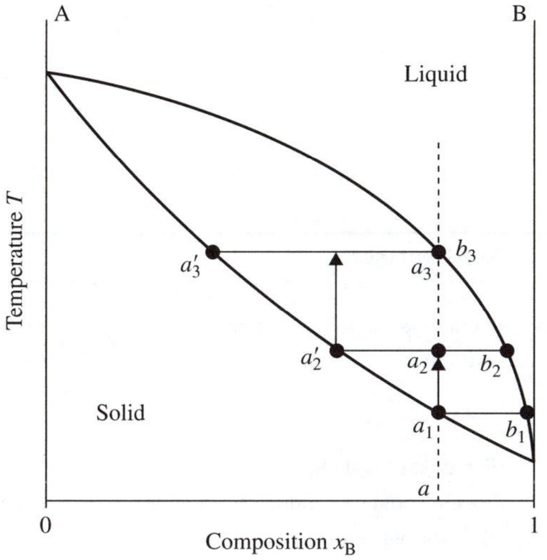

Fig B2

Box 6.2 *Exercise 2*

In the float zoning (FZ) method of silicon purification, a polycrystalline silicon rod is positioned atop a seed crystal and lowered through an electromagnetic coil (see Figure B3). The magnetic field generated by the coil creates electric currents, heating, and local melting in the rod. By slowly moving the coil upward impurities move with the melt zone. The lower surface of the melt zone solidifies to an ultrapure, single crystal as it slowly cools. Search <www.nrel.gov>.

Box 7.1 *Exercise 1*

In order to calculate the fractional saturation at the requested pressure we need to know the value of K in the Hill equation. K can be determined approximately from the figure shown in the Box. For $s = 0.5, p = p_{50}$ where p_{50} is the pressure of O_2 at 50% saturation.

$$\log\left(\frac{0.5}{1-0.5}\right) = \nu \log p_{50} - \nu \log K.$$

Therefore, $K = p_{50}$.

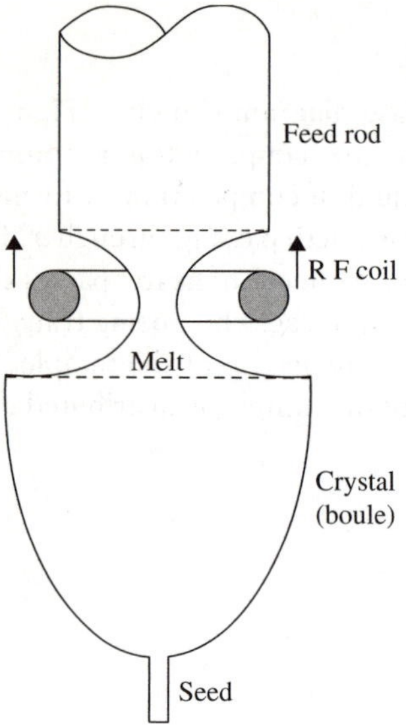

Fig B3

Advantages	Disadvantages
Produces ultrapure silicon for high efficiency photovoltaic cells and infrared detectors for space, defense, and environmental applications	Requires a smooth, uniform diameter, and crack-free feed rod
No crucible contamination	High cost of heating
Produces large boules (10 cm diameter)	Process must be conducted under helium or argon and 10^{-5} Torr vacuum
	Boron impurity is not removed from silicon
	Boule must be sliced with a diamond saw into thin wafers for microelectronic devices. This reduces the useful volume of the boule

From the graph we estimate that, for hemoglobin, $p_{50} = K = 26$ Torr; for myoglobin, $p_{50} = K = 5$ Torr. Solving the Hill equation for s we obtain

$$\frac{s}{1-s} = \left(\frac{p_{O_2}}{K}\right)^{\nu}$$

and

$$s = \frac{1}{\left(\frac{K}{p_{O_2}}\right)^{\nu} + 1}$$

Then for Hb we construct the following table.

p_{O_2}/Torr	5	10	20	30	60
s	9.8×10^{-3}	0.064	0.32	0.60	0.91

For Mb:

p_{O_2}/Torr	5	10	20	30	60
s	0.50	0.66	0.80	0.86	0.92

The values of s match well the values read from the graph.

Box 7.1 *Exercise 2*

Again we use $s = \dfrac{1}{\left(\dfrac{K}{p_{O_2}}\right)^{\nu} + 1}$ and construct the following tables.

For Hb:

p_{O_2}/Torr	5	10	20	30	60
s	1.4×10^{-3}	0.021	0.26	0.64	0.996

For Mb:

p_{O_2}/Torr	5	10	20	30	60
s	0.50	0.94	0.996	0.9992	0.99995

For $\nu = 4$, the curves lose their 'S', sigmoid shape.

Box 8.1 *Exercise 1*

$$H_2CO_3(aq) \rightleftharpoons H^+(aq) + HCO_3^-(aq).$$

$$K_{a1} = \frac{[H^+][HCO_3^-]}{[H_2CO_3]} = 7.9 \times 10^{-7} \text{at the physiological temperature of } 37°\text{C}.$$

$$\frac{[HCO_3^-]}{[H_2CO_3]} = \frac{7.9 \times 10^{-7}}{[H^+]} = \frac{7.9 \times 10^{-7}}{10^{-pH}}.$$

$$\text{Normal ratio (pH = 7.40): } \frac{[HCO_3^-]}{[H_2CO_3]} = \frac{7.9 \times 10^{-7}}{10^{-7.40}} = 20.\bar{0}.$$

$$\text{Ratio at onset of acidosis (pH = 7.35): } \frac{[HCO_3^-]}{[H_2CO_3]} = \frac{7.9 \times 10^{-7}}{10^{-7.35}} = \boxed{17.\bar{7}}.$$

$$\text{Ratio at onset of alkalosis (pH = 7.45): } \frac{[HCO_3^-]}{[H_2CO_3]} = \frac{7.9 \times 10^{-7}}{10^{-7.45}} = \boxed{22.\bar{3}}.$$

Box 8.1 *Exercise 2*

The Bohr effect observes that hemoglobin binds O_2 strongly when it is deprotonated at slightly higher blood pH. That is, the Hb saturation s, which is proportional to bound O_2, increases. The mathematic model

$$\log\left(\frac{s}{1-s}\right) = \nu \log p_{O_2} - \nu \log K \quad \text{where } \nu \text{ is the Hill coefficient}$$

defines the Hill coefficient. At high partial pressures, the last term is not the major influence. Ignoring it, we see that the left side of the equation increases as s and, consequently, the Hill coefficient must increase should this occur at constant pressure.

We conclude that the $\boxed{\text{Hill coefficient increases with a slight increase in pH}}$.

Box 9.1 *Exercise 1*

With the equilibrium approximation $\Delta G_m = 0$, the electrostatic potential difference for $[K^+]_{in}/[K^+]_{out} = 20$ is estimated by solving the following equation for $\Delta\phi$.

$$\Delta G_m = RT \ln \frac{[K^+]_{in}}{[K^+]_{out}} + zF\Delta\phi = 0 \qquad \text{or} \qquad \Delta\phi = -\frac{RT}{zF} RT \ln \frac{[K^+]_{in}}{[K^+]_{out}},$$

$$\Delta\phi = -\frac{\left(8.3145\,\text{J}\,\text{K}^{-1}\,\text{mol}^{-1}\right)(298\,\text{K})}{9.649 \times 10^4\,\text{C}\,\text{mol}^{-1}} \ln 20 = \boxed{-76.9\,\text{mV}} \quad [\text{Note}: 1\,\text{J}\,\text{C}^{-1} = 1\,\text{V}].$$

With the equilibrium approximation $\Delta G_m = 0$, the electrostatic potential difference for $[Na^+]_{in}/[Na^+]_{out} = 0.10$ is estimated by solving the following equation for $\Delta\phi$.

$$\Delta G_m = RT \ln \frac{[Na^+]_{in}}{[Na^+]_{out}} + zF\Delta\phi = 0 \qquad \text{or} \qquad \Delta\phi = -\frac{RT}{zF} RT \ln \frac{[Na^+]_{in}}{[Na^+]_{out}},$$

$$\Delta\phi = -\frac{\left(8.3145\,\text{J}\,\text{K}^{-1}\,\text{mol}^{-1}\right)(298\text{K})}{9.649 \times 10^4\,\text{C}\,\text{mol}^{-1}} \ln 0.10 = \boxed{+59.1\,\text{mV}} \quad [\text{Note}: 1\,\text{J}\,\text{C}^{-1} = 1\text{V}].$$

Only the $\Delta\phi$ value for K^+ is comparable to the observed resting potential of -62 mV. The $\Delta\phi$ value for Na^+ has the opposite sign. These computations are not expected to match the resting potential because the cell is not at equilibrium.

Box 9.1 *Exercise 2*

$$\Delta\phi = \frac{RT}{F} \ln \left(\frac{\sum_i P_i [M_i^+]_{out} + \sum_j P_j [X_j^-]_{in}}{\sum_i P_i [M_i^+]_{in} + \sum_j P_j [X_j^-]_{out}} \right) \qquad [\text{Goldman equation}].$$

$$\Delta\phi = \frac{\left(8.3145\,\text{J}\,\text{K}^{-1}\,\text{mol}^{-1}\right)(298\,\text{K})}{9.649 \times 10^4\,\text{C}\,\text{mol}^{-1}} \ln \left\{ \frac{(0.04 \times 440) + (1.0 \times 20) + (0.45 \times 50)}{(0.04 \times 50) + (1.0 \times 400) + (0.45 \times 560)} \right\}$$

$$= \boxed{-61.\bar{3}\,\text{mV}} \qquad [\text{Note}: 1\,\text{J}\,\text{C}^{-1} = 1\,\text{V}].$$

The Goldman equation is in good agreement with the observed resting potential of -62 mV.

Box 9.2 *Exercise 1*

(a) The cell reaction is

$$H_2(g) + \frac{1}{2}O_2(g) \rightarrow H_2O(l).$$

$$\Delta_r G^\ominus = \Delta_f G^\ominus (H_2O, l) = -237.13\,\text{kJ mol}^{-1} \quad [\text{TableD1.2}].$$

$$E^\ominus = -\frac{\Delta_r G^\ominus}{\nu F}[9.13] = \frac{+237.13\,\text{kJ mol}^{-1}}{(2) \times (96.485\,\text{kC mol}^{-1})} = \boxed{+1.23\,\text{V}}.$$

(b) The cell reaction is

$$C_6H_6(l) + \frac{15}{2}O_2(g) \rightarrow 6CO_2(g) + 3H_2O(l).$$

$$\begin{aligned}\Delta_f G^\ominus &= 6\Delta_f G^\ominus(CO_2, g) + \Delta_f G^\ominus(H_2O_2, l) - \Delta_f G^\ominus(C_6H_6, l) \\ &= (6) \times (-394.36) + (5) \times (-237.13) - (+124.3)]\,\text{kJ mol}^{-1} \\ &\quad [\text{Tables D1.1 and D1.2}] \\ &= -\ 3676.1\,\text{kJ mol}^{-1}\end{aligned}$$

In this reaction the number of electrons transferred, ν, is not immediately apparent as in part (a). To find ν we break the cell reaction down into half-reactions as follows

$$\text{R:}\quad \frac{15}{2}O_2(g) + 30e^- + 30H^+(aq) \rightarrow 15H_2O(l).$$

$$\text{L:}\quad 6CO_2(g) + 30e^- + 30H^+(aq) \rightarrow C_6H_6(l) + 12H_2O(l).$$

$$\text{R} - \text{L:}\quad C_6H_6(l) + \frac{15}{2}O_2(g) \rightarrow 6CO_2(g) + 3H_2O(l).$$

Hence, $\nu = 30$.

Therefore, $E = \dfrac{-\Delta G^\ominus}{\nu F} = \dfrac{+3676.1\,\text{kJ mol}^{-1}}{(30) \times (96.485\,\text{kC mol}^{-1})} = \boxed{+1.27\,\text{V}}.$

Box 9.2 *Exercise 2*

The simplest methane fuel cell oxidizes methane at the anode and reduces oxygen at the cathode.

Anode: $CH_4(g) + 8OH^-(aq) \rightarrow CO_2(g) + 6H_2O(l) + 8e^-$.

Cathode: $O_2(g) + 2H_2O(l) + 4e^- \rightarrow 4OH^-(aq)$.

Net reaction: $CH_4(g) + 2O_2(g) \rightarrow CO_2(g) + 2H_2O(l) \quad \nu = 8$.

$$\begin{aligned}\Delta G^\ominus &= \{1 \times (-394.36) + 2 \times (-237.13) - 1 \times (-50.72)\}\,\text{kJ mol}^{-1} \\ &= -817.9\,\text{kJ mol}^{-1}.\end{aligned}$$

$$E^\ominus = -\frac{\Delta G^\ominus}{\nu F} \quad [9.15] = -\frac{(-817.8 \times 10^3)\,\text{J mol}^{-1}}{8 \times \left(9.6485 \times 10^4\,\text{C mol}^{-1}\right)} = 1.06\,\text{V}.$$

However, we view the challenge of this exercise as a challenge to imagine a fuel cell in which methane is consumed at one electrode and produced at the other while being fed carbon monoxide as the cell oxidant. We imagine the oxidation of methane at the anode and the reduction of carbon monoxide at the cathode.

Anode: $CH_4(g) + 8OH^-(aq) \rightarrow CO_2(g) + 6H_2O(l) + 8e^-$.

Cathode: $CO(g) + 5H_2O(l) + 6e^- \rightarrow CH_4(g) + 6OH^-(aq)$.

Net reaction: $4CO(g) + 2H_2O(l) \rightarrow CH_4(g) + 3CO_2(g) \quad \nu = 24$.

$$\Delta G^\ominus = \{1 \times (-50.72) + 3 \times (-394.36) - [4 \times (-137.17) + 2\,(-237.13)]\}\ \mathrm{kJ\,mol^{-1}}$$

$$= -210.86\ \mathrm{kJ\,mol^{-1}}.$$

$$E^\ominus = -\frac{\Delta G^\ominus}{\nu F}\ [9.15] = -\frac{(-210.86 \times 10^3)\ \mathrm{J\,mol^{-1}}}{24 \times \left(9.6485 \times 10^4\ \mathrm{C\,mol^{-1}}\right)} = \boxed{0.09\ \mathrm{V}}.$$

This fuel cell produces methane spontaneously but it gives only a very small zero-current potential.

Box 10.1 *Exercise 1*

The time-scales of atomic processes are rapid indeed: according to the following table, a nanosecond is an eternity. Note that the times given here are in some way typical values for times that may vary over two or three orders of magnitude. For example, vibrational wavenumbers can range from about 4400 cm^{-1} (for H_2) to 100 cm^{-1} (for I_2) and even lower, with a corresponding range of associated times. Radiative decay rates of electronic states can vary even more widely. Times associated with phosphorescence can be in the millisecond and even second range. A large number of time-scales for physical, chemical, and biological processes on the atomic and molecular scale are reported in Figure 2 of A.H. Zewail, Femtochemistry: atomic-scale dynamics of the chemical bond. *J. Phys. Chem. A* **104**, 5660 (2000).

Radiative decay of excited electronic states can range from about 10^{-9} s to 10^{-4} s—even longer for phosphorescence involving 'forbidden' decay paths. Molecular rotational motion takes place on a scale of 10^{-12} to 10^{-9} s. Molecular vibrations are faster still, about 10^{-14} to 10^{-12} s. The mean time between collisions in liquids is similarly short, 10^{-14} to 10^{-13} s. Proton transfer reactions occur on a time-scale of about 10^{-10} to 10^{-9} s. Box 20.1 *Vision* describes several events in vision, including the 200-fs photoisomerization that gets the process started. Box 20.2 *Photosynthesis* lists time-scales of several energy-transfer and electron-transfer steps in photosynthesis. Initial energy transfer (to a nearby pigment) has a time-scale of around 10^{-13} to 10^{-11} s, with longer-range transfer (to the reaction center) taking about 10^{-10} s. Immediate electron transfer is also very fast (about 3 ps), with ultimate transfer (leading to oxidation of water and reduction of plastoquinone) taking from 10^{-10} to 10^{-3} s. Box 11.1 *Kinetics of protein folding* discusses helix–coil transitions, including experimental measurements of time-scales of tens or hundreds of microseconds (10^{-5} to 10^{-4} s) for formation of tightly packed cores. The rate-determining step for the helix–coil transition of small polypeptides has a relaxation time of about 160 ns in contrast to the faster 50 ns relaxation time of large protein.

Process	t/ns
Radiative decay of electronic excited state	1×10^{1}
Rotational motion	3×10^{-2}
Vibrational motion	3×10^{-5}
Proton transfer (in water)	2×10^{-5}
Initial chemical reaction of vision*	1×10^{-4}
Energy transfer in photosynthesis†	1×10^{-3}
Electron transfer in photosynthesis	3×10^{-3}
Polypeptide helix–coil transition	2×10^{2}
Collision frequency in liquids‡	4×10^{-4}
Harpoon reactions	5×10^{-2}

* Photoisomerization of retinal from 11-*cis* to all-*trans*.
† Time from absorption until electron transfer to adjacent pigment.
‡ Use formula for gas collision frequency at 300 K, parameters for benzene from data section, and density of liquid benzene.

Box 10.1 *Exercise 2*

[CO] changes little during the course of the reaction for the concentration given, so [Mb] follows pseudofirst-order kinetics.

$$[\mathrm{Mb}] = [\mathrm{Mb}]_0\, \mathrm{e}^{-k't}.$$

k' is the rate constant for the pseudofirst-order process. The rate constant, k, which is given, is for the second-order process. That is,

$$\text{rate} = \mathrm{k}\,[\mathrm{CO}]\,[\mathrm{Mb}] = k'\,[\mathrm{Mb}]\,.$$

$$k' = k\,[\mathrm{CO}]$$

$$= 5.8 \times 10^{5}\,\mathrm{dm^3\,mol^{-1}\,s^{-1}} \times 0.400\,\mathrm{mol\,dm^{-3}}$$

$$= 2.3 \times 10^{5}\,\mathrm{s^{-1}}.$$

A curve of [Mb] against time using this value of k' is shown in Figure B4.

Box 11.1 *Exercise 1*

(a) For the mechanism

$$hhhh\ldots \underset{k'_{\mathrm{a}}}{\overset{k_{\mathrm{a}}}{\rightleftarrows}} hchh\ldots$$

$$hchh\ldots \underset{k'_{\mathrm{b}}}{\overset{k_{\mathrm{b}}}{\rightleftarrows}} cccc\ldots$$

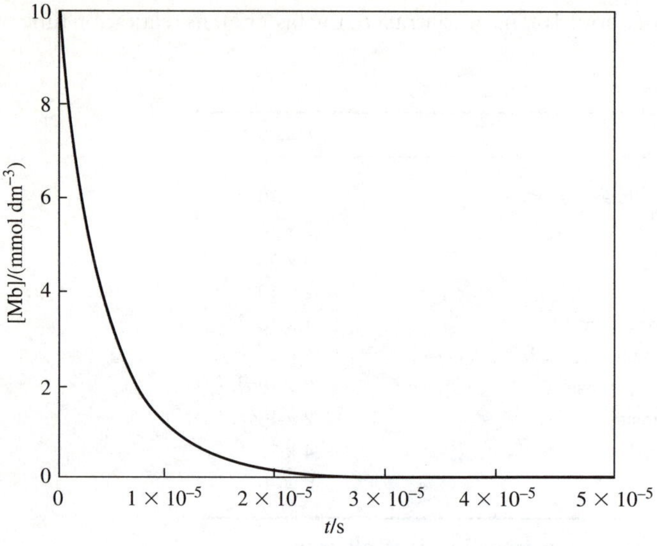

Fig B4

the rate equations are

$$\frac{\mathrm{d}[hhhh\ldots]}{\mathrm{d}t} = -k_\mathrm{a}[hhhh\ldots] + k'_\mathrm{a}[hchh\ldots],$$

$$\frac{\mathrm{d}[hchh\ldots]}{\mathrm{d}t} = k_\mathrm{a}[hhhh\ldots] - k'_\mathrm{a}[hchh\ldots] - k_\mathrm{b}[hchh\ldots] + k'_\mathrm{b}[cccc\ldots],$$

$$\frac{\mathrm{d}[cccc\ldots]}{\mathrm{d}t} = k_\mathrm{b}[hchh\ldots] - k'_\mathrm{b}[cccc\ldots].$$

(b) Apply the steady-state approximation to the intermediate.

$$\frac{\mathrm{d}[hchh\ldots]}{\mathrm{d}t} = k_\mathrm{a}[hhhh\ldots] - k'_\mathrm{a}[hchh\ldots] - k_\mathrm{b}[hchh\ldots] + k'_\mathrm{b}[cccc\ldots] = 0$$

so $[hchh\ldots] = \dfrac{k_\mathrm{a}[hhhh\ldots] + k'_\mathrm{b}[cccc\ldots]}{k'_\mathrm{a} + k_\mathrm{b}}$.

Therefore, $\dfrac{\mathrm{d}[hhhh\ldots]}{\mathrm{d}t} = -\dfrac{k_\mathrm{a}k_\mathrm{b}}{k'_\mathrm{a} + k_\mathrm{b}}[hhhh\ldots] + \dfrac{k'_\mathrm{a}k'_\mathrm{b}}{k'_\mathrm{a} + k_\mathrm{b}}[cccc\ldots]$.

This rate expression may be compared to that given in the text [Section 11.1] for the mechanism $\mathrm{A} \underset{k'}{\overset{k}{\rightleftarrows}} \mathrm{B}$.

Here $hhhh\ldots \underset{k'_\mathrm{eff}}{\overset{k_\mathrm{eff}}{\rightleftarrows}} cccc\ldots$ with $\quad k_\mathrm{eff} = \dfrac{k_\mathrm{a}k_\mathrm{b}}{k'_\mathrm{a} + k_\mathrm{b}} \quad k'_\mathrm{eff} = \dfrac{k'_\mathrm{a}k'_\mathrm{b}}{k'_\mathrm{a} + k_\mathrm{b}}$

Box 11.1 *Exercise 2*

It is difficult to make conclusive inferences about intermediates from kinetic data alone. For example, if rate measurements show formation of coils from helices with a single rate constant, they tell us nearly nothing about the mechanism. The rate law

$$\frac{d[cccc\ldots]}{dt} = k[hhhh\ldots]$$

is consistent with a single-step mechanism, with a two-step mechanism with a rate-determining second step, and with a two-step mechanism with a steady-state intermediate. Even if kinetic monitoring of the product shows production with two rate constants, the rate constants could belong to competing paths or to steps of a single reaction path. The best evidence for an intermediate's participation in a reaction is detection of the intermediate, or at least detection of structural features that can belong to a proposed intermediate but not reactant or product.

Box 11.2 *Exercise 1*

(a) The figure suggests that a chain-branching explosion does not occur at temperatures as low as 700 K. There may, however, be a thermal explosion regime at pressures in excess of 10^6 Pa.

(b) The lower limit seems to occur when

$$\log(p/\text{Pa}) = 2.1 \quad \text{so} \quad p = 10^{2.1}\,\text{Pa} = \boxed{1.3 \times 10^2\,\text{Pa}}.$$

There does not seem to be a pressure above which a steady reaction occurs. Rather the chain-branching explosion range seems to run into the thermal explosion range around

$$\log(p/\text{Pa}) = 4.5 \quad \text{so} \quad p = 10^{4.5}\,\text{Pa} = \boxed{3 \times 10^4\,\text{Pa}}.$$

Box 11.2 *Exercise 2*

The two equations that describe the time evolution of hydrogen radial concentration and the conditions under which they are applicable (i.e. make [H] a positive value) are

$$[\text{H}] = \frac{v_{\text{initiation}}}{k_{\text{termination}} - k_{\text{branching}}}\left\{1 - e^{-(k_{\text{termination}} - k_{\text{branching}})t}\right\}$$

when $k_{\text{termination}} > k_{\text{branching}}$ and $[O_2]$ is low

$$= \frac{v_{\text{initiation}}}{|\Delta k|}\left\{1 - e^{-|\Delta k|t}\right\} \quad \text{where } |\Delta k| = |k_{\text{termination}} - k_{\text{branching}}|$$

and

$$[\text{H}] = \frac{v_{\text{initiation}}}{k_{\text{termination}} - k_{\text{branching}}}\left\{1 - e^{-(k_{\text{termination}} - k_{\text{branching}})t}\right\}$$

when $k_{\text{branching}} > k_{\text{termination}}$ and $[O_2]$ is high

$$= \frac{-v_{\text{initiation}}}{k_{\text{branching}} - k_{\text{termination}}} \left\{1 - e^{(k_{\text{branching}} - k_{\text{termination}})t}\right\}$$

$$= \frac{v_{\text{initiation}}}{k_{\text{branching}} - k_{\text{termination}}} \left\{e^{(k_{\text{branching}} - k_{\text{termination}})t} - 1\right\}$$

$$= \frac{v_{\text{initiation}}}{|\Delta k|} \left\{e^{|\Delta k|t} - 1\right\}.$$

$[\mathrm{H}] \times |\Delta k| / v_{\text{initiation}}$ is plotted against $|k|\, t$ under both conditions in Figure B5. The conditions for the rapid growth of [H], and explosion, are $\boxed{k_{\text{branching}} > k_{\text{termination}} \text{ and high } [\mathrm{O_2}]}$. See Atkins and de Paula, *Physical Chemistry*, 8th edn, 2006, Section 23.2 for a detailed discussion of explosions.

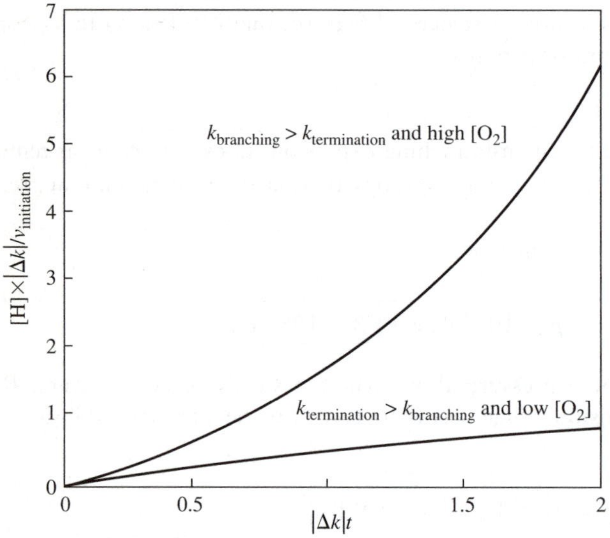

Fig B5

Box 13.1 *Exercise 1*

A stellar surface temperature of 3000 K–4000 K, (a 'red star') doesn't have the energetic particles and photons that are required for either the collisional or radiation excitation of a neutral hydrogen atom. Atomic hydrogen affects neither the absorption nor the emission lines of red stars in the absence of excitation. 'Blue stars' have surface temperature of 15000 K–20000 K. Both the kinetic energy and the black-body emissions display energies great enough to completely ionize hydrogen. Lacking an electron, the remaining proton cannot affect absorption and emission lines either.

In contrast, a star with a surface temperature of 8000 K–10000 K has a temperature low enough to avoid complete hydrogen ionization but high enough for black-body radiation to cause electronic transitions of atomic hydrogen. Hydrogen spectral lines are intense for these stars.

Simple kinetic energy and radiation calculations confirm these assertions. For example, a plot of black-body radiation against the radio photon energy and the ionization energy, I, is shown in Figure B6. It is clearly seen that at 25000 K a large fraction of the radiation is able to ionize the hydrogen ($h\nu/I$). It is likely that at such high surface temperatures all hydrogen is ionized and, consequently, unable to affect spectra.

Alternatively, consider the equilibrium between hydrogen atoms and their component charged particles:

$$H \rightleftharpoons H^+ + e^-.$$

The equilibrium constant is

$$K = \frac{p_+ p_-}{p_H p^\ominus} = \exp\left(\frac{-\Delta G^\ominus}{RT}\right) = \exp\left(\frac{-\Delta H^\ominus}{RT}\right) \times \exp\left(\frac{-\Delta S^\ominus}{R}\right).$$

Clearly $\Delta S^\ominus$ is positive for ionization, which makes two particles out of one, and $\Delta H^\ominus$, which is close to the ionization energy, is also positive. At a sufficiently high temperature, ions will outnumber neutral molecules. Using concepts developed in Chapter 22, one can compute the equilibrium constant; it turns out to be 60. Hence, there are relatively few undissociated H atoms in the equilibrium mixture, which is consistent with the weak spectrum of neutral hydrogen observed.

The details of the calculation of the equilibrium constant based on the methods of Chapter 22 follows. Consider the equilibrium between hydrogen atoms and their component charged particles:

$$H \rightleftharpoons H^+ + e^-$$

The equilibrium constant is

$$K = \frac{p_+ p_-}{p_H p^\ominus} = \exp\left(\frac{-\Delta G^\ominus}{RT}\right)$$

and, when written as a statistical thermodynamic analysis of a dissociation equilibrium, it becomes

$$K = \frac{q_+^\ominus q_-^\ominus}{q_H^\ominus N_A} e^{-\Delta E/RT}$$

$$\text{where } q^\ominus = \frac{RT}{g\, p^\ominus \Lambda^3} \quad \text{and} \quad \Lambda = \left(\frac{h^2}{2\pi kTm}\right)^{1/2}.$$

and where g is the degeneracy of the species. Note that $g_+ = 2, g_- = 2$, and $g_H = 4$. Consequently, these factors cancel in the expression for K.

$$\text{So} \quad K = \frac{RT}{p^\ominus N_A}\left(\frac{2\pi kT}{h^2}\right)^{3/2}\left(\frac{m_- m_+}{m_H}\right)^{3/2} e^{-\Delta E/RT}.$$

Note that the Boltzmann, Avogadro, and perfect gas constants are related ($R = N_A k$), and collect powers of kT; note also that the product of masses is the reduced mass, which is approximately equal to the mass of the electron. Note finally that the molar energy

ΔE divided by R is the same as the atomic ionization energy (2.179×10^{-18} J) divided by k:

$$K = \frac{(kT)^{5/2}\,(2\pi m_e)^{3/2}}{p^{\ominus}h^3} e^{-\Delta E/kT},$$

$$K = \frac{\left[(1.381 \times 10^{-23}\,\mathrm{J\,K^{-1}})\,(25000\,\mathrm{K})\right]^{5/2}\left[2\pi\,(9.11 \times 10^{-31}\,\mathrm{kg})\right]^{3/2}}{(10^5\,\mathrm{Pa})\,(6.626 \times 10^{-34}\,\mathrm{J\,s})^3}$$

$$\times \exp\left(\frac{-2.179 \times 10^{-18}\,\mathrm{J}}{\left(1.381 \times 10^{-23}\,\mathrm{J\,K^{-1}}\right)(25000\,\mathrm{K})}\right).$$

$$K = 60.$$

Thus, the equilibrium favors the ionized species, even though the ionization energy is greater than kT.

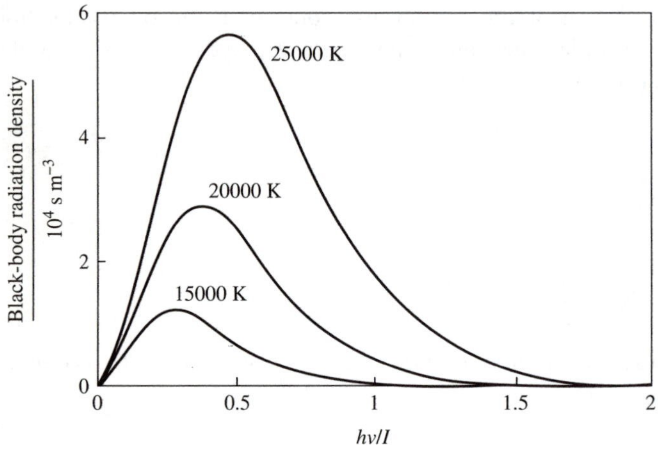

Fig B6

Box 13.1 *Exercise 2*

The wavenumber of a spectroscopic transition is related to the difference in the relevant energy levels. For a one-electron atom or ion, the relationship is

$$hc\tilde{\nu} = \Delta E = \frac{Z^2\mu_{\mathrm{He}}e^4}{32\pi^2\varepsilon_0^2\hbar^2 n_1^2} - \frac{Z^2\mu_{\mathrm{He}}e^4}{32\pi^2\varepsilon_0^2\hbar^2 n_2^2} = \frac{Z^2\mu_{\mathrm{He}}e^4}{32\pi^2\varepsilon_0^2\hbar^2}\left(\frac{1}{n_2^2} - \frac{1}{n_1^2}\right).$$

Solving for $\tilde{\nu}$, using the definition $\hbar = h/2\pi$ and the fact that $Z = 2$ for He, yields

$$\tilde{\nu} = \frac{\mu_{\mathrm{He}}e^4}{2\varepsilon_0^2 h^3 c}\left(\frac{1}{n_2^2} - \frac{1}{n_1^2}\right).$$

Note that the wavenumbers are proportional to the reduced mass, which is very close to the mass of the electron for both isotopes. In order to distinguish between them, we need to carry many significant figures in the calculation.

$$\tilde{\nu} = \frac{\mu_{\text{He}}(1.60218 \times 10^{-19}\,\text{C})^4}{2(8.85419 \times 10^{-12}\,\text{J}^{-1}\text{C}^2\,\text{m}^{-1})^2 \times (6.62607 \times 10^{-34}\,\text{Js})^3 \times (2.99792 \times 10^{10}\,\text{cm}\,\text{s}^{-1})}$$

$$\times \left(\frac{1}{n_2^2} - \frac{1}{n^2}\right),$$

$$\tilde{\nu}/\text{cm}^{-1} = 4.81870 \times 10^{35}(\mu_{\text{He}}/kg)\left(\frac{1}{n_2^2} - \frac{1}{n_1^2}\right).$$

The reduced masses for the ^{4}He and ^{3}He nuclei are

$$\mu = \frac{m_{\text{e}} m_{\text{nuc}}}{m_{\text{e}} + m_{\text{nuc}}}$$

where $m_{\text{nuc}} = 4.00260\,\text{u}$ for ^{4}He and 3.01603 u for ^{3}He or, in kg,

$$^4\text{He}\ m_{\text{nuc}} = (4.00260\,\text{u}) \times (1.66054 \times 10^{-27}\,\text{kg}\,\text{u}^{-1}) = 6.64648 \times 10^{-27}\,\text{kg},$$

$$^3\text{He}\ m_{\text{nuc}} = (3.01603\,\text{u}) \times (1.66054 \times 10^{-27}\,\text{kg}\,\text{u}^{-1}) = 5.00824 \times 10^{-27}\,\text{kg}.$$

The reduced masses are

$$^4\text{He}\ \mu = \frac{(9.10939 \times 10^{-31}\text{kg}) \times (6.64648 \times 10^{-27}\text{kg})}{(9.10939 \times 10^{-31} + 6.64648 \times 10^{-27})\,\text{kg}} = 9.10814 \times 10^{-31}\text{kg},$$

$$^3\text{He}\ \mu = \frac{(9.10939 \times 10^{-31}\text{kg}) \times (5.00824 \times 10^{-27}\,\text{kg})}{(9.10939 \times 10^{-31} + 5.00824 \times 10^{-27})\,\text{kg}} = 9.10773 \times 10^{-31}\,\text{kg}.$$

Finally, the wavenumbers for $n = 3 \rightarrow n = 2$ are

$$^4\text{He}\ \tilde{\nu} = (4.81870 \times 10^{35}) \times (9.10814 \times 10^{-31}) \times \left(\frac{1}{4} - \frac{1}{9}\right)\,\text{cm}^{-1}$$

$$= \boxed{60957.4\,\text{cm}^{-1}},$$

$$^3\text{He}\ \tilde{\nu} = (4.81870 \times 10^{35}) \times (9.10773 \times 10^{-31}) \times \left(\frac{1}{4} - \frac{1}{9}\right)\,\text{cm}^{-1}$$

$$= \boxed{60954.7\,\text{cm}^{-1}}.$$

The wavenumbers for $n = 2 \rightarrow n = 1$ are

$$^4\text{He}\ \tilde{\nu} = (4.81870 \times 10^{35}) \times (9.10814 \times 10^{-31}) \times \left(\frac{1}{1} - \frac{1}{4}\right)\,\text{cm}^{-1}$$

$$= \boxed{329170\,\text{cm}^{-1}},$$

$$^3\text{He}\ \tilde{\nu} = (4.81870 \times 10^{35}) \times (9.10773 \times 10^{-31}) \times \left(\frac{1}{1} - \frac{1}{4}\right)\,\text{cm}^{-1}$$

$$= \boxed{329155\,\text{cm}^{-1}}.$$

Box 15.1 *Exercise 1*

Single-walled carbon nanotubes (SWNT) may be either conductors or semiconductors depending upon the tube diameter and the chiral angle of the fused benzene rings with respect to the tube axis. Van der Waals forces cause SWNT to stick together in clumps, which are normally mixtures of conductors and semiconductors. SWNT stick to many surfaces and they bend, or drape, around nano-sized features that are upon a surface.

Only the semiconductor SWNT are suitable for the preparation of field-effect transistors (FET) so IBM researchers (*Science*, April 27, 2001) have developed a destructive technique for eliminating conducting tubes from conductor/semiconductor clumps with a current burst. The technique can also be used to remove the outer layers of multi-walled tubes that consist of multiple concentric tubes about a common axis. Bandgaps increase as the diameter of multi-walled tubes is decreased which means that the destructive technique can be used to tailor a semiconductor tube to specific requirements.

Here is a list of ideas for producing transistors with SWNT.

1 Cees Dekker and students (S.J. Tans *et al.*, *Nature*, **393**, 49 (1998)) have draped a semiconducting carbon nanotube over metal electrodes that are 400 nm apart atop a silicon surface coated with silicon dioxide. A bias voltage between the electrodes provides the source and drain of an FET. The silicon serves as a gate electrode. By adjusting the magnitude of an electric field applied to the gate, current flow across the nanotube may be turned on and off.
2 A section of a single nanotube may be exposed to potassium vapor to produce a p–n junction.
3 A single-electron transistor (SET) has been prepared by Cees Dekker and coworkers (*Science*, **293**, 76, (2001)) with a conducting nanotube. The SET is prepared by putting two bends in a tube with the tip of an AFM. Bending causes two buckles that, at a distance of 20 nm, serves as a conductance barrier. When an appropriate voltage is applied to the gate below the barrier, electrons tunnel one at a time across the barrier.
4 A semiconductor tube may be fused to a conductor tube to produce a SET similar to an SET.

Box 15.1 *Exercise 2*

Evaluate the sum of $\pm 1/r_i$, where r_i is the distance from the ion i to the ion of interest, taking $+1/r$ for ions of like charge and $-1/r$ for ions of opposite charge. The array has been divided into five zones, as shown in Figure B7. Zones B and D can be summed analytically to give $-\ln 2 = -0.69$. The summation over the other zones, each of which gives the same result, is tedious because of the very slow convergence of the sum. Unless you make a very clever choice of the sequence of ions (grouping them so that their contributions almost cancel), you will find the following values for arrays of different sizes.

10 × 10	20 × 20	50 × 50	100 × 100	200 × 200
0.259	0.273	0.283	0.286	0.289

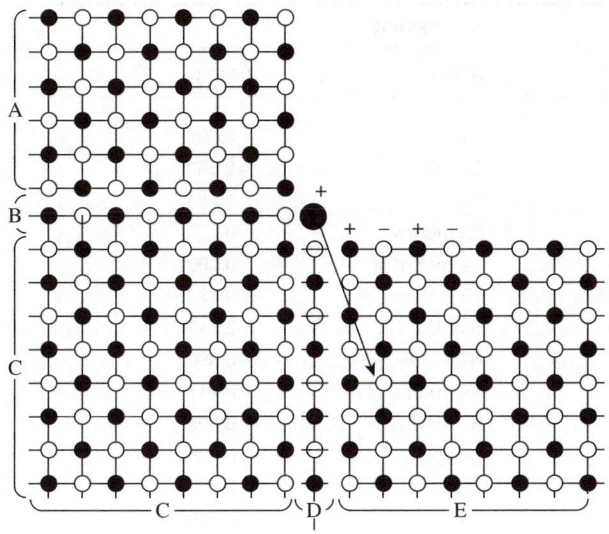

Fig B7

As the array becomes larger and larger the sum converges to 0.2892597... For a cation above a flat surface, the energy (relative to the energy at infinity, and in multiples of $e^2/4\pi\varepsilon_0 r_0$ where r_0 is the lattice spacing (200 pm), is Zone C + D + E = 0.29 – 0.69 + 0.29 = $\boxed{-0.11}$, which implies an attractive state.

Box 17.1 *Exercise 1*

(a) The table displays computed electrostatic charges (semi-empirical, PM3 level, PC Spartan Pro™of the DNA bases, modified by addition of a methyl group to the position at which the base binds to the DNA backbone. (That is, R = methyl for the computations displayed, but R = DNA backbone in DNA.) See Figure B8 for numbering.

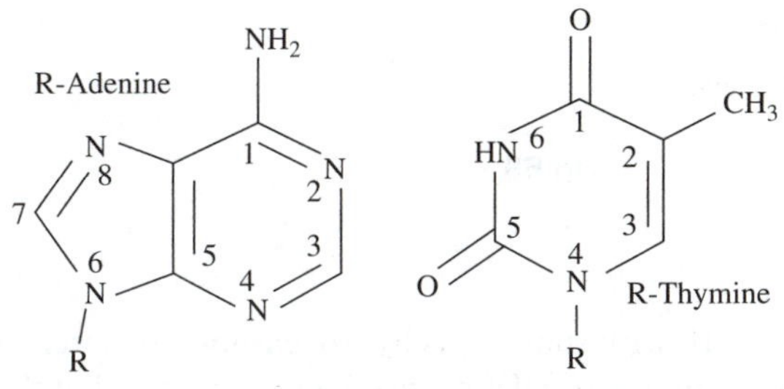

Fig B8

(b) and (c) On purely electrostatic grounds, one would expect the most positively charged hydrogen atoms of one molecule to bind to the most negatively charged atoms of another. The hydrogen atoms with a charge of at least +0.2 are candidates for hydrogen bonding. Atoms with the greatest negative charges (more negative than –0.400) and a lone pair of electrons are counterpart candidates to hydrogen bond with hydrogen atoms. Methyl groups may prevent hydrogen bonding when they are positioned for steric hindrance.

R = adenine		R = thymine	
Atom	Charge†	Atom	Charge
C1	−0.905	C1	−0.885
amino N	−0.656	O of C1	−0.580
amino H	−0.288	C2	−0.554
N2	−0.914	C2 methyl C	−0.180
C3	−0.785	C2 methyl H	−0.003
H of C3	−0.020	C3	−0.173
N4	−0.835	H of C3	−0.111
C5	−0.639	N4	−0.390
N6	−0.183	N4 methyl C*	−0.211
methyl C*	−0.113	N4 methyl H*	−0.002
methyl H*	−0.022	C5	−0.836
C7	−0.320	O of C5	−0.596
H of C7	−0.056	N6	−0.540
N8	−0.584	H of N6	0.264
C9	−0.268		

* Part of R group, so not really available for hydrogen bonding in DNA.
† Table displays average charge of atoms that are chemically equivalent.

(d) The naturally occurring pairs are shown in Figure B9. These configurations are quite accessible sterically, and they have the further advantage of multiple hydrogen bonds.

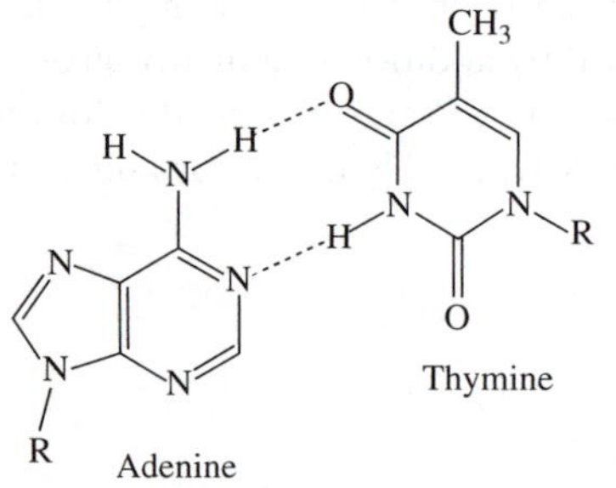

Fig B9

Box 17.1 *Exercise 2*

The drug Crixivan, shown in Figure B10 with some of its hydrogen-bonding interactions with HIV protease, is a competitive inhibitor of HIV protease and has several molecular features that optimize binding to the enzyme's active site. First, the highlighted hydroxyl group displaces a H_2O molecule that acts as the nucleophile in the hydrolysis of the substrate. Second, the carbon atom to which the key —OH group is bound has a tetrahedral geometry that mimics the structure of the transition state of the peptide hydrolysis reaction. However, the tetrahedral moiety in the drug is not cleaved by the enzyme. Third, the inhibitor is anchored firmly to the active site via a network of hydrogen bonds involving the carbonyl groups of the drug, a water molecule, and peptide NH groups from the enzyme.

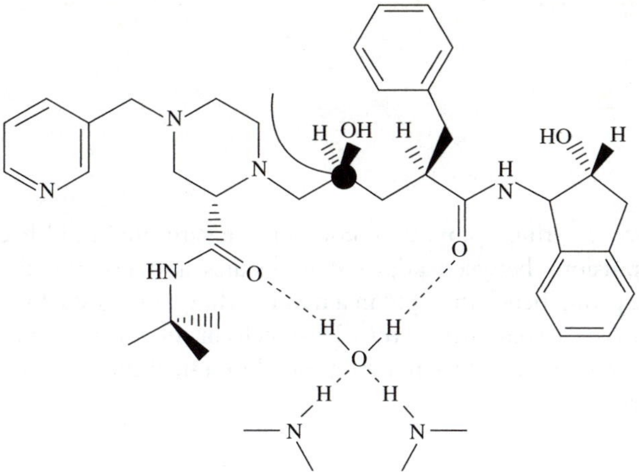

Fig B10

Box 18.1 *Exercise 1*

$$V_{\text{bend}} = \frac{1}{2}k_{\text{bend}}(\theta - \theta_e)^2$$

where $V_{\text{bend}} = 8.5\,\text{kJ mol}^{-1}/N_A$ and $\theta - \theta_e = (30 - 15)\text{deg} = 15\,\text{deg}$.

$$k_{\text{bend}} = \frac{2V_{\text{bend}}}{(\theta - \theta_e)^2} = \frac{2 \times 8.5 \times 10^3\,\text{J mol}^{-1}}{6.02 \times 10^{23}\,\text{mol}^{-1} \times (15\,\text{deg})^2}$$

$$= \boxed{1.3 \times 10^{-22}\,\text{J deg}^{-2}}$$

Box 18.1 *Exercise 2*

$$V_{\text{stretch}}(\text{per mole}) = N_A \times \frac{1}{2}k_{\text{stretch}}(R - R_e)^2$$

$$= 6.02 \times 10^{23}\,\text{mol}^{-1} \times \frac{1}{2} \times 400\,\text{N m}^{-1} \times [(165 - 152) \times 10^{-12}\,\text{m}]^2$$

$$= 2.04 \times 10^4\,\text{J mol}^{-1} = \boxed{20.4\,\text{kJ mol}^{-1}}.$$

Box 18.2 *Exercise 1*

In both cases, the diffusion is two-dimensional. Therefore,

$$l = (4Dt)^{1/2} \quad \text{and} \quad t = \frac{l^2}{4D}.$$

In a cell plasma membrane,

$$t = \frac{(1.0 \times 10^{-8}\,\text{m})^2}{4 \times 1.0 \times 10^{-8}\,\text{cm}^2\,\text{s}^{-1} \times 1\,\text{m}^2/10^4\,\text{cm}^2} = \boxed{2.5 \times 10^{-5}\,\text{s}}.$$

In a lipid bilayer,

$$t = \boxed{2.5 \times 10^{-6}\,\text{s}}.$$

Box 18.2 *Exercise 2*

Unsaturated lipids have lower melting points than comparable saturated lipid because of the alkene–alkene π repulsions between adjacent molecules and because the cis-conformation at each alkene group bends the lipid in a manner that reduces stacking and reduces the net strength of intermolecular attractions. Thus, cells produce a greater degree of unsaturated lipid at lower temperatures to maintain membrane-melting temperatures close to ambient temperature.

Box 19.1 *Exercise 1*

$$I = fI + M = fI + aT^4 \quad [12.3] \quad \text{so}$$

$$T = \left(\frac{I(1-f)}{a}\right)^{1/4} = \left(\frac{(343\,\text{W m}^{-2}) \times (1-0.30)}{5.67 \times 10^{-8}\,\text{W m}^{-2}\,\text{K}^{-4}}\right)^{1/4} = \boxed{255\,\text{K}}$$

where I is the incoming energy flux, f the albedo (fraction of incoming radiation absorbed), M the excitance, and a the Stefan–Boltzmann constant. Wien's displacement law relates the temperature to the wavelength of the most intense radiation

$$T\lambda_{\text{max}} = 2.9\,\text{mm K} \quad [12.2] \quad \text{so}$$

$$\lambda_{\text{max}} = \frac{2.9\,\text{mm K}}{T} = \frac{2.9\,\text{mm K}}{255\,\text{K}} = \boxed{11.4\,\mu\text{m}} \text{ in infared}.$$

Box 19.1 *Exercise 2*

(a) and (b) are infrared inactive because these normal modes do not give rise to a changing dipole moment.

(c) and (d) are infrared active because these normal modes do give rise to a changing dipole moment.

Box 20.1 *Exercise 1*

Percentage transmitted to the retina is

$$\begin{aligned} &70\% - 0.25 \times 70\% - 0.09 \times 0.75 \times 70\% - 0.43 \times (1 - 0.25 - 0.09 \times 0.75) \times 70\% \\ &= 70\% - (0.25 + 0.0675 + 0.390) \times 70\% \\ &= 20.5\%. \end{aligned}$$

Number of photons focused on the retina in 0.1 s is

$$0.205 \times 40\,\text{mm}^2 \times 0.1\,\text{s} \times 4 \times 10^3\,\text{mm}^{-2}\,\text{s}^{-1}$$

$$= \boxed{3 \times 10^3}.$$

More than what one might have guessed.

Box 20.1 *Exercise 2*

Trans-retinal has 10 conjugated carbon atoms. There are 9 carbon–carbon bonds, 4 single, and 5 double. There are 10 π-electrons associated with the double bonds. Two electrons occupy each level (by the Pauli principle), so levels E_1 to E_5 are filled. The minimum excitation energy is then $E_6 - E_5$.

$$E_n = \frac{n^2 h^2}{8\,mL^2}.$$

$$\Delta E = E_6 - E_5 = (36 - 25)\frac{h^2}{8mL^2} = 11\frac{h^2}{8mL^2}.$$

In calculating L, we can add an extra half-bond length at each end of the box, so that the length of the 'box' is $10 \times 140\,\text{pm} = 1.40 \times 10^{-9}$ m.

$$\Delta E = \frac{11 \times \left(6.63 \times 10^{-34}\,\text{J s}\right)^2}{8 \times 9.11 \times 10^{-31}\,\text{kg} \times \left(1.40 \times 10^{-9}\,\text{m}\right)^2} = 3.38 \times 10^{-19}\,\text{J}.$$

$$\Delta E = \frac{hc}{\lambda} \quad \text{or} \quad \lambda = \frac{hc}{\Delta E}.$$

$$\lambda = \frac{6.63 \times 10^{-34} \times \text{J s} \times 3.00 \times 10^8\,\text{m s}^{-1}}{3.38 \times 10^{-19}\,\text{J}} = 5.87 \times 10^{-7}\,\text{m}$$

$$= \boxed{587\,\text{nm}}.$$

This wavelength is in the visible region of the spectrum.

Box 20.2 *Exercise 1*

$$\underset{\text{Chlorophyll}}{\text{C}} + \underset{\text{Quinone}}{\text{Q}} \xrightarrow{h\nu} \text{C}^* + \text{Q} \xrightarrow{\text{electron transfer}} \text{C}^+ + \text{Q}^-$$

Direct electron transfer from the ground state of C is not spontaneous. It is spontaneous from the excited state. The difference between the ΔGs of the two processes is given by the expression

$$\Delta\,(\Delta G) = \Delta G_{\text{c}^*} - \Delta G_{\text{c}} \sim U_{\text{c}} - U_{\text{c}^*} \sim -(U_{\text{LUMO}} - U_{\text{HOMO}})$$

where U_{LUMO} and U_{HUMO} are energies of the LUMO and HOMO of chlorophyll. Since $\Delta(\Delta G) < 0$, we see that electron transfer is exergonic and spontaneous when the electron is transferred from the excited state of chlorophyll.

Box 20.2 *Exercise 2*

$$2H_2O \rightarrow O_2 + 4H^+(aq) + 4e^-$$

$$\underline{2NADP^+ + 2H^+(aq) + 4e^- \rightarrow 2NADPH}$$

$$2NADP^+ + 2H_2O \rightarrow O_2 + 2H^+(aq) + 2NADPH \quad \Delta_r G^\ominus = 438.0\,\mathrm{kJ\,mol^{-1}}$$

The net reaction for the generation of the reduced form of photoquinone (PQH_2),

$$H_2O + PQ \xrightarrow{\text{light,PSII}} \frac{1}{2}O_2 + PQH_2,$$

indicates that reduction requires 2 electrons. Yet, only 1 electron at a time is passed from excited P680 to PQ. To explain the facts, we hypothesize that the reduction occurs in stages with two stages requiring photon absorption by P680 and electron transfer to PQ. In order to regenerate the P680 after the initial electron transfer to PQ, the manganese–enzyme–water complex must partially oxidize the water and pass an electron to the oxidized form of P680. After photosystem II has absorbed 4 photons and generated $2PQH_2$ molecules, the manganese–enzyme complex has oxidized $2H_2O$ molecules to an O_2 molecule. The multiple stages of the process require that the oxygen atoms of water be complexed to the manganese in a succession of oxidation states.

Box 21.1 *Exercise 1*

We use $\nu = \dfrac{\gamma_N \mathcal{B}_{loc}}{2\pi} = \dfrac{\gamma_N}{2\pi}(1-\sigma)\mathcal{B}$ [21.16]

where $\mathcal{B}$ is the applied field.

Because shielding constants are quite small (a few parts per million) compared to 1, we may write for the purposes of this calculation

$$\nu = \frac{\gamma_N \mathcal{B}}{2\pi},$$

$$\nu_L - \nu_R = 100\,\mathrm{Hz} = \frac{\gamma_N}{2\pi}(\mathcal{B}_L - \mathcal{B}_R),$$

$$\mathcal{B}_L - \mathcal{B}_R = \frac{2\pi \times 100\,\mathrm{s^{-1}}}{\gamma_N}$$

$$= \frac{2\pi \times 100\,\mathrm{s^{-1}}}{26.752 \times 10^7\,\mathrm{T^{-1}\,s^{-1}}} - 2.35 \times 10^{-6}\,\mathrm{T}$$

$$= 2.35\,\mu\mathrm{T}.$$

The field gradient required is then

$$\frac{2.35\,\mu\mathrm{T}}{0.08\,\mathrm{m}} = \boxed{29\,\mu\mathrm{T\,m^{-1}}}.$$

Note that knowledge of the spectrometer frequency, applied field, and the numerical value of the chemical shift (because constant) is not required.

Box 21.1 *Exercise 2*

Assume that the radius of the disk is 1 unit. The volume of each slice is proportional to length of slice multiplied by δx (see Figure B11).

Length of slice at $x = 2\sin\theta$.

$x = \cos\theta$.

$\theta = \arccos x$.

x ranges from -1 to $+1$.

Length of slice at $x = 2$ sin (arcos x).

Plot $f(x) = 2$ sin (arcos x) against x between the limits -1 and $+1$. The plot is shown Figure B12.

The volume at each value of x is proportional to $f(x)$ and the intensity of the MRI signal is proportional to the volume, so the figure represents the absorption intensity for the MRI image of the disk.

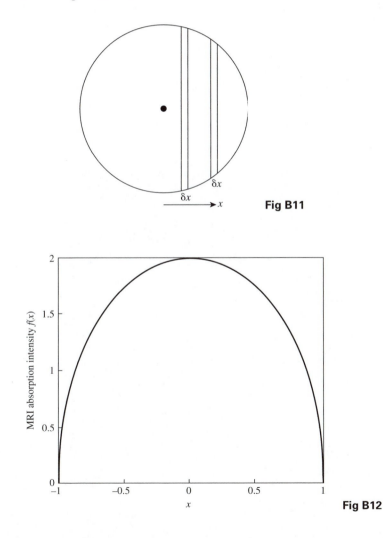

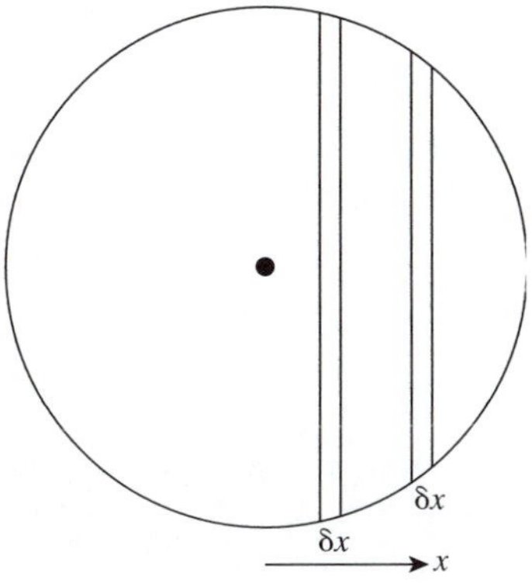

Fig B11

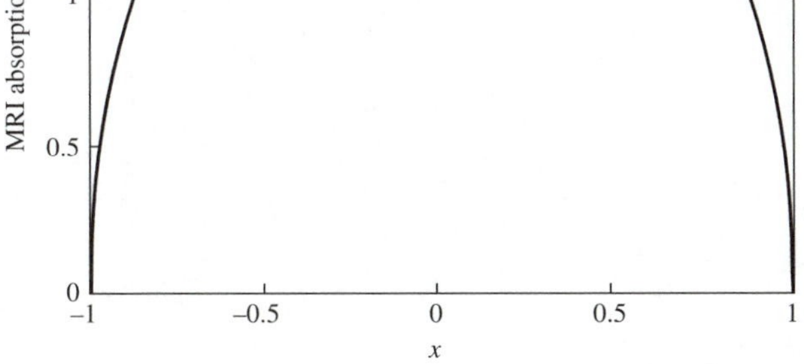

Fig B12